Energy Systems in Electrical Engineering

Series Editor

Muhammad H. Rashid, Florida Polytechnic University, Lakeland, USA

More information about this series at http://www.springer.com/series/13509

Neeraj Priyadarshi · Akash Kumar Bhoi ·
Ramesh C. Bansal · Akhtar Kalam
Editors

DC–DC Converters for Future Renewable Energy Systems

Springer

Editors
Neeraj Priyadarshi
Department of Energy Technology
Aalborg University
Esbjerg, Denmark

Akash Kumar Bhoi
Department of Computer Science
and Engineering
Sikkim Manipal Institute of Technology
Sikkim Manipal University
Majitar, Sikkim, India

Ramesh C. Bansal
Department of Electrical Engineering
University of Sharjah
Sharjah, United Arab Emirates

Akhtar Kalam
College of Engineering and Science
Victoria University
Footscray, VIC, Australia

ISSN 2199-8582 ISSN 2199-8590 (electronic)
Energy Systems in Electrical Engineering
ISBN 978-981-16-4387-3 ISBN 978-981-16-4388-0 (eBook)
https://doi.org/10.1007/978-981-16-4388-0

This Springer imprint is published by the registered company Springer Nature Singapore Pte Ltd.
The registered company address is: 152 Beach Road, #21-01/04 Gateway East, Singapore 189721, Singapore

Preface

This book describes modeling and analysis of DC–DC converters for future renewable energy systems. Chapter 1 will help researchers to get a thorough understanding about the design of key component of the DCMG (DC microgrid), i.e., DC–DC converter. It allows control operation to take place in interconnected renewable energy sources (RES), energy storage systems (ESS) and loads operating at different voltage levels to a common DC bus. In Chap. 2, the description of DC–DC high gain converters with different solar photovoltaic (PV)-based systems is presented, and then, an improved high gain buck–boost converter (IHGBBC) suitable for PV-based systems is demonstrated. A boost converter with the generalized structure of quadratic boosting cell is proposed in Chap. 3. The proposed structure having three quadratic boosting cells (n = 3) can produce a high magnitude of output DC voltage from a low-magnitude DC voltage by adjusting the number of quadratic cells in the generalized quadratic cell structure with an appropriate duty ratio (δ). Chapter 4 portrays the procedure to design a flyback converter (230 V DC–5 V DC) functioning under discontinuous conduction mode. The three-level and five-level of neutral point control inverter (NPC multilevel inverters) utilized for DC to AC control transformation are explained in Chap. 5. In Chap. 6, a bidirectional interleaved switched capacitor DC–DC converter is proposed as a power electronic interface for microgrids. Chapter 7 presents advanced LUO converters used for renewable energy applications in detail, and simulations of these converters are presented. The performance review of solar energy conversion systems (SECS) with the integration of the advanced DC/DC converter super-lift Luo converter (SLLC) is discussed in Chap. 8. Chapter 9 explains a detailed analysis of transformerless non-coupled inductor quadratic boost converters. Chapter 10 presents the design and modeling of photovoltaic (PV) emulator using DC–DC buck converter in MATLAB/Simulink and its hardware implementation utilizing Arduino controller. In Chap. 11, a smart low-voltage DC (LVDC) distribution system is designed in real time to utilize the power from hybrid DC microgrid made of solar photovoltaic (SPV) and wind energy conversion systems (WECS). In this work, a DC–DC converter plays a major role to achieve the proposed distribution system. Chapter 12 presents an approach using multiple input converter (MIC) to realize individual maximum power from photovoltaic modules. Chapter 13 presents the overview of control methods and parameter designing of the dual active bridge (DAB) power converter, applicable in the renewable energy system. Chapter 14 focuses on high-frequency signal-based fault diagnosis strategy for open-circuit (OCF) and short-circuit fault (SCF) identification on two-level three-phase voltage source power converter-based PMSM drive system. Chapter 15 describes modeling of different DC–DC converter for green energy applications. Chapter 16 focuses on the design and analysis of DC–DC buck converter with drift-free MPPT algorithm for a self-excited induction generator (SEIG)-based wind energy generation system. Chapter 17 explains design of roof-top photovoltaic system (RTPV) panels connected to the boost converter with an inverter and batteries. The design and implementation of the virtual synchronous machine (VSM) in microgrid (MG) are explained in Chap. 18. In Chap. 19, the dominant types of DC–DC power electronics converters used for green energy applications are discussed in detail. Chapter 20 proposes Internet of things (IoT)-based maximum power point tracker for brushless DC motor (BLDC)-driven photovoltaic water pumping system. Moreover, a dual-level boost converter is employed to obtain high output voltage with enhanced efficiency and has economical operation compared to classical boost converter. Switched mode fourth-order buck–boost converter using Type II and Type III controllers in DC grid applications has been discussed in Chap. 21. Fractional-order PI-lead controller design of DC–DC power converter for renewable energy applications is explained in Chap. 22. Chapter 23 explains power management of battery-integrated PV system with sliding mode controller (SMC)-controlled bidirectional converter. Overview of bidirectional DC–DC converters topologies for electric vehicle and renewable energy system has been discussed in Chap. 24. Chapter 25 examines conventional boost converter (CBC) and interleaved boost converter (IBC) with PV sources that are designed to attain maximum voltage from the converter for green energy applications.

Esbjerg, Denmark — Dr. Neeraj Priyadarshi
Majitar, India — Dr. Akash Kumar Bhoi
Sharjah, United Arab Emirates — Dr. Ramesh C. Bansal
Footscray, Australia — Prof. Dr. Akhtar Kalam

Contents

About the Editors

Neeraj Priyadarshi received M.Tech. degree in power electronics and drives in 2010 from Vellore Institute of Technology, Vellore, India, and a doctorate from Government College of Technology and Engineering, Udaipur, Rajasthan, India in 2015. His current research interests include power electronics, control systems, power quality, and solar power generation. Dr. Priyadarshi has a Post Doc from Bioenergy and Green Engineering, Department of Energy Technology, Aalborg University, Denmark. He is a Professor in the Department of Electrical Engineering, JIS Group Kolkata. He has published over 50 papers in journals and conferences. Dr. Priyadarshi is a reviewer of *IEEE Transactions Industrial Informatics*, *IEEE Transactions Industry Applications*, *IEEE Systems Journal*, *Journal of Cleaner Production*, *International Journal of Electrical Power and Energy Systems*, *IET Power Electronics*, and *Energy for Sustainable Development*.

Akash Kumar Bhoi is working as an Assistant Professor in the Department of Computer Science and Engineering at Sikkim Manipal Institute of Technology, India. He is also working as a Research Associate at Wireless Networks Research Laboratory, Institute of Information Science and Technologies, National Research Council (ISTI-CRN) Pisa, Italy. He is a University Ph.D. Course Coordinator for Research & Publication Ethics (RPE). He is a member of IEEE, ISEIS, and IAENG, an associate member of IEI, UACEE, and an editorial board member reviewer of Indian and international journals. He is also a regular reviewer of reputed journals, namely IEEE, Springer, Elsevier, Taylor and Francis, Inderscience, etc. His research areas are biomedical technologies, IoT, computational intelligence, antenna, and renewable energy. He has published several papers in national and international journals and conferences.

Ramesh C. Bansal is a Professor in the Department of Electrical Engineering at the University of Sharjah, UAE. Professor Bansal completed his M.E. from Delhi Technical University and Ph.D. from IIT Delhi, India, in 1996 and 2003, respectively. He has over 25 years of diversified experience in academic research and collaborating with industry and government organizations. He has supervised

20 Ph.D. researchers, four postdoctoral researchers and currently supervising several Ph.D. students. His research interests are in the areas of renewable energy and power systems, power electronics, electrical machines, and smart grids. Professor Bansal is a Fellow and Chartered Engineer IET-UK, Fellow Engineers Australia, Fellow Institution of Engineers (India), Fellow SAIEE, and Senior Member of IEEE-USA.

Akhtar Kalam has been associated with Victoria University (VU) since 1985. He is a former Deputy Dean of the Faculty of Health, Engineering, and Science and Head of Engineering of the College of Engineering and Science. Currently, he is the Head of External Engagement. He is also the current Chair of the Academic Board in the Engineering Institute of Technology, Perth, Australia, and the Editor in Chief of the *Australian Journal of Electrical and Electronics Engineering*. Further, he has a Distinguished Professorship position at the University of New South Wales, Sydney, Australia, three Indian and five Malaysian universities. He has also been appointed as the Editor in Chief of the *Australian Journal of Electrical and Electronic Engineering*. He received his B.Sc. and B.Sc. Engineering from Calcutta University and Aligarh Muslim University, India, respectively. He completed his M.S. and Ph.D. from the University of Oklahoma, USA, and the University of Bath, UK, respectively. His major interest areas are power system analysis, communication, control, protection, renewable energy, smart grid, IEC61850 implementation, and co-generation systems. Professor Kalam has conducted research, provided industrial consultancy, and published more than 542 publications in his area of expertise. He has written 26+ books in these areas. More than 35 Ph.D. students have graduated under his supervision and he is an external examiner of many external doctoral students in Australia and overseas.

Chapter 1
Design and Control of DC–DC Converters in a PV-Based LVDC Microgrid

Pradyumna Kumar Behera and Monalisa Pattnaik

Learning Objectives: Upon completion of this chapter, the readers will gain knowledge about

- Configuration of a PV-based LVDC microgrid
- Accurate mathematical modeling of photovoltaic system, battery and supercapacitor
- Understanding and importance of the key component of the DC microgrid (i.e., DC–DC converter)
- Design of DC–DC boost converter used for MPPT control
- Design of non-isolated-type buck–boost bidirectional converter for interfacing the energy storage systems
- Control strategy and proper power sharing among PV and HESS during various operating conditions

1.1 Introduction

The escalated global power consumption and also the adverse impact of extensive use of fossil fuel on the environment have expedited both the usage of renewable energy sources (RES) as well as development of cost-effective and smart microgrid technologies. Typically, microgrid is an autonomous low-voltage power system which encompasses multiple RES, energy storage systems (ESS), loads and various

P. K. Behera (✉) · M. Pattnaik
Department of Electrical Engineering, National Institute of Technology, Rourkela, Odisha 769008, India
e-mail: pattnaikm@nitrkl.ac.in

N. Priyadarshi et al. (eds.), *DC–DC Converters for Future Renewable Energy Systems*, Energy Systems in Electrical Engineering,
https://doi.org/10.1007/978-981-16-4388-0_1

power electronic devices and converters within clearly defined boundaries incorporating bidirectional energy exchange with the electric grid. Microgrid can operate seamlessly between islanded as well as grid-integrated mode. The operation of microgrid provides numerous advantages to the utility sector and customers, in terms of enhanced energy efficiency, decreased environmental impact, network operational advantages such as minimization of loss, voltage control and cost-effective electricity infrastructure replacement. DC microgrid (DCMG), one out of three possible microgrid topologies (AC, DC and hybrid AC–DC), has proliferated as an inevitable part of modern power distribution networks with an improved efficiency and reliability. A DCMG is a controllable system with group of distributed energy resources (DER) connected through a common DC bus and can supply power to the local loads. Furthermore, it is distinguished with a simple control and energy management with its counterpart, i.e., AC microgrid (ACMG) and overcomes various innate challenges such as reactive power compensation, phase synchronization and high inrush current. Also, the substantial growth of electronic loads like electric vehicles (EV), data centers, lighting system, battery chargers, cell phones, laptops, etc. points toward the adaptation of DCMG as a new standard in the LVDC power system. Numerous energy generation and storage technologies like solar photovoltaic (SPV), fuel cell (FC), battery, supercapacitor (SC), etc. are ingrained qualified for DCMG application because they supply DC power. In case of DC grid power system, the proper selection of suitable voltage level can supply most of the loads effectively. This also reduces various conversion stages making the design as well as control simpler than the AC system.

The idea of DCMG has been coined few decades ago, but numerous issues such as lack of proper standards and codes, misconceptions regarding safety and efficiency hindered its wide spread usage. DCMG is a promising solution for power generation because RES and ESS can be directly interfaced without conversion losses. The primary goal of control in DCMG is to maintain a constant voltage at the point of load (POL) and ensure power balance between generation and demand. But, the intermittent nature of RES creates voltage fluctuation at the DC bus. This voltage fluctuation can be easily suppressed by power exchange with the utility grid to fulfill the load demand in case of utility connected MG. However, the stand-alone DCMG rely mostly on energy storage to meet the generation-demand mismatch and ensure a stable DC bus voltage. Therefore, RES along with appropriate HESS and proper control mechanism can be considered as a reliable option for residential application or for remote locations where electricity supply from utility grid is practically infeasible.

This chapter seeks to address and solve power balance issues arising during dynamic conditions in a DCMG comprising of PV as the prime energy source, battery and SC as hybrid energy storage and the load. Further, this chapter will help researchers to get a thorough understanding about the design of key component of the DCMG (i.e., DC–DC converter). It allows control operation to take place in interconnected RES, ESS and loads operating at different voltage level to a common DC bus. The major challenges in design and control of LVDC microgrid involve development of control strategy to maintain a stiff DC-link voltage and proper power sharing within RES and ESS during various operating conditions. Therefore, in this chapter, an effort has been made to design the converters required for the PV-based LVDC microgrid. Also, the design of a coordinated controller and optimum power management scheme for a 48 V stand-alone LVDC microgrid typically for residential application is also presented.

1.2 Scientific Background and Motivation

During the last decade, the research on DCMG has gained major attention since they can supply DC loads directly avoiding notable conversion losses. Additionally, it is free from reactive power, phase synchronization issues and harmonics as compared to its AC counterpart (i.e., ACMG) [1]. Generally, DCMG is attributed as a low-voltage (LV) power system which encompasses several distributed energy resources (DER), energy storage devices and loads that can operate in stand-alone, islanded or grid-connected mode [2]. Low-voltage direct current (LVDC) distribution system with RES integration provides numerous advantages such as: (a) reduced AC–DC conversion stages improves energy saving due to high efficiency of DC equipment, (b) easily compatible with efficient home equipments like BLDC fans, LED lights etc., (c) better power quality (no power factor and harmonics related issues), (d) safe and secure voltage level which ensures human safety. LVDC systems are generally popular for their applications in marine, automotive and aerospace system. The selection of proper DC bus voltage is a crucial parameter for safety, efficiency and reliability of connected loads and generating units. Currently, more research focus is toward optimal selection of suitable voltage level for several applications. For example, a 380 V DC voltage standard has been developed for data centers with the combined effort of Lawrence Berkley Laboratory and Electric Power Research Institute (EPRI) [3, 4], a 24 V DCMG for interior lighting has been proposed by Emerge Alliance. So, the selection of suitable voltage level of a DCMG is application specific and depends upon the power requirement. In [5–7], it is shown that, a 48 V DCMG is considered suitable for residential applications. Further, for such cases, the operation of battery with or without direct contact is safe and efficient.

The primary concerns in designing and control of LVDC microgrid involve: (a) choice of suitable converter, (b) extraction of maximum power from RES, (c) voltage regulation and (d) power sharing among various sources and loads [7, 8]. The output power of PV is intermittent in nature and is affected due to change in climatic conditions. Thus, the system requires MPPT technique to provide maximum power irrespective of operating conditions. Efficient MPPT algorithms play a vital role in maximization of energy efficiency of PV array. A comparative research on several MPPT control techniques for PV system is clearly depicted in [9]. Among various MPPT control algorithms reported in literature, much focus is given to perturb and observe (P&O), incremental conductance (INC), adaptive step-size P&O, fractional open-circuit voltage (FOCV), fractional short-circuit current (FSCC) and several fuzzy inference MPPT techniques. However, P&O and INC techniques are widely used and well suited for applications seeking the true MPP due to low implementation complexity. The surplus power generated using these MPPT techniques can be stored in ESS and supplies back to the system whenever required [8, 10]. Unlike grid-connected microgrids that have reactive power support, stand-alone microgrids mostly leverage on ESS to balance the power consumption and generation mismatch. Also, the incorporation of ESS not only provides promising solution to triumph over several challenges such as voltage instability, power quality issues and load discrepancy but also ensures stable system operation. Hence, in recent years, many ESS technologies have been developed and are classified into electrical, electrochemical, mechanical and chemical-based storage systems. The extensively used energy storages are flywheel, supercapacitors (SC), superconducting magnetic energy storage (SMES), compressed air-based energy storage (CAES), batteries and hydrogen tanks. Among these, batteries have evolved as promising and notable energy storage medium for stable operation in case of utility interfaced system. Furthermore, they play an essential and much important role in managing momentary and uncertain power fluctuations in an off-grid system. The combination of renewable energy sources (RES) such PV, wind energy with ESS improves the reliability and resiliency of the system and reduces energy import dependence in addition to reduction in carbon footprints. Moreover, developing nations like India are actually moving toward building smart cities to attain energy sustainability, adopting e-vehicles for transportation where RES and ESS play an important role. However, it would be worth noting that, every ESS is confined to a particular application and has certain limitations based on power and energy density. In [11, 12], the authors developed a battery-integrated wind energy generation system to mitigate the power fluctuations arising due to wind speed change. However, the charge/discharge rate and power density of the battery are low to meet the load demand during peak operating periods. The integration of high power density supercapacitor with RES is proposed in [13] to eliminate the above demerits. But, SC having a lower energy density could not fulfill the long-term load demand. The SC has lower energy and higher power density in comparison to battery and vice versa [14]. Therefore, authors in [15] proposed an active configuration HESS to harness merits of both the energy storage devices. Several topologies of battery-SC HESS are discussed in [16]. They are classified mainly into passive, semi-active and active configurations. Among them, active configuration has independent control over battery and SC through bidirectional DC–DC converter. This particular topology not only improves the reliability of HESS but also enhances the efficiency, life cycle and overall system performance. As of now, numerous control schemes of HESS in electric vehicle (EV) applications are reported in literature. But, relatively less papers concentrated on voltage regulation of DC bus in microgrid. Authors in [17] developed a simple power sharing scheme in a hybrid AC/DC microgrid comprising of PV, wind energy generation system, SC and battery. However, internal power sharing among battery and SC is not considered. Another control technique proposed in [18] could not explain clearly about current sharing between SC and battery to deal with fluctuating load demand. In [19], a hysteresis controller for composite converter to control several battery packs in distributed power generation system is proposed. But, this controller fails to maintain desired grid voltage during rapid power fluctuations. The authors in [20] considered seven different voltage range to decide the operating mode and priority of power supply is set according to voltage level of the source. However, during transient condition, severe DC bus voltage variation is encountered. In [21], an improved voltage control loop for battery/SC HESS is presented which eradicates the high-frequency components of SC charging current and does not interfere with DC-link voltage control loop. To improve the controllability and reliability of HESS, several advanced control schemes such as artificial neural network (ANN) [22], fuzzy logic-based control (FLC) [23], optimization-based methods [24] and model predictive control (MPC) [25] are available in literature. However, such sophisticated methods require large storage space for data, rigorous data training and complex mathematical computations.

1.3 Architecture of PV-Based LVDC Microgrid with HESS

The architecture of a PV-based LVDC microgrid with parallel active HESS is shown in Fig. 1.1. The system configuration includes photovoltaic as the primary energy source, power electronic converters, SC and battery as HESS. In order to

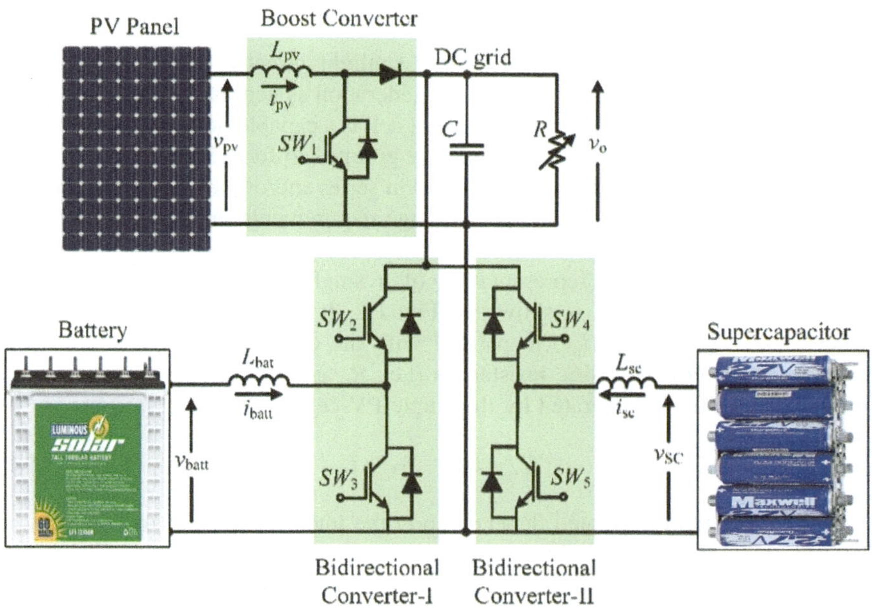

Fig. 1.1 Configuration of PV-based LVDC microgrid with hybrid energy storage system

improve the efficacy of the system, it is necessary to adopt an efficient MPPT algorithm for the PV generation system. The PV module is connected with the DC bus using a DC–DC converter. The battery and supercapacitor are interfaced with the DC bus to form a parallel active HESS using two different non-isolated half-bridge bidirectional converters. The HESS compensates the supply and demand mismatch and also helps in quick restoration of DC-link voltage (V_o) to desired level. Whenever the generation from PV system is more than the demand of the load, the HESS absorbs the excess power whereas in case of lesser generation, HESS supplies the deficit power to fulfill the load demand. The bidirectional power exchange between the DC grid and HESS is done using a buck–boost-type bidirectional converter. In Fig. 1.1. SW_1 represent the control switch for the boost converter. SW_2, SW_3 and SW_4, SW_5 represent the complementary control switches for bidirectional converters for battery bank and the SC pack, respectively. I_{pv}, I_{sc} and I_{bat} represent the PV module, SC and battery current, respectively, V_o is DC bus voltage. L_{pv}, L_{bat} and L_{sc} are PV panel, battery and SC filter inductance. C is DC-link side filter capacitance and R is the load resistance, respectively. The HESS not only maintains the power balance but also regulates the DC bus at desired level.

1.4 Accurate Mathematical Modeling of PV Array, Battery and Supercapacitor

1.4.1 Equivalent Mathematical Model of PV Panel

The photovoltaic (PV) system transforms sunlight directly into electricity. The fundamental unit of a photovoltaic energy generation system is PV cell. Depending upon the type of material used, an individual cell is a capable of producing 1 or 2 W of power approximately. Cells are generally grouped to form PV panel or module. These PV modules are ultimately connected in series and/or parallel arrangement to form PV array to meet the current and voltage requirements. All PV modules in an array are assumed identical for the sake of simplicity.

The equivalent circuit representation of a single-diode five-parameter exponential model of PV cell is shown in Fig. 1.2. It generally consists of a light generated current source (I_{ph}) in parallel with the diode (D) and also takes into account the effect of parasitic resistances (i.e., R_s and R_{sh}). Applying KCL at the node '*m*,' the current generated by the single PV cell can be expressed as:

$$I_{pv} = I_{ph} - I_D - I_{sh} \tag{1.1}$$

The current (I_{ph}) generated depends upon the intensity of light incident on the PV cell. Henceforth, the light generated current is also directly proportional to changes in temperature that corresponds to change in intensity of light. It can be expressed as:

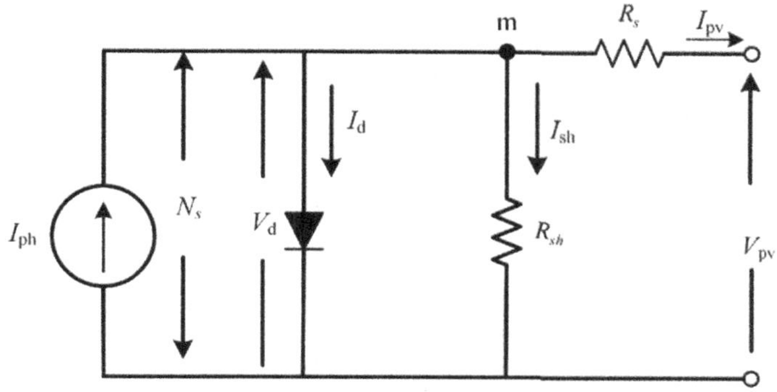

Fig. 1.2 Equivalent circuit representation of photovoltaic cell

$$I_{ph} = \left[I_{ph,n} + k_i(T - T_n)\right]{}^{S}\!/_{S_n} \tag{1.2}$$

According to semiconductor theory, the diode current (I_D) can be expressed as:

$$I_D = I_o\left[\exp\left(\frac{qV_d}{aKT}\right) - 1\right] \tag{1.3}$$

where

$$I_o = \frac{I_{sc,n} + k_i(T - T_n)}{\exp\left(\frac{V_{oc,n} + k_v(T - T_n)}{aN_s V_t}\right) - 1} \tag{1.4}$$

Substituting (1.2) and (1.3) in (1.1) we get,

$$I_{pv} = I_{ph} - I_o\left[\exp\left(\frac{V_{pv} + R_s I_{pv}}{aV_t}\right) - 1\right] - \frac{V_{pv} + R_s I_{pv}}{R_{sh}} \tag{1.5}$$

where

$$V_t = \frac{k.T}{q} \tag{1.6}$$

The equation governing the PV module current consisting of 'N_p' no. of parallel connected strings of PV cell as shown in Fig. 1.3 is expressed as:

$$I_{PV} = N_p I_{ph} - N_p I_o\left[e^{\frac{q}{akT}\left(\frac{V_{pv}}{N_s} + \frac{I_{pv}R_s}{N_p}\right)} - 1\right] - \frac{1}{R_{sh}}\left[\frac{N_p}{N_s}V_{pv} + I_{pv}R_s\right]$$

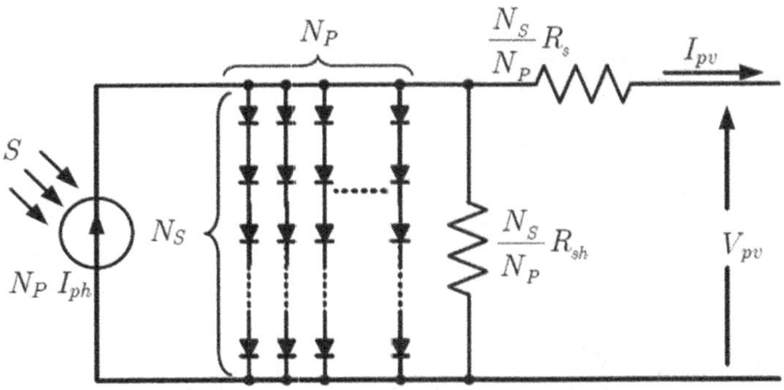

Fig. 1.3 Equivalent circuit representation of PV module consisting of 'N_p' number of strings of parallel PV cells

The current of the PV cell and the range of the operating voltage are determined from short-circuit and open-circuit conditions. From the previous equation, the short-circuit current and open-circuit voltages can determined as:

$$\begin{aligned} V_{oc}|_{I_{pv}=0} &= a.N_s.V_t \ln\left(\frac{I_{ph}}{I_o}+1\right) \\ I_{sc}|_{V_{oc}=0} &= I_{ph}\Big/\left(1+\frac{R_s}{R_{sh}}\right) \end{aligned} \tag{1.7}$$

Variables description from Sect. 3.1.2	
V_{pv}	Terminal voltage of PV panel (V)
I_{pv}	Terminal current of PV panel (A)
I_{ph}	Light generated or photon current (A)
I_D	Diode current (A)
I_{sh}	Shunt current (A)
I_o	Reverse saturation current (A)
V_{oc}	Open-circuit voltage (V)
I_{sc}	Short-circuit current (A)
$I_{ph,n}$	Photon current at STC (A)
$I_{sc,n}$	Short-circuit current at STC (A)
$V_{oc,n}$	Open-circuit voltage at STC (V)
T	Operating temperature (°C)
T_n	Nominal temperature (25 °C)
S	Operating irradiation (W/m^2)
S_n	Nominal irradiation (1000 W/m^2)
K_i	Short-circuit temperature coefficient
K_v	Open-circuit temperature coefficient
a	Ideality factor
N_s	Number of series connected cells
N_p	Number of parallel connected strings

(continued)

(continued)

Variables description from Sect. 3.1.2	
k	Boltzmann constant (1.38×10^{-23} J/K)
q	Electron charge (1.6023×10^{-19}C)
V_t	Thermal voltage imposed on the diode
V_d	Diode voltage (V)
R_s	Series resistance (Ω)
Rs_h	Shunt resistance (Ω)

1.4.2 Modeling of Battery

The equivalent circuit representation of the nonlinear model of battery is depicted in Fig. 1.4. It consists of a dependent voltage source (E) and a constant internal resistance (R_{int}). The model accounts SOC as its only state variable. The equations governing the battery model are given as the following equations:

$$E = E_o - K\frac{Q_0}{Q_0 - \int I_{batt}dt} + Ae^{-B\int I_{batt}dt} \quad (1.8)$$

$$V_{batt} = E - R_{int}I_{batt} \quad (1.9)$$

$$\%SOC = \left(1 - \frac{1}{Q}\int I_{batt}.dt\right) \times 100 \quad (1.10)$$

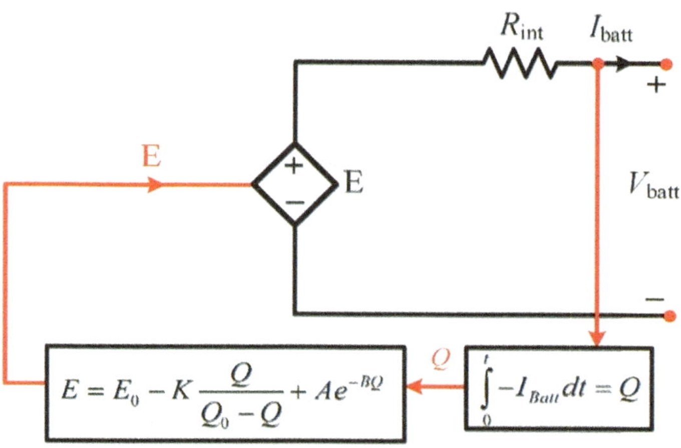

Fig. 1.4 Equivalent battery model

where

$$Q = \int I_{\text{batt}}.\mathrm{d}t.$$

Variables description from Fig. 1.4	
V_{batt}	Battery terminal voltage (V)
I_{batt}	Battery output current (A)
E	No-load voltage (V)
K	Polarization voltage (V/Ah)
Q	Actual battery charge (Ah)
Q_0	Rated battery capacity (Ah)
E_0	Constant voltage (V)
R_{int}	Internal resistance (Ω)
A	Exponential zone amplitude (V)
B	Exponential zone time constant (Ah^{-1})

1.4.3 *Modeling of Supercapacitor*

SC resembles a normal capacitor but provides significantly high capacitance. It has a fast charging and discharging capability. The equivalent circuit diagram of SC is shown in Fig. 1.5. It performs mid-way between conventional capacitors and electrochemical cells (i.e., battery) and has a high power density [20]. The output voltage of SC is generally expressed using Stern equation as follows:

$$V_{\text{sc}} = \frac{N_{\text{s}} Q_T d}{N_{\text{p}} N_{\text{e}} \varepsilon \varepsilon_0 A_i} + \frac{2 N_{\text{e}} N_{\text{s}} RT}{F} \sinh^{-1}\left\{\frac{Q_T}{N_{\text{p}} N_{\text{e}}^2 A_i \sqrt{8RT\varepsilon\varepsilon_0 c}}\right\} - R_{\text{int}}.I_{\text{sc}} \tag{1.11}$$

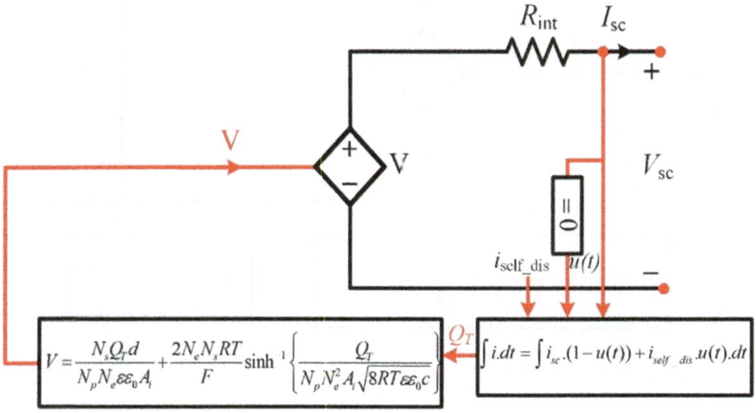

Fig. 1.5 Equivalent circuit diagram of supercapacitor

where

$$Q_T = \int I_{sc}.\mathrm{d}t \tag{1.12}$$

The SC electric charge equation can be modified considering $i_{sc} = 0$ to represent the self-discharge process as follows

$$Q_T = \int I_{self\_dis}.\mathrm{d}t \tag{1.13}$$

where

$$I_{self\_dis} = \begin{cases} \frac{C_T\alpha_1}{1+sR_{sc}C_T} & \text{if} \rightarrow t - t_{oc} \leq t_3 \\ \frac{C_T\alpha_2}{1+sR_{sc}C_T} & \text{if} \rightarrow t_3 < t - t_{oc} \leq t_3 \\ \frac{C_T\alpha_3}{1+sR_{sc}C_T} & \text{if} \rightarrow t - t_{oc} > t_3 \end{cases} \tag{1.14}$$

The constants α_1, α_2 and α_3 depict the rates of change of the SC voltage during different time intervals (t_{oc}, t_3), (t_3, t_4) and (t_4, t_5), respectively, as shown in Fig. 1.6.

Variables description from Fig. 1.5	
A_i	Area in between electrolyte and electrodes (m^2)
c	Molar concentration (mol/m^3)
r	Radius of the molecule (m)
F	Faraday constant
I_{sc}	Supercapacitor terminal current (A)
V_{sc}	Supercapacitor terminal voltage (V)
C_T	Rated capacitance (F)
R_{int}	Internal resistance of supercapacitor (Ω)
N_e	No. of electrode layers
N_A	Avogadro constant
N_p	No. of parallel supercapacitors
N_s	No. of series supercapacitors
Q_T	Electric charge (C)
R	Gas constant
T	Temperature (K)
ε	Permittivity of material
ε_0	Permittivity of free space

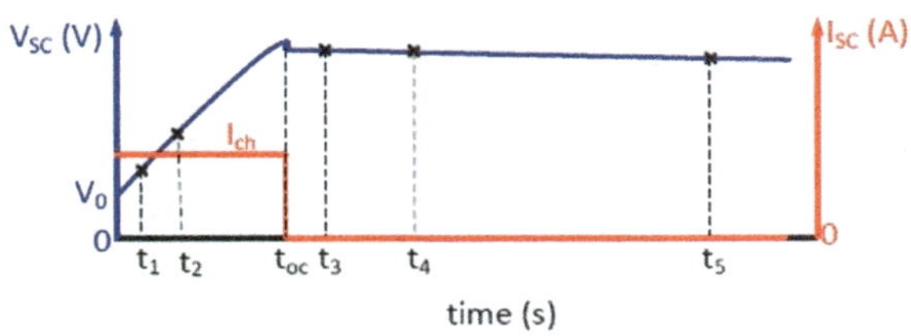

Fig. 1.6 Graph depicting the variation of SC voltage with respect to time

1.5 Design Considerations

Since the mathematical modeling and analysis of solar PV module, battery and supercapacitor have been discussed in Sect. 1.4, this section explains the design of DC–DC converters.

1.5.1 Design of DC–DC Boost Converter for MPPT Control

Boost converters are generally employed in industrial drives, automotive applications, adaptive control applications, battery power applications, etc. Where higher level of DC voltages are required with low level of input voltage. Particularly in case of PV application, the boost converter not only amplifies the output PV voltage to desired level but also performs the maximum power point tracking (MPPT) control. These converters are more popular because of their simple and rugged structure, less maintenance, easy operation and less cost.

Calculation of Parameters

The circuit diagram representation of the boost converter comprising MOSFET (SW_1), power diode (D), capacitor (C) and inductor (L_{pv}) is shown in Fig. 1.7. The converter during continuous current operation can be realized in two distinct modes in accordance with the status of the MOSFET.

The inductor current ripple (ΔI_L) is considered as 20% of the output1current and the capacitor voltage ripple (ΔV_c) is considered as 5% of the output1voltage and operating switching frequency (f_s) is 20 kHz to achieve the better performance of the converter. The switch and diode used in the converter are selected based on their voltage rating, current rating and operating characteristics.

The duty cycle is defined as the ratio of turn-on time of the switch to the total switching time.

$$D = \frac{T_{\text{on}}}{T_{\text{s}}} \tag{1.15}$$

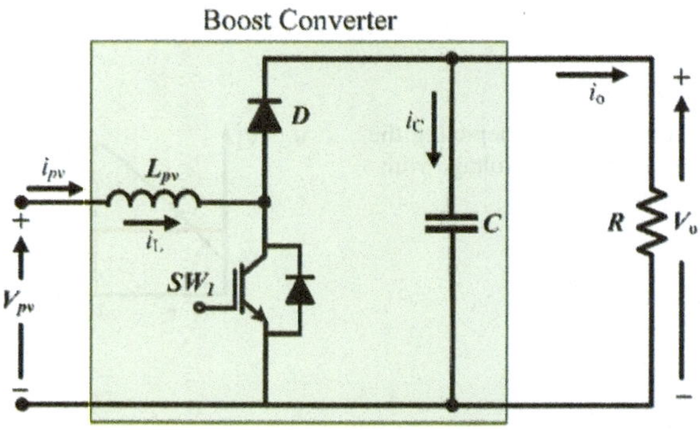

Fig. 1.7 Circuit schematic of DC–DC boost converter

The output voltage and output current can be calculated as:

$$\begin{aligned} V_o &= \frac{V_{pv}}{1-D} \\ I_o &= \frac{V_o}{R_L} \end{aligned} \tag{1.16}$$

The inductance (L_{pv}) and 1 capacitance (C_{pv}) can be calculated using 1.16 and 1.17

$$L_{pv} = \frac{V_{pv} \times D(1-D)}{\Delta I_L \times f_s} \tag{1.17}$$

$$C = \frac{V_{pv} \times D(1-D)}{8Lf_s^2 \Delta V_c} \tag{1.18}$$

1.5.2 Design of Bidirectional Converter

The parallel active configured battery-SC HESS provides the best utilization of SC capability as well as utilization of entire control of battery-SC currents. Therefore, battery-SC HESS in parallel active arrangement as shown in Fig. 1.1 has been used in the current work.

The operation and design of DC–DC non-isolated-type bidirectional converter is discussed right here since the functioning of both the battery and the supercapacitor converter will be identical. The battery energy storage system (BESS) as shown in Fig. 1.8 is developed using a battery stack and a DC–DC non-isolated-type buck–boost bidirectional converter comprising of two bidirectional switches (S_1 and S_2) with an input inductor (L_{bat}) and a filter capacitor (C) at the DC-link side. This particular converter can operate either in boost or buck mode in accordance with the duty cycle of the switches S_1 and S_2. The switches S_1 and S_2 are mutually exclusive to each other. Whenever the switch 'S_2' is switched on, the energy stored in the inductor through the battery increases. At the particular instant, 'S_2' is switched off and 'S_1' is switched on, the net current flows from the battery pack to the DC-link if $v_{batt} + v_l > v_o$ until 'S_1' remains in on condition. This is considered as the boost or battery discharging mode of operation. However, if the energy stored in the inductor decreases before 'S_2' is off or perhaps if $v_{batt} + v_l < v_o$ at end of the period when 'S_1' is switched on, the inductor stores energy in opposite direction, and therefore the current flows from DC-link to the battery pack. This mode of operation is considered as the buck. During this mode, the battery gets charged.

Filter design for bidirectional converter: The DC–DC bidirectional converter consists of two bidirectional switches with input inductor (L_{bat}) and output filter capacitor (C).

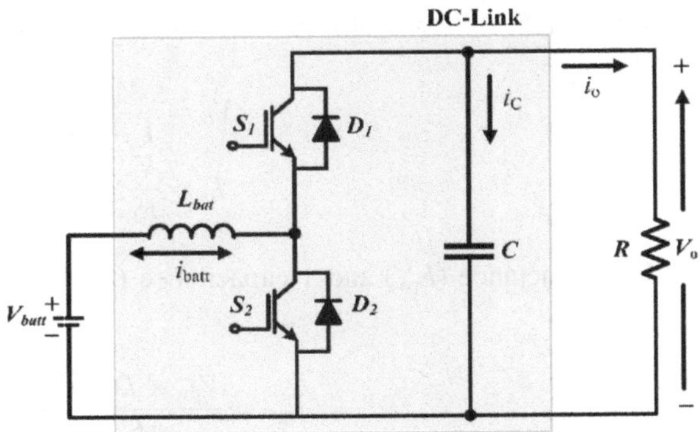

Fig. 1.8 Battery connected to the DC-link via buck–boost-type bidirectional converter

During the boost mode of operation, the inductor can be calculated as:

$$L_{bat\_boost} = \frac{V_{bat} D_{boost}}{f_{sw} \times \Delta I_{bat}} = \frac{V_o - V_{bat}}{f_{sw} \times \Delta I_{bat}} \cdot \frac{V_{bat}}{V_o} \tag{1.19}$$

During the buck mode of operation, the inductor can be calculated as:

$$L_{bat\_buck} = \frac{V_{bat}(1 - D_{buck})}{f_{sw} \times \Delta I_{bat}} = \frac{V_o - V_{bat}}{f_{sw} \times \Delta I_{bat}} \cdot \frac{V_{bat}}{V_o} \tag{1.20}$$

During continuous mode of operation, L_{bat} is selected as:

$$L_{bat} = \max[L_{b\_boost}, L_{b\_buck}] \tag{1.21}$$

DC bus capacitor is calculated as:

$$C = \frac{V_o D}{R \times \Delta V_o \times f_{sw}} \tag{1.22}$$

where V_{bat} is the battery voltage, D_{buck} and D_{boost} is the duty ratio during buck mode and boost mode, respectively, f_{sw} is the switching frequency, ΔV_o is the ripple content in DC bus voltage.

1.6 Control Strategy of PV-Based LVDC Microgrid with Parallel Active HESS Configuration

A stand-alone solar PV system with parallel active HESS arrangement is depicted in Fig. 1.1. It can be seen that the PV module is interfaced to the DC grid via a DC–DC boost converter. Controlling the duty ratio of converter using MPPT technique ensures the PV panel to work at its maximum power operating point (MPOP).

The HESS compensates the supply and demand mismatch and maintain a constant DC-link voltage (V_o). Whenever the generation from PV system is more than the load demand, the HESS absorbs the excess power whereas for less generation, HESS supplies the deficit power to fulfill the load demand. The bidirectional exchange of power between the DC grid and HESS is done using a buck–boost-type bidirectional converter. In Fig. 1.4b, SW_1 represent the control switch for the boost converter. SW_2, SW_3 and SW_4, SW_5 are the complementary control switches for bidirectional converters for battery bank and SC pack, respectively. I_{pv}, I_{sc} and I_{bat} represent the PV module, SC and battery current respectively, V_o is DC bus voltage. L_{pv}, L_{bat} and L_{sc} are PV panel, battery and SC filter inductance. C is filter capacitance and R is the load resistance, respectively. The parameters of all the converters are designed following the procedure mentioned in [17, 18] and the calculated values are listed in Table. 1.1.

The block diagram of the control strategy for HESS with parallel active topology is shown in Fig. 1.9. The main objective of the control strategy is to maintain and restore the DC bus voltage at the desired level under sudden variation in solar irradiation or load. The control strategy also aims to increase the battery life while ensuring minimal current stress upon the battery. The power necessary to balance the entire power flow at the DC bus during irradiation and load variation is segregated into two separate components: (a) average power component (P_{avg}) or the low-frequency component and (b) transient component (P_{tran}) or the high-frequency component.

The power balance can be represented as:

$$P_l(t) - P_{pv}(t) = P_b(t) + P_{sc}(t) = \overline{P}_{avg} + \hat{P}_{tran} \tag{1.23}$$

where $P_l(t)$, $P_{pv}(t)$, $P_b(t)$ and $P_{sc}(t)$ represent the load, PV, battery and supercapacitor power, respectively. The control scheme compares the DC-link voltage (V_o) with the reference voltage (V_{ref}) and the error signal is processed through a PI controller. The output of the PI controller is the total reference current (I_{tot_ref}) necessary to establish a constant DC-link voltage. So, the net power provided by the HESS can be expressed as:

Table. 1.1 Simulation parameters of PV-based LVDC microgrid with HESS

PV array at STC (1000 W/m^2, 25° C)
Maximum power = 960 W V_{oc} = 42 V, I_{sc} = 32 A V_{mpp} = 34 V, I_{mpp} = 28.4 A
Lead-acid Battery: 24 V, 14 Ah
Supercapacitor: 29 F, 32 V
DC–DC Boost Converter L_{pv} = 0.35 mH, C = 390 μF, f_s = 20 kHz
DC–DC Bidirectional Converter L_{bat} = 0.32 mH, L_{sc} = 0.355 mH, C = 390 μF, f_{sw} = 20 kHz
V_o = 48 V, R = 4.8 Ω

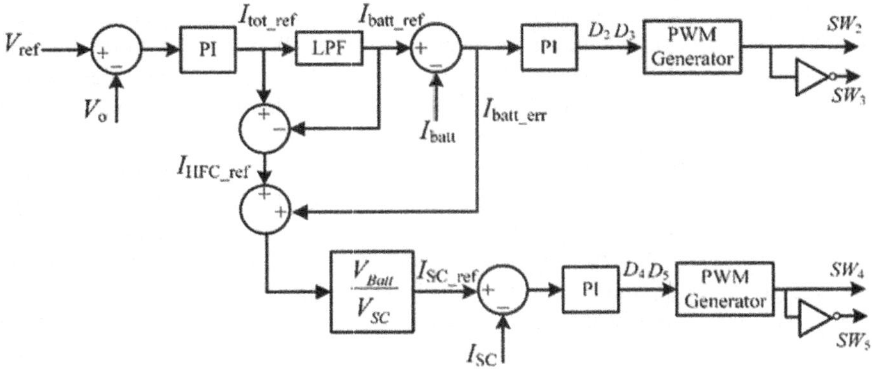

Fig. 1.9 Block diagram of control strategy of PV-based microgrid with HESS

$$\overline{P}_{avg}(t) + P_{tran}(t) = v_0 i_{tot\_ref} \tag{1.24}$$

I_{tot_ref} is decomposed into two frequency components (i.e., (I_{HFC_ref}) and ($I_{LFC\ \_ref}$)) using a low-pass filter (LPF). The low-frequency component is given as reference to the battery current controller.

$$I_{LFC\_ref} = I_{bat\_ref} = \mathrm{LPF}(I_{tot\_ref}) \tag{1.25}$$

$$I_{HFC\_ref} = I_{tot\_ref} - I_{LFC\_ref} \tag{1.26}$$

The difference between I_{bat_ref} and actual battery current (I_{bat}) is passed through another PI controller as shown in Fig. 1.10 to generate the duty ratios (D_2; D_3). For generation of switching pulses for operation of the battery converter switches (SW_2; SW_3), duty ratios (D_2; D_3) are provided to a PWM generator. Due to the slow dynamics of the battery, I_{bat} may not track the I_{bat_ref} instantly. Therefore, the equation of uncompensated battery power is given in Eq. 1.27.

$$P_{bat\_uncomp} = (I_{HFC\_ref} + I_{bat\_err}) \times V_{bat} \tag{1.27}$$

The supercapacitor compensates for the power fluctuations originating due to battery error component and the high-frequency part of the total reference current.

$$I_{sc\_ref} = \frac{P_{bat\_uncomp}}{V_{sc}} = \frac{(I_{HFC\_ref} + I_{bat\_err}) \times V_{bat}}{V_{sc}} \tag{1.28}$$

The reference component of SC I_{sc_ref} given Eq. 1.28, is compared with the actual current of the SC and the error given to the PI controller generates switching pulses for corresponding switches (SW_4; SW_5) of bidirectional converter interfaced with SC.

1.7 Simulation Results and Discussions

The PV system with hybrid energy storage system depicted Fig. 1.1 is simulated using MATLAB/Simulink. Initially, the PV panel is operated to work at MPPT mode. INC MPPT technique has been used to extract the peak power from the PV module. The control strategy depicted in Fig. 1.9 maintains the DC bus at desired level (i.e., 48 V) and achieves power sharing among PV, battery, SC and load. The control scheme employed for the overall system is evaluated under four different operating conditions: (i) Step increment in irradiation, (ii) step decrement in irradiation, (iii) step increment in load demand and (iv) step decrement in load demand.

1.7.1 Step Increment in Irradiation

In this case, initially the PV module is operated in MPPT mode with irradiation of 400 W/m^2 and an ambient temperature of 25 °C. The output power corresponding to this irradiation is shown in Fig. 1.10a. The PV output is increased at t = 3 s from 388 to 775 W upon irradiation increase from 400 to 800 W/m^2. During this condition, a constant load power of 480 W is maintained as shown in Fig. 1.10b. Due to a sudden increase in irradiation, the surplus power generated by the PV system is absorbed by HESS to maintain the DC-link voltage (V_o) at 48 V. SC absorbs the I_{HFC} of the surplus power momentarily, whereas the battery absorbs the average power component I_{LFC} as shown in Fig. 1.10c. As the PV power increases due to sudden irradiation change, the output voltage or the DC-link voltage increases in proportion instantly and again settles to its reference value quickly as shown in Fig. 1.10d. It can be observed from Fig. 1.10e that, at t = 3 s, the battery current makes a smooth transition from discharging mode to charging mode. Because, the SC takes care of high-frequency component momentarily which is shown in Fig. 1.10g. The % SOC of battery starts increasing as shown in Fig. 1.10f. The load current shown in Fig. 1.10h is maintained constant at 10 A throughout this operation.

1.7.2 Step Decrement in Irradiation

To check the transient performance of the control scheme under sudden decrease in PV generation, the input irradiation of the PV panel is suddenly decreased from 800 to 400 W/m^2 at t = 9 s. It is seen from Fig. 1.11a, PV power drops from 775 to 388 W and the load power is 480 W. So, the voltage decreases instantaneously as the PV generation decreases suddenly. To maintain the V_o at 48 V, the deficit power is delivered by the HESS as shown in Fig. 1.11c. The load power and DC-link voltage are maintained constant throughout the entire operation as shown in Fig. 1.11b, d, respectively. The battery current I_{batt} and %SOC of the battery are

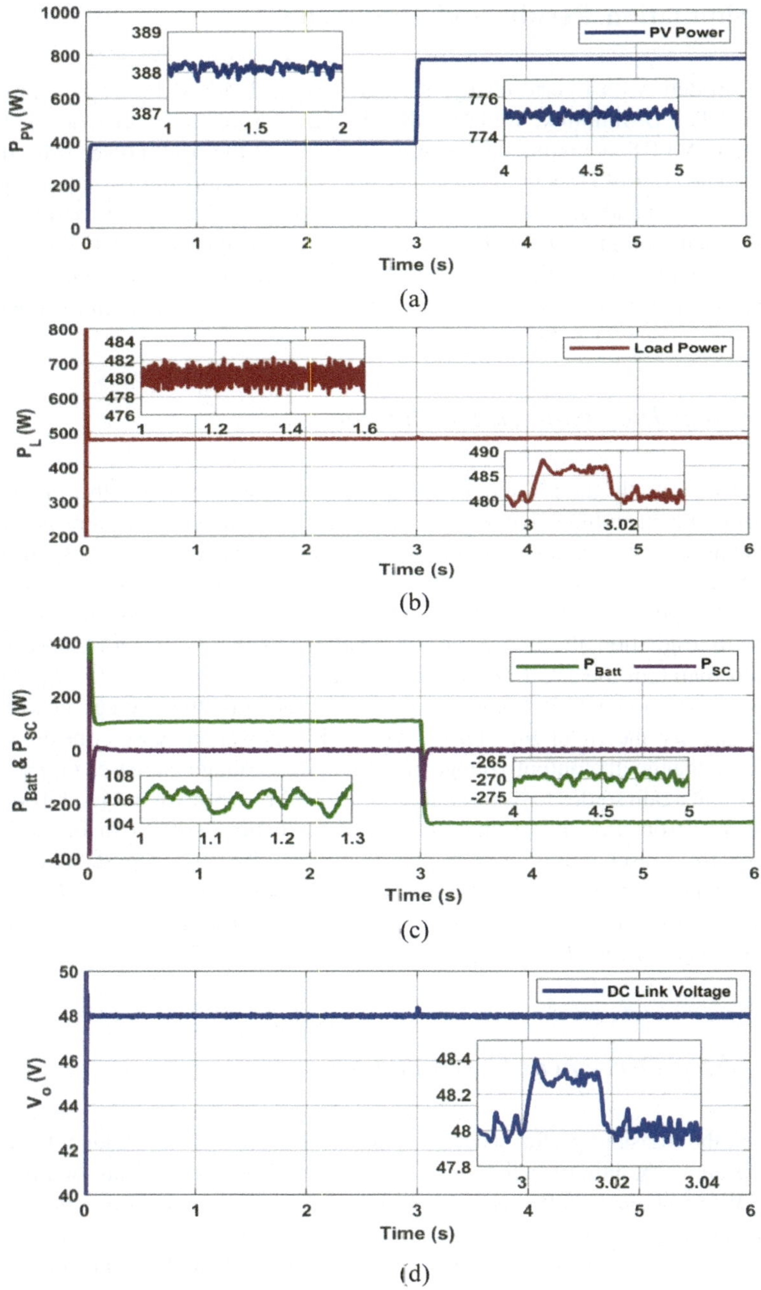

Fig. 1.10 Simulation results for step increment in irradiation **a** PV power, **b** load power, **c** battery and supercapacitor power, **d** DC-link voltage, **e** SOC of the battery, **f** battery current, **g** supercapacitor current, **h** load current

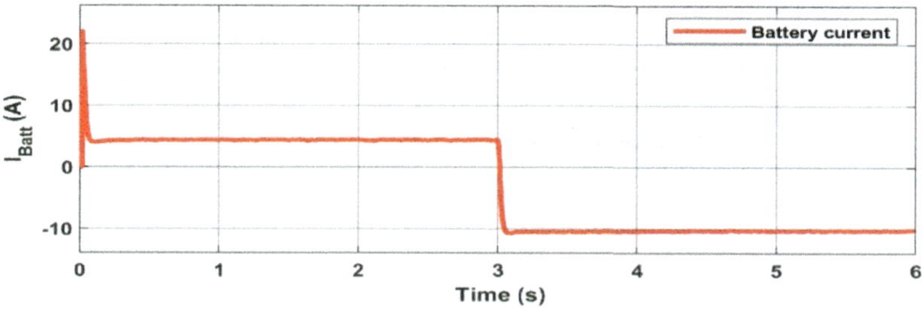

(e)

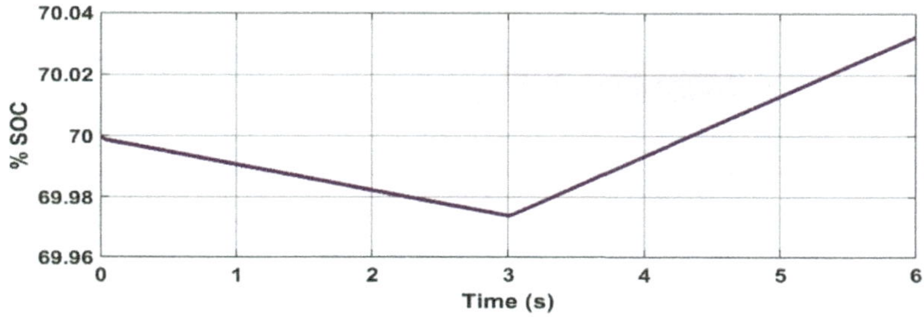

(f)

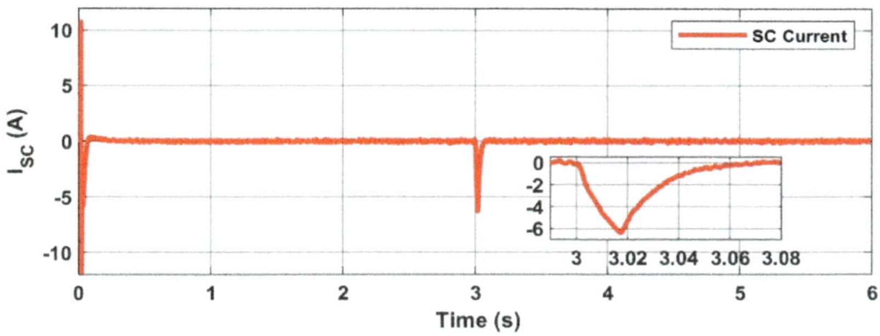

(g)

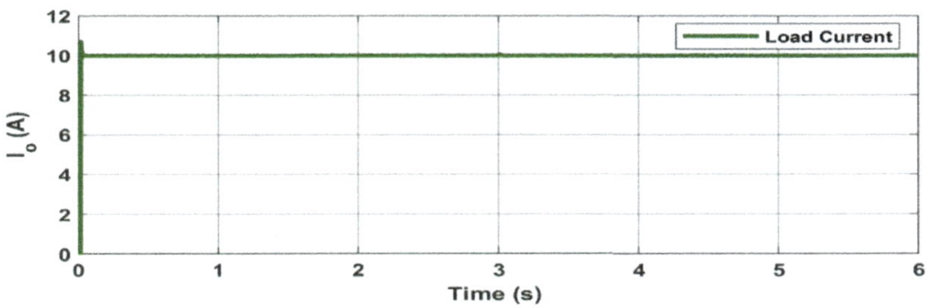

(h)

Fig. 1.10 (continued)

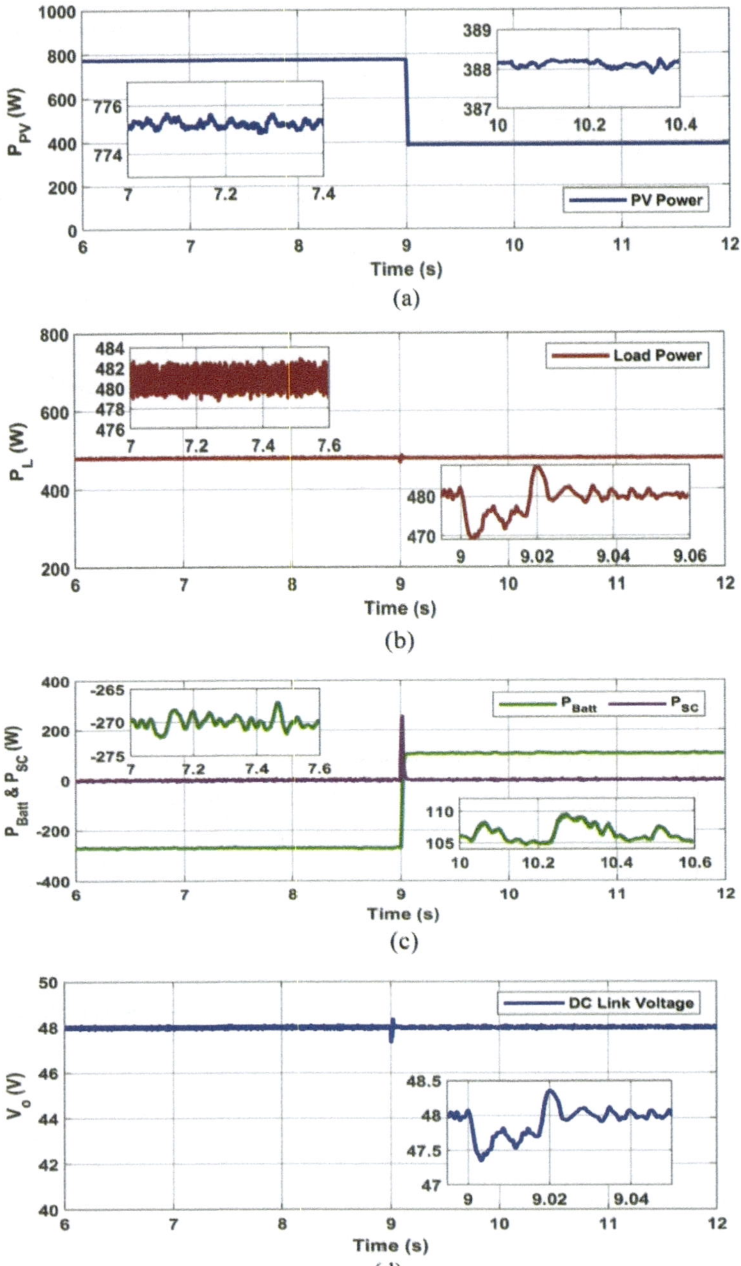

Fig. 1.11 Simulation results for step decrement in irradiation **a** PV power, **b** load power, **c** battery and supercapacitor power, **d** DC-link voltage, **e** SOC of the battery, **f** battery current, **g** supercapacitor current, **h** load current

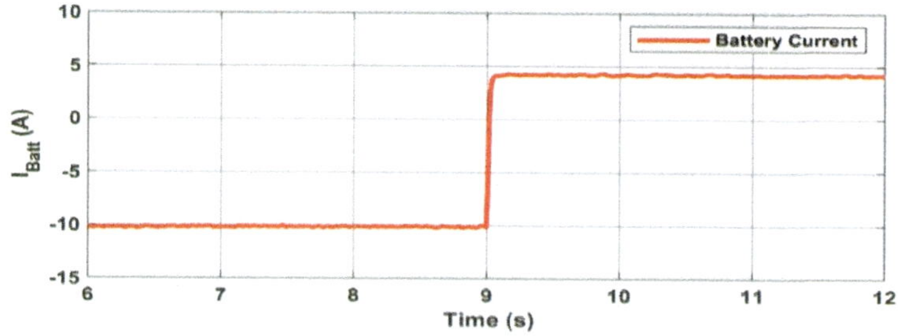

(e)

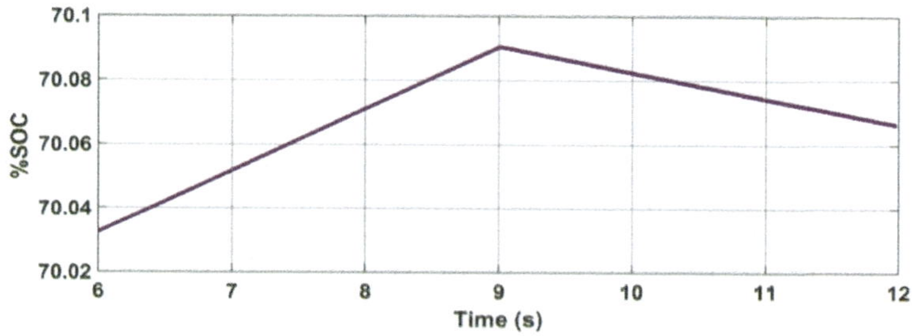

(f)

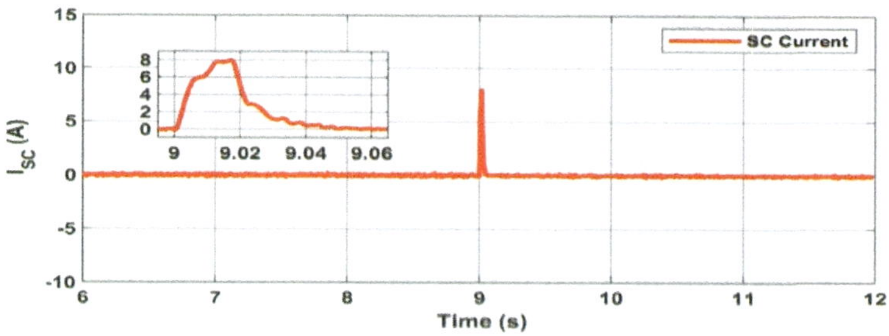

(g)

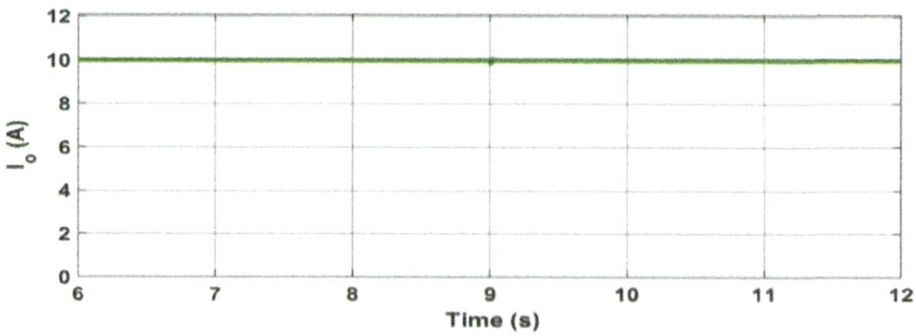

(h)

Fig. 1.11 (continued)

shown in Fig. 1.11e, f, respectively. The SC provides I_{HFC} of surplus demand along with the error component for short term as shown in Fig. 1.11g and diverts the steady-state component toward the battery. Therefore, due to low discharge current rate, the current stress on the battery reduces. The load current is maintained constant at 10 A as shown in Fig. 1.11h.

1.7.3 Step Increment in Load Demand

In this case, the simulation is performed under step increase in load demand at a constant irradiation of $S = 600$ W/m^2 and $T = 25$ °C. The PV module is operated at MPPT mode with $R = 4.8$ Ω. At $t = 3$ s, the load demand is suddenly increased by decreasing R to 2.4 Ω. At this instant, the PV panel supplies 583 W as shown in Fig. 1.12a, but the load demand is 960 W (Fig. 1.12b). To maintain the V_o at 48 V, the surplus demand is to be supplied by HESS as shown in Fig. 1.12c where SC supplies the high-frequency component until the battery provides the steady-state demand. It can be observed in Fig. 1.12d that the output voltage is stabilized quickly and a constant DC-link voltage of 48 V is maintained throughout. The waveforms of battery current, %SOC of battery, SC current and load current are shown Fig. 1.12e–h, respectively.

1.7.4 Step Decrement in Load Demand

The simulation results for sudden decrement in load demand are shown in Fig. 1.8 at a constant irradiance of $S = 600$ W/m^2 and $T = 25$ °C. With a constant PV output power of 583 W, at $t = 9$ s, the load demand is decreased suddenly from 960 to 480 W (Fig. 1.13a, b). Similarly, the HESS absorbs the excess power during sudden reduction in the load demand as shown in Fig. 1.13c in order to provide a stable system performance. The corresponding V_o, I_{batt}, %SOC, I_{sc} and I_o are shown Fig. 1.13d–h, respectively. From the corresponding results, it is found that the output voltage is stabilized quickly maintaining a constant value of 48 V.

1.8 Conclusion

The design, analysis and simulation of PV system with a parallel active HESS configuration are presented. The PV system is operated to run under MPP mode to meet the load requirement. A HESS consisting of SC and a battery pack helps to regulate the DC-link voltage and compensates the generation-demand mismatch. A simple control scheme for DC grid voltage regulation is employed. It is based on decomposition total reference current into low and high-frequency power components. The control scheme for voltage regulation and power sharing between battery and SC is successfully simulated using MATLAB/Simulink. The simulation results confirm that during sudden abrupt variation in irradiance and/or load demand, the steady-state component is managed by the battery storage system, whereas the transient component is taken care by SC momentarily. The advantages of control strategy can be attributed as better voltage regulation and less current stress on battery during transient conditions.

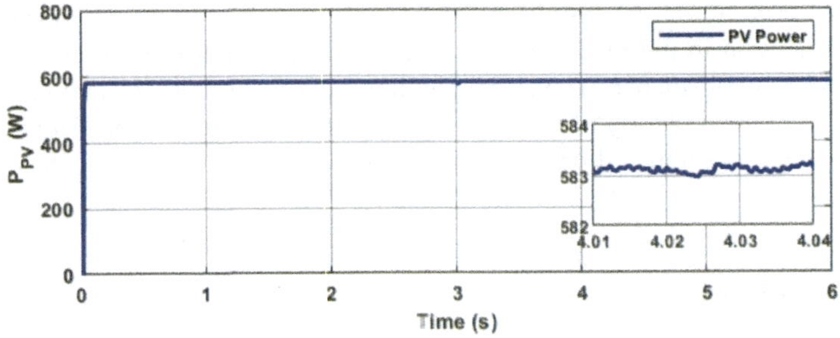

(a)

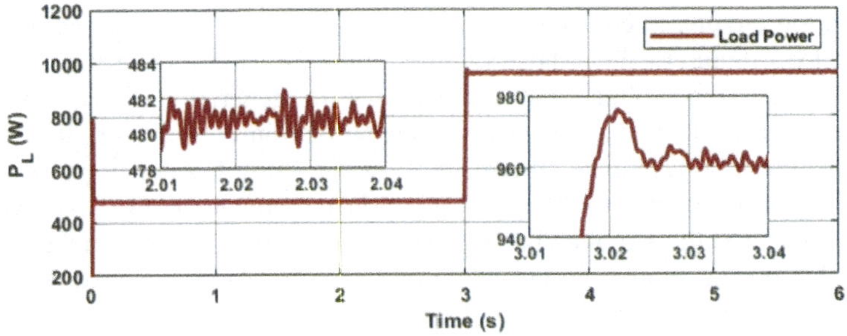

(b)

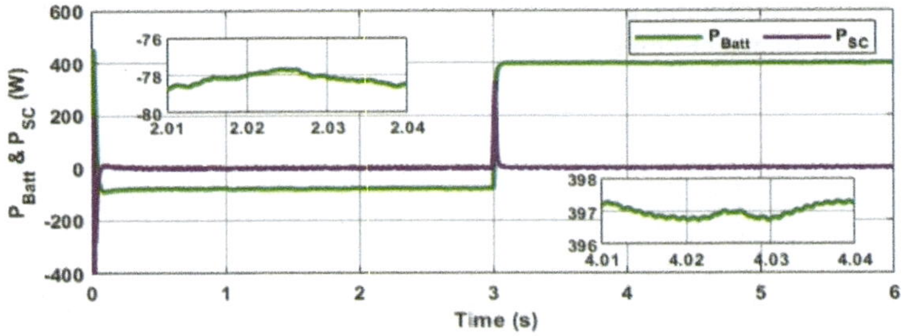

(c)

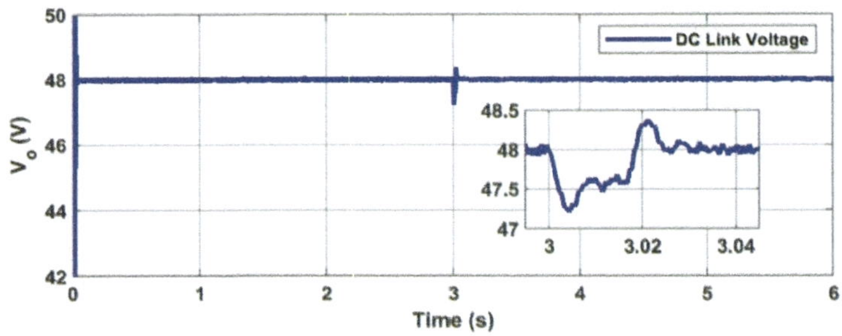

(d)

Fig. 1.12 Simulation results for step increase in load demand **a** PV power, **b** load power, **c** battery and supercapacitor power, **d** DC-link voltage, **e** SOC of the battery, **f** battery current, **g** supercapacitor current, **h** load current

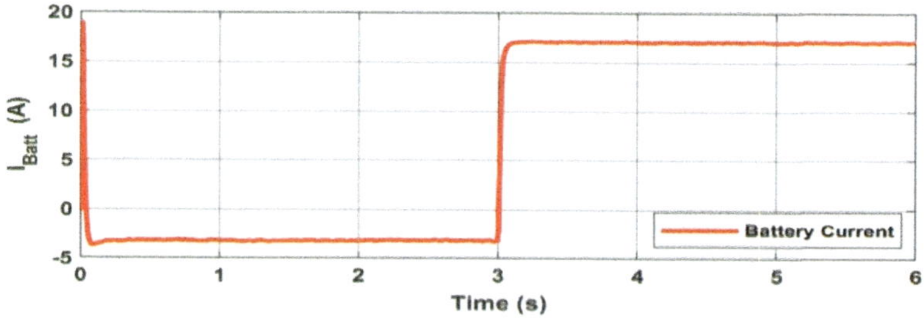

(e)

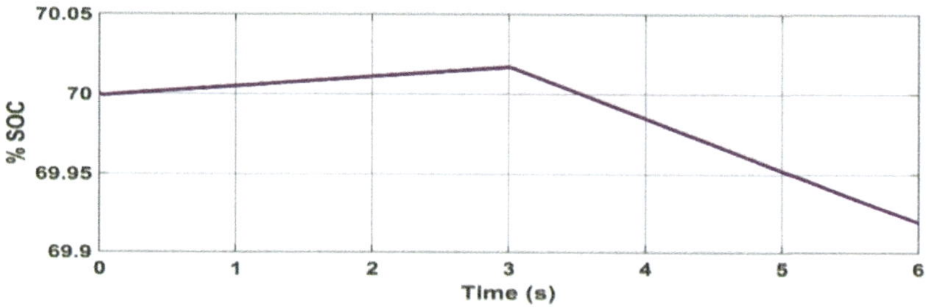

(f)

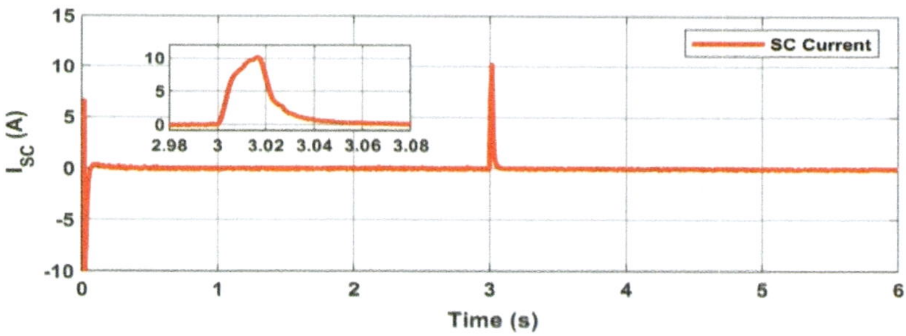

(g)

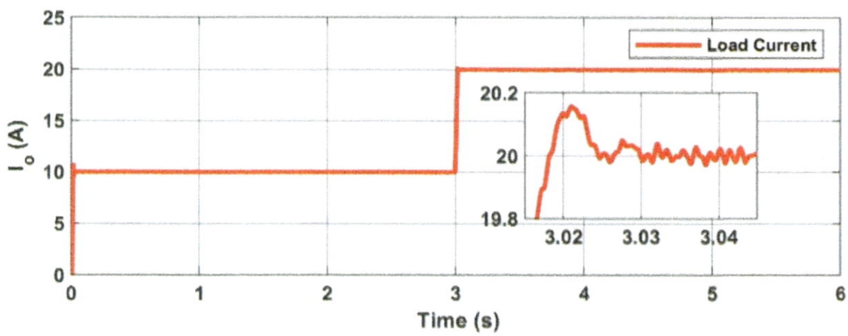

(h)

Fig. 1.12 (continued)

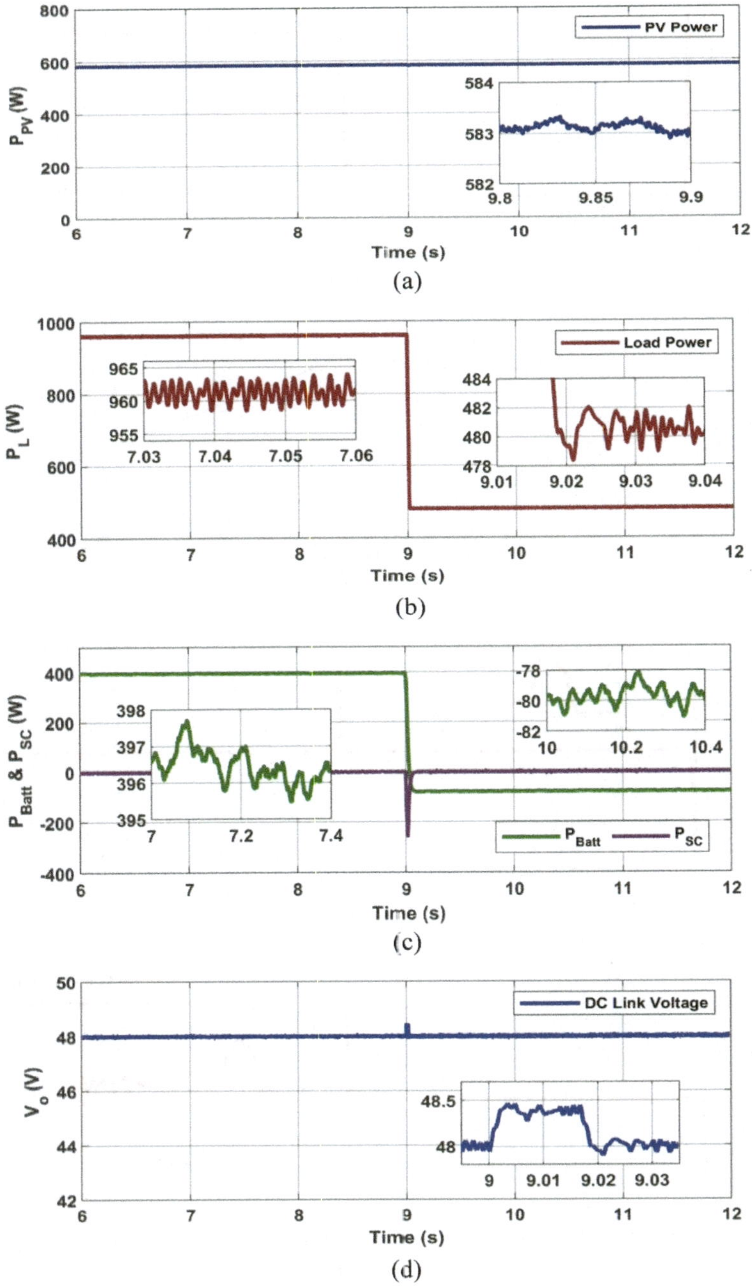

Fig. 1.13 Simulation results for step decrease in load demand **a** PV power, **b** load power, **c** battery and supercapacitor power, **d** DC-link voltage, **e** SOC of the battery, **f** battery current, **g** supercapacitor current, **h** load current

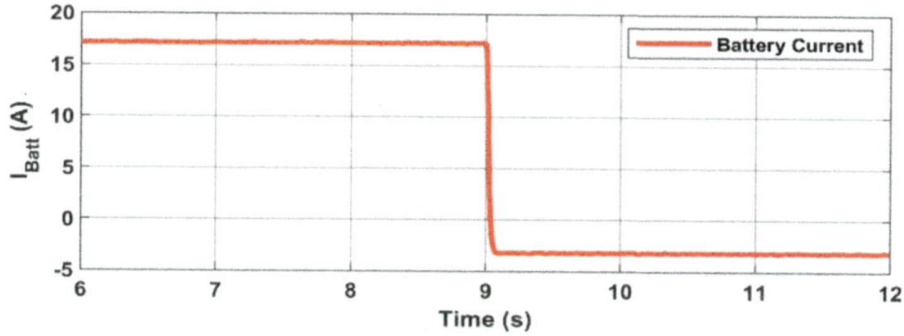

(e)

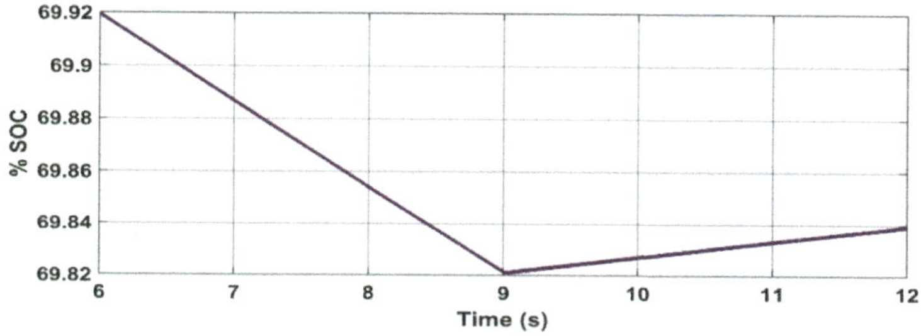

(f)

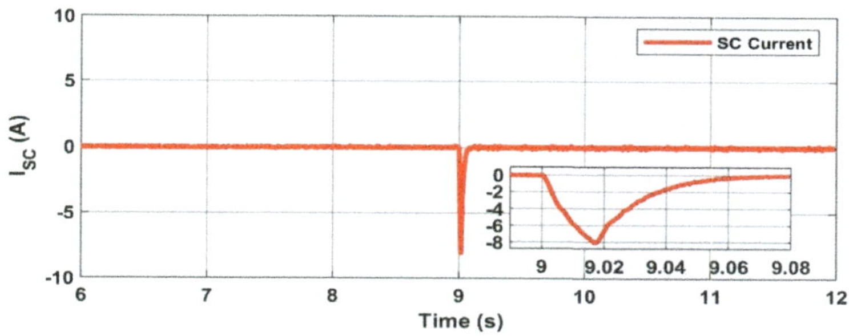

(g)

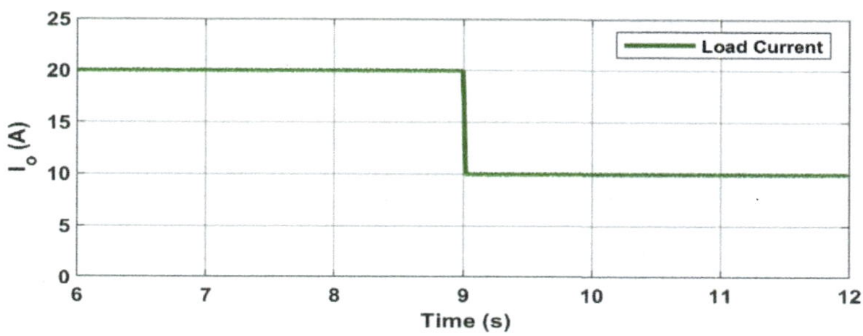

(h)

Fig. 1.13 (continued)

References

1. Olivares, D.E., Mehrizi-Sani, A., Etemadi, A.H., Canizares, C.A., Iravani, R., Kazerani, M., Hajimiragha, A.H., Gomis-Bellmunt, O., Saeedifard, M., Palma-Behnke, R., et al.: Trends in microgrid control. IEEE Trans. Smart Grid, **5**(4), 1905–1919 (2014)
2. Kumar, D., Zare, F., Ghosh, A.: Dc microgrid technology: system architectures, ac grid interfaces, grounding schemes, power quality, and communication networks, applications, and standardizations aspects. IEEE Access **5**(12), 230–256 (2017)
3. Mackay, L., Hailu, T.G., Mouli, G.C., Ramirez-Elizondo, L., Ferreira, J., Bauer, P.: From dc nano-and microgrids towards the universal dc distribution system-a plea to think further into the future. In: 2015 IEEE Power & Energy Society General Meeting. IEEE, pp. 1–5 (2015)
4. Li, W., Mou, X., Zhou, Y., Marnay, C.: On voltage standards for dc home microgrids energized by distributed sources. In: Proceedings of The 7th International Power Electronics and Motion Control Conference, vol. 3. IEEE, pp. 2282–2286 (2012)
5. Moussa, S., Ghorbal, M.J.-B., Slama-Belkhodja, I.: Bus voltage level choice for standalone residential dc nanogrid. Sustain. Cities Soc. **46**, 101431 (2019)
6. Anand, S., Fernandes, B.: Optimal voltage level for dc microgrids. In: IECON 2010–36th Annual Conference on IEEE Industrial Electronics Society. IEEE, pp. 3034–3039 (2010)
7. Dragicevic, T., Lu, X., Vasquez, J.C., Guerrero, J.M.: Dc microgrids part I: A review of control strategies and stabilization techniques. IEEE Trans. Power Electron. **31**(7), 4876–4891 (2015)
8. Augustine, S., Lakshminarasamma, N., Mishra, M.K.: Control of photovoltaic-based low-voltage dc microgrid system for power sharing with modified droop algorithm. IET Power Electron. **9**(6), 1132–1143 (2016)
9. Esram, T., Chapman, P.L.: Comparison of photovoltaic array maximum power point tracking techniques. IEEE Trans. Energy Convers. **22**(2), 439–449 (2007)
10. Jing, W., Lai, C.H., Wong, S.H.W., Wong, M.L.D.: Battery-supercapacitor hybrid energy storage system in standalone dc microgrids: a review. IET Renew. Power Gener. **11**(4), 461–469 (2016)
11. Shimizukawa, J., Iba, K., Hida, Y., Yokoyama, R.: Mitigation of intermittency of wind power generation using battery energy storage system. In: 45th International Universities Power Engineering Conference UPEC 2010. IEEE, pp. 1–4 (2010)
12. Serban, I., Marinescu, C.: Battery energy storage system for frequency support in microgrids and with enhanced control features for uninterruptible supply of local loads. Int. J. Electr. Power Energy Syst. **54**, 432–441 (2014)
13. Ding, M., Wang, B., Chen, Z., Chen, Z., Luo, Y., Zheng, G.: Stabilizing control strategy of complementary energy storage in renewable energy system. In: IEEE PES Innovative Smart Grid Technologies. IEEE, pp. 1–5 (2012)
14. Choi, M.-E., Kim, S.-W., Seo, S.-W.: Energy management optimization in a battery/supercapacitor hybrid energy storage system. IEEE Trans. Smart Grid **3**(1), 463–472 (2011)
15. Mendis, N., Muttaqi, K.M., Perera, S.: Management of battery-supercapacitor hybrid energy storage and synchronous condenser for isolated operation of pmsg based variable-speed wind turbine generating systems. IEEE Trans. Smart Grid **5**(2), 944–953 (2014)
16. Laldin, O., Moshirvaziri, M., Trescases, O.: Predictive algorithm for optimizing power flow in hybrid ultracapacitor/battery storage systems for light electric vehicles. IEEE Trans. Power Electron. **28**(8), 3882–3895 (2012)
17. Dong, B., Li, Y., Zheng, Z.: Composite converter of hybrid storage in distributed renewable energy generation system. In: 2011 International Conference on Electrical Machines and Systems. IEEE, pp. 1–4 (2011)
18. Cabrane, Z., Ouassaid, M., Maarou, M.: Management and control of storage photovoltaic energy using battery-supercapacitor combination. In: 2014 Second World Conference on Complex Systems (WCCS). IEEE, pp. 380–385 (2014)

19. Dong, B., Li, Y., Zheng, Z., Xu, L.: Control strategies of microgrid with hybrid dc and ac buses. In: Proceedings of the 2011 14th European Conference on Power Electronics and Applications. IEEE, pp. 1–8 (2011)
20. Joshi, M.C., Samanta, S.: Modified ultracapacitor voltage control loop for battery/uc hess. In: 2018 IEEE International Conference on Power Electronics, Drives and Energy Systems (PEDES). IEEE, pp. 1–5 (2018)
21. Ortuzar, M., Moreno, J., Dixon, J.: Ultracapacitor-based auxiliary energy system for an electric vehicle: implementation and evaluation. IEEE Trans. Industr. Electron. **54**(4), 2147–2156 (2007)
22. Feng, X., Gooi, H., Chen, S.: Hybrid energy storage with multimode fuzzy power allocator for pv systems. IEEE Trans. Sustain. Energy **5**(2), 389–397 (2014)
23. Guo, T., Liu, Y., Zhao, J., Zhu, Y., Liu, J.: A dynamic wavelet-based robust wind power smoothing approach using hybrid energy storage system. Int. J. Electr. Power Energy Syst. **116**, 105579 (2020)
24. Hu, Y., Chen, C., He, T., He, J., Guan, X., Yang, B.: Proactive power management scheme for hybrid electric storage system in evs: an mpc method. IEEE Trans. Intell. Transp. Syst. (2019)
25. Tsai, H.-L., Tu, C.-S., Su, Y.-J., et al.: Development of generalized photovoltaic model using matlab/Simulink. In: Proceedings of the World Congress on Engineering and Computer Science, vol. 2008. San Francisco, USA, pp. 1–6 (2008)
26. Tremblay, O., Dessaint, L.-A., Dekkiche, A.-I.: A generic battery model for the dynamic simulation of hybrid electric vehicles. In: 2007 IEEE Vehicle Power and Propulsion Conference. IEEE, pp. 284–289 (2007)
27. Farcas, C., Petreus, D., Ciocan, I., Palaghita, N.: Modeling and simulation of supercapacitors. In: 2009 15th International Symposium for Design and Technology of Electronics Packages (SIITME). IEEE, pp. 195–200 (2009)
28. https://in.mathworks.com/help/physmod/sps/powersys/ref/supercapacitor.html
29. Motahhir, S., El Ghzizal, A., Sebti, S., Derouich, A.: Modeling of photovoltaic system with modified incremental conductance algorithm for fast changes of irradiance. Int. J. Photoenergy **2018** (2018)
30. Joshi, M.C., Samanta, S.: Energy management with improved frequency sharing based control for battery/ultracapacitor hybrid energy system in the presence of delay. IET Power Electron. (2020)
31. Kollimalla, S.K., Mishra, M.K., Narasamma, N.L.: Design and analysis of novel control strategy for battery and supercapacitor storage system. IEEE Trans. Sustain. Energy **5**(4), 1137–1144 (2014)
32. Kollimalla, S.K., Mishra, M.K., et al.: A new control strategy for interfacing battery supercapacitor storage systems for pv system. In: 2014 IEEE Students' Conference on Electrical, Electronics and Computer Science. IEEE, pp. 1–6 (2014)
33. Zhang, F., Yang, Y., Ji, C., Wei, W., Chen, Y., Meng, C., Jin, Z., Zhang, G.: Power management strategy research for dc microgrid with hybrid storage system. In: 2015 IEEE First International Conference on DC Microgrids (ICDCM). IEEE, pp. 62–68 (2015)

Chapter 2
High Gain Buck–Boost Converter for Solar Photovoltaic (PV) System

Niraj Rana, Subrata Banerjee, and Kundan Kumar

2.1 Introduction

The rapid change in environmental conditions and gradually increasing electricity consumption led to require sustainable as well as a cheaper energy source. Due to the world energy resource shortage, the development of renewable energy sources (RES) is critical [1–3]. Nowadays, electricity production from the solar photovoltaic (PV) panel is a remarkable choice for power generation in industrial sectors due to its pollution-free characteristic [4]. The DC–DC power converters are extensively utilized in PV-based systems for interfacing between the PV panel and the connected load [5]. The converter must be designed to be connected directly to the PV panel and to perform operation for maximizing power extraction under all operating conditions employing MPPT controller [6].

The solar PV panels are extensively applied in both stand-alone and grid-connected systems and the block diagram representations of both systems are given in Fig. 2.1.

It is seen that the stand-alone system mainly involves of a PV panel, MPPT control system, DC–DC converter and load. But, the grid-connected PV-based system additionally requires solar inverter and the overall implementation requires more complex control.

However, the solar PV panel with low output voltage is the major drawback in solar power generation system. Therefore, to step-up the PV panel output voltage, the reliable and efficient converters are needed. The traditional DC–DC power converters such as boost converter (BC) and buck–boost converter (BBC) are

N. Rana (✉) · S. Banerjee
Department of Electrical Engineering, National Institute of Technology, Durgapur 713209, India

K. Kumar
Department of Electrical Engineering, National Institute of Technology, Manipur 795004, India

N. Priyadarshi et al. (eds.), *DC–DC Converters for Future Renewable Energy Systems*, Energy Systems in Electrical Engineering,
https://doi.org/10.1007/978-981-16-4388-0_2

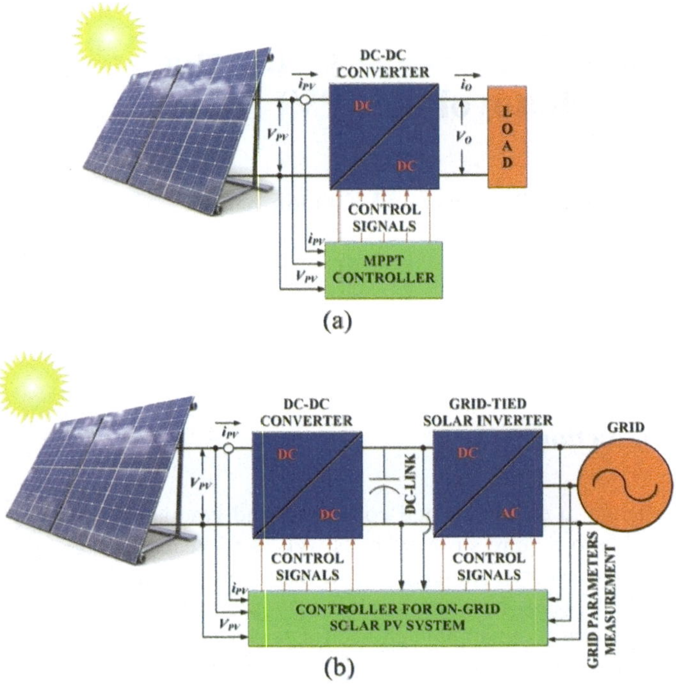

Fig. 2.1 Schematic representation of (**a**) stand-alone and (**b**) grid-connected PV-based system

employed with the MPPT-based controller at various places for maximum power extraction from the solar PV panel. But, there are certain shortcomings of the traditional converters such as limited voltage gain, high magnitude of load voltage ripple and input current ripple, poor dynamic performance and lesser power efficiency. Therefore, the applications of traditional DC–DC converters are restricted in some cases.

The isolated power electronic DC–DC converter (i.e., isolated buck–boost (flyback) converter, half/full-bridge converter) may be applied for high step-up the output voltage of the PV panel by increasing the isolated transformer turn ratio as well as duty cycle of power devices (i.e., MOSFETs and diodes). However, this technique for achieving high-voltage gain will cause high voltage and current stresses on the power devices, increases the size and weight of the overall system and poor power efficiency [7, 8]. So, a non-isolated class of DC–DC power electronics converter is preferred. The traditional non-isolated converters (i.e., BC and BBC) can be acquired high-voltage gain through extreme duty cycle operation of the power devices [9]. But, there are some unavoidable problems that happen during an extreme duty cycle operation such as the reverse recovery problem of converter diode, large ripple content, high electromagnetic interference (EMI) and low power efficiency [10].

Therefore, it is essential to design a non-isolated DC–DC power electronics converter that should have certain unique features for example high-voltage gain, lower ripple content, better power handling capability, enhanced power efficiency and improved dynamic characteristic [11].

Sometimes, the interleaving technique has been applied in the traditional converters to decrease the ripple content and to enrich the power handling capability. In this way, interleaved boost converter (IBC) as well as interleaved buck–boost converter (IBBC) have been developed and are reported in many literatures [11–15]. The interleaved DC–DC power electronics converters are applied in various industrial as well as portable electronics devices. Due to lesser ripple content in interleaved converters, their application in PV-based system is useful. However, the conventional interleaving technique cannot be able to raise the converter's voltage gain.

Some non-isolated DC–DC power electronics converter [16–18] are presented in Fig. 2.2, where converters produce high-voltage gain.

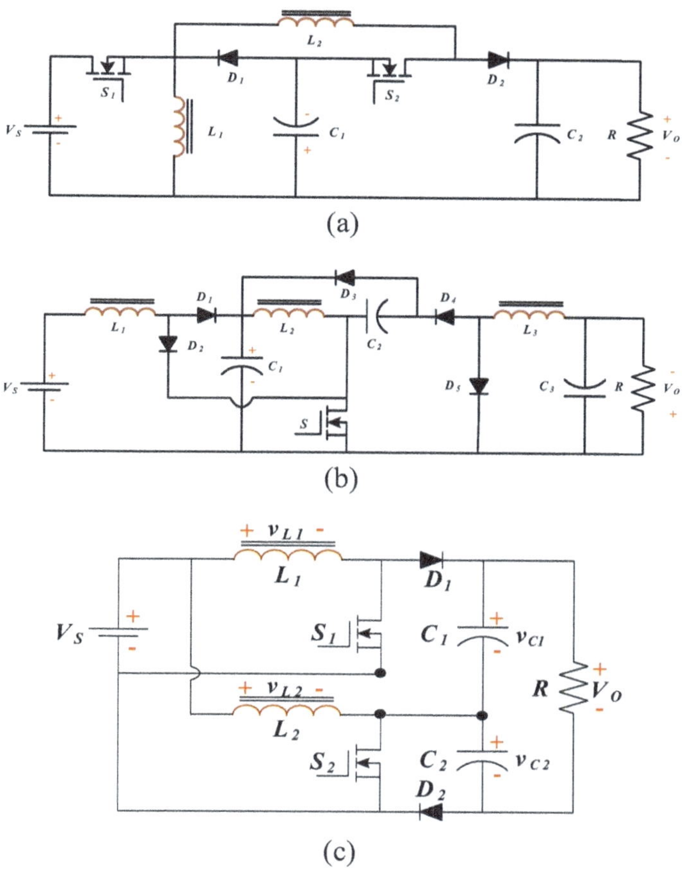

Fig. 2.2 High gain converters proposed in (a) [16], (b) [17] and (c) [18]

In [16] and [17] (Fig. 2.2a, b), two non-isolated high gain BBCs are demonstrated, where both converters produce square times voltage gain than the voltage gain of traditional BBC. However, these converters create more ripples with higher voltage gain so the conversion efficiency becomes poor.

The input parallel output series class of DC–DC power electronics boost converter [18] for enhancing the voltage gain as well as reducing the ripple content is shown in Fig. 2.2c. The converter produces the voltage gain, which is two times higher than the voltage gain of the traditional BC. But, the main drawback of the converter is that it cannot operate below the 50% duty cycle. Moreover, this converter has a serious EMI problem because the common ground bus is not present between the parallel–series-connected converter cells.

In this chapter, an improved high gain buck–boost converter (IHGBBC) suitable for PV-based systems has been demonstrated to overcome the above shortcomings of the converter systems. The IHGBBC is established by paralleling the input side and cascading the load side of two traditional BBCs cells. Consequently, IHGBBC will have double voltage gain as compared to a single TBBC cell. Due to the parallel connection of two TBBCs cells and 180° phase shifting operation (interleaving technique) between them, the overall rating is increased as well as ripple content is also reduced as compared to a single TBBC cell. The duty-to-output small-signal transfer function (TF) of IHGBBC is developed through small-signal averaging of state-space equations to investigate the dynamic characteristics of the converter. It is seen from the derived TF of IHGBBC that there exists a right-half-plane (RHP) zero, so it is a non-minimum phase system like TBBC. As a result, the dynamic performance of the converter during transient conditions is sluggish in nature. However, the closed-loop performance of IHGBBC can be improved by Type-III compensator which is established through the K-factor approach [19–22] and integrated with closed-loop IHGBBC. The closed-loop IHGBBC with Type-III compensator has been fabricated in the laboratory, and the performance has been tested under different operating conditions. It is observed that IHGBBC demonstrates outstanding performances under all operating circumstances.

2.2 Proposed Improved High Gain Buck–Boost Converter (IHGBBC)

2.2.1 Circuit Topology and Working Principle

The proposed IHGBBC is formed by combining two different traditional topologies of BBCs, and the developed circuit of power converter is bestowed in Fig. 2.3. Two TBBCs are paralleled in the input side and are cascaded in the load side as well as interleaving technique (i.e., operating phase displacement between two TBBCs is 180° with identical switches duty cycle in a switching interval) is applied between them.

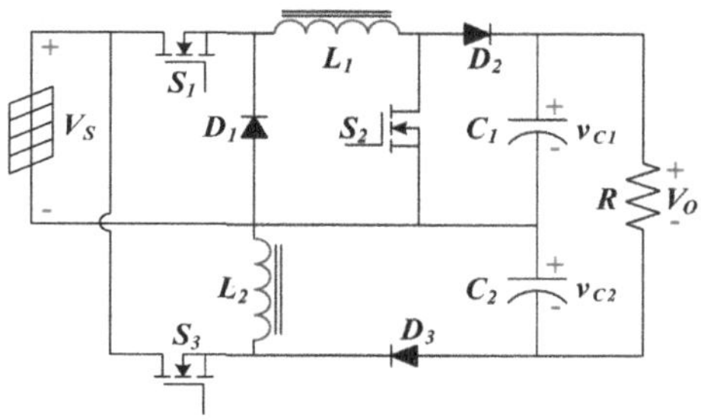

Fig. 2.3 Power circuit of IHGBBC

In this manner, IHGBBC produces lesser ripples with improved performances along with it produces high-voltage gain as compared to voltage gain of a single TBBC cell. Consequently, IHGBBC demonstrates higher power handling capability with greater than 90% power conversion efficiency as compared to single TBBC cell.

The circuit components of IHGBBC (shown in Fig. 2.3) are indicated as follows: source voltage or output voltage of PV panel (V_S), inductors (L_1 & L_2), power MOSFETs (S_1, S_2 & S_3), capacitors (C_1 & C_2), load resistance (R), power diodes (D_1, D_2, & D_3) and output or load voltage (V_O). The voltages across C_1 & C_2 are denoted by v_{C1} & v_{C2}, respectively. There are two distinct working modes of the IHGBBC during continuous conduction mode (CCM). As per the working states (Mode-I & Mode-II) of IHGBBC, the corresponding circuits schematic are demonstrated in Fig. 2.4a, b. Figure 2.5 illustrates the several key waveforms of IHGBBC. Where v_{L1}, v_{L2}, i_{L1} & i_{L2} are the inductors voltage and current; V_{S1}, V_{S2} & V_{S3} are the gate pulses for switches S_1, S_2 & S_3; δ_1 & δ_2 are duty cycle of respective TBBC cells; T is one switching interval.

During Mode-I of IHGBBC, both S_1 & S_2 will be in ON state and S_3 will be in OFF state. So, both D_1 & D_2 will be reverse bias and D_3 will be forward bias. V_S is supplying energy to L_1, L_2 is supplying energy to C_2. Finally, both C_1 & C_2 are supplying energy to R as well as maintain the required voltage across R.

Similarly, during Mode-II of IHGBBC, both S_1 & S_2 will be in OFF state and S_3 will be in ON state. Consequently, both D_1 & D_2 will be forward bias and D_3 will be reverse bias. Here, V_S is supplying energy to L_2, L_1 is supplying energy to C_1. In this mode also, both C_1 & C_2 are supplying energy to R as well as maintain the required voltage across R.

2.2.2 Voltage Gain

The steady-state voltages expressions of L_1, L_2, C_1 & C_2 of IHGBBC in Mode-I can be derived by adopting KVL in the equivalent circuit as shown in Fig. 2.4a and are presented as follows.

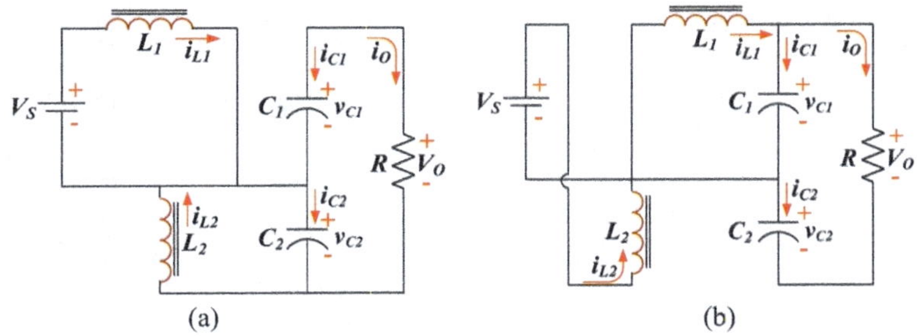

Fig. 2.4 State circuits of IHGBBC (**a**) Mode-I & (**b**) Mode-II

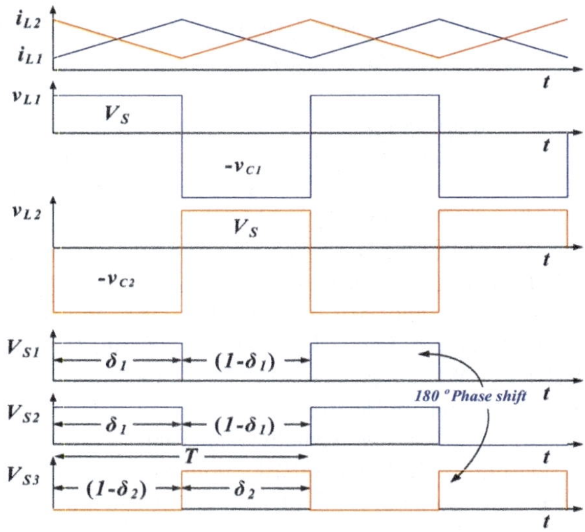

Fig. 2.5 Key waveforms of IHGBBC during CCM operation

$$v_{L1} = V_S \tag{2.1}$$

$$v_{L2} = -v_{C2} \tag{2.2}$$

$$V_O = v_{C1} + v_{C2} \tag{2.3}$$

Similarly, for Mode-II (Fig. 2.4b) operation of IHGBBC, the steady-state voltages expressions of L_1, L_2, C_1 & C_2 are given below.

$$v_{L1} = -v_{C1} \tag{2.4}$$

$$v_{L2} = V_S \tag{2.5}$$

$$V_O = v_{C1} + v_{C2} \tag{2.6}$$

Now, by utilizing the equal volt-sec balance principle across L_1, one can establish (2.7) and it can be written as follows.

$$v_{C1} = \frac{\delta_1 V_S}{(1 - \delta_1)} \tag{2.7}$$

Similarly, (2.8) is deduced through equal volt-sec balance approach across L_2, and it is given below.

$$v_{C2} = \frac{\delta_2 V_S}{(1 - \delta_2)} \tag{2.8}$$

It is to be noted that two connected TBBCs are assumed as identical as well as lossless system. All parameters of the converters are identical (i.e., $L_1 = L_2$, $C_1 = C_2$ and $\delta_1 = \delta_2 = \delta$).

Now, by adding (2.7) and (2.8), the final voltage gain of IHGBBC is presented in (2.9).

$$\frac{V_O}{V_S} = \frac{2\delta}{(1 - \delta)} \tag{2.9}$$

However, a single TBBC cell provides voltage gain which is given below in the following equation.

$$\frac{V_O}{V_S} = \frac{\delta}{(1 - \delta)} \tag{2.10}$$

From (2.9) and (2.10), it is found that the derived voltage gain of IHGBBC is two times higher as compared a single TBBC cell.

2.2.3 Parameters Design

This subsection presents the selection procedure of inductors (L_1 & L_2) and capacitors (C_1 & C_2) of IHGBBC.

2.2.3.1 Inductors (L_1 & L_2)

The values of L_1 & L_2 are chosen based on (2.11) and (2.12), which essential to satisfy the following criteria to enable the converter can work in CCM.

$$L_1 > \frac{1}{4} RT(1 - \delta)^2 \tag{2.11}$$

$$L_2 > \frac{1}{4} RT(1 - \delta)^2 \tag{2.12}$$

2.2.3.2 Capacitors (C_1 & C_2)

The capacitors (C_1 & C_2) are taken based on (2.13) and (2.14) through proper consideration of some parameters of IHGBBC like source voltage (V_S), output voltage (V_O), capacitors ripple voltages (Δv_{C1} & Δv_{C2}) and switching interval (T).

$$C_1 = \frac{\delta T V_O}{R \Delta v_{C1}} \tag{2.13}$$

$$C_2 = \frac{\delta T V_O}{R \Delta v_{C2}} \tag{2.14}$$

2.3 Small-Signal Modeling of IHGBBC

The IHGBBC is a kind of time-variant and highly nonlinear switching converter. Nevertheless, it can be possible to convert a continuous time-invariant linear converter model. In this process, the averaging of state equations is utilized to deduce the duty-to-output TF of IHGBBC. There are two distinct working states (Mode-I and Mode-II) of the IHGBBC during CCM. Hence, the differential state equations for Mode-I operation (Fig. 2.4a) of IHGBBC are given below in the following equations.

$$\frac{\mathrm{d}i_{L1}}{\mathrm{d}t} = \frac{V_S}{L_1} \tag{2.15}$$

$$\frac{\mathrm{d}i_{L2}}{\mathrm{d}t} = \frac{-v_{C2}}{L_2} \tag{2.16}$$

$$\frac{\mathrm{d}v_{C1}}{\mathrm{d}t} = \frac{-V_O}{RC_1} \tag{2.17}$$

$$\frac{\mathrm{d}v_{C2}}{\mathrm{d}t} = \frac{i_{L2}}{C_2} - \frac{V_O}{RC_2} \tag{2.18}$$

Similarly, the differential state equations for Mode-II (Fig. 2.4b) operation of IHGBBC can be written as follows.

$$\frac{\mathrm{d}i_{L1}}{\mathrm{d}t} = \frac{-v_{C1}}{L_1} \tag{2.19}$$

$$\frac{\mathrm{d}i_{L2}}{\mathrm{d}t} = \frac{V_S}{L_2} \tag{2.20}$$

$$\frac{\mathrm{d}v_{C1}}{\mathrm{d}t} = \frac{i_{L1}}{C_1} - \frac{V_O}{RC_1} \tag{2.21}$$

$$\frac{\mathrm{d}v_{C2}}{\mathrm{d}t} = \frac{-V_O}{RC_2} \tag{2.22}$$

It is assumed that two TBBCs are identical (i.e., $\delta_1 = \delta_2 = \delta$, $L_1 = L_2 = L$ & $C_1 = C_2 = C$) so the average differential state equations can be written as follows.

$$\frac{\mathrm{d}i_{L1}}{\mathrm{d}t} = \frac{V_S}{L_1}\delta - \frac{v_{C1}}{L_1}(1-\delta) \tag{2.23}$$

$$\frac{\mathrm{d}i_{L2}}{\mathrm{d}t} = \frac{V_S}{L_2}\delta - \frac{v_{C2}}{L_2}(1-\delta) \tag{2.24}$$

$$\frac{\mathrm{d}v_{C1}}{\mathrm{d}t} = \frac{-V_O}{RC_1}\delta + \left(\frac{i_{L1}}{C_1} - \frac{V_O}{RC_1}\right)(1-\delta) \tag{2.25}$$

$$\frac{\mathrm{d}v_{C2}}{\mathrm{d}t} = \frac{-V_O}{RC_2}\delta + \left(\frac{i_{L2}}{C_2} - \frac{V_O}{RC_2}\right)(1-\delta) \tag{2.26}$$

Hence, the final duty-to-output TF of IHGBBC is found by solving the above expressions (2.23)–(2.26), and it is given below in the following equation.

$$\frac{V_O(s)}{\delta(s)} = \frac{2V_S}{(1-\delta)^2} \times \frac{\left\{1 - \frac{2\delta L}{R(1-\delta)^2}s\right\}}{\left\{\frac{LC}{(1-\delta)^2}s^2 + \frac{2L}{R(1-\delta)^2}s + 1\right\}} \tag{2.27}$$

The converter (Fig. 2.3) parameters are selected by utilizing the design criteria described in Sect. 2.2 and are shown in Table 2.1. The mentioned parameters are used for the simulation as well as experimental study. After inserting the design parameters values into (2.27), the duty-to-output TF for a specific working state of IHGBBC is established and is given in (2.28).

$$\frac{V_O(s)}{\delta(s)} = \frac{-11832(s - 4.708 \times 10^4)}{(s^2 + 42.55s + 1.211 \times 10^6)} \tag{2.28}$$

It is interesting to note that IHGBBC is having one RHP zero in its output to control transfer function (2.28), and this is non-minimum phase system. So design

Table 2.1 Parameters of IHGBBC

Sl. no.	Parameters	Value
1	Supply voltage (V_S)	36 V
2	Load voltage (V_O)	20–110 V
3	Switching frequency (*f*)	50 kHz
4	Inductors (L_1 & L_2)	275 μH
5	Load resistance (*R*)	30–100 Ω
6	Capacitors (C_1 & C_2)	470 μF

of a suitable phase lead compensation is required to overcome the shortcomings in its closed-loop operation.

2.4 Results and Discussions

To obtain excellent transient performance of IHGBBC in closed loop, the Type-III compensator is established through the traditional K-factor procedure [19–22] and applied for closed-loop control of the proposed converter. The TF of the developed compensator by using K-factor is given in (2.29).

$$G_C(s) = \frac{2.85 \times 10^5 (s + 353.1)^2}{s(s + 1.64 \times 10^5)^2} \tag{2.29}$$

The stability performance of IHGBBC is analyzed through the frequency-domain analysis. The bode diagram of IHGBBC with and without compensator is shown in Fig. 2.6. The performance specifications of IHGBBC with the developed compensator are shown in Table 2.2. It can be observed that IHGBBC with advanced compensator demonstrates excellent stability during closed loop.

To verify the closed-loop dynamic performances of IHGBBC with developed compensator, extensive experimental analyses have been carried out by source voltage (V_S) variation as well as load resistance (*R*) variation in both step-down and step-up states of converter. The output voltage responses of IHGBBC under all working circumstances have been captured and are presented in Figs. 2.7 and 2.8. It is perceived that the output voltage responses of IHGBBC exhibit outstanding performances with faster rise time, minimum settling time and minimum overshoot in all cases. Figure 2.9 demonstrates the key experimental waveforms of IHGBBC during CCM operation for both step-down and step-up modes.

The comparative ripple and efficiency analyses of the proposed IHGBBC and traditional BBC have been performed and the graphical representations are demonstrated in Figs. 2.10 and 2.11, respectively. It is found that the proposed IHGBBC has lesser ripple than traditional BBC and due to lesser ripple the proposed converter has higher power efficiency as compared to traditional one.

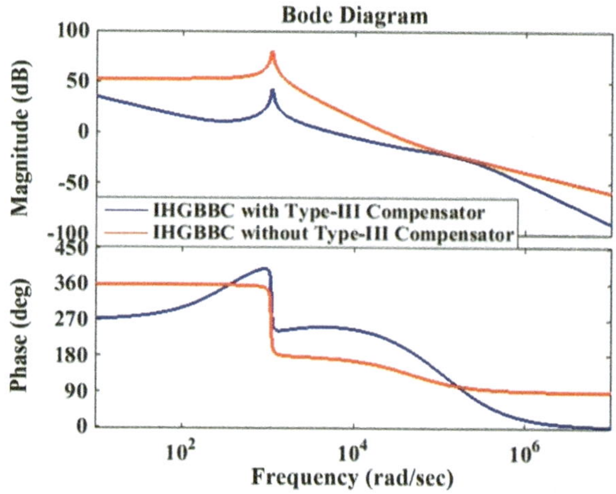

Fig. 2.6 Bode diagram of IHGBBC with and without compensator

Table 2.2 Performance specifications of IHGBBC with Type-III compensator

Specifications	Value
Gain margin	16.8 dB
Gain crossover frequency	6170 rad/s
Phase margin	72.1°
Phase crossover frequency	57,500 rad/s

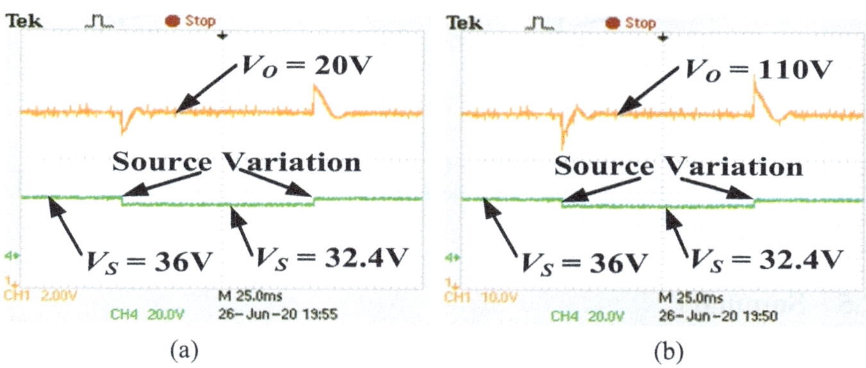

Fig. 2.7 Dynamic output voltage (V_O) responses of IHGBBC against ±10% step variations are given in source voltage (V_S) during (**a**) buck state and (**b**) boost state (CH1: Output voltage (V_O) & CH4: Source voltage variation)

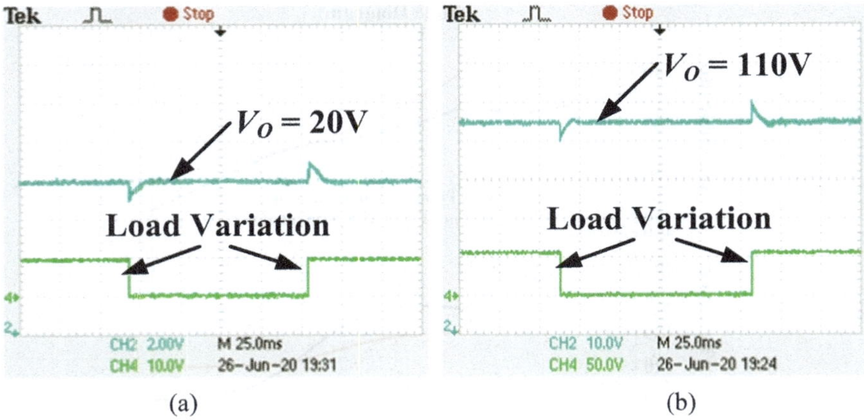

Fig. 2.8 Dynamic output voltage (V_O) responses of IHGBBC against ±50% step variations are given in load resistance (R) during (**a**) buck state and (**b**) boost state (CH2: Output voltage (V_O) & CH4: Load resistance variation)

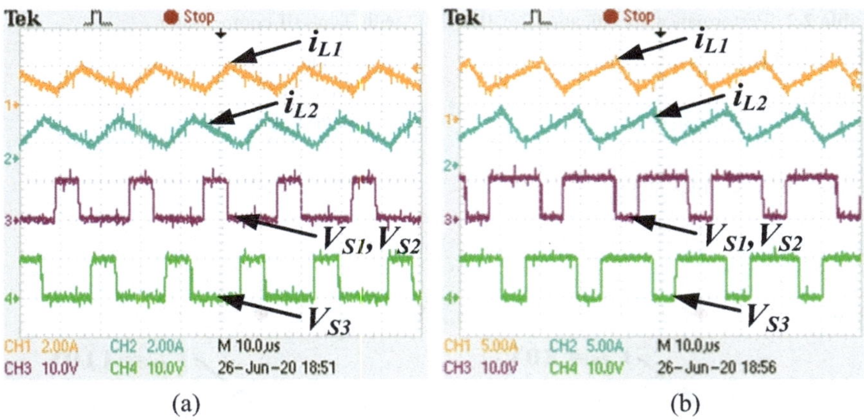

Fig. 2.9 Key waveforms of IHGBBC during CCM operation (**a**) buck state and (**b**) boost state [CH1: i_{L1}, CH2: i_{L2}, CH3: gate pulse for switches S_1 & S_2 (V_{S1} & V_{S2}) & CH4: gate pulse for switch S_3 (V_{S3})]

2.5 Summary

The power converter with high-voltage gain is considered to be an important aspect for both stand-alone and grid-connected PV-based system. In this chapter, one such high gain buck–boost converter named IHGBBC has been discussed. This converter is originated by connecting in parallel at the input side and cascading in the load side of two TBBC cells. The converter has been fabricated in the laboratory,

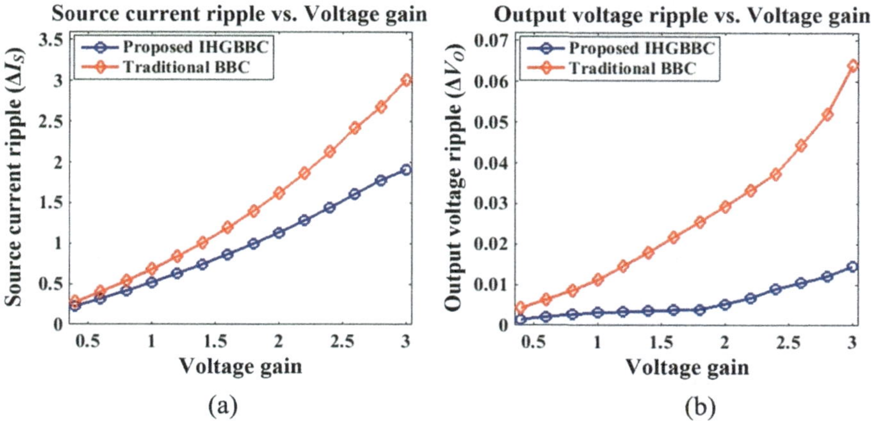

Fig. 2.10 Experimental ripples comparison between proposed IHGBBC and traditional BBC (**a**) ΔI_S against voltage gain (**b**) ΔV_O against voltage gain

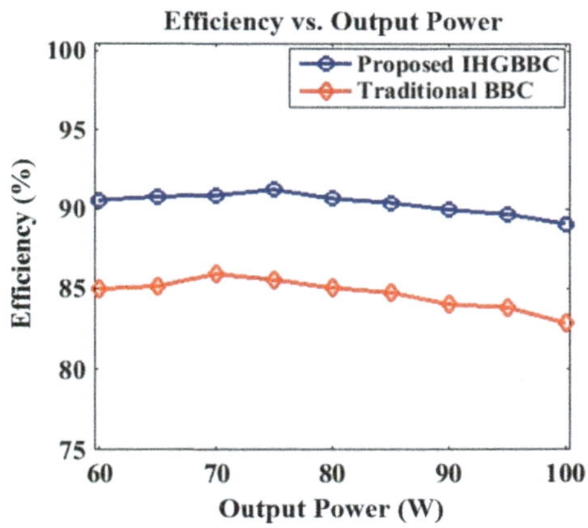

Fig. 2.11 Experimental efficiency (%) comparison between proposed IHGBBC and traditional BBC

and its closed control operation has been tested with it. The detailed working principle with different modes of operation, mathematical modeling, controller design, simulation and experimental results of IHGBBC has been presented. The non-inverting load voltage with a wide range can be generated through IHGBBC. The proposed power converter is compact, durable and easy to develop in real time with a simple analog components-based control circuit. The superiority of IHGBBC is substantiated through the obtained results from simulation as well as experimental studies. From the above analyses, it is clear that the proposed IHGBBC has some unique features like high-voltage gain, faster dynamic response, high power efficiency and smaller ripple. Thus, IHGBBC can be applied in solar PV systems due to its uniqueness, simplicity and improved performances as well as its easy to implement in real time.

The IHGBBC can be integrated with some practical applications of solar systems such as PV-based electric vehicles, PV-based street lighting systems and PV-based irrigation systems.

References

1. Vighetti, S., Ferrieux, J.-P., Lembeye, Y.: Optimization and design of a cascaded DC/DC converter devoted to grid-connected photovoltaic systems. IEEE Trans. Power Electron. **27**(4), 2018–2027 (2012)
2. Freitas, A.A.A., Tofoli, F.L., Júnior, E.M.S., Daher, S., Antunes, F.L.M.: High-voltage gain dc–dc boost converter with coupled inductors for photovoltaic systems. IET Power Electron. **8**(10), 1885–1892 (2015)
3. Moradpour, R., Ardi, H., Tavakoli, A.: Design and implementation of a new SEPIC-based high step-up DC/DC converter for renewable energy applications. IEEE Trans. Industr. Electron. **65**(2), 1290–1297 (2018)
4. Shen, C.-L., Chiu, P.-C.: Buck-boost-flyback integrated converter with single switch to achieve high voltage gain for PV or fuel-cell applications. IET Power Electron. **9**(6), 1228–1237 (2016)
5. Forouzesh, M., Shen, Y., Yari, K., Siwakoti, Y.P., Blaabjerg, F.: High-efficiency high step-Up DC-DC converter with dual coupled inductors for grid-connected photovoltaic systems. IEEE Trans. Power Electron. **33**(7), 5967–5982 (2018)
6. Urtasun, A., Lu, D.D.-C.: Control of a single-switch two-input buck converter for MPPT of two PV strings. IEEE Trans. Ind. Electron. **62**(11), 7051–7060 (2015)
7. Zeng, J., Qiao, W., Qu, L.: An isolated three-port bidirectional DC-DC converter for photovoltaic systems with energy storage. IEEE Trans. Ind. Appl. **51**(4), 3493–3503 (2015)
8. Silveira, G.C., Tofoli, F.L., Bezerra, L.D.S., Bascopé, R.P.T.: A nonisolated DC-DC boost converter with high voltage gain and balanced output voltage. IEEE Trans. Ind. Electron. **61**(12), 6739–6746 (2014)
9. Banaei, M.R., Bonab, H.A.F.: A novel structure for single-switch nonisolated transformerless buck-boost DC-DC converter. IEEE Trans. Ind. Electron. **64**(1), 198–205 (2017)
10. Wai, R.-J., Lin, C.-Y., Duan, R.-Y., Chang, Y.-R.: High-efficiency DC-DC converter with high voltage gain and reduced switch stress. IEEE Trans. Ind. Electron. **54**(1), 354–364 (2007)
11. Shin, H.-B., Jang, E.-S., Park, J.-G., Lee, H.-W., Lipo, T.A.: Small-signal analysis of multiphase interleaved boost converter with coupled inductors. IEE Proc. Electric Power Appl. **152**(5), 1161–1170 (2005)
12. J.S. Anu Rahavi, T. Kanagapriya, R. Seyezhai.: Design and analysis of interleaved boost converter for renewable energy source. In: Proceeding IEEE International Conference on Computing, Electronics and Electrical Technologies (ICCEET), pp. 447–451 (2012)
13. Tseng, K.-C., Huang, C.-C.: High step-up high-efficiency interleaved converter with voltage multiplier module for renewable energy system,. IEEE Trans. Ind. Electron. **61**(3), 1311–1319 (2014)
14. Sivanagaraju, G., Samata, S., Kunzler, L.M., Feistel, K.R., Rathore, A.K., Lopes, L.A.: PFC interleaved buck-boost converter for telecom power application. In: Proceeding 43rd Annual Conference of the IEEE Industrial Electronics Society (IECON), pp. 2299–2304 (2017)
15. Luna, T.-R.G., Vargas, I.A., Forsyth, A.J., Pulido, K.C., Elizondo, P.-E V., Cervantes, I., Olguín, F.G., Parra, A.V.: Two-phase, dual interleaved buck-boost DC-DC converter for automotive applications. IEEE Trans. Ind. Appl. **56**(1), 390–402 (2020)
16. Miao, S., Wang, F., Ma, X.: A new transformerless buck-boost converter with positive output voltage. IEEE Trans. Ind. Elect. **63**(5), 2965–2975 (2016)

17. N. Zhang, G. Zhang, K. W. See, B. Zhang.: A single-switch quadratic buck-boost converter with continuous input port current and continuous output port current. IEEE Trans. Power Electron. **33**(5), 4157–4166 (2018)
18. Hu, X., Gong, C.: A high gain input-parallel output-series DC/DC converter with dual coupled inductors. IEEE Trans. Power Electron. **30**(3), 1306–1317 (2015)
19. Venable, H.D.: The K-factor: a new mathematical tool for stability analysis and synthesis. In: Proceeding Powercon, San Diego, USA (1983)
20. Chan, K.-F., Lam, C.-S., Zeng, W.-L., Zheng, W.-M., Sin, S.-W., Wong, M.-C.: Generalized type III controller design interface for DC-DC converters. In: Proceeding IEEE Region 10 Conference (TENCON-2015), Macao, China (2015)
21. Hwu, K.I., Jiang, W.Z., Shieh, J.-J.: Implementation of type III controller for KY converter based on PSIM. In: Proceeding 41st Annual Conference of the IEEE Industrial Electronics Society (IECON-2015), Yokohama, Japan (2015)
22. Bamgboje, D.O., Harmon, W., Tahan, M., Hu, T.: Low cost high performance LED driver based on a self-oscillating boost converter. IEEE Trans. Power Electron. **34**(10), 10021–10034 (2019)

Chapter 3
Developed Boost Converter with Generalized Quadratic Boosting Cell to Minimize Capacitor Voltage Stresses and Reduce the Cost of Utilized Capacitors

Lipika Nanda and Adyasha Acharya

3.1 Introduction

The new era of electrical technology depends on the development of power electronic devices. A DC-to-DC converter is one of the most commonly used power electronic devices that converts a fixed input DC voltage to variable DC voltage output. These converters are widely implemented in applications like automobiles, traction motors, marine hoists, mine haulers, trolley care and renewable energy systems [1]. These converters can also be used as switched mode power converters (SMPS) that consist of a switch reactive circuit elements like inductor and capacitor for attenuating the switching ripples in the output power. These devices have a numerous advantages over the other linear regulators. Voltage of such devices can be controlled by adjusting the duty ratio. The new era of electrical technology mostly relies on the non-conventional energy sources whose output is insufficient for high voltage applications [1]. The best solution for the mentioned problem is general boost converter, but the maximized output gain of the boost converter is practically not feasible to achieve due to the dependency of voltage gain on duty ratio range [2–4]. Increasing the duty ratio for maximum output gain, the converter efficiency gets compromised [5]. Various other issues like switch and capacitor voltage stresses also tend to deteriorate the converters efficiency. The solution of such demerits is the newly derived topology such as cascaded boost converter [6, 7], quadratic boost converter [8], multidevice boost converter [9] and modified quadratic converters [10–13]. A cascaded boost converter steps up the voltage without wide variation in duty ratio, but decreases the overall efficiency of converter due to multiple switching losses. A quadratic boost converter is developed to

L. Nanda (✉) · A. Acharya
School of Electrical Engineering, KIIT University, Bhubaneswar, India
e-mail: lnandafel@kiit.ac.in

N. Priyadarshi et al. (eds.), *DC–DC Converters for Future Renewable Energy Systems*, Energy Systems in Electrical Engineering,
https://doi.org/10.1007/978-981-16-4388-0_3

maximize the output voltage gain at low duty ratio and shows minimal power losses. The switch and capacitor voltage stresses are the main problems in quadratic boost topology, that was solved by the application of a capacitor-inductor-diode (CLD) [14]. CLD cell is added to a multidevice boost converter such that a maximized output voltage gain and minimal stress voltage across the switch is obtained, but due to extra switch the losses increase. A new quadratic boost converter is introduced that is much better in terms of reduced capacitor voltage stress on comparison with quadratic boost topology but the input current is discontinuous in nature [15]. The topology-I attains a maximized output voltage gain for a low duty ratio range reducing the discontinuity in input current, but it requires extra inductors and capacitors [16–18]. In this paper, the quadratic cell has been generalized. The proposed structure contains quadratic boosting cell (n = 3). Each cell contains a capacitor, two inductors and two diodes. The aim of the cell is to maximize the output voltage gain for a low range of duty cycle. The inductors used in cells charge in a parallel manner and discharge in serial manner. The capacitors used in each cell have low stress voltage across them that minimizes their cost and size. Also, the voltage stress further reduces if the duty ratio is minimized. So as per the application the voltage gain can be increased by selecting boosting cells for a proper duty cycle. Hence, the dependency of output gain on number of boosting cells is prominent.

3.2 Proposed Topology

The proposed topology is developed in this section. The generalized quadratic boosting cell structure is developed and implemented. Proposed converter's operating modes and different wave forms are discussed in details. The theoretical calculations are presented in order to find the values of inductance and capacitance. Finally, the merits and demerits are discussed.

3.2.1 *Quadratic Boosting Cell and Converter*

The largest output voltage gain of boost converter is practically difficult to achieve, followed by other issues like switch and capacitor voltage stresses that deteriorate the efficiency of converter [19–21]. In Fig. 3.1, a quadratic boosting structure is shown that consist of a capacitor C, two diodes (D_1 and D_2) and two inductors (L_1 and L_2). Inductors in ON state of switch charge in parallel manner and discharge in serial manner in switch's OFF state thereby increasing the output gain [22, 23]. Adding single boosting cell (n = 1) in replacement of the inductor L in a boost converter in Fig. 3.2, the output voltage gain expression is given in Eq. (3.1):

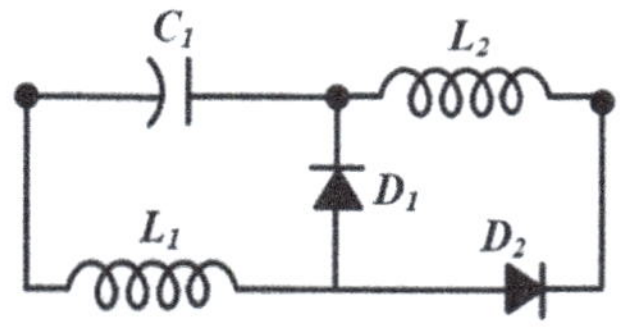

Fig. 3.1 Basic boosting cell

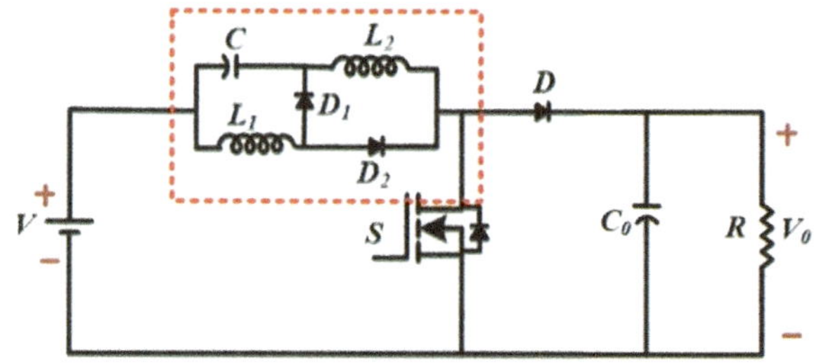

Fig. 3.2 Boost converter with single quadratic boosting cell (n = 1)

$$\frac{V_0}{V} = \frac{1}{(1-\delta)^2} \tag{3.1}$$

3.3 Generalized Quadratic Boosting Cell Structure

Addition of an extra capacitor C_2, inductor L_3 and two diodes (D_3, D_4) to the boosting cell forms a two quadratic boosting cell structure (n = 2) presented in Fig. 3.3 and output voltage gain is given in Eq. (3.2) as

$$\frac{V_0}{V} = \frac{1}{(1-\delta)^3} \tag{3.2}$$

Similarly addition of another capacitor C_3, inductor L_4 and diodes (D_5, D_6) in the previous structure shown in Fig. 3.4 forms a three quadratic boosting cell structure (n = 3) and the voltage gain is given in Eq. (3.3):

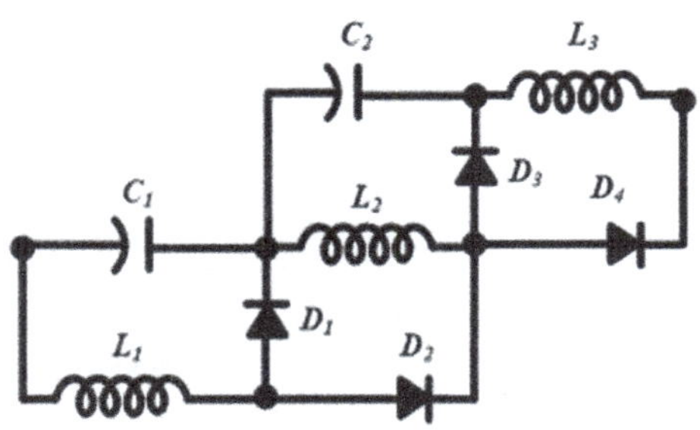

Fig. 3.3 Two quadratic boosting cells (n = 2)

$$\frac{V_0}{V} = \frac{1}{(1-\delta)^4} \tag{3.3}$$

Therefore, on addition of number of capacitors, inductors and diodes in same manner, n number of quadratic boosting cell structure are developed as shown in Fig. 3.5 and the generalized voltage gain expression is equated in Eq. (3.4). Hence, the output voltage gain dependency on the number of quadratic boosting cells n is figured out.

$$\frac{V_0}{V} = \frac{1}{(1-\delta)^{n+1}} \tag{3.4}$$

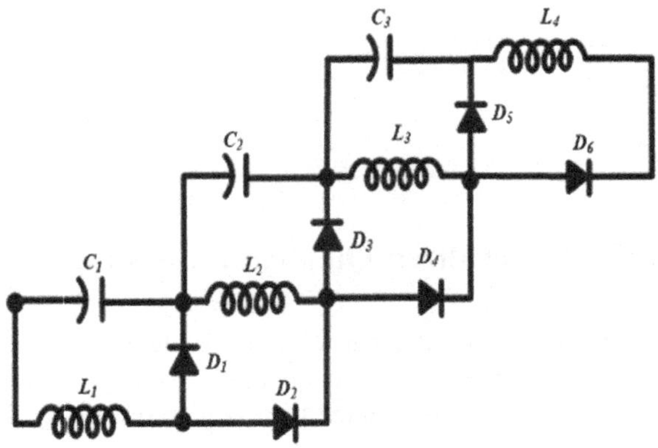

Fig. 3.4 Three quadratic boosting cell (n = 3)

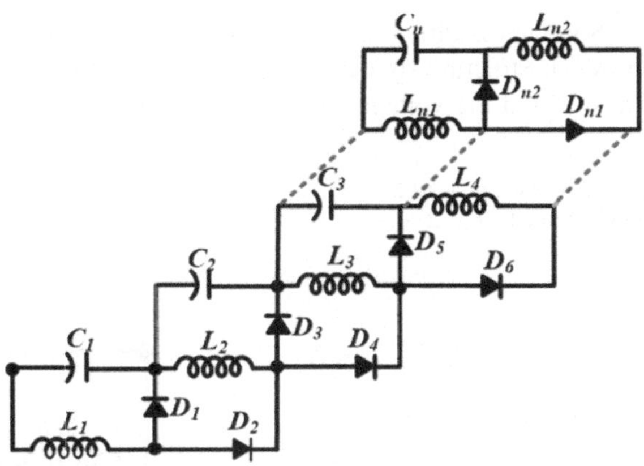

Fig. 3.5 n quadratic boosting cells

3.4 Proposed Converter and Operating Modes

The proposed converter with a generalized quadratic boosting cell structure ($n = 3$) is shown in Fig. 3.6. First cell contains two inductors (L_1, L_2), capacitor C_1 and two diodes (D_1, D_2). Second cell contains two inductors (L_3, L_2), capacitor C_2 and two diodes (D_3, D_4). Third cell contains two inductors (L_4, L_3), capacitor C_3 and two diodes (D_5, D_6). Each component in the proposed converter is assumed ideal.

3.4.1 Mode1

In mode1 the switch is ON as shown in Fig. 3.8. Diodes (D_1, D_3, D_5, D_7) turned reverse biased and diodes (D_2, D_4, D_6) turned forward biased. The input supply V charges the inductor L_1, inductor L_2 is charged from capacitor C_1 and V, similarly inductors L_3 and L_4 get charged from capacitors (C_2, C_1, V) and (C_3, C_1 and V). So the inductor currents (I_{L1}, I_{L2}, I_{L3}, I_{L4}) rise linearly as shown in Fig. 3.7. Hence, it is observed that inductors (L_1, L_2) and (L_3, L_4) charge parallely and the capacitor C_0 discharges energy to load.

3.4.2 Mode 2

In mode 2 the switch is OFF as shown in Fig. 3.9. Diodes (D_1, D_3, D_5, D_7) turned forward biased and diodes (D_2, D_4, D_6) turned reverse biased. The inductors (L_1, L_2) discharge in serial manner and inductor currents (I_{L1}, I_{L2}) fall linearly. Similarly the inductors (L_3, L_4) also discharge serially and inductor currents

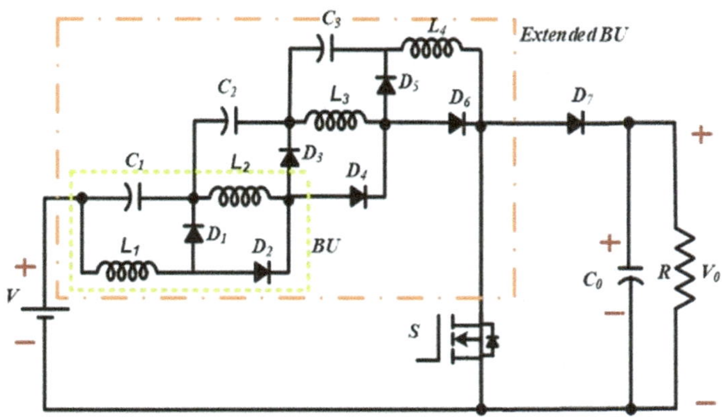

Fig. 3.6 Equivalent circuit of proposed topology using three boosting cells ($n = 3$)

(I_{L3}, I_{L4}) fall linearly as shown in Fig. 3.7, supplying energy to capacitor C_0 and load. Applying volt sec balance to inductor voltage equations, we get following capacitor voltage equations:

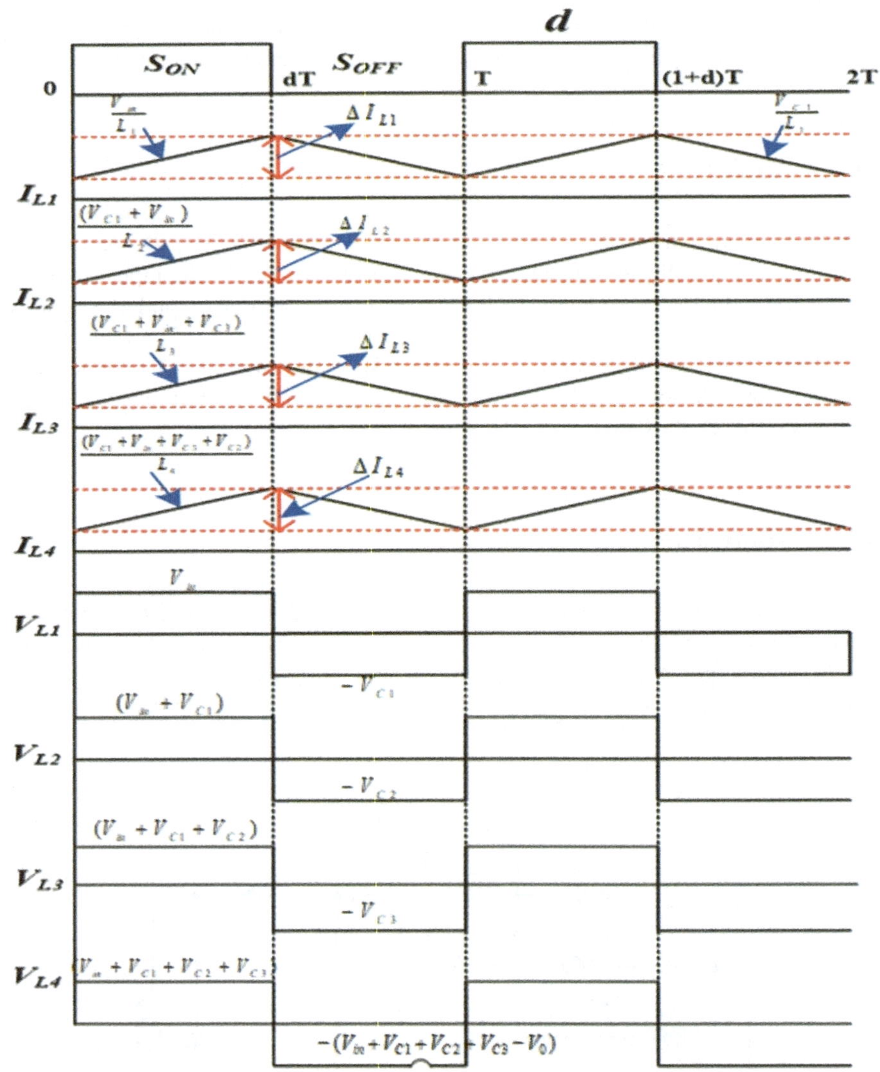

Fig. 3.7 Inductor current and voltage waveforms

3.4.3 Voltage Equation

The proposed structure is operated in both switch ON and OFF mode. The voltage equations in different modes are as follows.

3.4.3.1 Mode 1

During the switch ON mode that is visible in Fig. 3.8, the voltage across the four inductors (V_{L1}, V_{L2}, $V_{L3,}V_{L4}$) is determined. The voltage across inductor L_1 is given in Eq. (3.5) which is equal to the supply voltage *V*. Similarly, the voltage across inductor L_2 is given in Eq. (3.6) that is given as sum of input voltage *V* and capacitor voltage V_{C1}. Also, the voltage across the inductor L_3 is sum of input voltage *V* and capacitor voltage V_{C1} and V_{C2} that is given in Eq. (3.7). Finally, the

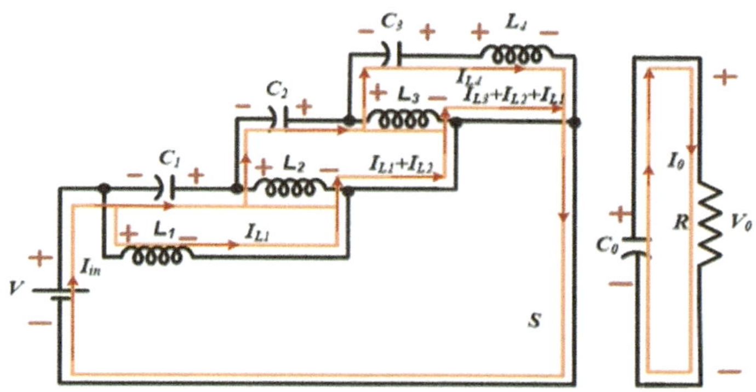

Fig. 3.8 Structure of proposed topology while switch is ON

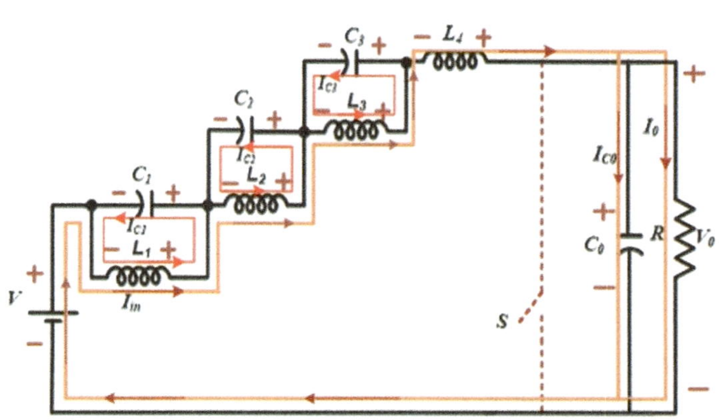

Fig. 3.9 Structure of proposed topology while the switch is OFF

voltage across the inductor L_4 is the sum of input voltage V and capacitor voltage V_{C1}, V_{C2} and V_{C3} shown in Eq. (3.8).

$$V_{L1} = V \quad (3.5)$$

$$V_{L2} = V + V_{C1} \quad (3.6)$$

$$V_{L3} = V_{C2} + V_{C1} + V \quad (3.7)$$

$$V_{L4} = V_{C3} + V_{C2} + V + V_{C1} \quad (3.8)$$

3.4.3.2 Mode 2

During the switch OFF mode as seen in Fig. 3.9, the voltage across the four inductors (V_{L1}, V_{L2}, V_{L3}, V_{L4}) is determined. The voltage across inductor L_1 is given in Eq. (3.9) which is equal to the capacitor voltage V_{C1}. Similarly, the voltage across inductor L_2 is given in Eq. (3.10) that is equal to capacitor voltage V_{C2}. Also, the voltage across the inductor L_3 is equal to capacitor voltage V_{C3} that is given in Eq. (3.11). Finally, the voltage across the inductor L_4 is shown in Eq. (3.12).

$$V_{L1} = V_{C1} \quad (3.9)$$

$$V_{L2} = V_{C2} \quad (3.10)$$

$$V_{L3} = V_{C3} \quad (3.11)$$

$$V_{L4} = V_0 - V - V_{C1} - V_{C2} - V_{C3} \quad (3.12)$$

Applying voltage second balance to inductor voltage equations, the capacitor voltage equations are determined (where T_{on} is on time and T_{off} is off time of switch). Voltage second balance of the inductor voltage V_{L1} during both modes is given in Eq. (3.13). Similarly, voltage second balance in inductor voltages (V_{L2}, V_{L3}, V_{L4}) is given in Eqs. (3.14)–(3.16), respectively.

$$V \times T_{on} = V_{C1} \times T_{off} \quad (3.13)$$

$$(V + V_{C1}) \times T_{on} = V_{C2} \times T_{off} \quad (3.14)$$

$$(V_{C2} + V_{C1} + V) \times T_{on} = V_{C3} \times T_{off} \quad (3.15)$$

$$(V + V_{C1} + V_{C2} + V_{C3}) \times T_{on} = (V_0 - V - V_{C1} - V_{C2} - V_{C3}) \times T_{off} \quad (3.16)$$

After dividing T_{on} on both sides of Eq. (3.13), the following capacitor voltage V_{C1} is obtained that is given in Eq. (3.17), where δ is the duty ratio and is given as ratio of turn on (T_{on}) time to total time interval T and ($T = T_{on} + T_{off}$).

$$V_{C1} = \left[\frac{V \times \delta}{(1-\delta)}\right] \tag{3.17}$$

Similarly dividing T_{on} on both the sides in Eq. (3.14), the capacitor voltageV_{C2} is determined in Eq. (3.18)

$$V_{C2} = \left[\frac{V \times \delta}{(1-\delta)^2}\right] \tag{3.18}$$

Similarly, the capacitor voltage V_{C3} is obtained that is given in Eq. (3.19)

$$V_{C3} = \left[\frac{V \times \delta}{(1-\delta)^3}\right] \tag{3.19}$$

Finally, dividing Eq. (3.16) by T_{on} on both sides and putting Eqs. (3.17)–(3.19) in Eq. (3.16), the capacitor voltage V_0 that is the output voltage is given in Eq. (3.20) as

$$\left[V_0 = \frac{V}{(1-\delta)^4}\right] \tag{3.20}$$

Hence, Eq. (3.20) is the obtained output voltage of proposed structure.

3.4.4 Current Equations

During ON mode that is visible in Fig. 3.8, the capacitor current I_{C0} is equal to output current I_0 that is given in Eq. (3.21). Similarly, in Eq. (3.22), the capacitor current I_{C3} is equal to inductor current I_{L4}. Also, the capacitor current I_{C2} is equal to the sum of inductor currents (I_{L4}, I_{L3}) that is presented in Eq. (3.23). Finally, the capacitor current I_{C4} is the sum of inductor currents (I_{L2}, I_{L3}, I_{L4}) given in Eq. (3.24).

$$I_{C0} = I_0 \tag{3.21}$$

$$I_{C3} = I_{L4} \tag{3.22}$$

$$I_{C2} = I_{L4} + I_{L3} \tag{3.23}$$

$$I_{C1} = I_{L2} + I_{L3} + I_{L4} \tag{3.24}$$

During OFF mode as seen in Fig. 3.9, the capacitor current I_{C0} is given in Eq. (3.25). Similarly, the capacitor current I_{C3} is seen in Eq. (3.26). Also, the capacitor current I_{C2} is presented in Eq. (3.27). Finally, the capacitor current I_{C1} is given in Eq. (3.28). The average current across capacitor C_0 is given as:

$$I_{C0} = I_{L4} - I_0 \tag{3.25}$$

$$I_{C3} = I_{L3} - I_{L4} \tag{3.26}$$

$$I_{C2} = I_{L2} - I_{L4} \tag{3.27}$$

$$I_{C1} = I_{L1} - I_{L4} \tag{3.28}$$

The average capacitor currents over a switching period should be zero at steady state, so current second balance is applied in capacitor currents (I_{C1}, I_{C2}, I_{C3}, I_{C0}) equations given as follows (where T_{on} is on time and T_{off} is off time of switch and δ is the duty ratio).

$$I_0 \times T_{on} = (-I_0 + I_{L4}) \times T_{off} \tag{3.29}$$

$$I_{L4} \times T_{on} = (I_{L3} - I_{L4}) \times T_{off} \tag{3.30}$$

$$(I_{L4} + I_{L3}) \times T_{on} = (I_{L2} - I_{L4}) \times T_{off} \tag{3.31}$$

$$(I_{L4} + I_{L3} + I_{L2}) \times T_{on} = (I_{L1} - I_{L4}) \times T_{off} \tag{3.32}$$

Equation (3.29) shows the net charge second balance of the capacitor current I_{C0}. Similarly, Eqs. (3.30)–(3.32) depict the net charge balance of the capacitor currents(I_{C3}, I_{C2} I_{C1}) simultaneously. After dividing T_{on} on both sides of Eq. (3.29), the average current across capacitor C_0 is given in Eq. (3.33),where δ is the duty ratio and is given as ratio of turn on (T_{on}) time to total time interval T and ($T = T_{on} + T_{off}$).

$$I_0 \times \delta = (I_{L4} - I_0) \times (1 - \delta) \tag{3.33}$$

Similarly, on dividing T_{on} on both the sides of Eq. (3.30), the average current across capacitor C_3 is given in Eq. (3.34) as

$$I_{L4} \times \delta = (I_{L3} - I_{L4}) \times (1 - \delta) \tag{3.34}$$

Hence, dividing T_{on} on both the sides of Eq. (3.31), the average current across capacitor C_2 is given in Eq. (3.35):

$$(I_{L4} + I_{L3}) \times \delta = (I_{L2} - I_{L4}) \times (1 - \delta) \tag{3.35}$$

Finally, dividing T_{on} on both the sides of Eq. (3.32), the average current across capacitor C_1 is given in Eq. (3.36):

$$(I_{L4} + I_{L3} + I_{L2}) \times \delta = (I_{L1} - I_{L4}) \times (1 - \delta) \tag{3.36}$$

Hence, on solving Eq. (3.33), the expression of inductor current I_{L4} w.r.t. output current I_0 is given in Eq. (3.37):

$$I_{L4} = \frac{I_0}{(1 - \delta)} \tag{3.37}$$

Similarly, placing Eq. (3.37) in Eq. (3.34), the expression of inductor current I_{L3} w.r.t. output current is given in Eq. (3.38) as

$$I_{L3} = \frac{I_0}{(1 - \delta)^2} \tag{3.38}$$

Also, placing Eq. (3.38) in (3.35) and solving the expression of inductor current I_{L2} w.r.t. output current is given in Eq. (3.39) as

$$I_{L2} = \frac{I_0}{(1 - \delta)^3} \tag{3.39}$$

Finally, placing Eq. (3.39) in (3.36) and further solving the expression of inductor current I_{L1} w.r.t. output current is given in Eq. (3.40)

$$I_{L1} = \frac{I_0}{(1 - \delta)^4} \tag{3.40}$$

3.4.5 Inductance Equations

Figure 3.7 presents the waveforms of inductor currents in CCM by taking the boundary conditions in which the minimum inductor current becomes zero. Hence, the minimum inductor current Eqs. (3.41)–(3.44) are given in as follows:

$$I_{L1,\min} = I_{L1} - \frac{\Delta I_{L1}}{2} = 0 \tag{3.41}$$

$$I_{L2,\min} = I_{L2} - \frac{\Delta I_{L2}}{2} = 0 \tag{3.42}$$

$$I_{L3,\min} = I_{L3} - \frac{\Delta I_{L3}}{2} = 0 \tag{3.43}$$

$$I_{L4,\min} = I_{L4} - \frac{\Delta I_{L4}}{2} = 0 \tag{3.44}$$

From Fig. 3.9. the voltage across the inductor L_1 is obtained by integrating it over T_{on} period of switch and is given in Eq. (3.45). After further solving and rearranging the terms, final inductor current ripple ΔI_{L1} expression is given in (3.46):

$$V_{L1} = V \int_0^{\delta t} dt = L_1 \times \Delta I_{L1} \tag{3.45}$$

$$\Delta I_{L1} = \frac{V \times \delta}{L_1 f_s} \tag{3.46}$$

Similarly from Fig. 3.9 the voltage across the inductor L_2 is obtained by integrating it over T_{on} period of switch and is given in Eq. (3.47). After further solving and rearranging the terms, final inductor current ripple ΔI_{L2} expression is given in (3.48):

$$V_{L2} = (V + V_{C1}) \int_0^{\delta t} dt = \Delta I_{L2} \times L_2 \tag{3.47}$$

$$\Delta I_{L2} = \frac{V \times \delta}{f_s(1 - \delta)L_2} \tag{3.48}$$

Similarly, the voltage across the inductor L_3 is obtained by integrating it over T_{on} period of switch and is given in Eq. (3.49). After further solving and rearranging the terms, final inductor current ripple ΔI_{L3} expression is given in (3.50) as

$$V_{L3} = (V + V_{C1} + V_{C2}) \int_0^{\delta t} dt = L_3 \times \Delta I_{L3} \tag{3.49}$$

$$\Delta I_{L3} = \frac{V \times \delta}{f_s(1 - \delta)^2 L_3} \tag{3.50}$$

Finally, the voltage across the inductor L_4 is obtained by integrating it over T_{on} period of switch and is given in Eq. (3.51). After further solving and rearranging the terms, final inductor current ripple ΔI_{L4} expression is given in (3.52):

$$V_{L4} = (V + V_{C1} + V_{C2} + V_{C3}) \int_0^{\delta t} dt = \Delta I_{L4} \times L_4 \tag{3.51}$$

$$\Delta I_{L4} = \frac{V \times \delta}{f_s(1-\delta)^3 L_4} \tag{3.52}$$

Substituting Eqs. (3.35), (3.41) in (3.36) the inductance equation is as follows:

$$L_1 = \frac{V \times \delta(1-\delta)^4}{2f_s I_O} \tag{3.53}$$

Substituting Eqs. (3.34), (3.43) in (3.37), the inductance obtained is as follows:

$$L_2 = \frac{V \times \delta(1-\delta)^2}{2f_s I_O} \tag{3.54}$$

Substituting Eqs. (3.33), (3.45) in (3.38) the following inductance is determined:

$$L_3 = \frac{V \times \delta}{2f_s I_O} \tag{3.55}$$

Finally, putting Eqs. (3.32), (3.47) in (3.38) the inductance equation is:

$$L_4 = \frac{V \times \delta}{2f_s I_O(1-\delta)^2} \tag{3.56}$$

3.5 Theoretical Comparison

The proposed converter is estimated with existing converters as shown in Table 3.1 on various parameters. The obtained output gain of proposed structure is largest on estimation with other existing converters for same duty ratio. It is mainly registered that output gain of the derived structure relies on the number of quadratic boosting cell structure for an appropriate duty ratio.

Figure 3.10 presents the dependency of output voltage gain on the number of quadratic boosting cell structure (n) for a range of duty ratio ($0 < \delta < 5$). It is registered that the output voltage gain keeps on increasing as the number of boosting cells (n) increases. So it is concluded that the voltage gain of the structure can be adjusted as per application by selecting proper duty ratio and number of quadratic cells. Further on increasing the cells beyond ($n = 2$) the number of components will increase that eventually increases the power losses in converter structure which is not recommended.

Table 3.1 Theoretical estimation of different converter with proposed topology

Converter	Boost	Quadratic boost [16]	Cascaded boost [3]	Topology-I [12]	Proposed converter (n = 3)
Output voltage	$V_O = \frac{V}{1-\delta}$	$V_O = \frac{V}{(1-\delta)^2}$	$V_O = \frac{V}{(1-\delta)^4}$	$V_O = \frac{V}{(1-\delta)^3}$	$V_O = \frac{V}{(1-\delta)^4}$
Switch	1	1	4	1	1
Frequency	F	F	F	F	F
Capacitors	1	2	4	3	4
Inductors	1	2	4	3	4
Diodes	1	2	4	5	7
V_{C1}	–	$V_{C1} = \frac{V}{(1-\delta)}$	$V_{C1} = \frac{V}{1-\delta}$	$V_{C1} = \frac{V}{1-\delta}$	$V_{C1} = \frac{V\times\delta}{1-\delta}$
V_{C2}	–	$V_{C2} = \frac{V}{(1-\delta)^2}$	$V_{C2} = \frac{V}{(1-\delta)^2}$	$V_{C2} = \frac{V\times\delta}{(1-\delta)^2}$	$V_{C2} = \frac{V\times\delta}{(1-\delta)^2}$
V_{C3}	–	–	$V_{C3} = \frac{V}{(1-\delta)^3}$	–	$V_{C3} = \frac{V\times\delta}{(1-\delta)^3}$

3.6 Capacitor Stress Voltage Comparison

The proposed structure is estimated with cascaded boost converter on the basis of voltage stresses across each capacitor for different range of duty ratio considering the same output gain. Figure 3.11 presents the reduced capacitor stress voltage (V_{C1}) of proposed structure given as blue curve as compared to the capacitor stress voltage (V_{C1}) of cascaded converter given as red curve (over a wide range of duty ratio). For example, considering (δ = 0.5, f = 5 kHz, V_O = 480 V, R = 460 Ω) the capacitor stress voltage (V_{C1}) of proposed structure is 27 V whereas the capacitor stress voltage (V_{C1}) of cascaded converter is 58 V.

Similarly, Fig. 3.12 presents the suppression in capacitor stress voltage (V_{C2}) of proposed structure given as blue curve as compared to the capacitor stress voltage (V_{C2}) of cascaded converter that is given as red curve over a range of duty ratio. For example, considering (δ = 0.5, f = 5 kHz, V_O = 480 V, R = 460) the capacitor stress voltage (V_{C2}) of proposed structure is 55 V whereas the capacitor stress voltage (V_{C2}) of cascaded converter is 117 V.

Finally, Fig. 3.13 shows the reduction in capacitor stress voltage (V_{C3}) of proposed converter given as blue curve as compared to the capacitor stress voltage (V_{C3}) of cascaded converter that is given as red curve over a wide range of duty ratio. For example, considering (δ = 0.5, f = 5 kHz, V_O = 480 V, R = 460 Ω) the capacitor stress voltage (V_{C3}) of proposed structure is 110 V whereas the capacitor stress voltage (V_{C3}) of cascaded converter is 235 V.

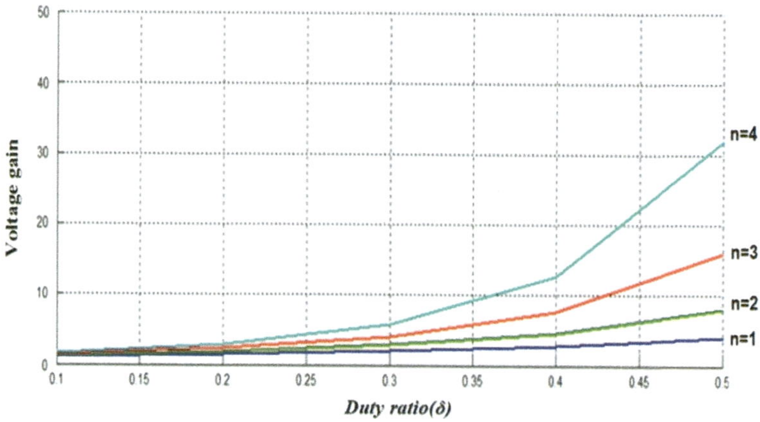

Fig. 3.10 Variation in output voltage gain with n boosting cell

Fig. 3.11 Reduction in capacitor stress voltage (V_{C1}) of proposed converter than the cascaded boost converter's capacitor stress voltage (V_{C1})

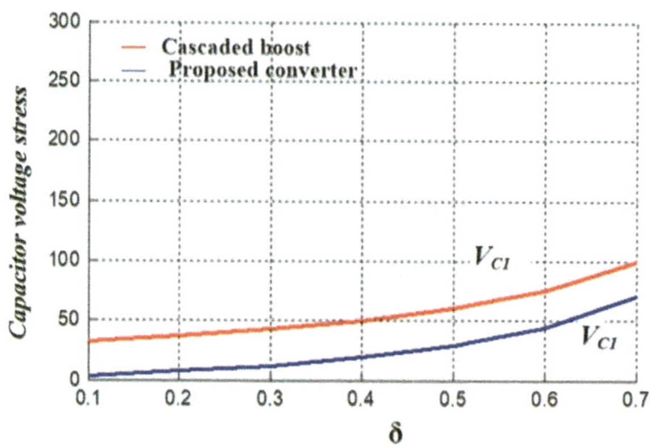

3.7 Simulation Study

The developed converter structure is designed using following specifications given in Table 3.2 to determine the values of inductance ($L = 4$) and finally simulated in MATLAB/Simulink platform. Hence, the calculated values of inductance ($L = 4$) from following parameters are obtained. The standard values of capacitors ($C_1 = C_2 = C_3 = 470\ \mu\text{F}$) are selected.

$$L_1 = \frac{V \times \delta(1-\delta)^4}{2f_s I_0} = \frac{30 \times 0.5 \times (1-0.5)^4}{2 \times 5000 \times 1.041} = 0.09\ \text{mH}$$

Fig. 3.12 Reduction in capacitor stress voltage (V_{C2}) of proposed converter than the cascaded boost converter's capacitor stress voltage (V_{C2})

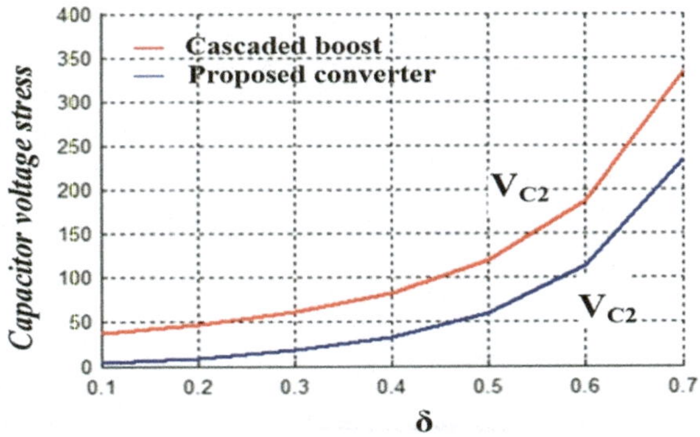

$$L_2 = \frac{V \times \delta(1-\delta)^2}{2f_s I_0} = \frac{30 \times 0.5 \times (1-0.5)^2}{2 \times 5000 \times 1.041} = 0.36 \text{ mH}$$

$$L_3 = \frac{V \times \delta}{2f_s I_0} = \frac{30 \times 0.5}{2 \times 5000 \times 1.041} = 1.44 \text{ mH}$$

$$L_4 = \frac{V \times \delta}{2f_s I_0(1-\delta)^2} = \frac{30 \times 0.5}{2 \times 5000 \times 1.041 \times (1-0.5)^2} = 5.769 \text{ mH}$$

Considering the parameters given in Table 3.2, the output voltage (V_O = 444.5 V) of proposed topology and the output current (I_O = 0.96 A) are shown in Figs. 3.14 and 3.15, respectively. Simultaneously after simulation in MATLAB/Simulink platform, the ripple currents across inductors are also determined. Figure 3.16 presents the ripple current across the inductor L_1 obtained. Similarly, inductor current ripples across inductors (L_2, L_3, L_4) are shown in Figs. 3.17, 3.18 and 3.19 along with their upper and lower ripple values simultaneously. Also, Fig. 3.20 shows the voltage stress across the switch (V_{SW} = 444.1 V) in the proposed topology.

Fig. 3.13 Reduction in capacitor stress voltage (V_{C3}) of proposed converter than the cascaded boost converter's capacitor stress voltage (V_{C3})

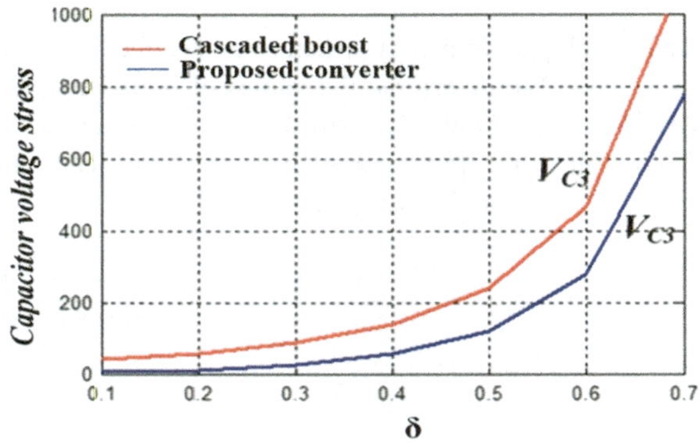

Table 3.2 Parameters to design the proposed structure

Parameter	Values
Input voltage (V)	30 V
Output voltage (V_0)	480 V
Duty ratio (δ)	0.5
Frequency (f)	5 kHz
Power (P)	500 W
Input current (I_{in})	16.66 A
Output current (I_0)	1.041 A
Load resistance (R)	460 Ω

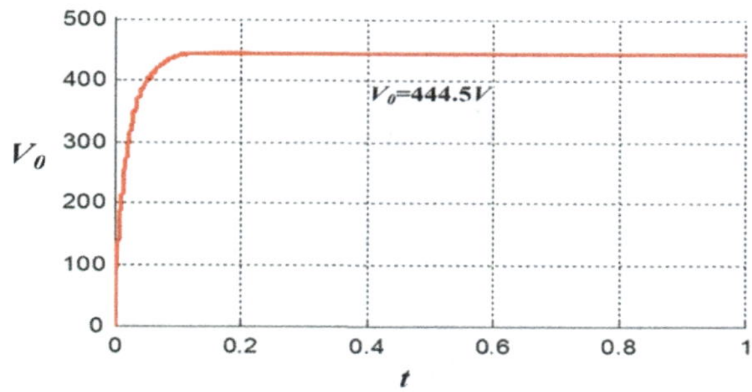

Fig. 3.14 Output voltage (V_0) waveform

3.8 Capacitor Stress Voltage Comparison Results

The simulated results of proposed topology on estimation with the classical cascaded boost converter on the basis of stress voltage of the various capacitors are determined. It is evident that the stress voltages are minimal in case of proposed topology as compared to the existing cascaded boost converter as shown in Table 3.3. This table shows the simulated values of various stress voltages across the different utilized capacitor considering following parameters ($\delta = 0.5$, $f = 5$ kHz, $V_0 = 480$ V, $R = 460$ Ω).

It is analyzed that the capacitor stress voltage ($V_{C1} = 27$ V) in proposed converter presented in Fig. 3.21 is much reduced than the capacitor stress voltage ($V_{C1} = 58$ V) in cascaded boost converter presented in Fig. 3.22. Similarly the capacitor stress voltage ($V_{C2} = 55$ V) in proposed converter presented in Fig. 3.23 is much reduced than the capacitor stress voltage ($V_{C2} = 117$ V) in cascaded boost converter presented in Fig. 3.24. Finally, the capacitor stress voltage ($V_{C3} = 110$ V) in proposed converter presented in Fig. 3.25 is much reduced than the capacitor stress voltage ($V_{C1} = 235$ V) in cascaded boost converter presented in Fig. 3.26.

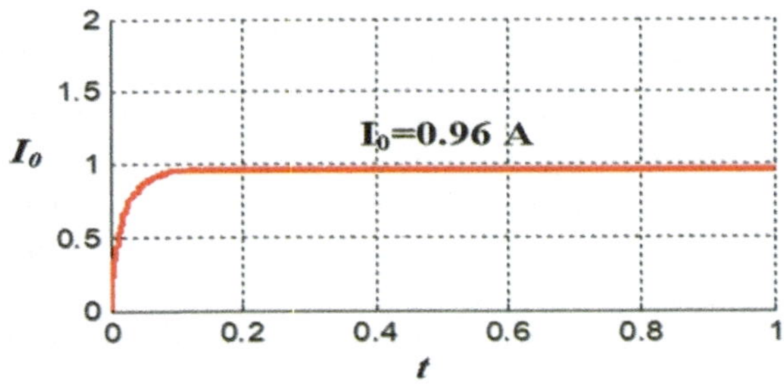

Fig. 3.15 Output current (I_0) waveform

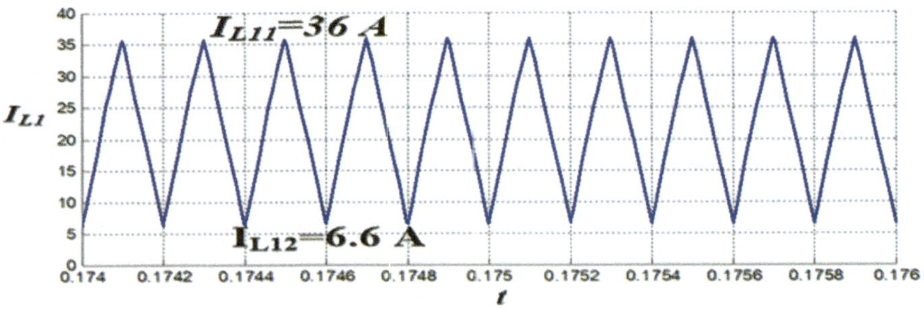

Fig. 3.16 Inductor current (I_{L1}) waveform where upper limit inductor current (I_{L11} = 36 A) and lower limit inductor current (I_{L12} = 6.6 A)

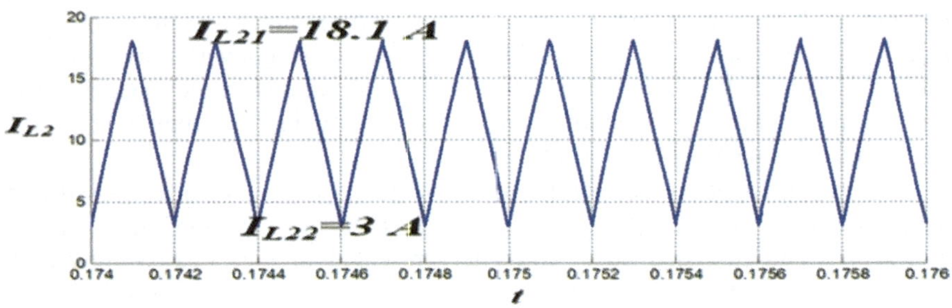

Fig. 3.17 Inductor current (I_{L2}) waveform where upper limit inductor current (I_{L21} = 18.1 A) and lower limit inductor current (I_{L22} = 3 A)

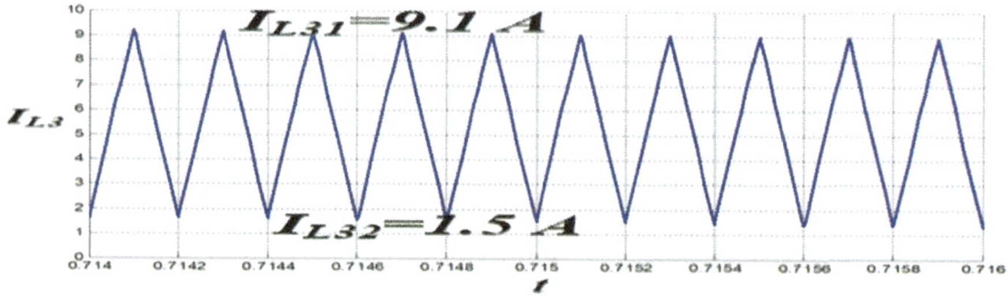

Fig. 3.18 Inductor current (I_{L3}) waveform where upper limit inductor current(I_{L31} = 9.1 A) and lower limit inductor current (I_{L31} = 1.5 A)

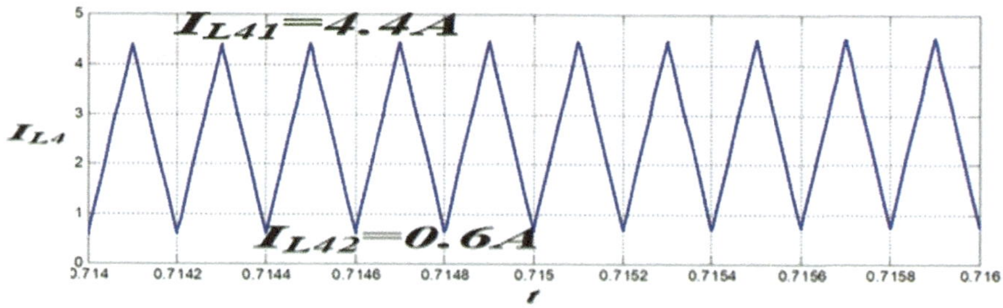

Fig. 3.19 Inductor current (I_{L4}) waveform where upper limit inductor current (I_{L41} = 4.4 A) and lower limit inductor current (I_{L42} = 0.6 A)

Fig. 3.20 Voltage across switch (V_{sw})

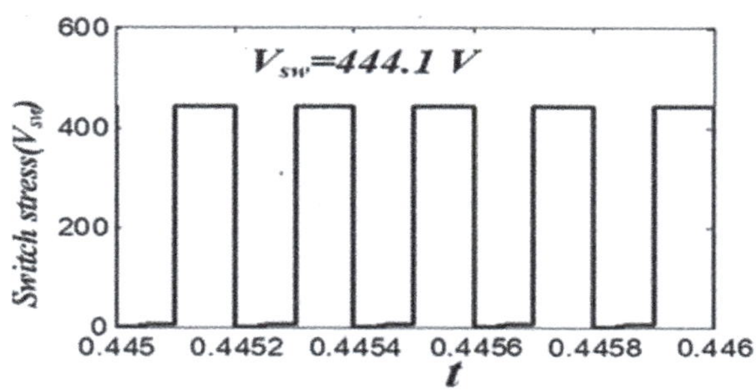

Table 3.3 Comparison table on the basis of capacitor stress voltages

Converters	Cascaded boost (V)	Proposed topology (V)
V_{C1}	58	27
V_{C2}	117	55
V_{C3}	235	110

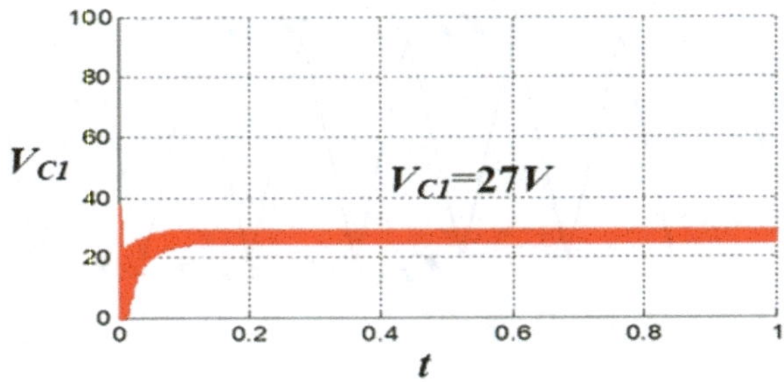

Fig. 3.21 Capacitor voltage stress (V_{C1}) of the proposed topology

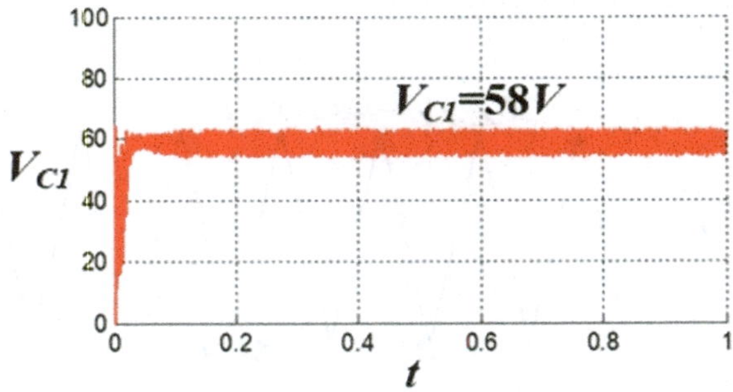

Fig. 3.22 Capacitor voltage stress (V_{C1}) of the cascaded boost converter

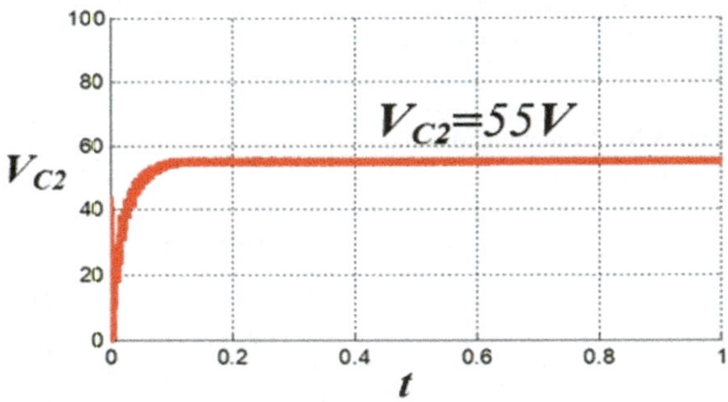

Fig. 3.23 Capacitor voltage stress (V_{C2}) of the proposed topology

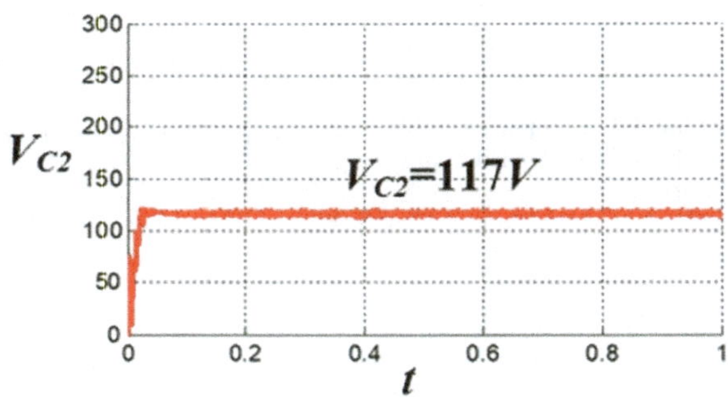

Fig. 3.24 Capacitor voltage stress (V_{C2}) of the cascaded boost converter

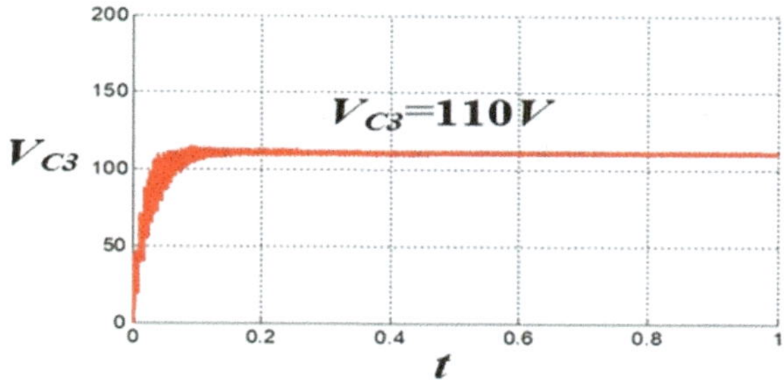

Fig. 3.25 Capacitor voltage stress (V_{C3}) of the proposed topology

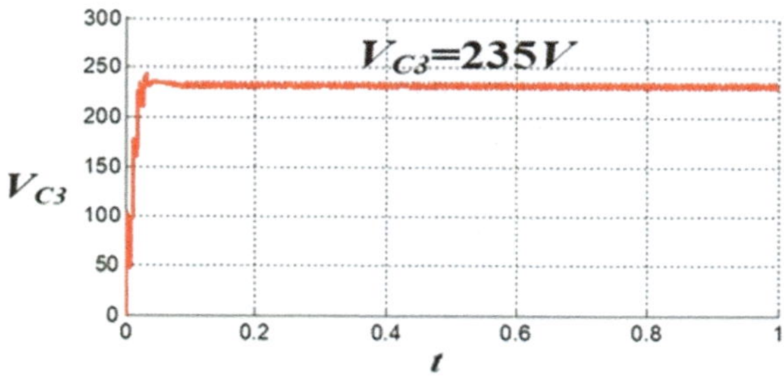

Fig. 3.26 Capacitor voltage stress (V_{C3}) of the cascaded boost converter

3.9 Conclusion

A complete simulation of the proposed converter with generalized structure of quadratic boosting cell (n = 3) is performed in MATLAB/Simulink. The proposed converter structure shows an increase in its output voltage (V_0 = 444.5 V) gain even for minimal range of duty ratio (δ = 0.5). The structure having maximized voltage gain at output terminals justifies its dependency on number of quadratic boosting cells (n = 3). The converter structure also on comparison with other various topology shows the maximum output voltage gain for a low input voltage (V = 30 V) value. Hence, as per the application, the output voltage can be obtained by adjusting the boosting cells n for a particular duty ratio range (δ). The reduction in voltage stresses is mainly observed across the utilized capacitors when the proposed converter is estimated with the cascaded boost converter for the consistent duty ratio (δ = 0.5) and same switching frequency (f = 5 kHz), power (P = 500 W) and input supply (V = 30 V). The simulated results and output wave shapes of capacitor stress voltages (V_{C1}, V_{C2}, V_{C3}) between the proposed structure and cascaded boost converter are discussed in detail. This feature reduces the capacitor voltage ratings as well as cost of the capacitors utilized in proposed converter making the converter cost effective. Hence, the converter is better as compared to other topology with respect to the output voltage gain and capacitor stress voltages.

References

1. Maksimovic, D., Cuk, S.: Switching converters with wide range dc conversion range. IEEE Trans. Power Electron. **6**(1), 369–378 (1996)
2. Middlebrook, R.D.: Transformerless dc-dc converters with large conversion ratios. IEEE Trans. Power Electron. 484–488 (1988)
3. Huber, L., Jovanovic, M.M.: A design approach for server power supply for networking applications. In: IEEE Applied Power Electronics Conference and Exposition, pp. 1163–1169 (2000)
4. Ortiz Lopez, G., Leyvia Ramos, J.: Modelling and analysis of switch mode cascaded converters with a single active switch. IEEE Power Electron. **1**, 478 (2000)
5. Tseng, K.C., Liang, T.J.: High efficiency step up converter. IEEE Power Electron. 182–190 (Apr 2004)
6. De Novaes, Y.R., Rufer, A., Barbi, I.: A new quadratic three level dc/dc converter suitable for fuel cell applications. IEEE Power Conversion Conference, pp. 601–607 (2007)
7. Choe, G.Y., Kang, H.S., Lee, B.K., Lee, W.Y.: Design consideration of interleaved converters for fuel cell applications. In: Proceeding of International Conference on Electrical Machines and System, Seoul, Korea (2007)
8. Acharya, A.., Nanda, L., Roy, T., Misra, B.: Boost converter with generalized quadratic boosting cell with reduced capacitor voltage stresses. Lecture notes in Electrical Engineering, 691, pp. 79–81 (2021)
9. Kadari, R., Gaubert, J.P., Champenois, G.: Performance analysis of transformer less single switch quadratic boost converter for grid connected photovoltaic systems. In: Proc. Int. Conf. ICEM'10, Rome, Italy, Sept 2010, pp.1–7
10. Li, W., He, X.: Review of non isolated high step up dc-dc converters in photo-voltaic grid connected applications. IEEE Trans. Ind. Electron. **58**, 1239–1250 (Apr 2011)
11. Yang, P., Xu, J., Zhou, G., Zang, S.: A new quadratic boost converter with high voltage step up ratio and reduced voltage stresses. In: 2012 IEEE 7th International Power Electronics and Motion Control Conference, Aaia, June 2012, harbin, China
12. Wijerantne, D.S., Moschopoulos, G.: Quadratic power conversion for power electronics. IEEE Trans. Circ. Syst. 426–438 (2012)
13. Khoucha, F., Benrabah, A., Herizi, O., Kheloui, A.: An improved MPPT interleaved boost converter for solar electric vehicle application. In: 2013 IEEE 4th International Conference on Power Engineering, Energy and Electrical Drives
14. Nanda, L., Acharya, A.: Evaluation on various DC–Dcnon-isolated maximized voltage gain converter topology. In: 11th International Conference on advances in computing, control and telecommunication technologies, ACT, 2020, pp. 173–178 (2020)
15. Tattiwong, K., Bunlasksananusorn, C.: Analysis Design and Experimental Verification of Quadratic Boost Converter. IEEE (2014)
16. Ye, Y., Eric Chang, K.W.: Quadratic boost converter with buffer capacitor voltage stress. IET Power Electron. **7**(5), 1162 (2014)
17. Blaabjerg, F., Yang, Y., Ma, K., Wang, X.: Power electronics the key technology for the renewable energy system integration. In: International Conference on Renewable Energy and Research Applications,Palermo, 2015,pp. 1618–1626
18. Priyadarshi, N., Padmanaban, S., Maroti, P.K., Sharma. A.: An extensive practical investigation of FPSO-based MPPT for grid integrated PV system under variable operating conditions with anti-islanding protection. IEEE Syst. J. 1–11 (2018)
19. Priyadarshi, N., Padmanaban, S., Bhaskar, M.S., Blaabjerg, F., Sharma, A.: A fuzzy SVPWM based inverter control realization of grid integrated PV-wind system with FPSO MPPT algorithm for a grid-connected PV/wind power generation system: hardware implementation. IET Electr. Power Appl. 1–12 (2018)

20. Padmanaban, S., Priyadarshi, N., Holm-Nielsen, J.B., Bhaskar, M.S., Azam, F., Sharma, A. K.: A novel modified sine-cosine optimized MPPT algorithm for grid integrated PV system under real operating conditions. IEEE Access **7**, 10467–10477 (2019). https://doi.org/10.1109/ACCESS.2018.2890533
21. Padmanaban, S., Priyadarshi, N., Holm-Nielsen, J.B., Bhaskar, M.S., Hossain, E., Azam, F.: A hybrid photovoltaic-fuel cell for grid integration with jaya-based maximum power point tracking: experimental performance evaluation. IEEE Access **7**, 82978–82990 (2019). https://doi.org/10.1109/ACCESS.2019.2924264
22. Kamalapathi, K., Priyadarshi, N., Padmanaban, S., Holm-Nielsen, J.B., Azam, F., Umayal, C., Ramachandaramurthy, V.K.: A hybrid moth-flame fuzzy logic controller based integrated Cuk converter fed brushless DC motor for power factor correction. Electronics **7**, 288 (2018)
23. Das, I., Roy, T.: A New Multi device Boost Converter Topology with reduced Switching Stresses and High Voltage Applications, pp. 77, MFIIS, Sept 2015

Lipika Nanda is currently working as Assistant professor in KIIT Deemed to be University (Institute of Eminence). She has completed her M.Tech in Power Electronics and Drives. She has completed her PhD research on designing various Multilevel Inverter Topologies. Her area of interest is multilevel inverter, DC-DC converters, Renewable energy sources etc. She has published in many national and international journals and conferences. Currently she is an active member of IEEE (PES), ISTE, ISCA and ISLE.

Adyasha Acharya completed M.Tech in the field of Power Electronics and Drives in KIIT Deemed to be University (Institute of Eminence). She has published research papers in IEEE and Springer conferences. Her area of interest is study of various DC-DC converters, their designing and applications.

Chapter 4
Comparative Analysis of Nonlinear SMC Controller with Linear PID Controller for Flyback Converter

Ashutosh Gupta and Dheeraj Joshi

4.1 Introduction

Switching converters have developed an important component of numerous military and commercial applications. Switching power converters that play major important role in power supply have become pervasive in many power equipment. The DC–DC power converters have created a remarkable research curiosity in the power electronics for their analysis, control and modeling.

Being small size, simple structure and high efficiency and so on, switched mode DC–DC converters has been broadly used in aerospace, IT, naval ships, power supplies and everyday usages [1]. Switched mode DC–DC power converters are nonlinear as well as time varying systems [2]. With recent advancement of power electronics applications and the increase in battery-powered electronic device, most of the power electronics or the battery systems are powered through DC–DC converters in recent years [3]. Numerous researches are going on to design the control strategies model, and numerous innovative models have been proposed.

In traditional controlling methods, there are vast innovation for the analysis and control of the various linear time-invariant (LTI) systems whereas actual systems are nonlinear and time invariant. DC–DC converters can be measured as time-variant as well as nonlinear systems due to their characteristic switching process and designing depends on instantaneous states of the power switches in the converters. The model with nonlinearity and disturbance, the conventional techniques are not fruitful. The conventional PWM technique-based controllers are only suited for specific conditions and cannot be applicable in large scale as they are based on small-signal analysis. That is why their modeling is a complex task. However, accurate analytical modeling of pulse width modeling (PWM) of DC–DC converters is extremely important in system for the analysis and design in numerous

A. Gupta (✉) · D. Joshi
IEEE, Department of Electrical Engineering, Delhi Technological University, Delhi, India

N. Priyadarshi et al. (eds.), *DC–DC Converters for Future Renewable Energy Systems*, Energy Systems in Electrical Engineering,
https://doi.org/10.1007/978-981-16-4388-0_4

applications, e.g., submarines, mainframe computers, aeronautics, medical, automobiles, naval ships, aerospace and telecommunications equipment.

However, the researchers are going on to develop its robust stability for the DC–DC converters. Converter's control design is usually implemented using small-signal analysis around its operating point, so that one can apply linear control strategies to study its performance characteristics [2]. There are lot of nonlinear controllers like backstepping control, fuzzy control, sliding mode control, model predictive control and so on [4–7]. Among these methods, SMC has some privilege of robustness against parameter uncertainties, fast dynamic response and load disturbances. Sliding mode control theory is a derived using variable structure system theory. For a flyback converter, one of the most powerful robust control strategies can be sliding mode controller. Various researches are going on to develop sliding mode controller for flyback converters [8–11].

4.2 Buck–Boost Converter

A DC–DC converter is used usually to transfer DC voltage at one level to another level where input voltage is chosen by different voltage source [12–14]. Buck–boost converter shown in Fig. 4.1 is capable of generating an output voltage higher or lower value than the input voltage and has property to step down or step up the output voltage [15–17]

Its voltage gain equation is stated as

$$\frac{V_{out}}{V_{in}} = \frac{D}{1 - D} \tag{4.1}$$

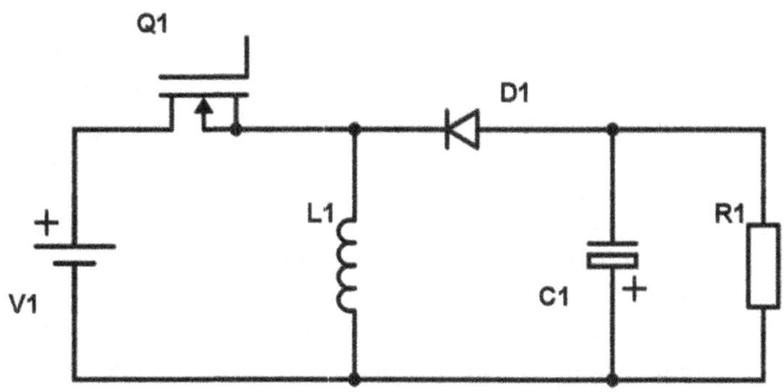

Fig. 4.1 Buck–boost converter topology

V_{out} = Output voltage over C.

V_{in} = Input source voltage across V_1.

D = duty cycle of switching pulse.

4.2.1 Working

A buck–boost converter has the same configuration as that of flyback converter but there is no transformer. It consists of a switch (Q_1), a voltage source V_1, an inductor L_1, a power diode D_1, a capacitor C_1 and a resistive load R_1 working in continuous conduction mode (CCM) for operation. Assuming fixed state condition capacitor is initially charged having V_{out} voltage across it and a finite current flow through the inductor. Let us consider that the switch Q_1 is ON, so input DC voltage V_1 charges the inductor L_1 magnetically as the diode D_1 is reversed biased. This leads to surge in inductor current expressed as

$$V = L \times \frac{di}{dt} \tag{4.2}$$

V = Voltage over L inductor.

L = Inductance value in mH.

i = Inductor current in A.

And as soon as switch Q_1 is switched OFF, the current in the inductor L_1 is going to transfer energy to the capacitor C_1 in the same direction, making diode D_1 forward bias and starts conducting according to Faradays law of electromagnetic induction. The capacitor will charge in opposite direction so the output will be inverted.

The state that average voltage over inductor ought to be equal to zero to derive voltage gain expression is,

During ON state,

$$V_{L1} = V_{in} \tag{4.3}$$

and during OFF state,

$$V_{L1} = V_{out} \tag{4.4}$$

Including average voltage conditions norms,

$$V_{in} \times D \times T - V_{out} \times (1 - D) \times T = 0 \tag{4.5}$$

$$\frac{V_{out}}{V_{in}} = \frac{D}{1 - D} \tag{4.6}$$

T = Time period of switching pulse.
So, the final Eq. 4.6 represents the voltage expression for buck–boost converter.

4.3 Flyback Converter

Converters with the implementation of isolation transformer can have numerous outputs of various magnitude and polarities. A flyback converter shown in Fig. 4.2 is a simple switched mode converter used in both AC–DC and DC–DC conversion with isolated input and output [8].

Flyback converter here is a DC–DC converter which is an advance form of buck–boost converter where a transformer is placed by eliminating inductor for galvanic separation between input and output so that voltage ratios are multiplied with an additional advantage of isolation. They are widely used in various applications such as PV system, DC power supply, power adapters and so on. Flyback converter ensures good output regulation and fast response over wide frequency bandwidth because the flyback converter is a non-minimum phase system with a right-half-plane zero in its voltage transfer function [4]. Voltage gain expression for a flyback converter is defined as

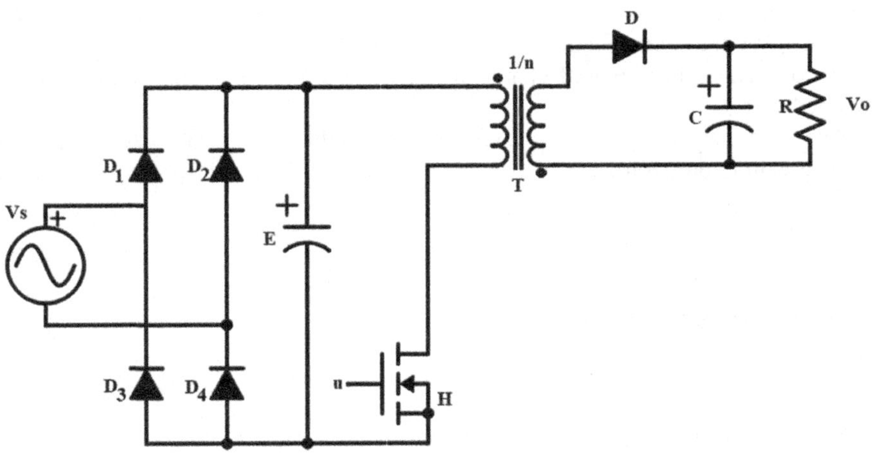

Fig. 4.2 Flyback converter topology

$$V_{\text{in}} DT - V_{\text{out}} \frac{N_1}{N_2}(1 - D)T = 0 \tag{4.7}$$

$$\frac{V_{\text{out}}}{V_{\text{in}}} = \frac{N_1}{N_2} \times \frac{D}{1 - D} \tag{4.8}$$

N_1 = Primary side turns ratio of transformer.

N_2 = Secondary side turns ratio of transformer.

4.4 Dynamics and Modeling of Flyback Converter

Consider a flyback converter having boost topology shown in Fig. 4.3 which is an electrical model where transformer is modeled as equivalent inductor L. The parameters E, L, H, D, C, R, n, i_Landv_C represent input voltage, transformer inductance, MOSFET switch, diode, output filter capacitor, load resistor, transformer turns ratio, inductor current and output capacitive voltage of the flyback converter, respectively.

Figure 4.4 illustrates the voltage and current waveform for ON and OFF state of switch (H) considering the continuous conduction mode.

In any converter, there are basic two states of a switch, which are used in expressing the voltage gain equation.

During the switch H is in ON state, the current builds up in the primary side of the transformer T as the input DC voltage power the primary side. In result to that energy gets stored in the magnetizing inductance L of transformer. Diode D will be

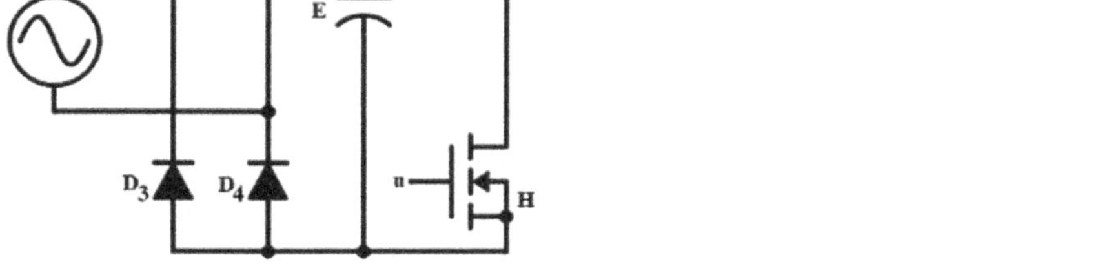

Fig. 4.3 Electrical model of flyback converter including model of transformer

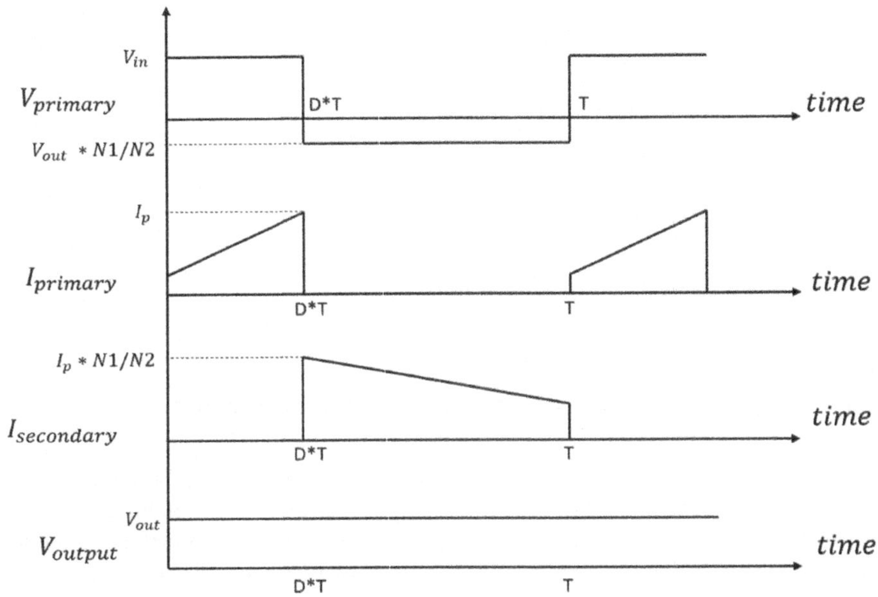

Fig. 4.4 Voltage and current waveform for PWM switching of flyback converter

reversed bias due to dot polarity of the transformer *T*, so secondary side will be open circuited, and it will not conduct. During this condition, output capacitor (*C*) provides current to the load.

And when the switch *H* is switched OFF, as per the dot polarity of the transformer, the current in the inductor *L* is going to transfer energy to the capacitor *C* in the same direction, making diode *D* forward bias on the secondary side and starts conducting according to Faradays law of electromagnetic induction. With this cycle gets complete and DC voltage is received across load resistance.

Considering negligible winding resistance and leakage inductance, whereas it has nonzero core reluctance transformer can be modeled as two winding inductors with magnetizing inductance *L* on primary side of ideal transformer. The system has two state variables, current i_L and voltage v_C, whose dynamics are described as:

$$u = 1 : \begin{cases} L\dfrac{\mathrm{d}\, i_L}{\mathrm{d}t} = E \\ C\dfrac{\mathrm{d}\, v_c}{\mathrm{d}t} = -\dfrac{v_c}{R} \end{cases} \quad u = 0 : \begin{cases} L\dfrac{\mathrm{d}\, i_L}{\mathrm{d}t} = -\dfrac{v_c}{n} \\ C\dfrac{\mathrm{d}\, v_c}{\mathrm{d}t} = \dfrac{i_L}{n} - \dfrac{v_c}{R} \end{cases} \tag{4.9}$$

Above Eq. (4.3) can be represented in single switched model of flyback converter as (Table 4.1).

Table 4.1 Specification for flyback converter

Description	Parameter	Value
Input voltage	V_s	220 V (RMS)
Output voltage	V_o	5 V (DC)
Frequency	f_s	100 kHz
Inductance	L	1 mH
Capacitance	C	250 μF
Load resistance	R	2.5 Ω
Turn ratio	N	6/320

$$\begin{cases} L \cdot i_L = -(1-u) \cdot \dfrac{v_c}{n} + u \cdot E \\ C \cdot v_c = (1-u) \cdot \dfrac{i_L}{n} - \dfrac{v_c}{R} \end{cases} \tag{4.10}$$

4.5 Proportional Plus Integral Plus Derivative Controller

Most of the desired performance of a system can be achieved by suitable combination of proportional, integral and derivative control action. The PID controller is widely implemented because it is easy to understand and is quite operative. The transfer function of a PID controller is expressed by:

$$C(S) = K_p\left(1 + \tau_d + \frac{1}{\tau_{iS}}\right)$$

It is a second-order controller, but it has versatile applicability. Any type of SISO system can use this controller, e.g., linear, nonlinear, time delay, etc. For MIMO system, it is first decoupled into many SISO system and PID controller is implemented in each SISO system. Though, for suitable implementation, a controller has to be tuned for a precise process, i.e., choice of P, I, D constraints is very important and process dependent. Unless the parameters are precisely selected, a controller may reason variability to the closed-loop system.

The general variation on a closed-loop system with each controller parameter is tabulated below (Table 4.2).

In order to obtain a desired overall response, adjust each of the gains K_p, K_i and K_d. It is not always essential that all the combination of proportional, derivative and integral actions should be combined in the controller. In most of the cases, a simple P-I assembly will aid the objective. This project also deals with the PID controller implementation on flyback converter in subsequent section for comparative study.

Figure 4.5 shows the basic configuration of PID control structure for flyback converter. Simulation model designed for the flyback converter is simulated in

Table 4.2 Controller parameter variation

Close-loop response	Rise time	Overshoot	Settling time	S–S error
K_p	Fall	Rise	Small change	Fall
K_i	Fall	Rise	Rise	Fall
K_d	Small change	Fall	Fall	No change

Simulink of MATLAB under resistive load condition to study its performance. A simulation model designed for the flyback converter is simulated in Simulink of MATLAB under different load condition to study its performance. Figure 4.10 shows the flyback converter with closed-loop control using PID control algorithm in Simulink. The reason behind using PID controller is its application to different processes and its simple structure. A source of 220 V is first rectified using a bridge rectification and then passed to the input side of flyback converter which include a high-frequency transformer and protection circuit and a closed-loop PID control is performed with one of its multiple outputs of 5 V as a reference voltage. Proper tuning of the PID controller requires wide trials. PID block parameters like K_p, K_i and K_d are chosen after repetitive trial and error method. With the value of 1, 15 and 0 for K_p, K_i and K_d, respectively, are entered in the PID block. Initially voltage error is given to the PID controller, then error generated is forwarded to a pulse width modulation generator for producing switching pulse for the MOSFET switch. After all the calculation for the flyback transformer and parameter for converter, performance of flyback converter is studied for closed-loop PID controller. The model is simulated on MATLAB–Simulink, and various outputs are studied and verified. The parameters for flyback converter are shown in Table 4.1.

4.6 Sliding Mode Controller

4.6.1 Principle of Sliding Mode Controller

Variable controller-based sliding mode control is introduced to make the controlled system to hold robust dynamic response to the variable system parameter [9] (Fig. 4.6).

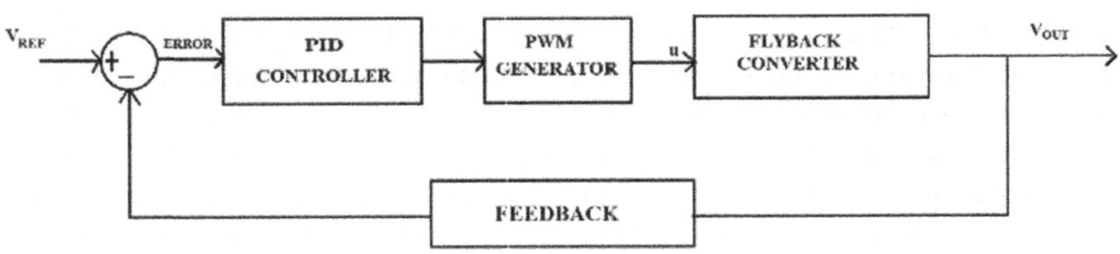

Fig. 4.5 Basic PID controller topology

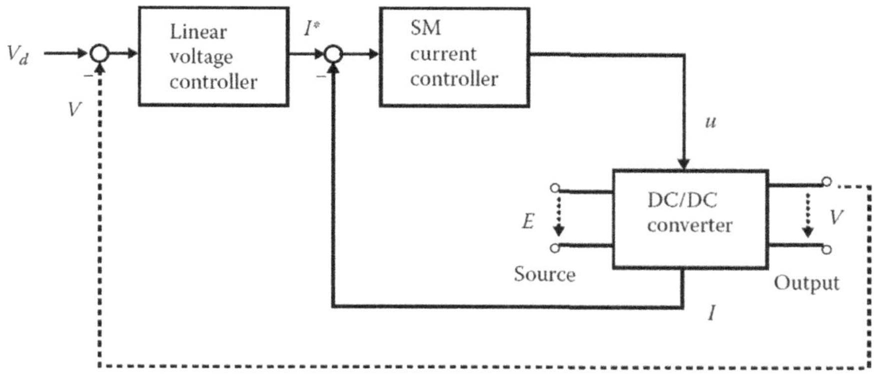

Fig. 4.6 A topology of cascaded control design of general DC–DC converters (Ref: sliding mode control in electromechanical system)

The basic concept of SMC is to have a sliding surface in relations to system variables as reference path to trace, so that systems state trajectory reaches that sliding surface and lock itself to that surface to attain the equilibrium position. Also, it can be said that the controlled system reaches the designed steady state [14]. This phenomenon can be realized with the agreement to some desired condition with infinite switching frequency.

- Hitting condition
- Existence condition
- Stability condition

As soon as these conditions are achieved in the process, then the system becomes immune to internal and external disturbances and the system will be in equilibrium with minimum regulation error along with fast transient response.

4.6.2 *Sliding Mode Control of Flyback Converter*

As shown Fig. 4.7 is diagram of flyback converter.

The differential equation for flyback converter under switching condition (ON/OFF) is expressed below:

$$\begin{cases} L \cdot \dot{i}_L = -(1-u) \cdot \dfrac{v_c}{n} + u \cdot E \\ C \cdot \dot{v}_c = (1-u) \cdot \dfrac{i_L}{n} - \dfrac{v_c}{R} \end{cases} \tag{4.11}$$

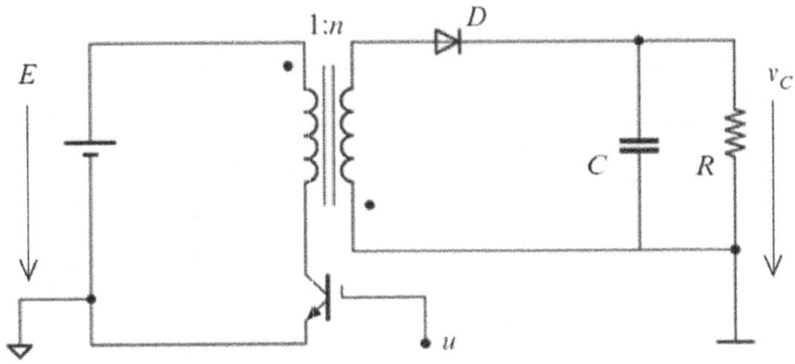

Fig. 4.7 Flyback converter schematics

where the typical representations have been used to describe the parameters for converter and the transformer ratio is represented using n [16], $x = [i_L v_c]^T$ defines the state vector for flyback converter and $s(\mathbf{x}) = 0$ defines the surface, where

$$S(\mathbf{x}) = v_c - v_c^* \tag{4.12}$$

with v_c^* constant is an accurate selection for starting the design of a SMC required in regulating the output voltage which is stated underneath:

$$u = \begin{cases} u^+ = 1 & \text{if } s(\mathbf{x}) < 0 \\ u^- = 0 & \text{if } s(\mathbf{x}) > 0 \end{cases} \tag{4.13}$$

The initial stage of the designing a SMC is to check if the transversality condition is reached. This means to verify if the aspect multiplying u in the equation of $\dot{s}$ is nonzero. Considering the state Eq. (4.11), it then outcomes:

$$S(X) = v_c = (1-u)\frac{i_L}{nC} - \frac{v_c}{RC} = \frac{i_L}{nC} - \frac{v_c}{RC} - \frac{i_L}{nC} \cdot u \tag{4.14}$$

where the term $i_L/(nC)$ multiplying u is nonzero since $i_L > 0$.

In a next step, it is essential to compute the corresponding control. This is found by resolving the expression $S(\mathbf{x}) = 0$ aimed at u. Rendering to (4.14), it provides $(1-u) \cdot {}^{i_L}/_{n} - {}^{v_c}/_{R} = 0$, hereafter:

$$u_{eq} = 1 - v_c \cdot {}^{n}/_{(R \cdot i_L)} \tag{4.15}$$

Since imposing that $0 = u^- < u_{eq} < u^+ = 1$ it can presume the area of the sliding mode's presence. The constrain $u_{eq} > 0$ provides $v_c \cdot n/(R \cdot i_L) > 0$, which is correct since $v_c > 0$ and $i_L > 0$ in all working condition. The inequality $u_{eq} < 1$ primes to $i_L > v_c \cdot {}^{n}/_{R}$, which confines the shaded area shown in Fig. 4.8.

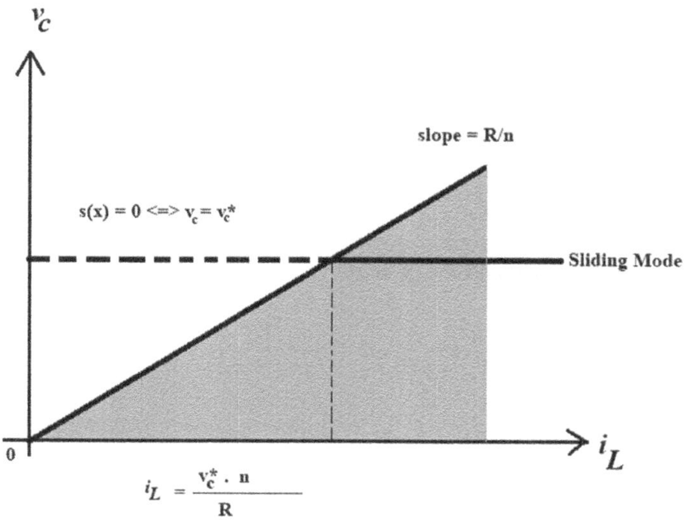

Fig. 4.8 Switching surface for sliding mode controller to control flyback converter output

The third design stage entails in construing the corresponding dynamic when in sliding mode. To this conclusion, Eq. 4.15 is replaced in the flyback converter Eq. 4.11, in the meantime seeing that $S(\mathbf{x}) = 0(v_c = v_c^*)$.

Also $S(\mathbf{x}) = 0$. Therefore, the corresponding dynamic of the inductor current expressed below:

$$L \cdot i_L = -(1 - u_{eq}) \cdot v_c^*/n + u_{eq} \cdot E \tag{4.16}$$

Using Eq. 4.15 one can find after some simple operation:

$$L \cdot i_L = \frac{v_c^*(nE - v_c^*)}{R \cdot i_L} + E \tag{4.17}$$

Above expression is a nonlinear function of i_L which defines that the dynamic of current and expressed as $i_L = h(i_L, v_c^*)$. One can continue by linearization by considering an assumed operating point to evaluate the nature of this dynamic. Considering i_{eq0} be the equivalent value of current to the steady-state operating point confirmed by regulating the capacitor voltage v_c to a positive set point value v_{co}^*, i_{eq0} is consequently found by equating to zero with the current time derivative and putting $v_c = v_{co}^*$. The outcome will be

$$i_{eq0} = \frac{v_{co}^*(v_{co}^* - nE)}{ER} \tag{4.18}$$

A positive value of i_{eq0} is confirmed unless $v_{co}^{*} > nE$. Consider $\tilde{i_L} = i_L - i_{eq0}$ and $\tilde{v_c^*} = v_c^* - v_{co}^*$ be insignificant variances of the state variables near operative point (i_{eq0}, v_{co}^*). Linearizing the system Eq. 4.17 near this point provides

$$\tilde{i_L} = \left(\frac{\partial h}{\partial i_L}\right) \Big| (i_{eq0}, v_{co}^*) \cdot \tilde{i_L} \Big| + \left(\frac{\partial h}{\partial v_c^*}\right) \Big|_{(i_{eq0}, v_{co}^*)} \cdot \tilde{v}_c^* \tag{4.19}$$

where the magnitude of a and b can be derived using some algebraic calculation as

$$a = \frac{E^2}{L v_{co}^* (v_{co}^* - nE)}, b = \frac{E}{L(v_{co}^* - nE)} \left(\frac{nE}{v_{co}^*} - 2\right) \tag{4.20}$$

It is note that $a > 0$ as $v_{CO}^* > nE$, so the inductor current i_L is not stable since the transfer function $\frac{\tilde{V}_C^*(s)}{\tilde{I}_L(s)}$—with $\tilde{V}_C^*(s)$ and $\tilde{I}_L(s)$ being the Laplace transforms of $\tilde{v}_C^*$ and $\tilde{i}_L$, correspondingly—has a pole located in right half-plane. Thus, the matching dynamic is not stable.

To achieve stable equivalent dynamic on the sliding surface for the system, one can figure a state feedback K such that the closed-loop dynamic of the system is steady. Furthermore, this dynamic can also be adjusted to ensure anticipated performance, i.e., settling time. Therefore, by imposing stabilizing time constant T_0, the value of K outcomes as:

$$K = -\frac{a + 1/T_0}{b} \tag{4.21}$$

Figure 4.9 shows the closed-loop sliding mode control of the converter including the stabilization of the corresponding process.

It can be imagined from the diagram that the final expression of the control law depicts a switching surface that includes the inductor current as one of the parameters of sliding surface $i_L : s(\mathbf{x}) = v_C - v_C^* + K \cdot (i_L - i_{eq})$, where the magnitude of i_{eq} is specified by Eq. (4.18). After carefully examining the last

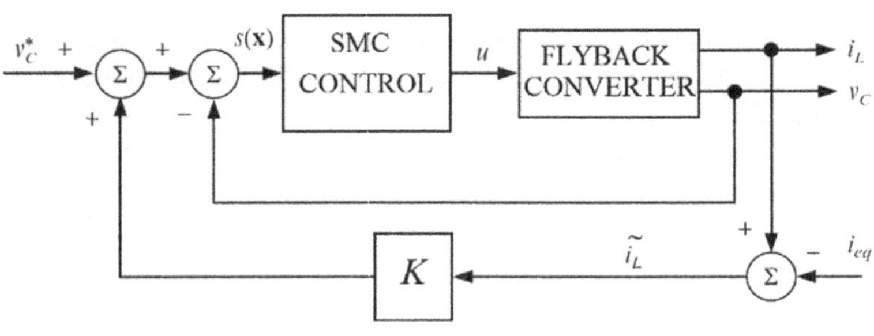

Fig. 4.9 Control scheme for flyback converter using SMC and stabilizing the feedback

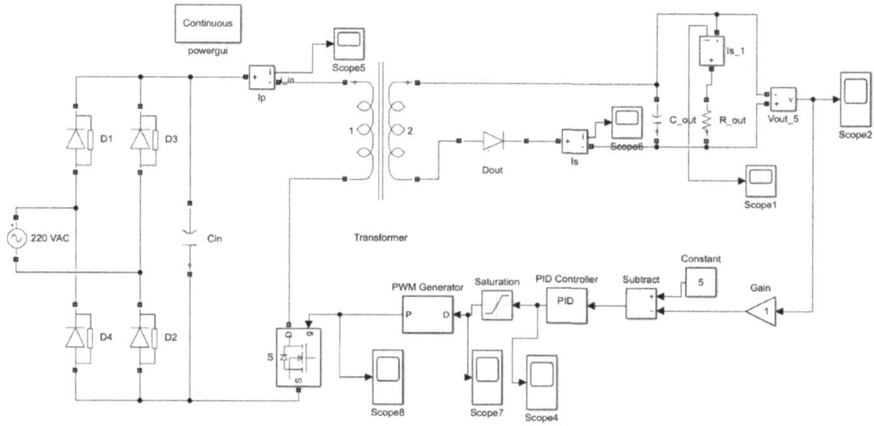

Fig. 4.10 Flyback converter with closed-loop PID control in MATLAB

expression, the value of i_{eq} is load resistance R dependant, which needs to be calculated. Hence, Fig. 4.9 shows a load estimator being calculated for the system based on some dependable procedures from circuit while implementing the control diagram.

4.7 Simulation Model and Result

4.7.1 Effect of Linear PID Controller

Simulation model for flyback converter using PID control is designed considering the parameters mentioned in paper and the value of K_p, K_i and K_d is entered as calculated. Figure 4.10 shows the simulation model of PID controller for flyback converter. Figures 4.11 and 4.12 show the output waveforms for voltage and current for flyback converter under PID controller. Peak output voltage for flyback converter in case of PID controller is 5.042 V. Rise time to reach desire state in case of output voltage of 5 V is 49.438 μs. Output voltage and current have ripple in case of PID controller. Steady-state error for flyback converter is 5.16%.

4.7.2 Effect of SMC Controller

Following the calculation in Section VI for flyback converter, simulation model is designed using the proposed design described in the earlier Fig. 4.9. Figure 4.13 shows the Simulink model for the flyback converter under sliding mode controller. Result for the same is shown in the subsequent section. After designing the

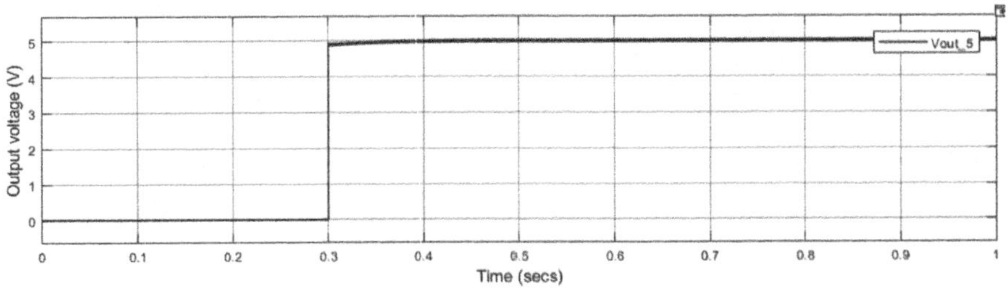

Fig. 4.11 Simulation waveform of PID-controlled flyback converter for 5 V output voltage

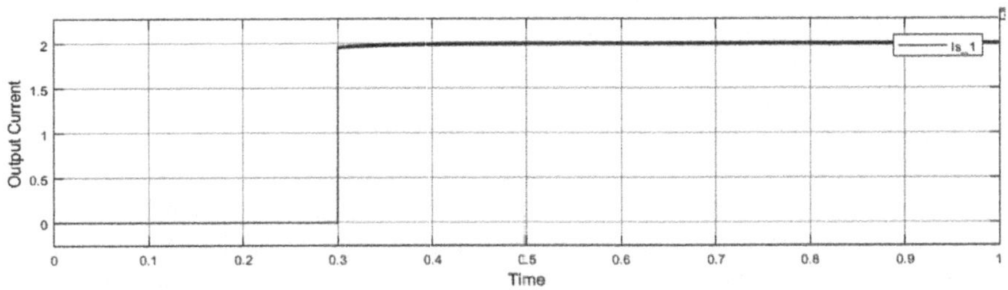

Fig. 4.12 Simulation waveform of PID-controlled flyback converter for output current

Simulink model for flyback convertor for unregulated 5 V output voltage, sliding mode controller is initiated for closed-loop control to get regulated 5 V output voltage. Initially a sliding surface is designed to minimize the voltage and current error as per the sliding mode control technique, i.e., $s(\mathbf{x}) = 0$. Then a multiplying factor u $(-1 < u < 1)$ is added to the sliding surface, and it is checked that the derivative of sliding surface must be nonzero to proceed further. In the next step, equivalent control law is developed by equating derivative of sliding surface with zero which is:

$$u_{eq} == 1 - v_C \cdot \frac{n}{R \cdot i_L}$$

By imposing that $0 = u^- < u_{eq} < u^+ = 1$, one can believe the area of the sliding mode's existence for the system.

Next step is to deduce the equivalent dynamic of the system when in sliding mode. The closed-loop SMC topology involving the stabilizing effect on this dynamic model is shown in Fig. 4.13.

Result for the same is studied. After designing the Simulink model for flyback convertor for unregulated 5 V output voltage, sliding mode controller is initiated for closed-loop control to get regulated 5 V output voltage. Initially a sliding surface is designed to minimize the voltage and current error as per the sliding mode control technique, i.e., $s(\mathbf{x}) = 0$.

Peak output voltage for flyback converter in case of SMC controller is 4.997 V. Rise time to reach desire state in case of output voltage of 5 V is 470.15 ns. Steady-state error for flyback converter under close-loop control using sliding mode is 0.4% (Figs. 4.14, 4.15 and 4.16).

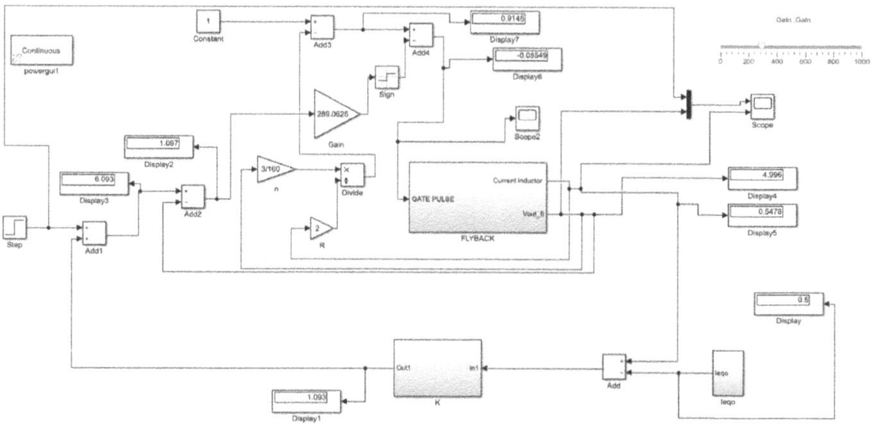

Fig. 4.13 Flyback converter with closed-loop SMC control in MATLAB

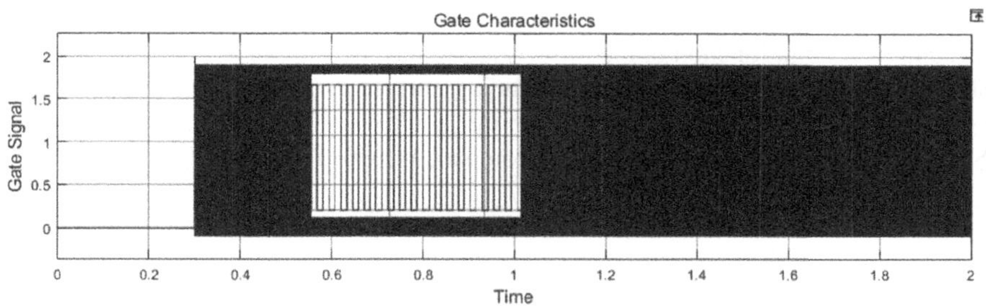

Fig. 4.14 Simulation for gate switching under SM controller

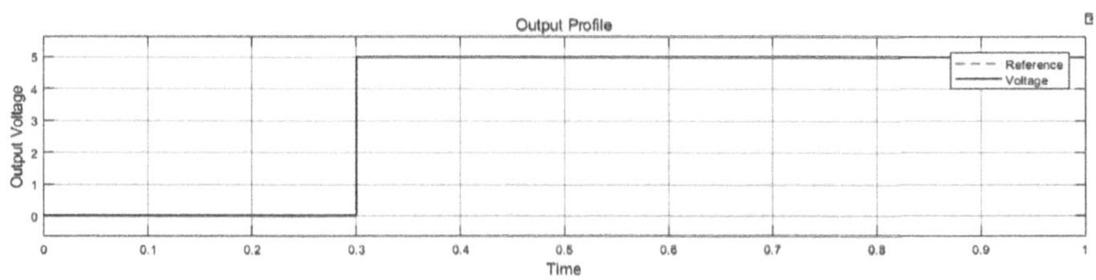

Fig. 4.15 Simulation waveform of SMC controlled flyback converter for 5 V output voltage

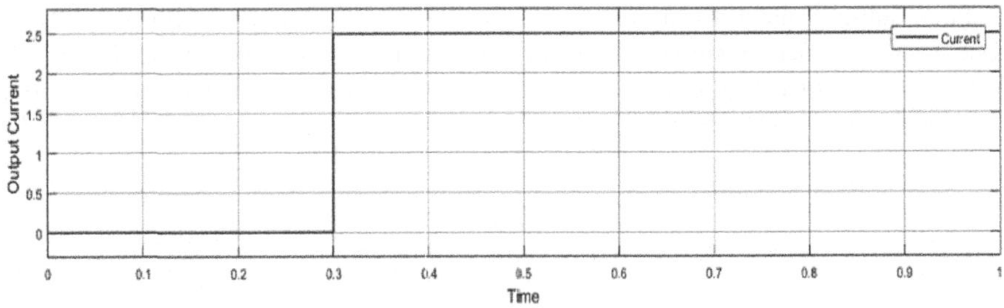

Fig. 4.16 Simulation waveform of SMC controlled flyback converter for output current

4.8 Conclusion

For DC–DC converter as in this case, flyback converter was controlled using both linear and nonlinear controller. In this paper, the basic principle and design procedure of the robust sliding mode controller are discussed. For controlling a nonlinear converter, most suitable controllers are nonlinear as it reduces multiple perturbated calculation which was tedious work in linear controller. Performance of flyback converter is more robust in sliding mode controller with steady-state error of less than 1% with no peak overshoot. Even the settling time is much faster in SMC controller. Sliding mode is a robust controller but the main problem associated with this is chattering effect. This effect could damage the physical part of the system. However, the chattering effect can be suppressed by SMC control with a boundary layer in which the discontinuous sign function of the SMC is substituted by saturation function. Here, the sliding variable is the function of the sliding surface and states which represents a connection between state variables. From the result waveform, it can be said that the SMC is having more robust control in comparison with PID controller.

Acknowledgements I would like to express thanks to Prof. Dheeraj Joshi for his numerous discussions as well as suggestions to explore various aspect of problem statement.

I would like to acknowledge Prof. Uma Nangia, Head of the Dept. of Electrical Engineering, DTU, Delhi for extending their support during Course of this study.

I would be failing in my responsibility if I do not acknowledge the cooperation rendered during numerous stages of Lab Equipment requirement by Miss. Vandana Rani (Technical Assistant), Miss. Renu (Technical Assistant) and Mr. Raju (Staff Power Electronics).

References

1. Zhou, C., Zhang, Q., Ezechias, D.D., Gao, Y., Deng, H., Qu, S.: A general digital PID controller based on PWM for buck converter. Proc. World Congr. Intell. Control Autom. **2015**, 4596–4599 (Mar 2015). https://doi.org/10.1109/WCICA.2014.7053488
2. Seshachalam, D., Tripathi, R.K., Chandra, D.: Practical implementation of sliding mode control for boost converter. In: IEEE Asia-Pacific Conf. Circuits Syst. Proceedings, APCCAS, vol. 00, no. 2, pp. 650–653 (2006). https://doi.org/10.1109/APCCAS.2006.342091

3. Li, H.: Sliding-Mode PID Control of DC-DC Converter, pp. 730–734 (2010)
4. Padmanaban, S., Priyadarshi, N., Holm-Nielsen, J.B., Bhaskar, M.S., Azam, F., Sharma, A. K.: A novel modified sine-cosine optimized MPPT algorithm for grid integrated PV . IEEE Access **7**, 10467–10477 (2019). https://doi.org/10.1109/ACCESS.2018.2890533
5. Padmanaban, S., Priyadarshi, N., Holm-Nielsen, J.B., Bhaskar, M.S., Hossain, E., Azam, F.: A hybrid photovoltaic-fuel cell for grid integration with jaya-based maximum power point tracking: experimental performance evaluation. IEEE Access **7**, 82978–82990 (2019). https://doi.org/10.1109/ACCESS.2019.2924264
6. Priyadarshi, N., Padmanaban, N., Holm-Nielsen, J.B., Blaabjerg, F., Bhaskar, M.S.: An experimental estimation of hybrid ANFIS–PSO-based MPPT for PV grid integration under fluctuating sun irradiance. IEEE Syst. J. **14**(1), 1218–1229 (2020). https://doi.org/10.1109/JSYST.2019.2949083
7. Priyadarshi, N., Padmanaban, N., Bhaskar, M.S., Blaabjerg, F., Holm-Nielsen, J.B., Azam, F., Sharma, A.K.: A hybrid photovoltaic-fuel cell-based single-stage grid integration with Lyapunov control scheme. IEEE Syst. J. https://doi.org/10.1109/JSYST.2019.2948899
8. Lin, M.T.: A second order sliding mode controller for the flyback converter. In: Conf. Proc.—IEEE Int. Conf. Syst. Man Cybern., vol. 2014, January, pp. 2289–2292 (2014). https://doi.org/10.1109/smc.2014.6974267
9. Gupta, D., Rana, A., Joshi, K.: Convertor under Continuous and Discontinuous Conduction mode using PI Controller (2020)
10. Bacha, S., Antoneta, I.M., Bratcu, I.: Advanced Textbooks in Control and Signal Processing Power Electronic Converters Modeling and Control (2013)
11. Utkin, V., Guldner, J., Shi, J.: Sliding Mode Control in Electro-mechanical Systems, 2nd edn (2017)
12. Chetty, P.: Current injected equivalent circuit approach to modelling switching dc-dc converters. Aerosp. Electron. Syst. IEEE Trans. AES-**17**(6), 501–505 (Nov 1981)
13. Davoudi, A., Jatskevich, J., De Rybel, T.: Numerical state-space average-value modelling of PWM dc-dc converters operating in DCM and CCM. Power Electron. IEEE Trans. **21**(4), 1003–1012
14. Cúk, S., Middlebrook, R.: A general unified approach to modelling switching dc to dc converters in discontinuous conduction mode. In: Proc. IEEE PESC'77, 1977, pp. 36–57 (1977)
15. Iqbal, H.K., Abbas, G.: Design and analysis of SMC for second order DC-DC flyback converter. In: 17th IEEE International Multi Topic Conference 2014, Karachi, 2014, pp. 533–538. https://doi.org/10.1109/INMIC.2014.7097398
16. Mohan, N., Undeland, T., Robbins, W.: Power Electronics. Wiley, New York (1995)
17. Rashid, M.H.: Power Electronics: Circuit, Devices and Applications, 2nd edn. Prentice-Hall, Englewood Cliffs, NJ (1993)
18. Vardia, M., Priyadarshi, N., Ali, I., Azam, F., Bhoi, A.K.: Maximum power point tracking for wind energy conversion system. In: Bhoi, A., Sherpa, K., Kalam, A., Chae, G.S. (eds.) Advances in Greener Energy Technologies. Green Energy and Technology (2020). Springer, Singapore. https://doi.org/10.1007/978-981-15-4246-6_36

Chapter 5
Effect of Irradiance on THD of Neutral Point Clamped Inverter Fed from PV Cell

Rohit Kumar Gautam, Subrat Behera, and Ranjeeta Patel

5.1 Introduction

In the present scenario, the whole world is confronting a massive increase in energy consumption. So, most of the industrialized processes need to build efficiency and diminish generation expenses. And this is accomplished by expanding establishment's size also expanding the power of every single electrically related machine and components. The DC schemes are operated at lower voltages in case of traction systems, shipboard power systems and telecommunication equipment [1].

The main focus is to present and review an analysis of NPC MLI of five-phase three-level compared with five-phase two-level NPC inverter at similar load performing for wider modulation index [2, 3]. In the recommended method, the selection of the switching is done in order to reduce the common-mode voltage of the inverter and also the THD of the current get minimized. And after simulation of the model, the result observed is that the five-phase three-level NPC inverter holds best performance than two-level NPC. In this paper, validation of performance parameters is done with a hardware setup [4].

The other researcher worked in related areas to my topic. In this paper, the development in speed control techniques of the IMD has led to a wide scope of usage in most electrical drives. The paper focuses on reduction of the harmonics distortions and common-mode voltage (CMV) by using five-level NPC inverter in case of three phases IMD. Here, both levels of inverters are compared for better performance and after comparison five-level NPC has been found better [5].

The other paper audits various kinds of medium- and high-voltage power converters utilized in sustainable power source and smart grid which consist of IMD of variable frequency. This paper entails the topology for converters and analyzes their specific utilizations [6].

R. K. Gautam · S. Behera (✉) · R. Patel
School of Electrical Engineering, Kalinga Institute of Industrial Technology, Deemed To Be University, Bhubaneswar, India

N. Priyadarshi et al. (eds.), *DC–DC Converters for Future Renewable Energy Systems*, Energy Systems in Electrical Engineering,
https://doi.org/10.1007/978-981-16-4388-0_5

There are different modulation techniques present to govern the multilevel inverters and the methods are (1) selective harmonic elimination pulse width modulation (SHE-PWM) and (2) sinusoidal pulse width modulation (SPWM). These techniques can be classified as indicated according to the switching frequency utilized for respective applications. The most popular modulation methods is SPWM which is widely used in much industrial application [7]. There are several multicarrier techniques developed on the basis of classical SPWM technique for multilevel inverter. It helps in diminishing the output voltage harmonics. Here the study of MATLAB SIMULATION models of both levels NPC multilevel inverter is done with SPWM control technique [8].

The attractive highlights of multilevel inverters are (i) The generated output voltage having lower d*v*/d*t* with very low distortion, (ii) it extracts input current along with low distortion, (iii) operation is done with a lower switching frequency, and (iv) generate smaller common-mode (CM) voltage.

The neutral point clamped (NPC) inverter has the voltage level at midpoint which is known as the neutral point of the inverter. The NPC inverter efficiently doubles the level of voltage of the device at the output side if proper switching pattern is introduced. Some of the main applications are laminators, process plants, conveyors, AC pumps, fans, blowers, compressors and many more advanced applications also [9].

In this paper, the multilevel inverter topology deals with the input PV solar voltage source. Power electronics module is designed for maximum extraction of power from solar. PV cell can produce limited quantity of electric power even at good weather conditions and power conditioners that utilize MPPT technique [10].

The advantages associated with PV systems are (i) Power delivered by PV system or solar cell is clean and silent. (ii) PV systems do not discharge any dangerous water or air contamination in the earth exhaust normal assets, or imperil creature or human safety. (iii) A PV system of any scale can be developed depending on energy demand. (iv) The PV cells are used in space effectively for emergency systems. (v) PV powers each satellite surrounding the earth as it works dependably for long periods of time with ignorable maintenance [11].

The literature survey of another paper explains the optimization and comparison of neutral point clamped dual-output inverters. Here, the harmonic analysis is being carried out and found to be lower among the converters [12–14]. To upgrade the performance of multilevel inverters, use of numerous switches becomes mandatory. Previously, neutral point clamped three-level inverter requires twelve switches to regulate the three-phase load [15, 16]. The main motive of the paper is to reduce the cost of the system, weight and numbers of gate driver circuits; it reduces the switch count with a minimum number of switches of three-level inverter topology that is suggested for dual input and output [17].

In all photovoltaic (PV) power systems, maximum power point tracking (MPPT) is an essential feature. In several papers and applications, the classical "hill climbing" and "incremental conductance" MPPT algorithms are commonly applied [18–20]. Both algorithms perturb the operating conditions of the PV array, and the changes in the power generated are detected [21, 22]. This paper, based on the single-diode model, begins to address the scale of the disturbance of the operating conditions for both algorithms. In order not to run away under such complex conditions, the result is used to pick the updating frequency for the two algorithms. The two calculations are executed in an inverter and tested over 16 days of simultaneous operation. Simple statistical techniques, the paired t-test, were applied to the data with the result that both algorithms perform equally well.

The paper has been arranged as Sect. 5.2 describes the design and modelling of PV module, Sect. 5.3 describes about the conventional and PV integrated three-level and five-level with their switching schemes. Section 5.4 depicts the simulation results with detail results description, Sect. 5.5 represents a clear comparison about the PV integrated three-level and five-level NPC inverter and without PV and finally concluded at last.

5.2 Modelling of PV Cell

As we know that the power electronics devices absorb light from the source and coverts part of the energy into electrical energy, the unit of PV source is PV cell, which is simple P–N junction diode formed by semiconductor materials [11–13] (Fig. 5.1).

At point X, we are applying Kirchhoff current law:

$$I = I_{ph} - I_d \tag{5.1}$$

Diode current:

$$I_d = I_o(\mathrm{e}^{(qv/nkT-1)}) \tag{5.2}$$

Output current:

$$I = I_{ph} - I_o(\mathrm{e}^{(qv/nkT-1)}) \tag{5.3}$$

The standard parameters for design are: Saturation current (I_o), parallel resistance (R_p), photo-cell current (I_{ph}), series resistance (R_s) and quality factor (n) [11].

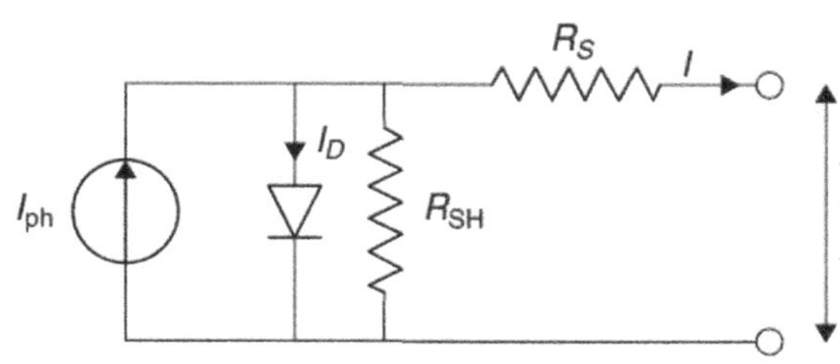

Fig. 5.1 Single-diode equivalent circuit of solar PV cell

Now, we are going to apply Kirchhoff current law at point X:

$$I = I_{ph} - I_d - I_{sh} \tag{5.4}$$

Photo-cell Current:

$$I_{ph} = [I_{sc} + ki(T - 298)] * \frac{I_r}{1000} \tag{5.5}$$

Reverse saturation current:

$$I_{rs} = \frac{I_{sc}}{\left[\left(e^{\frac{qV_{oc}}{nkT}}\right) - 1\right]} \tag{5.6}$$

Saturation current:

$$I_o = I_{rs} * \left[\frac{T}{T_n}\right]^3 * e^{\left[q.E_{go}/nk\left(\frac{1}{T}-\frac{1}{T_n}\right)\right]} \tag{5.7}$$

Shunt current:

$$I_{sh} = \frac{V + I + R_s}{R_p} \tag{5.8}$$

Output voltage:

$$I = I_{ph} - I_o\left(e^{\frac{q(V+I*R_S)}{nkT}-1}\right) - \frac{V + I * R_s}{R_p} \tag{5.9}$$

5.2.1 MPPT Algorithm

Photovoltaic (PV) usually have low efficiency initially. In order to improve its efficiency, maximum power point tracking is used. In this perturb and observe method, minor perturbations are introduced for causing power variations. If an increase in voltage results in increase in power, then the operating point of the PV module is set on the left of the MPP. Maximum power point is located at zero slope point on the power curve of the PV array. The resistance equals the negative deferential resistance [14]. PV array voltage is always adjusted as per the algorithm set for maximum power. The flowchart of the adopted P&O algorithm for the charge controller is shown in Fig. 5.2.

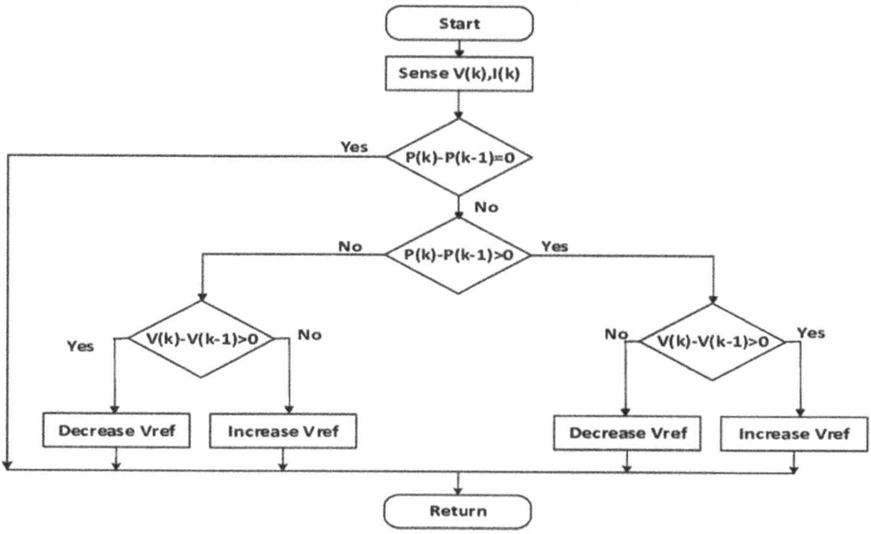

Fig. 5.2 Flowchart of the adopted P&O algorithm for the charge controller

5.2.1.1 Constants Used in PV Model

The charge controller such as series and shunt type are only able to protect the battery but do not get a better result in an optimum and efficient use of the PV source. In this paper, we are going to use boost converter rather than the series and shunt type of controller to interface the load combination and battery with the PV array because it gives out optimum use and smooth control of the PV source. The boost converter is used for the regulation of the output of PV array for feeding the load. A typical configuration or a circuit diagram is shown below in Fig. 5.3 (Table 5.1).

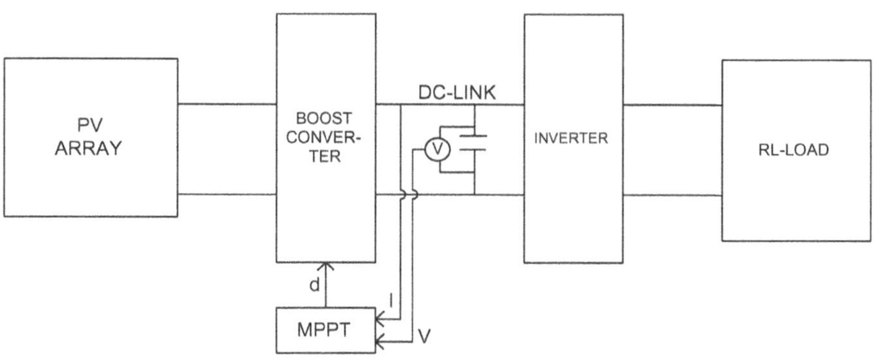

Fig. 5.3 Circuit diagram of PV cell

Table 5.1 Different parameters of components used for simulation

Name of components	Ratings
Inductance	1.15 mH
Capacitance	4.03 mF
Load resistance	100 Ω
K_c	0.032 A
Q	1.69 e^{-19} C
K	1.38 e^{-23} JK^{-1}
N	1.3
E_{go}	1.1
R_s	0.221 Ω
R_{sh}	415.405 Ω
T_n	298 °C
V_{oc}	32.9 V
I_{sc}	8.21 A
N_s	54
T	Operating temperature

The control schemes of the boost converter that we are using in this paper is the classic control used to control the boost converter. As we know that the PV cell must charge the battery in more efficient. Between the PV array and the battery, a DC–DC converter is used to extract the maximum power from it. The duty cycle of the boost converter is normally controlled for maximum voltage conversion. The voltage reference of the external control loops is acquired from an old style that is perturb and observe (P&O) MPPT calculation. By using PI controller, the PV voltage error is regulated, and internal PI control loop is additionally used to manage the current of an inductor. After voltage get boost up through boost converter, the voltage are supplied to the multilevel inverter through which the RL load is connected.

5.3 Multilevel NPC Inverter with PV Integration and SPWM Switching Schemes

5.3.1 Three-Phase Three-level NPC Inverters

Here, three-phase, three-level neutral point clamped inverter is demonstrated in Fig. 5.4. In this topology, the capacitors C_1 and C_2 are in series which split input voltage DC from PV system into three different levels. Vdc, 0 and –Vdc are the three states of the output voltage. The diode which is connected in each phase of multilevel inverter is the key component which distinguishes this circuit from other conventional inverters. The basic operation to compress the magnitude of voltage

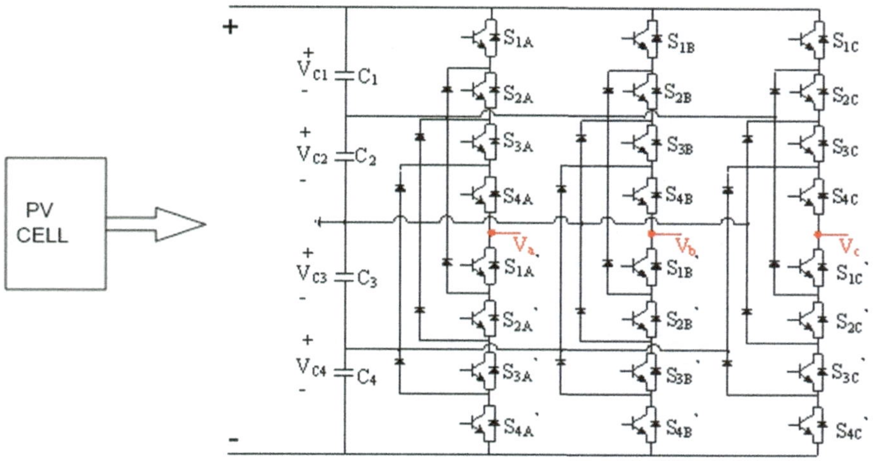

Fig. 5.4 Circuit diagram of three-level NPC inverter using PV cell

through the switches into half of the DC voltage is done by the help of diodes which are connected across each legs of the inverter. By following the circuit diagram of the three-level inverter, the simulation is done on MATLAB.

5.3.1.1 Switching Patterns

Table 5.2 presents the three switching patterns of three-level MLI along with output voltage.

5.3.2 Switching Scheme for Three-level NPC MLI

Figure 5.5 shows the circuital explanation of above tabulation and also pictorial diagram shows the flow of current in the multilevel inverter which are essential to get the required output voltages.

The Advancement of MLI's topography seems a challenge to expand the current modulation techniques in the case of multilevel inverter. The strategies that are

Table 5.2 Switching patterns of 3 level MLI

Mode	S_{1A}	S_{2A}	S'_{1A}	S'_{2A}	Voltage
Mode-I	1	1	0	0	Vdc
Mode-II	0	1	1	0	0
Mode-III	0	0	1	1	−Vdc

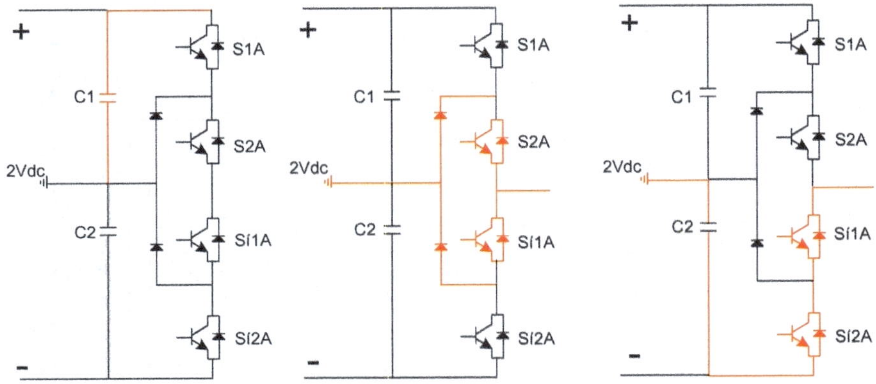

Fig. 5.5 Switching states of three-level NPC MLI

created required studying the extra complication of the inverter having to control the extra switching states of semiconductor devices.

Among them, the modulation technique which is utilized is SPWM, because it is very simple and provides high performance to reduce the harmonic content by using phase-shifting method.

In most of the industries, the commonly used switching technique is SPWM switching technique. The strategies are described by steady amplitude pulses with various duty cycles for each and every period. Modulation of the pulses is done to minimize the higher-order harmonics improving THD. The most common method in the case of motors and inverters is SPWM, and it is most famous due to its simplicity. The operation of this method includes sine waves and triangular carrier wave with high frequency which is used for the generation of PWM signal. So by comparing both waves, the pulse or switching signals are generated.

5.3.3 *Five-level Neutral Point Clamped Inverter*

As we know that by improving the level of a multilevel inverter, the efficiency of the system gets increased with low harmonic distortion. Essentially, for applications with high-control, it is conceivable to utilize direct converters and indirect converters. In recent years, the regular progress of IGBTs with high-voltage and the application of these semiconductor switches in most of the voltage source converter (VSC) topography has stimulated an extreme increment of power ratings and nominal voltage of self-commutated converters. In the topology of five-level neutral point clamped multilevel inverter, the supply is provided by PV cell with MPPT controller using boost converter to boost up the supply voltage which is input to the inverter. Here, the voltage clamping diodes are essential for blocking the reverse voltage. Basically, even number of bulk capacitor is used for dividing the mutual DC bus, build upon the number of voltage levels of the inverter.

At the middle of the line which is called neutral point, the capacitors are in sequence. The $(m - 1)$ pairs of valve are connected to the clamping diodes with neutral point and also capacitor from the DC bus in which 'm' is amount of voltage magnitudes of inverter. For the higher levels of voltage, the required number of clamping diodes is relatively high for NPC topology. The purpose behind the holding diodes is to compress the voltage of DC bus and connecting both the midpoints of the inverter, so it is called diode clamped inverter. The output voltage levels of five-level inverter which is generated among point R and neutral point N are Vdc/2, Vdc/4, 0, −Vdc/4 and −Vdc/2, and according to this, the whole switches get ON and OFF. The waveforms are closer to sinusoidal signals as we are increasing the voltage level. There are some advantages and also disadvantages of this topology. The advantages are: (I) One isolated DC link is required. (II) Voltage of capacitors in main DC link is identical. (III) It provides low THD at output. And the disadvantages are: (I) It does not have self-balancing property. (II) By increasing the number of levels, the number of clamping diodes grows rapidly. (III) To balance the voltage of DC link capacitor, complex control method is used.

The given illustration Fig. 5.6 is showing the circuital diagram of five-level multilevel inverter with PV cell along with RL load, and MATLAB is done by following this circuit diagram.

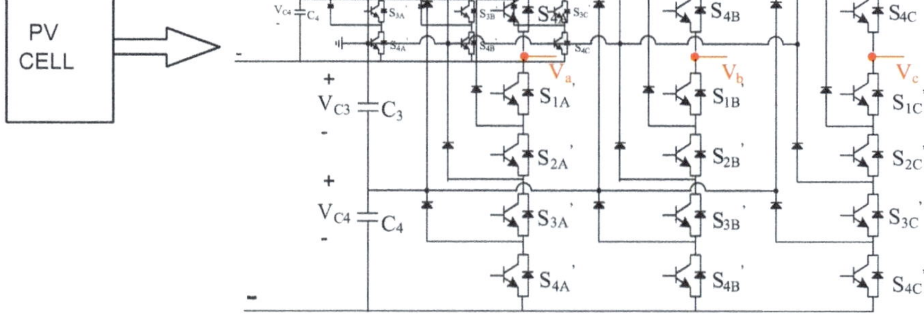

Fig. 5.6 Five-level NPC multilevel inverter using PV cell

5.3.3.1 Various States of Switching for the Above MLIs Along with Outputs Voltage

Figure 5.7 is explanation of Tabulation 2 and showing the flow of current in MLI's inverter which is necessary to get required output voltage V_{C4} (Table 5.3).

5.3.4 Switching Scheme for Five-level NPC Inverter

See Fig. 5.7.

5.4 Simulation and FFT Results:

Carrier waves used for pulse generation is shown in Figs. 5.8 and 5.9.

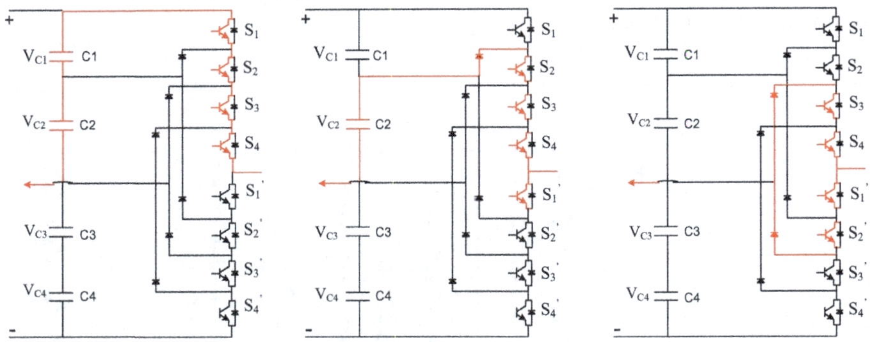

Fig. 5.7 Five-level NPC inverter switching states

Table 5.3 Output voltage and switching states

Mode	S_{1A}	S_{2A}	S_{3A}	S_{4A}	S'_{1A}	S'_{2A}	S'_{3A}	S'_{4A}	Voltage
Mode-I	1	1	1	1	0	0	0	0	Vdc/2
Mode-II	0	1	1	1	1	0	0	0	Vdc/4
Mode-III	0	0	1	1	1	1	0	0	0
Mode-IV	0	0	0	1	1	1	1	0	−Vdc/2
Mode-V	0	0	0	0	1	1	1	1	−Vdc/4

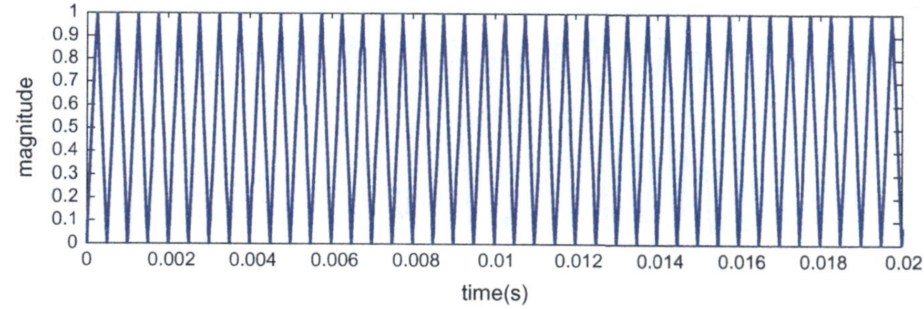

Fig. 5.8 Triangular or carrier wave

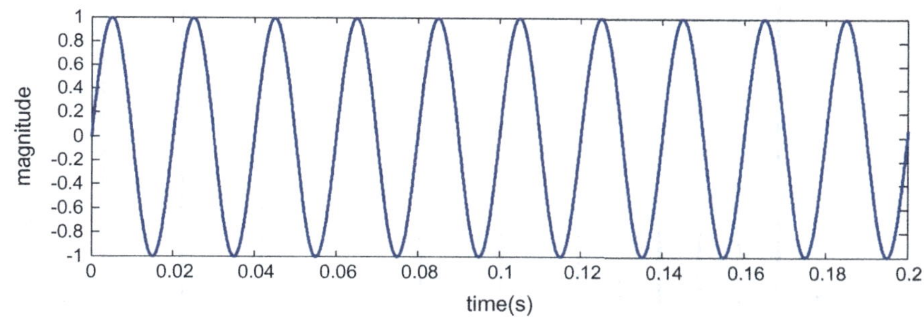

Fig. 5.9 Sinusoidal wave or control wave

5.4.1 Results Without PV Integration

Here, we are showing the simulation results of the both NPC multilevel inverters without PV integration under the similar condition. Fig. 5.10a, b, c, d demonstrated shows the current waveforms and THD analysis of both the levels of inverter with the help of MATLAB software (Fig. 5.11).

The fast Fourier transform (FFT) study with RL load and along with PV cell is presented in Fig. 5.12. There are five cycles for 50 Hz power system for discrete Fourier transform techniques. The total FFT analysis is done on MATLAB software, and after analysis, the THD is 7.60% along with frequency (50 Hz) (Figs. 5.13 and 5.14).

The fast Fourier transform (FFT) study of five-level NPC with RL load along with PV cell is shown in Fig. 5.15. There are five cycles for 50 Hz power system for discrete Fourier transform techniques. The total FFT analysis is done on MATLAB software, and after analysis, the total harmonic distortion (THD) of the output waveform is 4.57% along its fundamental frequency (50 Hz).

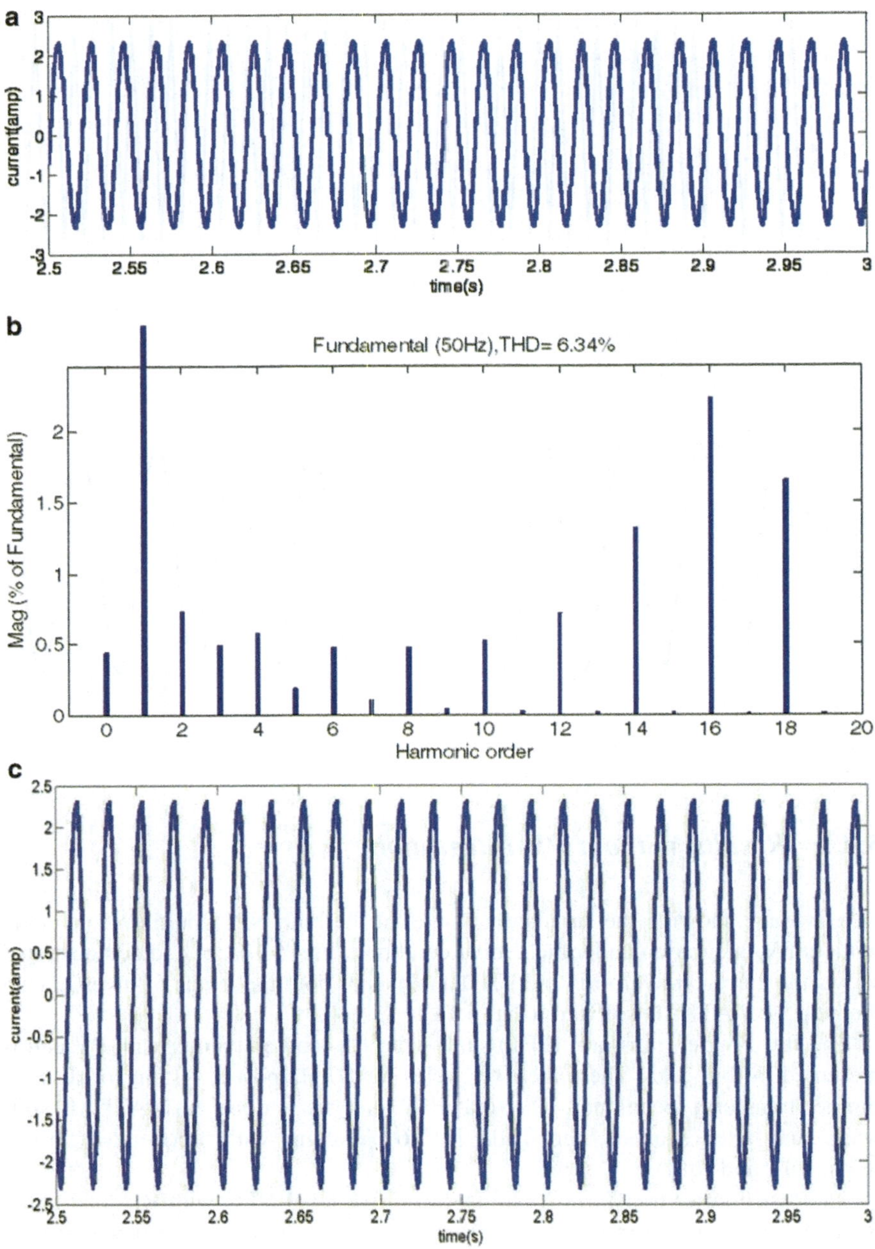

Fig. 5.10 **a** Current waveform of three-level NPC inverter without PV cell. **b** THD analysis of three-level NPC inverter without PV cell. **c** Current waveform of five-level NPC inverter without PV cell. **d** THD analysis of five-level NPC inverter without PV cell

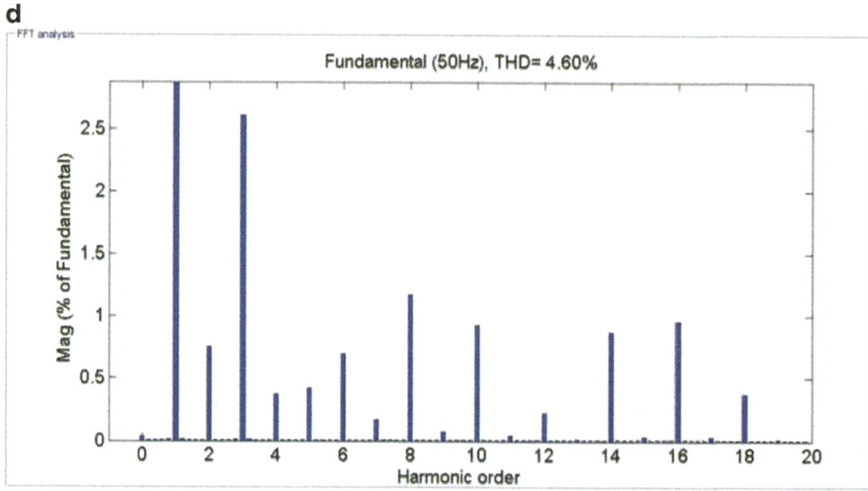

Fig. 5.10 (continued)

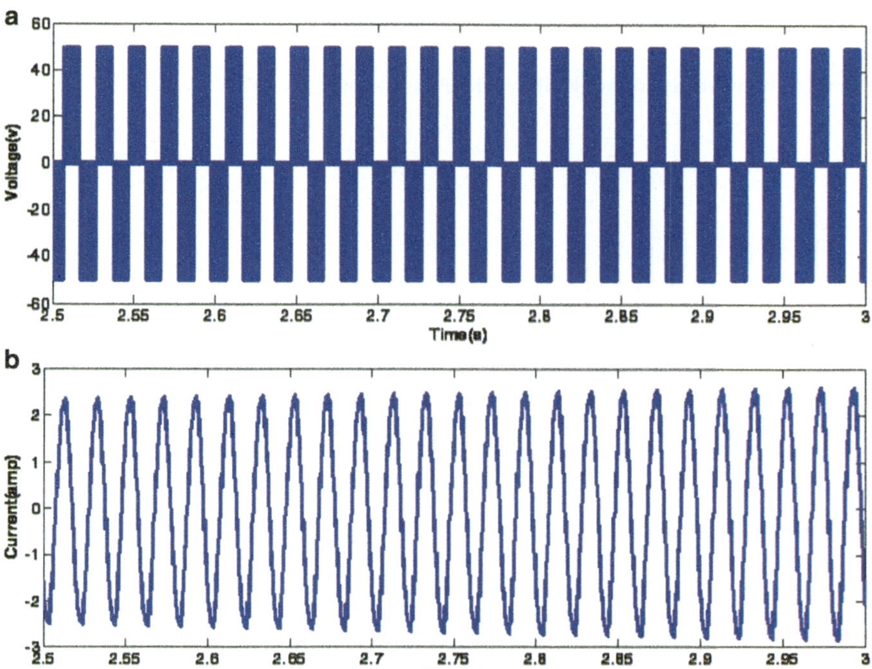

Fig. 5.11 **a** Three-level phase voltage waveform with PV. **b** Current waveform of three-level NPC inverter with PV cell

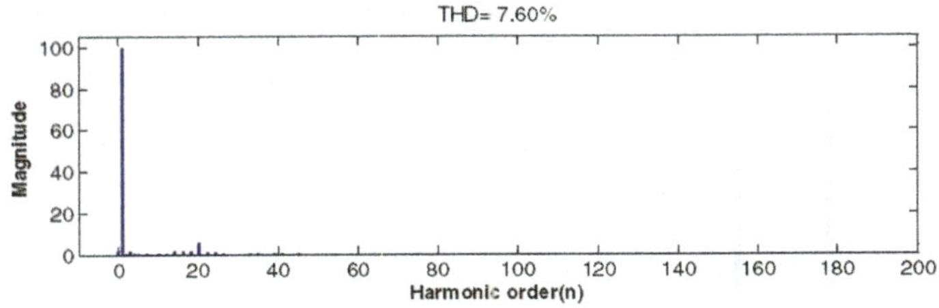

Fig. 5.12 FFT analysis of current a three-level NPC inverter with PV cell

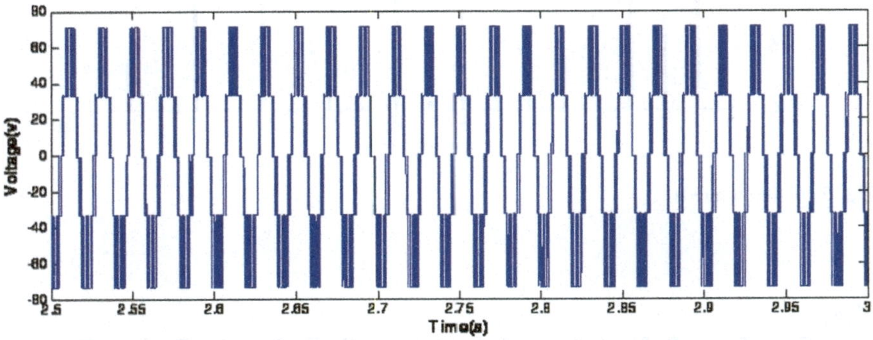

Fig. 5.13 Phase voltage of five-level multilevel inverter with PV cell

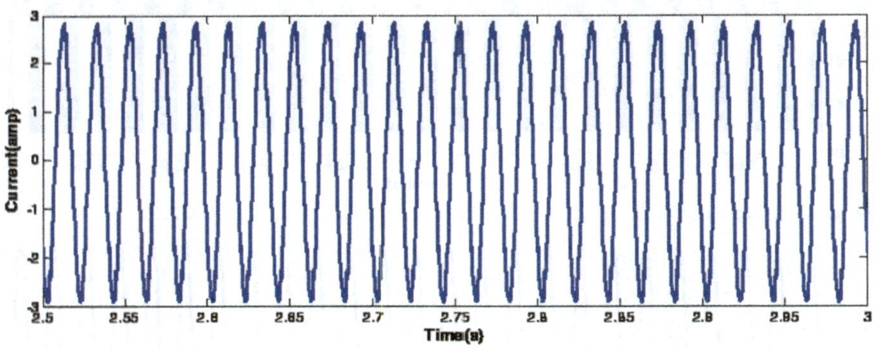

Fig. 5.14 Current waveform of five-level NPC inverter with PV cell

5.4.2 Comparison of Inverters With PV and Without PV

The comparative study of all the modules is done utilizing MATLAB software. Each multilevel inverter of the above structures (three-level NPC and five-level NPC) is incorporated with two different supply system in which one set of

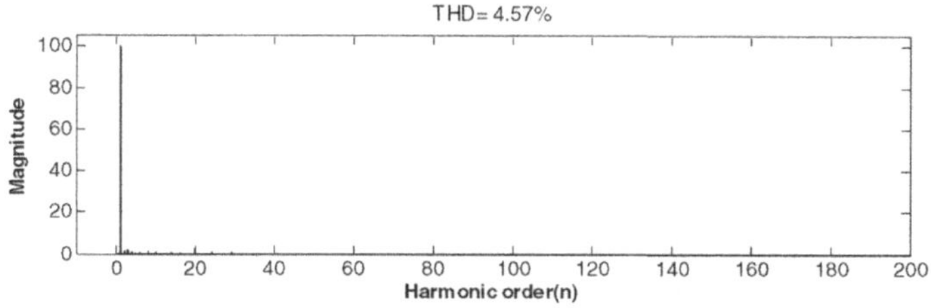

Fig. 5.15 FFT of current of a five-level NPC MLI with PV cell

three-level NPC and five-level NPC is fed by DC source and one set of three-level and five-level NPC MLIs is fed by solar PV cell. First of all, both levels of NPC MLIs with PV cell and with DC source are discussed. Owing to occurring losses in the inverter model IGBT switches are used in order to simulate and obtain the results close to practical case. The supply voltage which is provided to the three-level NPC with DC source is about 100v with RL load in which the value of L and R is taken from the experimental analysis, and the output phase voltage that we get is also −100v, 0v and 100v. The current waveform that we are getting is about 2.5A, and after the FFT, the total harmonic distortion has come about 6.34% without PV.

Three-level NPC inverter with PV cell and the supply voltage that is provided to the inverter is about 140v with RL load, and the output phase voltage obtained after simulation is approximately 140v, 0v and −140v. The FFT analysis is done, and total harmonic distortion has been observed as 7.60%. So, the comparison of three-level NPC inverter is completed.

Here, the five-level NPC inverter with and without PV has been discussed. In first case, we considered five-level NPC without PV cell. In this model, DC supply of 140v is supplied to the inverter with RL load, and the ratings of load are taken from experimental value. After the simulation, the phase voltage obtained is also approximately 140v, 0v and −140v, the current is nearly 2.5A, and after the FFT analysis, the THD is 4.04%.

Secondly five-level NPC with PV cell is discussed. The supplied voltage provided is 140v with RL load and the output phase voltage that has been observed is −140v, 0v and +140v which is nearly about supplied voltage and the current we observed is about 2.5A. After the FFT analysis, the THD has been found about 4.30%. To better calculate the performance of the three and five levels of NPC with or without PV cell, a complete simulation is achieved and notes that all the inverters are simulated under the same conditions.

The comparison of both levels of NPC inverter is done and observed with the result analysis. In this paper, three-level inverter output waveform levels is −300 V, 0 V and +300 V, and from FFT analysis, total harmonic distortion is found as 42.22% in line voltage, and for five-level NPC, the obtained THD is 34.21% in line voltage. In this paper, the THD is done for load current which is observed less as compared to line voltage. Here, the supply voltage is provided by both DC voltage source and by solar PV cell to the both three- and five-level NPC inverters. But in few papers, only DC voltage is supplied to the MLIs which makes the analysis a new improvement.

5.4.3 Comparison of Three-level and Five-level NPC With Without and With PV Cell

See Table 5.4.

5.5 Effect of Change of Irradiance on NPC Inverter Keeping Load Parameters Same

The effect of irradiance on the input side is studied with the THD keeping load parameters constant. Both the NPC Inverters are supplied by solar PV cell. The irradiance is varied, and the comparison is conducted for the THD analysis of both the inverters. Table 5.5 shows the comparison at different value of irradiance. It is

Table 5.4 Comparison of three-level and five-level NPC without and with PV cell

Component	Three-level NPC inverter	Five-level NPC inverter
DC bus capacitors	2	4
No. of switches	12	24
THD with RL load without PV cell (%)	6.34	4.04
THD with RL load with PV cell (%)	7.36	4.30
Phase voltage without PV cell (v)	140	140
Phase voltage with PV cell (v)	140	140
Rating of R load (ohms)	145	145
Rating of L load (mH)	150	150

Table 5.5 Irradiance versus THD of load current table with PV cell

Si. no	Irradiance	THD for three-level	THD for five-level
1	100	8.47	15.3
2	400	7.83	5.58
3	600	7.78	5.32
4	1000	7.70	4.58
5	2000	7.6	4.5

observed that five-level NPC inverter showed better results and THD was better than three-level NPC. The FFT analysis for both the inverters at one particular value of irradiance (=1000) has been demonstrated in Table 5.5.

5.5.1 Three-level Load Current of NPC MLI with PV

See Figs. 5.16 and 5.17.

5.5.2 Five-level Load Current of NPC MLI with PV

See Figs. 5.18 and 5.19.

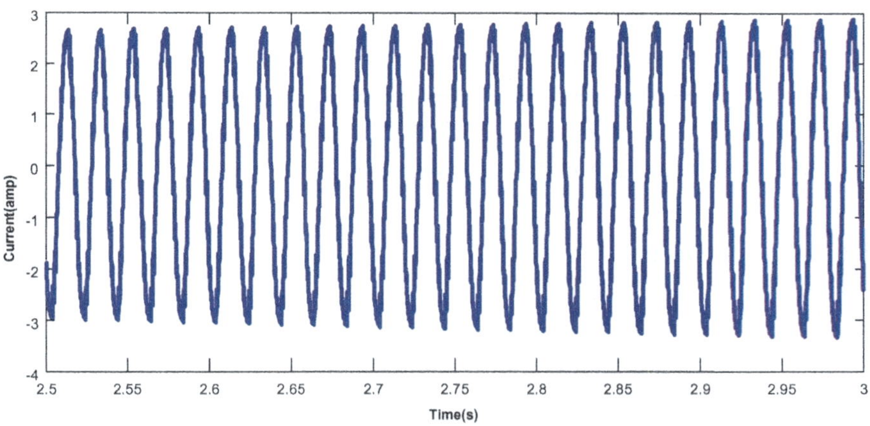

Fig. 5.16 Load current of a three-level NPC MLI with PV cell having irradiance 1000 W/m^2

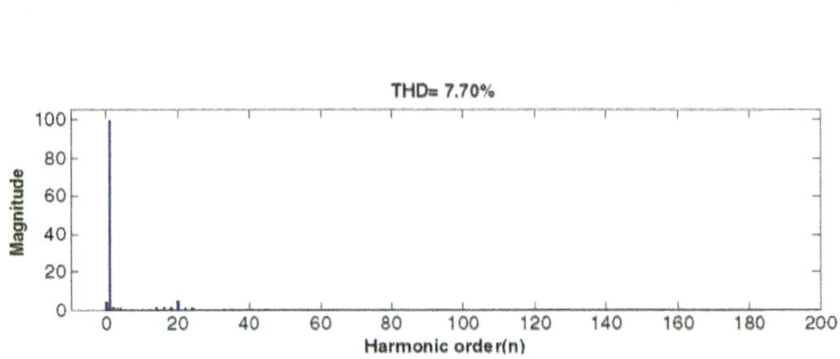

Fig. 5.17 FFT of current of a three-level NPC MLI with PV cell having irradiance 1000 W/m^2

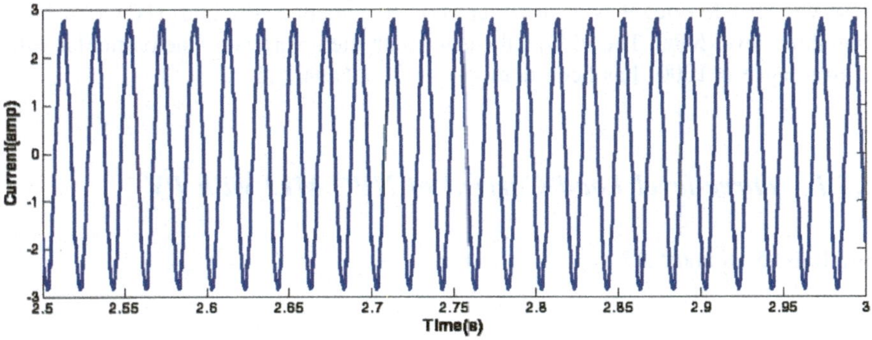

Fig. 5.18 Load current of a five-level NPC MLI with PV cell having irradiance 1000 W/m^2

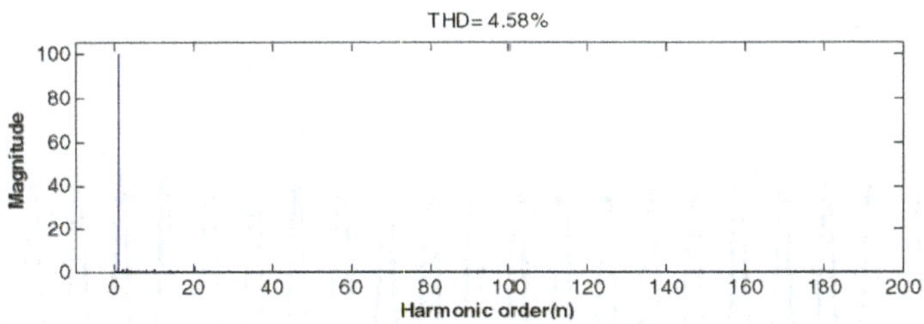

Fig. 5.19 THD of five-level With NPC-PV having irradiance 1000 W/m^2

5.6 Conclusion

Here, the detailed performance analysis of neutral point clamped MLI fed from PV cell has been presented. Along with this, the comparison of both three- and five-level NPC inverter with PV cell and without PV cell is depicted which describes the suitability of the devices in different applications. Here it is well explained about the effectiveness of multilevel inverter utilized for DC to AC control transformation with simulation results. The performances of each type of these multilevel inverters are simulated and analyzed under the same conditions. This work additionally displays a multilevel three-level and five-level inverter utilizing PV cell in the source side interchanging conventional DC source. All the parameters or constants which are used for PV cell are mentioned properly in a tabulation form for the smooth operation of the system. The THD analysis of the current waveforms of both the three-level NPC and five-level NPC is done with PV cell and without PV cell which have been found to be quite nominal and within the limits and the THD of five-level NPC inverter in both the cases found to be better than three-level NPC. And the effect of irradiance on THD is also observed to be better in case of five-level compared to three-level NPC inverter.

References

1. Adhikary, R.D., Halder, T., Balow, W.: Diode & neutral point clamped five-level inverter for the power quality issues. 978–1–5386–6722–4/19/$31.00 ©2019 IEEE
2. Pillay, T.L., Saha, A.K.: A comparative analysis of three phase npc multi-level inverter IEEEPES/IAS (2018)
3. Ch, B., Muduli, U.R., Behera, R.K.: Performance comparison of five-phase three-level NPC to five-phase two-level voltage source inverter. 978–1–5386–9316–2/18/$31.00 2018 IEEE (2018)
4. Mistry, J.N., Vakharia, K., Tailor, D., Patel, B.: Software and hardware implementation of five level neutral point clamped inverter. 978–1–4673–9939–5/16/$31.00 ©2016 IEEE (2016)
5. Ch, B., Muduli, U.R., Behera, R.K.: Performance comparison of five-phase three-level NPC to five-phase two-level voltage source inverter. 2018 IEEE (2018)
6. AkAgi, H.: Multilevel Converters: Fundamental Circuits and Systems, vol. 105, no. 11 (2017)
7. Mane, J.J., Muley, S.P., Aware, M.V.: Performance of 5-level NPC inverter with multi-multicarrier multi-modulation technique. In: Proceedings of 2012 IEEE International Conference on Drives and Energy Systems (PEDES), Bangalore, pp.1–5 (2012)
8. Rodriguez, J., Bernet, S., Steimer, P.K.: A survey on neutral-point-clamped inverters. IEEE Trans. Ind. Electron. **57**(7) (2010)
9. Choi, U.-M.: Method to minimize the low-frequency neutral-point voltage oscillations with time-offset injection for neutral-point-clamped inverters. IEEE Trans. Ind. Appl. **51**(2) (2015)
10. Arunachalam, K., Vairakannu, S.K., Nallathambi, G.: Comparison and optimisation of three-level neutral point clamped dual output inverter. IET Circuits Devices Syst. **14**(1) (2020)
11. Shongwe, S.: Comparative analysis of different single-diode PV modeling methods. IEEE J. Photovoltaics **5**(3) (2015)
12. Jadli, U., Thakur, P., Shukla, R.D.: A new parameter estimation method of solar photovoltaic. IEEE J, Photovoltaics **8**(1) (2018)
13. Rahman, S.A., Varma, R.K., Vanderheide, T.: Generalized model of a photovoltiac panel. IET Renew. Power Gener. **8** (2014)
14. Priyadarshi, N., Padmanaban, S., Maroti,P.K., Sharma. A.: An extensive practical investigation of FPSO-Based MPPT for grid integrated PV system under variable operating conditions with anti-islanding protection. IEEE Syst. J. 1–11 (2018)
15. Priyadarshi, N., Padmanaban, S., Bhaskar, M.S., Blaabjerg, F., Sharma, A.: A fuzzy SVPWM based inverter control realization of grid integrated PV-wind system with FPSO MPPT algorithm for a grid-connected PV/wind power generation system: hardware implementation. IET Electric Power Appl. 1–12 (2018)
16. Azam, F., Priyadarshi, N., Nagar, H., Kumar, S., Bhoi, A.K.: An overview of solar-powered electric vehicle charging in vehicular Adhoc network. In: Electric Vehicles. Green Energy and Technology. Springer, Singapore (2021)
17. Padmanaban, S., Priyadarshi, N., Holm-Nielsen, J.B., Bhaskar, M.S., Azam, F., Sharma, A. K.: A novel modified sine-cosine optimized MPPT algorithm for grid integrated PV system under real operating conditions. IEEE Access **7**, 10467–10477 (2019). https://doi.org/10.1109/ACCESS.2018.2890533
18. Padmanaban, S., Priyadarshi, N., Holm-Nielsen, J.B., Bhaskar, M.S., Hossain, E., Azam, F.: A hybrid photovoltaic-fuel cell for grid integration with jaya-based maximum power point tracking: experimental performance evaluation. IEEE Access **7**, 82978–82990 (2019). https://doi.org/10.1109/ACCESS.2019.2924264
19. Modi, S., Kevin, K., Usha, P.: Mathematical modeling, simulation and performance analysis of solar cell. In: 2018 International Conference on Power Energy, Environment and Intelligent Control (PEEIC)
20. Salmi, Z.T., Bouzguenda, M., Gastli, A., Masmoudi, A.: MATLAB/Simulink modelling of solar photovoltaic cell. Int. J. Renew. Energy Res. **2**(2) (2012)

21. Patel, H., Gupta, M., Bohre, A.K.: Mathematical modeling and performance analysis of MPPT based solar PV system. In: 2016 International Conference on Electrical Power and Energy Systems (ICEPES)
22. Ali, A.N.A., Saied, M.H., Mostafa, M.Z., Abdel-Moneim, T.M.: A Survey of Maximum PPT techniques of PV Systems. 978-1-4673-1835-8/12/$31.00 ©2012 IEEE

Chapter 6
Bidirectional Interleaved Switched Capacitor DC–DC Converter for Renewable Energy Applications

B. S. Nalina, V. Kamaraj, M. Chilambarasan, and M. Ramesh Babu

6.1 Introduction

Fossil fuel, coal oil and gas are used for the conventional generation of electricity. These energy sources are depleting, very expensive and cause pollution which damages the environment [1]. The introduction of renewable energy sources such as solar, wind, biogas and ocean for the generation of electricity helps in overcoming the issues created by the conventional electricity generation [2]. The advantages of renewable energy sources are they are free of cost as they are available in nature and do not cause any damage to the environment unlike conventional energy sources [3]. The major drawback of these renewable energy sources is they cannot be obtained continuously as they are seasonal [4, 5]. Energy storage devices aid in overcoming the drawbacks of renewable energy sources. When various power requirements from different locations are needed, individual distributed generating units are used. The usage of individual generating units has certain limitations which is overcome by the concept of microgrids [6]. Microgrid is an interconnection of various distributed generation units like solar PV systems, wind turbines, microturbine, different types of loads, energy storage devices like battery, flywheel and supercapacitor. In a microgrid system, the power is generated at the load site by which the transmission losses are reduced and enhance unin-

B. S. Nalina (✉) · V. Kamaraj
Department of Electrical and Electronic Engineering, SSN College of Engineering, Kalavakkam, Chennai, Tamilnadu, India
e-mail: kamarajv@ssn.edu

M. Chilambarasan · M. R. Babu
Department of Electrical and Electronic Engineering, St. Joseph's College of Engineering, OMR, Chennai, Tamilnadu, India
e-mail: chilambarasanm@stjosephs.ac.in

M. R. Babu
e-mail: rameshbabum@stjosephs.ac.in

N. Priyadarshi et al. (eds.), *DC–DC Converters for Future Renewable Energy Systems*, Energy Systems in Electrical Engineering,
https://doi.org/10.1007/978-981-16-4388-0_6

terrupted supply of energy. The power variations between the renewable energy source and the load are balanced by energy storage devices [7]. Power electronic devices are required to maintain the power flow between the source and load and also to interface the renewable energy source with the DC bus and load. DC–DC converters are used as an interface in microgrid. Traditional DC–DC converters are not able to fulfill the requirements in the field of renewable energy sources as they are not available all the time and they transfer power in only one direction [8]. Renewable energy sources require flow of power in both the directions [4, 5]. The introduction of renewable energy brings a difference to this concept from unidirectional power flow to bidirectional power flow [9–12]. Bidirectional DC–DC converters are used to ensure the flow of power in both the directions [13]. Based on the propulsive isolation between the input side and the output side, bidirectional DC–DC converter is classified as

(i) isolated bidirectional DC–DC converter
(ii) non-isolated bidirectional DC–DC converter

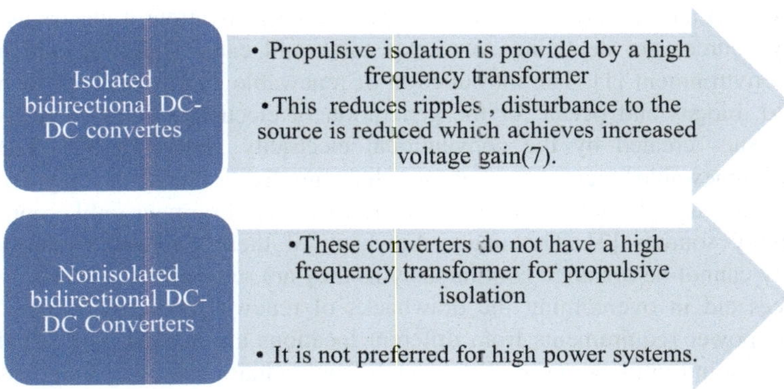

The main disadvantage of isolated bidirectional DC–DC converters is increased voltage spikes on the devices caused by the leakage inductance of the transformer. The different topologies of isolated converters are [14–17], forward-flyback converter half-bridge converters, flyback converters and full-bridge converters. The different topologies of non-isolated bidirectional DC–DC converters are buck–boost, multilevel converters, sepic/zeta, switched capacitor and coupled inductor. In case of converters using coupled inductor, the voltage gain can be increased by adjusting the turns ratio of the inductor [18]. The main drawback is the power transferring potential of the converter is reduced by the capacity of the magnetic core of the coupled inductor. A converter with high efficiency and voltage gain is proposed [19]. However, the coupled inductor used to produce high voltage gain increases the cost of the system. An interleaved bidirectional dual active bridge (DAB) DC–DC converter is proposed as a power electronic interface in microgrid [20]. The converter increases the power density by regulating the output voltage. The converter also aids in power sharing even in variation among the converters.

A modular interleaved boost converter is first proposed by integrating a forward energy-delivering circuit in which the high voltage ratio is achieved by a voltage doubler circuit. The converter can also be used for higher power rating by connecting in parallel [21]. Traditional buck converter and boost converters are best suited low-voltage applications but have the disadvantage of low efficiency and are not eligible for energy storage applications. The voltage ripples at the input side can be reduced by interleaved bidirectional DC–DC converter in [22] but it has increased voltage stress on the switches. This drawback is overcome by bidirectional three-level DC–DC converters. Based on the literature review on different types of isolated and non-isolated bidirectional DC–DC converters, a converter topology is proposed in which the features of two-phase interleaved bidirectional DC–DC converter and three-phase bidirectional DC–DC converter are combined. The proposed converter has the following advantages: (i) reduction of ripple current by the implementation of interleaving technique; (ii) reduced voltage stress; and (iii) increase or decrease in voltage gain by the switched capacitor and improved efficiency. The proposed converter operates two different modes boost (step-up) operation and buck (step-down) operation and ensures power flow in both directions. In boost operation, the capacitors are charged in parallel and they are discharged in series. In buck operation, the capacitors are charged in series and discharged in parallel. The proposed converter is interfaced with a solar-powered microgrid.

The organization of the work is as follows. The topology of the proposed converter is explained in Sect. 6.2. Section 6.3 describes the modes of operation of the converter. The proposed bidirectional interleaved switched capacitor DC–DC converter interfaced with solar-powered microgrid is discussed in Sect. 6.4. The simulation results are discussed in Sect. 6.5. Section 6.6 briefly describes about the efficiency calculation of the system.

6.2 Topology of Bidirectional Interleaved Switched Capacitor DC–DC Converter

Figure 6.1 shows the basic topology of the proposed bidirectional interleaved switched capacitor DC–DC converters. The topology of the converter consists of five switches S_1, S_2, S_3, S_4 and S_5, capacitors C_1, C_2, C_3 and inductors L_1 and L_2. The input voltage and current are represented as V_{low} and I_{low}, respectively. The output voltage and current are denoted as V_{high} and I_{high}. The input capacitor is C_{low}. The inductors L_1, L_2 and the switches S_1 and S_2 perform the interleaving operation whereas the other switches and capacitors act as a switching capacitor. The converter operates in step-up (boost) mode and step-down (buck) mode. A PID controller is implemented to control the voltage and currents for bidirectional power flow. The converter operates in step-up (boost) mode for a duty cycle greater than

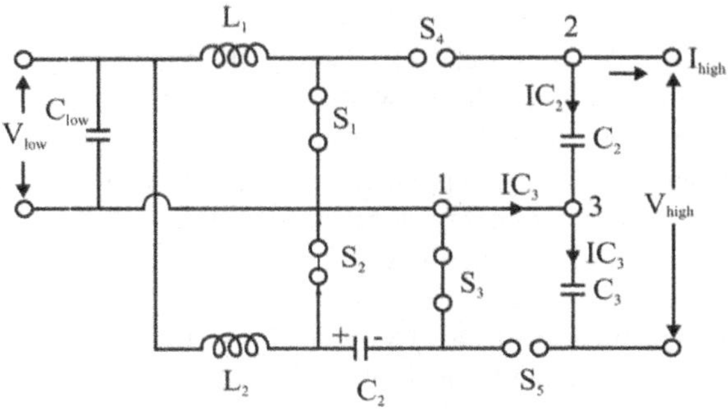

Fig. 6.1 Topology of bidirectional interleaved switched capacitor DC–DC converter

Table 6.1 Specifications of bidirectional interleaved switched capacitor DC–DC converter

Specifications	Values
V_{low}	50–150 V
V_{out}	400 V
L_1	353 μH
L_2	347 μH
C_{low}	520 μF
C_1	520 μF
C_2	520 μF
Rated power	1 Kw
Switching frequency	120 kHz

0.5 and in step-down (buck) mode for duty cycle 0.5–1. The inductor and capacitor values of the converter are calculated from the following equation (Table 6.1)

$$L_l = L_2 = \text{dboost}(1 - \text{dboost})T_s * V_{\text{high}}/2\Delta IL_1 \tag{6.1}$$

$$C_3 = C_1 = C_2 = (1/\Delta V_c)I_{\text{high}} * DT_s \tag{6.2}$$

6.3 Modes of Operation

6.3.1 Step-Up (Boost) Modes of Operation

In step-up (boost) mode, power flows from source (input side) to load (output side). The battery is energized in this mode of operation. The operation of the converter in this mode is carried out for duty cycle greater than 0.5.

6.3.1.1 Mode 1

In mode 1, the switches S_1 and S_2 are turned ON and the other three switches S_3, S_4 and S_5 are in turn OFF condition. The flow of energy in Fig. 6.2a. The capacitors C_1 and C_3 which are connected in series get charged. As the capacitors are connected in parallel to the load (load 2), the load is energized with the voltage drop across the capacitors which are connected in parallel to it. The capacitor C_{low} supplies energy to the load at the input side (load 1). The relationship between the input and output voltage and currents is obtained by the following analysis.

By applying Kirchoff's voltage law in Fig. 6.2a, we get

$$\frac{di_{L_1}}{dt} = \frac{1}{L_1} V_{\text{in}} \tag{6.3}$$

$$\frac{di_{L_2}}{dt} = \frac{1}{L_2} .V_{\text{in}} \tag{6.4}$$

$$\frac{dV_{C_{\text{low}}}}{dt} = \frac{1}{C_{\text{low}} r} V_{C_{02}} - \frac{1}{C_{\text{low}} r} V_{C_{\text{low}}} \tag{6.5}$$

$$\frac{dV_{C_{01}}}{dt} = -\frac{1}{RC_{01}} V_{C_{01}} - \frac{1}{RC_{01}} V_{C_{02}} \tag{6.6}$$

$$\frac{dV_{C_{02}}}{dt} = -\frac{1}{RC_{02}} V_{C_{02}} - \frac{1}{RC_{02}} V_{C_{01}} + \frac{V_{C_{02}}}{C_{02} r} \tag{6.7}$$

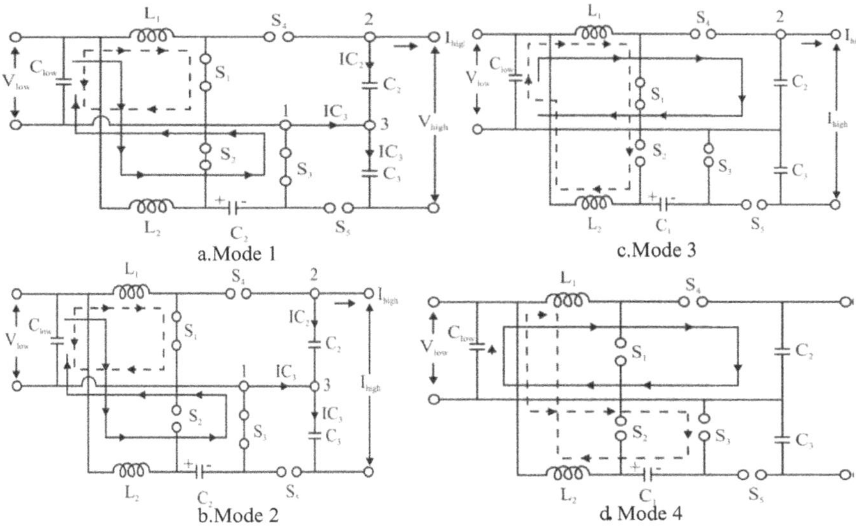

Fig. 6.2 Modes of operation is step-up mode

Equations (6.3–6.7) are written as:

$$\begin{bmatrix} \frac{di_{L_1}}{dt} \\ \frac{di_{L_2}}{dt} \\ \frac{dV_{C_{low}}}{dt} \\ \frac{dV_{C_{01}}}{dt} \\ \frac{dV_{C_{02}}}{dt} \end{bmatrix} = \underbrace{\begin{bmatrix} 0 & 0 & 0 & 0 & 0 \\ 0 & 0 & 0 & 0 & 0 \\ 0 & 0 & -\frac{1}{C_{low}r} & 0 & \frac{1}{C_{low}r} \\ 0 & 0 & 0 & -\frac{1}{RC_{02}} & -\frac{1}{C_{low}r} \\ 0 & 0 & 0 & -\frac{1}{RC_{02}} & -\frac{1}{RC_{02}} + \frac{1}{C_{02}r} \end{bmatrix}}_{A_1} \begin{bmatrix} i_{L_1} \\ i_{L_2} \\ V_{C_{low}} \\ V_{C_{01}} \\ V_{C_{02}} \end{bmatrix} + \underbrace{\begin{bmatrix} \frac{1}{L_1} \\ \frac{1}{L_2} \\ 0 \\ 0 \\ 0 \end{bmatrix}}_{B} V_{in} \quad (6.8)$$

6.3.1.2 Mode 2

The switch S_1 is in ON condition and the switch S_2 is in OFF condition. The switches S_4 and S_5 are in turn OFF condition. Both the switches S_2 and S_3 are connected in parallel. The switch S_3 gets turned ON as it is connected with the switch S_2. The flow of energy is shown in Fig. 6.2b. The capacitors C_1, C_2 and load 2 form a loop. The capacitors C_1 and C_2 charged in mode 1. These capacitors discharge energy in this mode which energizes load 2. The voltage across load 1 is equal to the voltage drop across the capacitor C_{low}. Applying Kirchoff's voltage law in Fig. 6.2b,

$$\frac{di_{L_1}}{dt} = \frac{1}{L_1} V_{in} \quad (6.9)$$

$$\frac{di_{L_2}}{dt} = \frac{1}{L_2} V_{low} - \frac{1}{L_2} V_{C_{low}} \quad (6.10)$$

$$\frac{dV_{C_{low}}}{dt} = \frac{i_{L_2}}{C_{low}} - \frac{V_{C_{02}}}{RC_{low}} \quad (6.11)$$

$$\frac{dV_{C_{01}}}{dt} = -\frac{V_{C_{01}}}{RC_{01}} - \frac{V_{C_{02}}}{RC_{01}} \quad (6.12)$$

$$\frac{dV_{C_{02}}}{dt} = -\frac{V_{C_{01}}}{RC_{02}} - \frac{V_{C_{02}}}{RC_{02}} \quad (6.13)$$

Equations (6.9–6.13) are written as:

$$\begin{bmatrix} \frac{di_{L_1}}{dt} \\ \frac{di_{L_2}}{dt} \\ \frac{dV_{C_{low}}}{dt} \\ \frac{dV_{C_{01}}}{dt} \\ \frac{dV_{C_{02}}}{dt} \end{bmatrix} = \underbrace{\begin{bmatrix} 0 & 0 & 0 & 0 & 0 \\ 0 & 0 & -\frac{1}{L_2} & 0 & 0 \\ 0 & \frac{1}{C_{low}} & 0 & 0 & \frac{1}{RC_{low}} \\ 0 & 0 & 0 & -\frac{1}{C_{01}} & -\frac{1}{RC_{01}} \\ 0 & 0 & 0 & -\frac{1}{RC_{02}} & -\frac{1}{RC_{02}} \end{bmatrix}}_{A_2} \begin{bmatrix} i_{L_1} \\ i_{L_2} \\ V_{C_{low}} \\ V_{C_{01}} \\ V_{C_{02}} \end{bmatrix} + \underbrace{\begin{bmatrix} \frac{1}{L_1} \\ \frac{1}{L_2} \\ 0 \\ 0 \\ 0 \end{bmatrix}}_{B} V_{in} \quad (6.14)$$

6.3.1.3 Mode 3

In this mode, the switch S_1 is turned OFF and the switches S_2 and S_4 are in turned ON condition. The capacitor C_1 is charged and the capacitor C_2 gets charged by the voltage drop across the capacitor C_1. The capacitors C_1, C_2 and load 2 form a loop and the load 2 is energized by the voltage drop across the capacitors C_1 and C_2. The load 1 is supplied by the voltage drop across the capacitor C_{low}.

Applying Kirchoff's voltage law in Fig. 6.2c,

$$\frac{di_{L_1}}{dt} = \frac{1}{L_1} V_{in} - \frac{1}{L_1} V_{C_{01}} \quad (6.15)$$

$$\frac{di_{L_2}}{dt} = \frac{1}{L_2} V_{in} \quad (6.16)$$

$$\frac{dV_{C_{low}}}{dt} = -\frac{V_{C_{low}}}{C_{low} r} + \frac{V_{C_{02}}}{C_{low} r} \quad (6.17)$$

$$\frac{dV_{C_{01}}}{dt} = \frac{1}{C_{01}} i_{L_1} - \frac{V_{C_{02}}}{RC_{01}} - \frac{V_{C_{01}}}{RC_{01}} \quad (6.18)$$

$$\frac{dV_{C_{02}}}{dt} = -\frac{V_{C_{02}}}{RC_{02}} - \frac{V_{C_{01}}}{RC_{02}} - \frac{V_{C_{02}}}{C_{02} r} \quad (6.19)$$

Equations (6.15–6.19) are written as

$$\begin{bmatrix} \frac{di_{L_1}}{dt} \\ \frac{di_{L_2}}{dt} \\ \frac{dV_{C_{low}}}{dt} \\ \frac{dV_{C_{01}}}{dt} \\ \frac{dV_{C_{02}}}{dt} \end{bmatrix} = \underbrace{\begin{bmatrix} 0 & 0 & 0 & -\frac{1}{L_1} & 0 \\ 0 & 0 & 0 & 0 & 0 \\ 0 & 0 & -\frac{1}{C_{low}r} & 0 & \frac{1}{C_{low}r} \\ \frac{1}{C_{01}} & 0 & 0 & -\frac{1}{RC_{01}} & -\frac{1}{RC_{low}} \\ 0 & 0 & 0 & -\frac{1}{RC_{02}} & -\frac{1}{RC_{02}} + \frac{1}{C_{02}r} \end{bmatrix}}_{A_3} \begin{bmatrix} i_{L_1} \\ i_{L_2} \\ V_{C_{low}} \\ V_{C_{01}} \\ V_{C_{02}} \end{bmatrix} + \underbrace{\begin{bmatrix} \frac{1}{L_1} \\ \frac{1}{L_2} \\ 0 \\ 0 \\ 0 \end{bmatrix}}_{B_3} V_{in} \tag{6.20}$$

6.3.1.4 Mode 4

The switches S_1 and S_2 are in turn OFF condition whereas the switches S_3 and S_4 are in turn ON condition. The input voltage V_{low} charges the capacitor C_1 and the voltage drop across C_1 charges the capacitor C_2. The capacitors C_1 and C_2 supplies energy to the load 2. The capacitor C_{low} is charged in mode 3 and voltage drop across C_{low} supplies energy to load 1.

By applying Kirchoff's voltage law to Fig. 6.2d,

$$\frac{di_{L_1}}{dt} = \frac{1}{L_1} V_{in} - \frac{1}{L_1} V_{C_{01}} \tag{6.21}$$

$$\frac{di_{L_2}}{dt} = \frac{1}{L_2} V_{C_{low}} + \frac{V_{in}}{L_2} \tag{6.22}$$

$$\frac{dV_{C_{low}}}{dt} = \frac{1}{C_{low}} i_{L_2} - \frac{1}{RC_{low}} V_{C_{02}} \tag{6.23}$$

$$\frac{dV_{C_{01}}}{dt} = -\frac{1}{RC_{01}} V_{C_{01}} - \frac{1}{RC_{01}} V_{C_{02}} + \frac{1}{C_{01}} i_{L_1} \tag{6.24}$$

$$\frac{dV_{C_{02}}}{dt} = \frac{-V_{C_{01}}}{R} - \frac{V_{C_{02}}}{RC_{02}} \tag{6.25}$$

Equations (6.21–6.25) are written as

$$\begin{bmatrix} \frac{di_{L_1}}{dt} \\ \frac{di_{L_2}}{dt} \\ \frac{dV_{C_{low}}}{dt} \\ \frac{dV_{C_{01}}}{dt} \\ \frac{dV_{C_{02}}}{dt} \end{bmatrix} = \underbrace{\begin{bmatrix} 0 & 0 & 0 & 0 & 0 \\ 0 & 0 & -\frac{1}{L_2} & 0 & 0 \\ 0 & \frac{1}{C_1} & 0 & 0 & -\frac{1}{RC_{low}} \\ \frac{1}{C_{01}} & 0 & 0 & -\frac{1}{RC_{01}} & -\frac{1}{RC_{01}} \\ 0 & 0 & 0 & -\frac{1}{RC_{02}} & -\frac{1}{RC_{02}} \end{bmatrix}}_{A_4} \begin{bmatrix} i_{L_1} \\ i_{L_2} \\ V_{C_{low}} \\ V_{C_{01}} \\ V_{C_{02}} \end{bmatrix} + \underbrace{\begin{bmatrix} \frac{1}{L_1} \\ \frac{1}{L_2} \\ 0 \\ 0 \\ 0 \end{bmatrix}}_{B_4} V_{in} \tag{6.26}$$

Let us assume the converter operates in continuous conduction mode, the switches S_1 and S_2 are turned ON individually or together for a time duration of d_1 and d_2 and turned OFF for time $(1-d_1)T_{S_1}$ and $(1-d_2)T_{S_2}$, where d_1 and d_2 are the duty cycles of switch S_1 and S_2, T_{S_1}, T_{S_2} are the switching periods of switches S_1 and S_2.

The averaged state-space equations of IBSC converter in step-up mode are derived from equations [6, 11, 13] (Table 6.2).

$$[\dot{X}] = A_x + B_v$$

$$A = A_1 + A_2 + A_3 + A_4$$

$$B = B_1 + B_2 + B_3 + B_4$$

$$-\frac{1}{L_1}(1-d_1)V_{C_{01}} + \frac{1}{L_1}V_{in} = 0 \tag{6.27}$$

$$-\frac{1}{L_2}(1-d_2)V_{C_{low}} + \frac{1}{L_2}V_{in} = 0 \tag{6.28}$$

$$\frac{(1-d_1)}{C_{01}}i_{L_2} - \frac{1}{RC_{low}}(1-d_2)V_{C_{02}} - \frac{1}{C_{low}r}d_2 = 0 \tag{6.29}$$

$$\frac{1}{C_{01}}(1-d_1)\,i_{L_1} - \frac{1}{RC_{01}}V_{C_{01}} - \frac{1}{RC_{01}}V_{C_{02}} = 0 \tag{6.30}$$

Table 6.2 Modes of operation of bidirectional interleaves switched capacitor DC–DC converter in step-up mode

Modes	S_1	S_2	S_3	S_4	S_5	C_1	C_2	Load 1	Load 2
Mode 1	ON	ON	ON	OFF	OFF	Charging	Charging	Energized	Energized
Mode 2	ON	OFF	ON	OFF	OFF	Discharging	Discharging	Energized	Energized
Mode 3	ON	OFF	OFF	OFF	OFF	Charging	Charging	Energized	Energized
Mode 4	OFF	OFF	ON	ON	OFF	Discharging	Discharging	Energized	Energized

$$-\frac{1}{RC_{02}}V_{C_2}-\frac{1}{RC_{02}}V_{C_{02}}+\frac{d_2}{C_{02}r}V_{C_{02}} \quad (6.31)$$

6.3.2 Step-Down Mode

In step-down mode, the converter operates for a duty cycle less than 0.5. The battery discharges energy in this mode of operation. The flow of energy is from the output-to-input side. The switches S_3 and S_4 are controlled in step-down mode. The different modes of operation are shown in the Fig. 6.3.

6.3.2.1 Mode 1

The switches S_3 and S_4 are in turn ON condition in mode 1. The power flow direction is shown in Fig. 6.3a. The capacitor C_1 is already charged in mode 4 of

Fig. 6.3 Step-down modes of operation

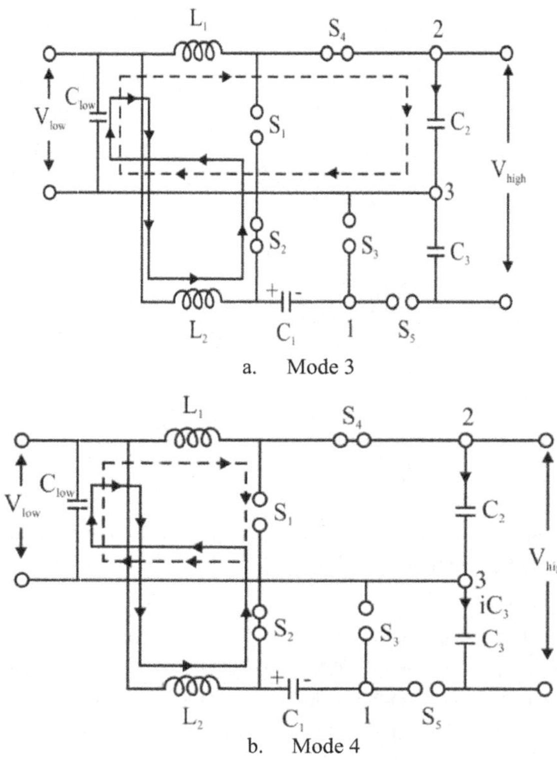

step-up mode. The capacitor discharges in this mode and charges capacitor C_{low} and inductor L_1. As both the capacitors C_1 and C_2 are connected in parallel to the load 2, the load 2 is supplied by both the capacitors. The load 1 is energized by the voltage drop across the capacitor C_{low}.

Applying Kirchoff's voltage law in Fig. 6.3a,

$$\frac{di_{L_1}}{dt} = \frac{V_{C_{02}}}{L_1} - \frac{V_{C_{\text{low}}}}{L_1} \tag{6.32}$$

$$\frac{di_{L_2}}{dt} = \frac{1}{L_2} V_{C_{01}} - \frac{1}{L_2} V_{C_{\text{low}}} \tag{6.33}$$

$$\frac{dV_{C_{01}}}{dt} = -\frac{1}{C_{01}} i_{L_2} \tag{6.34}$$

$$\frac{dV_{C_{02}}}{dt} = -\frac{1}{C_{02}r} V_{C_{02}} - \frac{1}{C_{02}r} V_{C_{03}} - \frac{1}{C_{02}} i_{L_1} \tag{6.35}$$

$$\frac{dV_{C_{03}}}{dt} = -\frac{1}{C_{03}r} V_{C_{02}} - \frac{1}{C_{03}r} V_{C_{03}} \tag{6.36}$$

$$\frac{dV_{C_{\text{low}}}}{dt} = \frac{1}{C_{\text{low}}} i_{L_1} + \frac{1}{C_{\text{low}}} i_{L_2} + \frac{1}{RC_{\text{low}}} V_{C_{\text{low}}} \tag{6.37}$$

Equations (6.23–6.28) are written as:

$$\begin{bmatrix} \frac{di_{L_1}}{dt} \\ \frac{di_{L_2}}{dt} \\ \frac{dV_{C_{01}}}{dt} \\ \frac{dV_{C_{02}}}{dt} \\ \frac{dV_{C_{03}}}{dt} \\ \frac{dV_{C_{\text{low}}}}{dt} \end{bmatrix} = \underbrace{\begin{bmatrix} 0 & 0 & 0 & -\frac{1}{L_1} & 0 & -\frac{1}{L_1} \\ 0 & 0 & \frac{1}{L_2} & 0 & 0 & -\frac{1}{L_2} \\ 0 & -\frac{1}{C_{01}} & 0 & 0 & 0 & 0 \\ -\frac{1}{C_{02}} & 0 & 0 & -\frac{1}{C_{02}r} & -\frac{1}{C_{02}r} & 0 \\ 0 & 0 & 0 & -\frac{1}{C_{03}r} & -\frac{1}{C_{03}r} & 0 \\ \frac{1}{C_{\text{low}}} & \frac{1}{C_{\text{low}}} & 0 & 0 & \frac{1}{RC_{\text{low}}} & 0 \end{bmatrix}}_{A_1} \begin{bmatrix} i_{L_1} \\ i_{L_2} \\ V_{C_{01}} \\ V_{C_{02}} \\ V_{C_{03}} \\ V_{C_{\text{low}}} \end{bmatrix} + \underbrace{\begin{bmatrix} 0 \\ 0 \\ 0 \\ \frac{1}{C_{02}r} \\ \frac{1}{C_{03}r} \\ 0 \end{bmatrix}}_{B_1} V_{\text{low}} \tag{6.38}$$

6.3.2.2 Mode 2

The switch S_3 is turned ON, and switch S_4 is in turn OFF condition. The direction of power flow is shown in Fig. 6.3b. The charged capacitor C_1 in mode 1 and the energy stored in the inductor L_1 supply energy to the load 1. The capacitors C_1 and C_2 discharge and energize load 2.

Applying Kirchoff's voltage law to Fig. 6.3b,

$$\frac{\mathrm{d}i_{L_1}}{\mathrm{d}t} = \frac{V_{C_{\mathrm{low}}}}{L_1} \tag{6.39}$$

$$\frac{\mathrm{d}i_{L_2}}{\mathrm{d}t} = \frac{1}{L_2}V_{C_{01}} - \frac{V_{C_{\mathrm{low}}}}{L_2} \tag{6.40}$$

$$\frac{\mathrm{d}V_{C_{01}}}{\mathrm{d}t} = -\frac{1}{C_{01}}i_{L_2} \tag{6.41}$$

$$\frac{\mathrm{d}V_{C_{02}}}{\mathrm{d}t} = -\frac{V_{C_{02}}}{C_{02}r} - \frac{V_{C_{03}}}{C_{02}r} \tag{6.42}$$

$$\frac{\mathrm{d}V_{C_{03}}}{\mathrm{d}t} = \frac{-V_{C_{02}}}{C_{03}r} - \frac{V_{C_{03}}}{C_{03}r} \tag{6.43}$$

$$\frac{\mathrm{d}V_{C_{\mathrm{low}}}}{\mathrm{d}t} = \frac{1}{C_{\mathrm{low}}}i_{L_1} + \frac{1}{C_{\mathrm{low}}}i_{L_2} + \frac{1}{C_{\mathrm{low}}R}V_{C_{\mathrm{low}}} \tag{6.44}$$

Equations (6.29–6.34) are written as:

$$\begin{bmatrix} \frac{\mathrm{d}i_{L_1}}{\mathrm{d}t} \\ \frac{\mathrm{d}i_{L_2}}{\mathrm{d}t} \\ \frac{\mathrm{d}V_{C_{01}}}{\mathrm{d}t} \\ \frac{\mathrm{d}V_{C_{02}}}{\mathrm{d}t} \\ \frac{\mathrm{d}V_{C_{03}}}{\mathrm{d}t} \\ \frac{\mathrm{d}V_{C_{\mathrm{low}}}}{\mathrm{d}t} \end{bmatrix} = \underbrace{\begin{bmatrix} 0 & 0 & 0 & 0 & 0 & -\frac{1}{L_1} \\ 0 & 0 & \frac{1}{L_2} & 0 & 0 & -\frac{1}{L_2} \\ 0 & -\frac{1}{C_{01}} & 0 & 0 & 0 & 0 \\ 0 & 0 & 0 & -\frac{1}{C_{02}r} & -\frac{1}{C_{02}r} & 0 \\ 0 & 0 & 0 & -\frac{1}{C_{03}r} & -\frac{1}{C_{03}r} & 0 \\ \frac{1}{C_{\mathrm{low}}} & \frac{1}{C_{\mathrm{low}}} & 0 & 0 & 0 & \frac{1}{RC_{\mathrm{low}}} \end{bmatrix}}_{A_2} \begin{bmatrix} i_{L_1} \\ i_{L_2} \\ V_{C_{01}} \\ V_{C_{02}} \\ V_{C_{03}} \\ V_{C_{\mathrm{low}}} \end{bmatrix} + \underbrace{\begin{bmatrix} 0 \\ 0 \\ 0 \\ \frac{1}{C_{02}r} \\ \frac{1}{C_{03}r} \\ 0 \end{bmatrix}}_{B_2} V_{\mathrm{low}} \tag{6.45}$$

6.3.2.3 Mode 3

The switch S_3 is in turn OFF condition and S_4 is turn ON condition. The flow of power is shown in Fig. 6.3c. The capacitor C_1 charges the capacitor C_{low} and energizes the inductor L_1. The load 1 is energized by the voltage drop across the capacitor C_{low}. The capacitors C_1 and C_2 supply energy to load 2.

Applying Kirchoff's voltage law to Fig. 6.3c,

$$\frac{\mathrm{d}i_{L_1}}{\mathrm{d}t} = -\frac{1}{L_1}V_{\mathrm{low}} - \frac{1}{L_1}V_{C_{02}} \tag{6.46}$$

$$\frac{\mathrm{d}i_{L_2}}{\mathrm{d}t} = -\frac{1}{L_2}V_{\mathrm{low}} \tag{6.47}$$

$$\frac{dV_{C_{01}}}{dt} = -\frac{1}{C_{01}r}V_{C_{01}} + \frac{1}{C_{01}r}V_{C_{03}} \tag{6.48}$$

$$\frac{dV_{C_{02}}}{dt} = -\frac{1}{C_{02}}i_{L_1} - \frac{1}{C_{02}r}V_{C_{02}} - \frac{1}{C_{02}r}V_{C_{03}} \tag{6.49}$$

$$\frac{dV_{C_{03}}}{dt} = \frac{1}{C_{03}r}V_{C_{01}} - \frac{1}{C_{03}r}V_{C_{02}} - \frac{1}{C_{03}(r+r_1)}V_{C_{03}} \tag{6.50}$$

$$\frac{dV_{C_{low}}}{dt} = \frac{1}{C_{low}}i_{L_1} + \frac{1}{C_{low}}i_{L_2} + \frac{1}{C_{low}R}V_{C_{low}} \tag{6.51}$$

Equations (6.40–6.45) are written as

$$\begin{bmatrix} \frac{di_{L_1}}{dt} \\ \frac{di_{L_2}}{dt} \\ \frac{dV_{C_{01}}}{dt} \\ \frac{dV_{C_{02}}}{dt} \\ \frac{dV_{C_{03}}}{dt} \\ \frac{dV_{C_{low}}}{dt} \end{bmatrix} = \underbrace{\begin{bmatrix} 0 & 0 & 0 & -\frac{1}{L_1} & 0 & -\frac{1}{L_1} \\ 0 & 0 & 0 & 0 & 0 & -\frac{1}{L_2} \\ 0 & 0 & -\frac{1}{C_{01}r} & 0 & \frac{1}{C_{01}r} & 0 \\ -\frac{1}{C_{02}} & 0 & 0 & -\frac{1}{C_{02}r} & -\frac{1}{C_{02}r} & 0 \\ 0 & 0 & \frac{1}{C_{03}r} & -\frac{1}{C_{03}r} & -\frac{1}{C_{03}(r+r_1)} & 0 \\ \frac{1}{C_{low}} & \frac{1}{C_{low}} & 0 & 0 & 0 & \frac{1}{RC_{low}} \end{bmatrix}}_{A_3} \begin{bmatrix} i_{L_1} \\ i_{L_2} \\ V_{C_{01}} \\ V_{C_{02}} \\ V_{C_{03}} \\ V_{C_{low}} \end{bmatrix} + \begin{bmatrix} 0 \\ 0 \\ 0 \\ \frac{1}{C_{02}r} \\ \frac{1}{C_{03}r} \\ 0 \end{bmatrix} V_{high} \tag{6.52}$$

6.3.2.4 Mode 4

In this mode of operation, both the switches S_3 and S_4 are in turn OFF condition whereas the switches S_1 and S_2 are in turn ON condition. The flow of energy is shown in Fig. 6.3d. The already charged capacitor C_{low} energizes load 1. The capacitors C_1 and C_{out2} are connected in parallel to load supplies energy to the load. Applying Kirchoff's voltage law to Fig. 6.3d

$$\frac{di_{L_1}}{dt} = -\frac{V_{low}}{L_1} \tag{6.53}$$

$$\frac{di_{L_2}}{dt} = -\frac{V_{low}}{L_2} \tag{6.54}$$

$$\frac{dV_{C_{01}}}{dt} = \frac{V_{C_{03}}}{C_{01}r_1} - \frac{V_{C_{01}}}{C_{01}r_1} \tag{6.55}$$

$$\frac{dV_{C_{02}}}{dt} = -\frac{1}{C_{02}r}V_{C_{02}} - \frac{1}{C_{02}r}V_{C_{03}} \tag{6.56}$$

$$\frac{dV_{C_{03}}}{dt} = \frac{1}{C_{03}r_1}V_{C_{01}} - \frac{1}{C_{03}r}V_{C_{02}} + \frac{1}{C_{03}(r+r_1)}V_{C_{03}} \tag{6.57}$$

$$\frac{dV_{C_{low}}}{dt} = \frac{1}{C_{low}}i_{L_1} + \frac{1}{C_{low}}i_{L_2} + \frac{1}{RC_{low}}V_{C_{low}} \tag{6.58}$$

Equations (6.41)–(6.46) are written as

$$\begin{bmatrix} \frac{di_{L_1}}{dt} \\ \frac{di_{L_2}}{dt} \\ \frac{dV_{C_{01}}}{dt} \\ \frac{dV_{C_{02}}}{dt} \\ \frac{dV_{C_{03}}}{dt} \\ \frac{dV_{C_{low}}}{dt} \end{bmatrix} \underbrace{\begin{bmatrix} 0 & 0 & 0 & 0 & 0 & -\frac{1}{L_1} \\ 0 & 0 & 0 & 0 & 0 & -\frac{1}{L_2} \\ 0 & 0 & -\frac{1}{C_{01}r_1} & 0 & \frac{1}{C_{01}r_1} & 0 \\ 0 & 0 & 0 & -\frac{1}{C_{02}r} & -\frac{1}{C_{02}r} & 0 \\ 0 & 0 & \frac{1}{C_{03}r} & -\frac{1}{C_{03}r} & -\frac{1}{C_{03}(r+r_1)} & 0 \\ \frac{1}{C_{low}} & \frac{1}{C_{low}} & 0 & 0 & 0 & \frac{1}{RC_{low}} \end{bmatrix}}_{A_3} \begin{bmatrix} i_{L_1} \\ i_{L_2} \\ V_{C_{01}} \\ V_{C_{02}} \\ V_{C_{03}} \\ V_{C_{low}} \end{bmatrix} + \begin{bmatrix} 0 \\ 0 \\ 0 \\ \frac{1}{C_{02}r} \\ \frac{1}{C_{03}r} \\ 0 \end{bmatrix} V_{high} \tag{6.59}$$

The averaged state-space equation of IBSC converter in step-down mode is given by (Table 6.3)

$$[\dot{X}] = [A]x + [B]v$$

$$[A] = A_1 + A_2 + A_3 + A_4$$

$$[B] = B_1 + B_2 + B_3 + B_4$$

$$\frac{1}{L_1}d_4V_{C_{01}} - \frac{1}{L_1}V_{low} = 0 \tag{6.60}$$

Table 6.3 Modes of operation of bidirectional interleaved switched capacitor converter in step-down mode

Modes	S_1	S_2	S_3	S_4	S_5	C_1	C_2	Load 1	Load 2
Mode 1	OFF	OFF	ON	ON	OFF	Discharging	Discharging	Energized	Energized
Mode 2	OFF	OFF	ON	OFF	OFF	Discharging	Discharging	Energized	Energized
Mode 3	OFF	OFF	OFF	ON	OFF	Discharging	Discharging	Energized	Energized
Mode 4	ON	ON	OFF	OFF	OFF	Discharging	Discharging	Energized	Energized

$$\frac{1}{L_2}d_3V_{C_{02}} - \frac{1}{L_2}V_{\text{low}} = 0 \tag{6.61}$$

$$-\frac{1}{C_{01}}d_3i_{L_2} - \frac{1}{C_{01}r_1}(1-d_3)V_{C_{01}} + \frac{1}{C_{01}r_1}(1-d_3)V_{C_{03}} = 0 \tag{6.62}$$

$$-\frac{1}{C_{01}}d_4i_{L_1} - \frac{1}{C_{01}r}V_{C_{02}} + \frac{1}{C_{01}r}V_{C_{03}} = \frac{1}{C_{02}r}V_{\text{high}} \tag{6.63}$$

$$-\frac{1}{C_{02}r}(1-d_3)V_{C_{01}} + \frac{1}{C_{02}r}V_{C_{02}} - \frac{1}{C_{02}r}V_{C_{03}} - \frac{(1-d_3)}{C_{01}r}V_{C_{03}} = \frac{1}{C_{03}r}V_{\text{high}} \tag{6.64}$$

$$\frac{1}{C_{\text{low}}}i_{L_1} + \frac{1}{C_{\text{low}}}i_{L_2} + \frac{1}{RC_{\text{low}}}V_{C_{\text{low}}} = 0 \tag{6.65}$$

6.4 Operation of Bidirectional Interleaved Switched Capacitor DC–DC Converter with Solar Microgrid

6.4.1 Solar System

The proposed microgrid includes two or more renewable energy as a source. In this work, two solar PV arrays with an output power of 1.5 kW each are considered. The mathematical modeling of the solar PV array is done [1]. The equivalent circuit model of the solar PV array is shown in Fig. 6.4.

By applying Kirchoff's current law to the equivalent circuit, the output current from the PV cell is calculated,

$$I = I_L - I_d$$

The diode equation is as in Eq. 6.2

$$I_d = I_s(e^{qV_d/kT} - 1)$$

Fig. 6.4 Equivalent circuit of solar cell

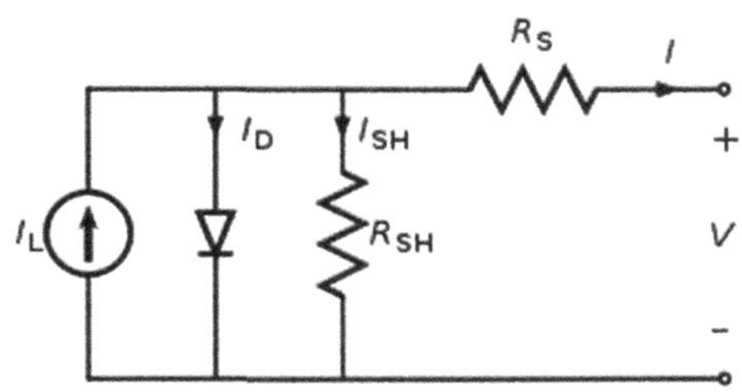

Depending on the solar insolation and the cell's working temperature, the photocurrent is calculated by the following expression

$$I_L = [I_{SC} + K_i(T - 298)]\frac{\lambda}{1000}$$

The saturation current of the module is expressed as 4

$$I_L = [I_{SC} + K_i(T - 298)]\frac{\lambda}{1000}$$

$$I_s = I_{rs}[(T/T_r)^3 \exp\{q^{E_g}/Ak\{(1/T_r) - (1/T)\}]$$

The output current of the PV module is given by 5

$$I_{pv} = N_p IL - N_p I_s[\exp\{q(R_s + I_{pv}R_s)/N_s AkT\} - 1]$$

where

I_L Photon current
I_d diode current
I_s saturation current
T solar cells actual temperature
T_r reference temperature
E_g bandgap energy
K Boltzmann's constant
N_p number of cells in parallel
N_s Number of cells in series
Q Electron charge.

The solar parameters are calculated from the above equations. The V-I and P–V characteristic curves obtained are dependent only on the solar irradiance and the temperature.

The maximum power point tracking (MPPT) is done to obtain the point at which the maximum power is obtained so that the efficiency is also increased [2]. The maximum power is obtained when the source impedance and the load impedance matches. Perturb and observe (P&O) MPPT algorithm is used in this work [3]. The advantage of P&O technique over other MPPT techniques are low implementation cost, only current and voltage values are sufficient.

The VI characteristics of the simulated solar cell with 900 W/m^2 irradiance are shown in Fig. 6.5.

The output solar power, voltage and current in both step-up and step-down mode are shown in Figs. 6.6, 6.7 and 6.8. 0–0.5 indicates step-up mode and 0.5–1 indicates step-down mode.

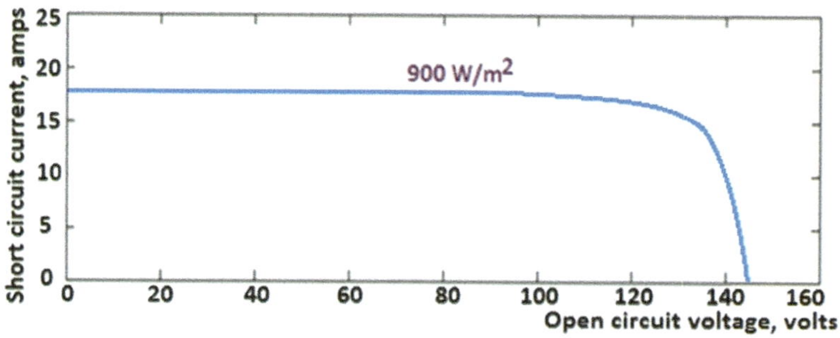

Fig. 6.5 VI characteristics of solar cell

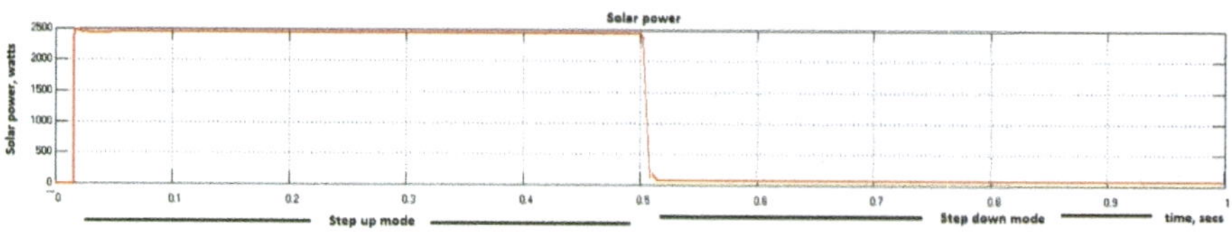

Fig. 6.6 Solar output power

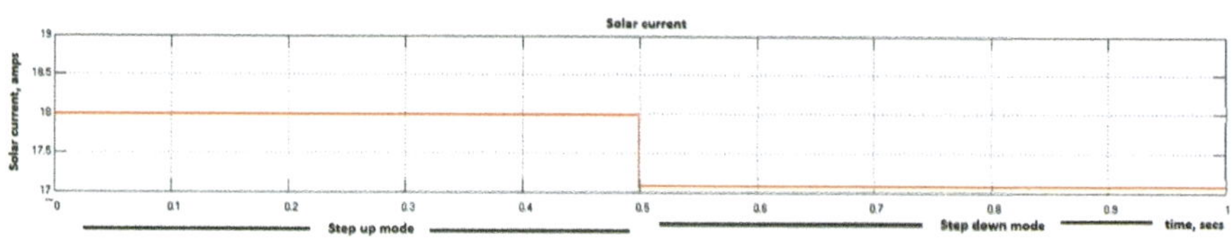

Fig. 6.7 Solar output current

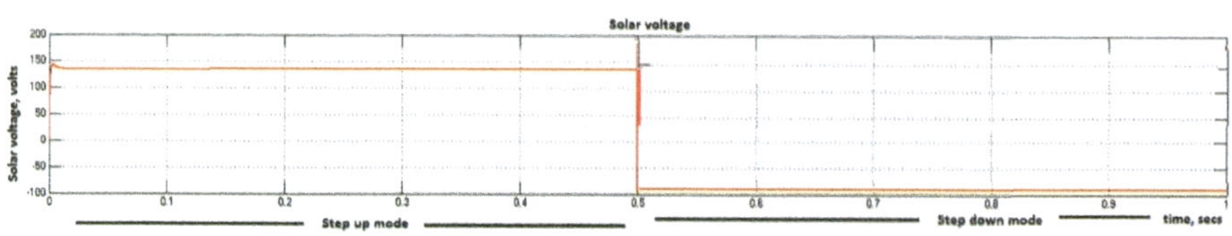

Fig. 6.8 Solar output voltage

6.4.2 Energy Storage Devices

The energy storage devices are essential in microgrids as the renewable energy sources are not available continuously. In this work, lead acid battery is used as an energy storage device. The battery gets charged when the source is available and

Table 6.4 Design specifications of battery

S. no.	Parameters	Values
1	Battery power	1.6 kW
2	Battery voltage	400 V
3	Battery current	4A
4	Number of hours used	3
5	Number of days used	1
6	Battery efficiency	85%
7	Depth of discharge (DoD)	0.8

discharges the stored energy when the source is not available. This supports the load to get continuous supply of energy. The state of health (SOH) of the battery determines the maximum capacity of the battery. The capacity of the battery is understood by the depth of discharge (DOD), charging and discharging cycles. The parameters and specifications of the battery are modeled as in Table 6.4.

Battery Capacity = (Total Power * No. of hours used * No. of days used)/(Battery efficiency * DoD * Battery voltage)

$$\begin{aligned} \text{Cb} &= (1600 * 3 * 1)/(0.85 * 0.8 * 400) \\ &= 17.6\,\text{Ah} \end{aligned}$$

$$\begin{aligned} \text{Charging current} \quad \text{Ib} &= 0.2 * \text{Cb} \\ &= 0.2 * 16 \\ \text{Ib} &= 3.2\,\text{A} \end{aligned}$$

Battery discharging current

$$\begin{aligned} \text{Ib discharge} &= (\text{Total power})/(\text{Battery voltage*Battery efficiency}) \\ &= (1600)/(400 * 0.85) \\ &= 4.7\text{A} \end{aligned}$$

The battery voltage and current in charging and discharging mode are shown in Fig. 6.9. The battery charges in step-up mode and discharges in step-down mode which can be seen in the simulation result are shown in Fig. 6.9.

6.4.3 Load

Two resistive loads with different power capacities are considered. One load is connected across the source and other load is connected across the battery. The solar system and the battery are designed as per the load requirements. The load specifications are given in the following Table 6.5.

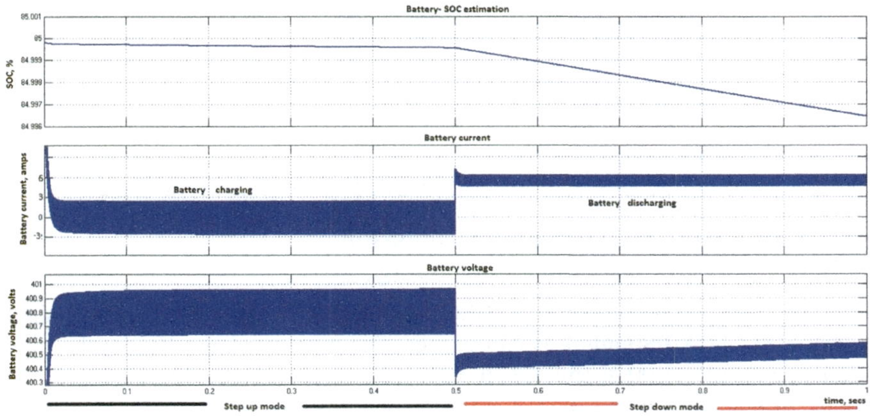

Fig. 6.9 Battery output power, current and voltage

Table 6.5 **a** Parameters of Load 1 **b** Parameters of Load 2

a	
Voltage	400 V
Current	25 A
Power	1000 W
b	
Voltage	140 V
Current	3 A
Power	420 W

In step-up mode of operation charging of battery takes place. The solar energy source is available and the input power to the system is only from solar. The solar energizes battery and the load. The input and output power are calculated

$$\text{Input Power (From Solar)} = 3\text{kW}\ (1.5\text{kW} + 1.5\text{kW})$$

$$\begin{aligned}\text{Output power } &= \text{Battery charging power} + \text{Load 1} + \text{Load 2} \\ &= (400 * 3.2) + 1000 + 420 \\ &= 2700\,\text{W} \\ &= 2.7\,\text{kW}\end{aligned}$$

In step-down mode of operation discharging of battery takes place. The solar energy source is not available, and the energy stored in the battery acts as an input source. The battery energizes both the loads. The input and output power are calculated

$$\text{Discharging battery power} = 400 * 4 = 1600\,\text{W}$$

$$\begin{aligned}\text{Output power} &= \text{Load 1} + \text{Load 2}\\ &= 1420\,\text{W}\end{aligned}$$

6.5 Simulation Results and Discussion

To understand the performance of IBSC DC–DC converter, a detailed simulation model in MATLAB/Simulink is carried out. The output voltage V_{low} at the input side is regulated as 100 V and the output voltage at the output side as 400 V. The output current at the input side is designed to be 2.5 A, and the output current at the output side is regulated as 2.3 A. Figures 6.10 and 6.11 show the output voltage and current waveforms at the output side of the converter. It is observed that the output voltage as 400 V and the output current as 2.5 A.

Output voltage (Load side).

Figures 6.12 and 6.13 show the output voltage and output current at the input side of the converter. It is observed that the output voltage as 100 V and output current as 2.3 A. Figure 6.14 shows the voltage and current of inductor L_1 in step-up and step-down modes of operation. The step-up mode for 0–0.5 and step-down mode for 0.5–1.

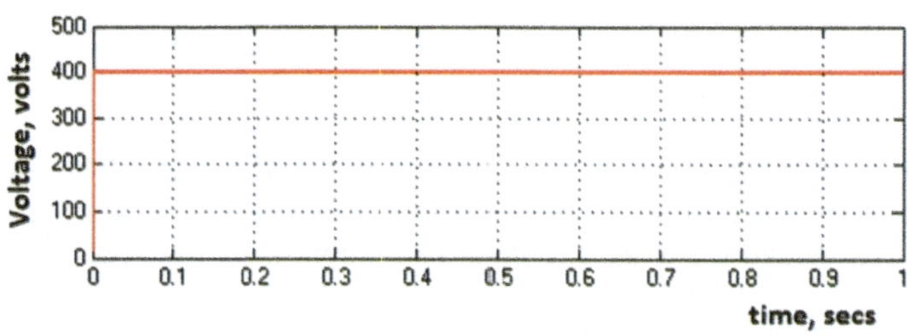

Fig. 6.10 Output voltage across the output side output current (load side)

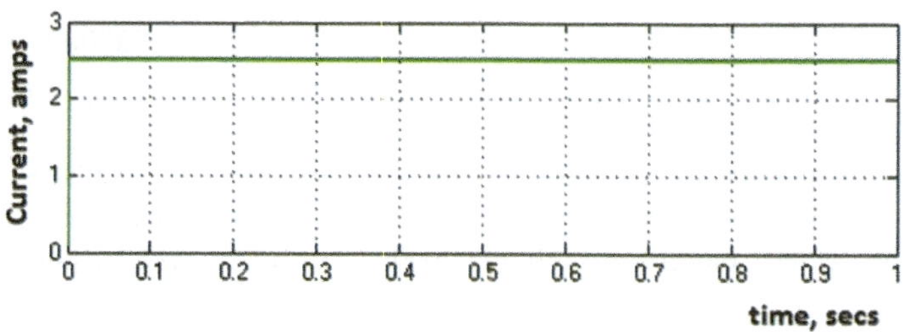

Fig. 6.11 Output current through the output side

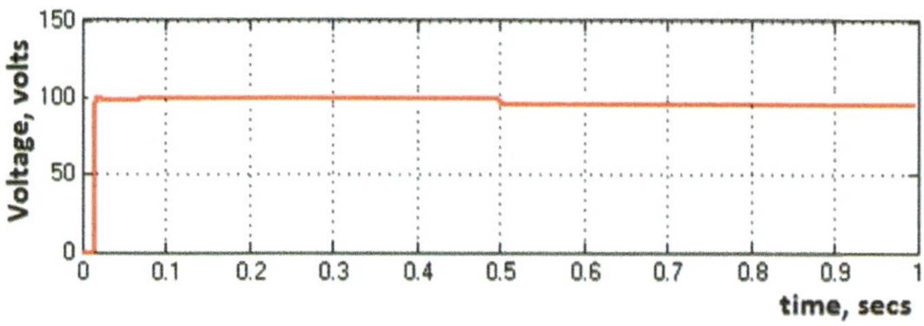

Fig. 6.12 Output voltage across the input side

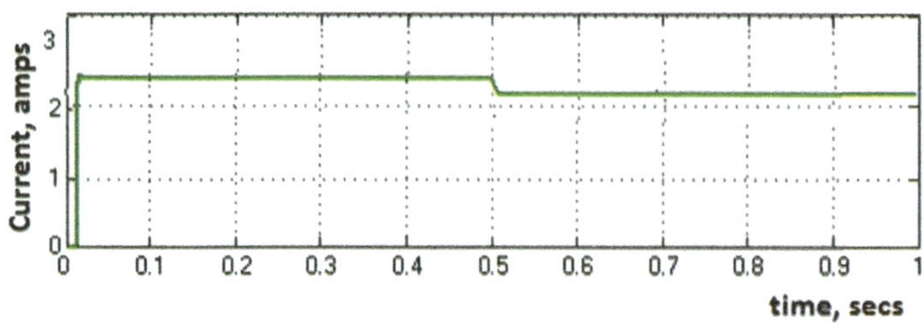

Fig. 6.13 Output current through the Input side

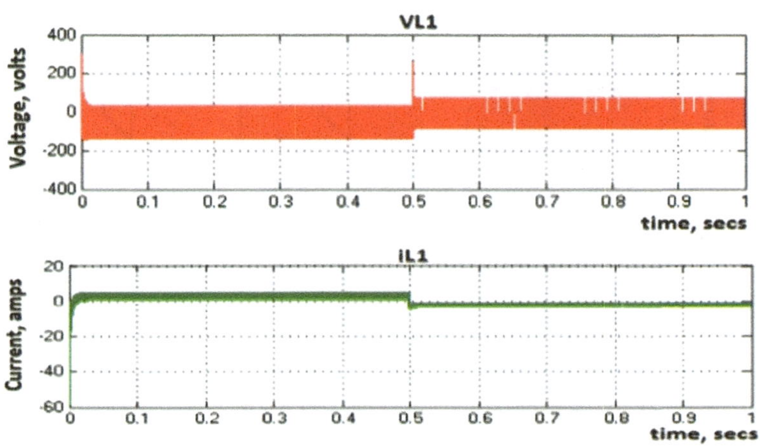

Fig. 6.14 Voltage and currents of Inductor L_1 in step-up and step-down modes of operation

Figure 6.15 shows the output voltage and current of inductor L_2 in step-up and step-down modes of operation. Figures 6.16, 6.17, 6.18, 6.19, 6.20, 6.21, 6.22, 6.23 and 6.24 indicate the voltage across the switches S_1, S_2, S_3, S_4 and S_5 and the currents flowing through them are observed.

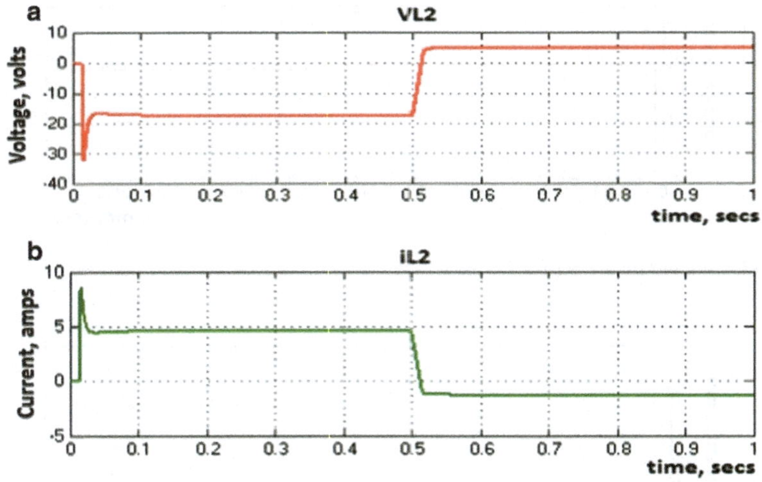

Fig. 6.15 **a** Inductor current (L_2): Step-up: (Time 0–0.5 s), Step-down: (Time 0.5–1 s). **b** voltage and current of inductor L_2 in step-up and step-down mode of operation

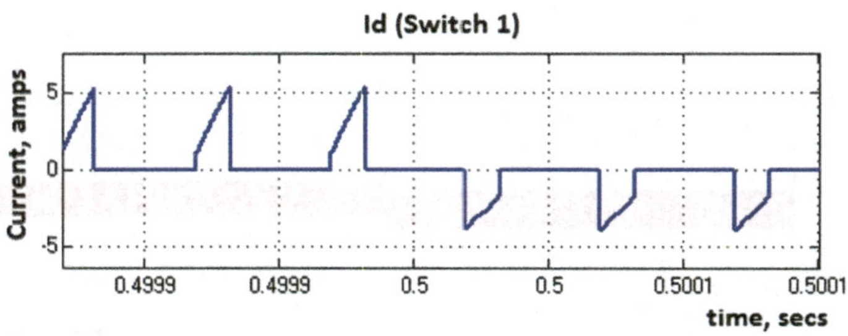

Fig. 6.16 Current through the switch S_1

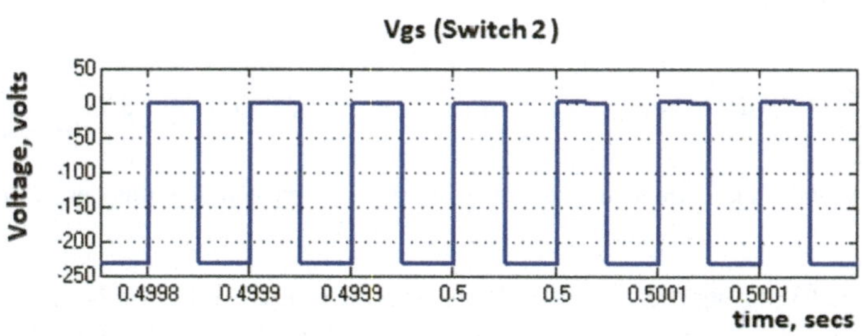

Fig. 6.17 Voltage across the switch S_2

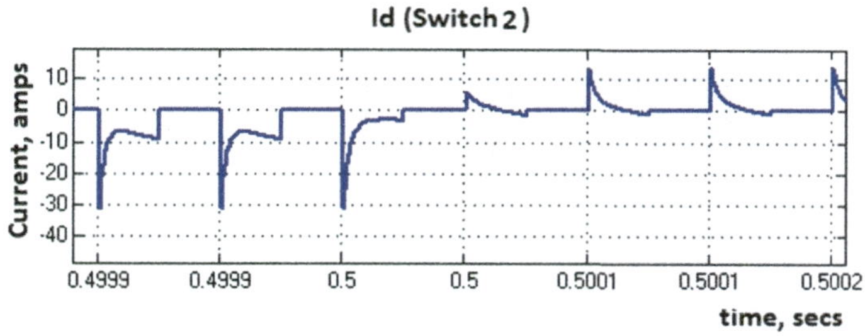

Fig. 6.18 Current through the switch S_2

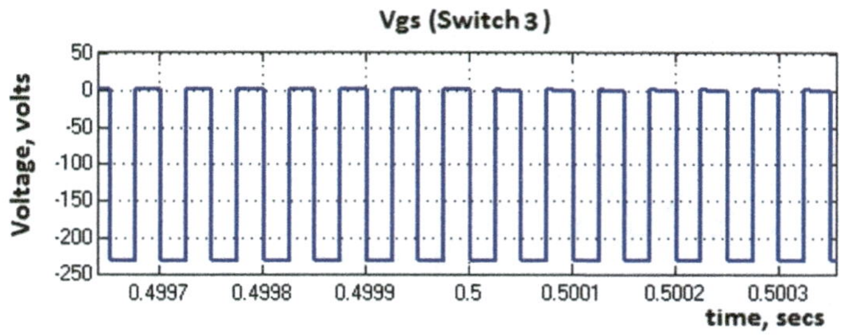

Fig. 6.19 Voltage across the switch S_3

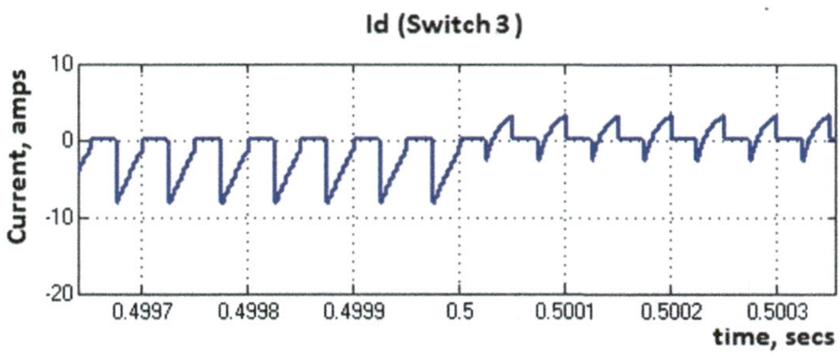

Fig. 6.20 Current through the switch S_3

6.6 Efficiency

Efficiency of a system is an important factor in determining the performance of the system. It determines the efficient power delivered to the load from the system. The efficiency of the system is calculated from the input and output power including the losses. In this work, solar power is considered as input power and the battery power is considered as the output power. The efficiency calculation is as follows

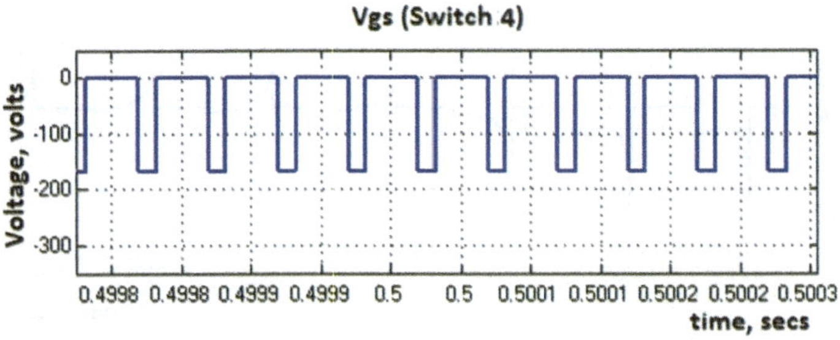

Fig. 6.21 Voltage across the switch S_4

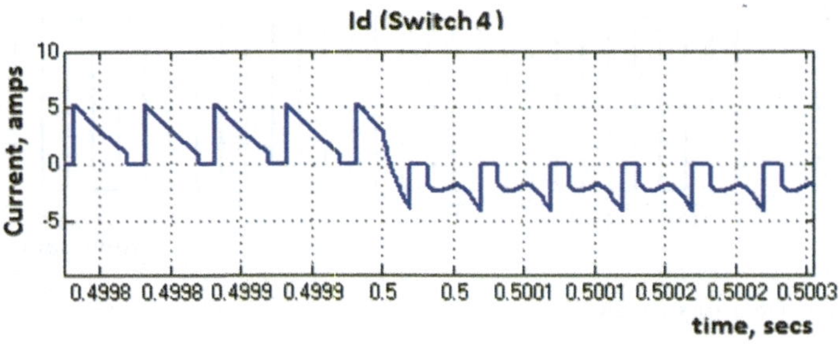

Fig. 6.22 Current through the switch S_4

Fig. 6.23 Voltage across the switch S_5

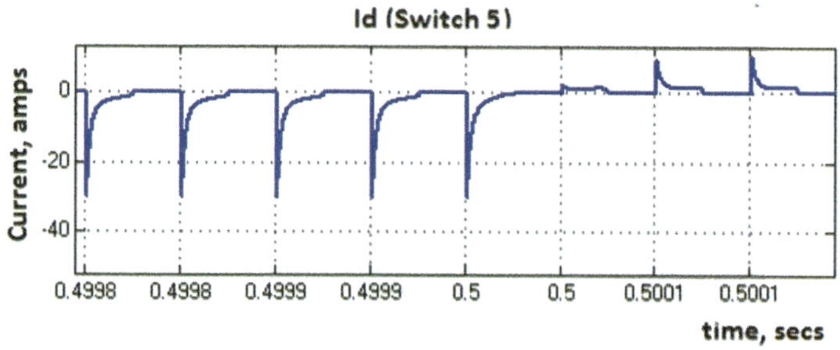

Fig. 6.24 Current through the switch S_5

$$
\begin{aligned}
\text{Efficiency} &= \text{Output power}/\text{output power} + \text{losses}\\
\text{Input power} &= \text{Output power} + \text{losses}\\
\text{Efficiency} &= \text{Output power}/\text{Input power}\\
\text{Solar voltage} &= 140\,\text{V}\\
\text{Solar current} &= 2.5\,\text{A}\\
\text{Solar power} &= 140 * 2.5 = 2520\,\text{W}\\
\text{Battery voltage} &= 400\,\text{V}\\
\text{Battery current} &= 2.5\ (\text{charging})\\
\text{Battery current} &= 3.5\ (\text{discharging})\\
\text{Battery power} &= 400.6 * 2.5 = 1051\ (\text{Charging})\\
\text{Battery power} &= 400.4 * 3.5 = 1401\ (\text{Discharging})\\
\text{Output power} &= \text{Load 1 (input side)} + \text{Load 2 (output side)}\\
&= (400 * 2.5) + (100 * 2.5)\\
&= (1000) + (250) = 1250\,\text{W}
\end{aligned}
$$

Step-up mode (charging):

$$\text{Efficiency} = 1250 + 1051 = 2301/2520 = 91.3\%$$

Step-down mode (discharging):

$$\text{Efficiency} = 1250/1401 = 90.2\%$$

From the above calculations, the efficiency of the system is calculated as 91.3% in step-up mode and 90.2% in step-down mode.

6.7 Conclusion

Microgrids which include renewable energy are becoming popular nowadays. These microgrids require a power electronic device as an interface. The proposed bidirectional interleaved switched capacitor DC–DC converter has many advantages such as bidirectional power flow, reduced ripples, increased voltage gain and efficiency. The converter also operates in both step-up and step-down mode. The proposed system uses battery as an energy storage device. As the proposed converter operates in both the directions and includes an energy storage device, continuous supply of energy to the load is ensured. The modes of operation of the proposed configuration are tested and verified using MATLAB/Simulink, and a hardware prototype of 1 kW prototype has been developed. The proposed configuration is interfaced with solar photovoltaic system and proved to be efficient for microgrids with solar as a renewable energy source.

References

1. Cardell, J., Tabors, R.: Operation and control in a competitive market: distributed generation in a restructured industry. In: The Energy Journal Special Issue: Distributed Resources: Toward a New Paradigm of the Electricity Business, The International Association for Energy Economics, Clevland, Ohio, USA, pp. 111–135 (1998)
2. Guerrero, J.M., Blaabjerg, F., Zhelev, T., Hemmes, K., Monmasson, E., Jemei, S., Comech, M.P., Granadino, R., Frau, J.I.: Distributed generation: Toward a new energy paradigm. IEEE Ind. Electron. Mag. **4**, 52–64 (2010)
3. Jamil, M., Hussain, B., Abu-Sara, M., Boltryk, R.J., Sharkh, S.M.: Microgrid power electronic converters: state of the art and future challenges. In: 2009 44th International Universities Power Engineering Conference (UPEC) 11172188 (2009)
4. Ravi, D., Reddy, B.M., Shimi, S.L., Samue, P.: Bidirectional DC to DC converters: an overview of various topologies, switching schemes and control techniques. Int. J. Eng. Technol. **7**(4.5), 360–365 (2018)
5. Reddy, B.M., Samue, P.: Analysis of secluded bi-direction DC/DC converters for the performance enhancement of photo-voltaic system and energy storage system, Conference Paper, October 2017 (2017). https://doi.org/10.1109/POWERI.2016.8077310
6. Zhou, H., Duong, T., Sing, S.T., Khambadkone, A.M.: Interleaved bi-directional dual active bridge DC–DC converter for interfacing ultracapacitor in micro-grid application. 2010 IEEE, pp. 2229–2234 (2010)
7. Thomas, J., Binu, A.B., Alex, J., Babu, A., Varghese, B.M.: Wide conversion ratio bidirectional DC–DC converter for DC microgrid applications. Int. Res. J. Eng. Technol. (IRJET) **07**(06) (2020). e-ISSN: 2395-0056
8. Zhang, Y., Gao, Y., Li, J., Sumner, M.: Interleaved switched-capacitor bidirectional DC–DC converter with wide voltage-gain range for energy storage systems. IEEE Trans. Power Electron. **33**(5)
9. Priyadarshi, N., Padmanaban, S., Maroti, P.K., Sharma. A.: An extensive practical investigation of FPSO-based MPPT for grid integrated PV system under variable operating conditions with anti-islanding protection. IEEE Syst. J. 1–11 (2018)

10. Priyadarshi, N., Padmanaban, S., Bhaskar, M.S., Blaabjerg, F., Sharma, A.: A fuzzy SVPWM based inverter control realization of grid integrated PV-Wind system with FPSO MPPT algorithm for a grid-connected PV/Wind Power generation system: hardware implementation. IET Electric Power Appl. 1–12 (2018)
11. Padmanaban, S., Priyadarshi, N., Holm-Nielsen, J.B., Bhaskar, M.S., Azam, F., Sharma, A. K.: A novel modified sine-cosine optimized MPPT algorithm for grid integrated PV system under real operating conditions. IEEE Access **7**, 10467–10477 (2019). https://doi.org/10.1109/ACCESS.2018.2890533
12. Padmanaban, S., Priyadarshi, N., Holm-Nielsen, J.B., Bhaskar, M.S., Hossain, E., Azam, F.: A hybrid photovoltaic-fuel cell for grid integration with Jaya-based maximum power point tracking: experimental performance evaluation. IEEE Access **7**, 82978–82990 (2019). https://doi.org/10.1109/ACCESS.2019.2924264
13. Kamalapathi, K., Priyadarshi, N., Padmanaban, S., Holm-Nielsen, J.B., Azam, F., Umayal, C., Ramachandaramurthy, V.K.: A hybrid moth-flame fuzzy logic controller based integrated Cuk converter fed brushless DC motor for power factor correction. Electronics **7**, 288 (2018)
14. Kashif, M.: Bidirectional flyback DC–DC converter for hybrid electric vehicle: Utility, working and PSPICE computer model. Asia Pacific Conference on Postgraduate Research in Microelectronics and Electronics, December, pp. 61–66 (2012)
15. Li, H., Peng, F.Z., Lawler, J.S.: A natural ZVS medium-power bidirectional DC–DC converter with minimum number of devices. IEEE Trans. Ind. Appl. **39**(2), 525–535 (2003)
16. Zhang, Z., Ouyang, Z., Thomsen, O.C., Andersen, M.A.E.: Analysis and design of a bidirectional isolated DC–DC converter for fuel cells and supercapacitors hybrid system. IEEE Trans. Power Electron. **27**(2), 848–859 (2012)
17. Lee, J.-Y., Jeong, Y.-S., Han, B.-M.: A Two-stage isolated/bidirectional DC/DC converter with current ripple reduction technique. IEEE Trans. Industr. Electron. **59**(1), 644–646 (2012)
18. Wai, R.J., Duan, R.Y.: High-efficiency bidirectional converter for power sources with great voltage diversity. IEEE Trans. Power Electron. **22**(5), 1986–1996 (2007)
19. Chung, H.S.H., Chow, W.C., Hui, S.Y.R.: Development of a switched-capacitor dc–dc converter with bi-directional power flow. IEEE Trans. Circuits Syst. **47**(9), 1383–1390 (2000)
20. Chung, H.S., Ioinovici, A., Cheung, W.L.: Generalized structure of bi-directional switched-capacitor dc/dc converters. IEEE Trans. Circuits Syst. I Fundam. Theory Appl. **50**(6), 743–753 (2003)
21. Zou, K., Scott, M.J., Wang, J.: A switched-capacitor voltage tripler with automatic interleaving capability. IEEE Trans. Power Electron. **27**(6), 2857–2868 (2012)
22. Cornea, O., Andreescu, G.D., Muntean, N., Dan, H.: Bidirectional power flow control in a dc microgrid through a switched-capacitor cell hybrid dc–dc converter. IEEE Trans. Ind. Electron. **64**(4), 3012–3022 (2017)

Chapter 7
LUO Converters for Renewable Energy Applications

P. S. V. Kishore, Jayachandra Bogineni, J. Rajesh, Sukanta Halder, and N. Jayaram

7.1 Introduction

Nowadays, fossil fuels, namely coal, gas, oil and nuclear, etc., are the sources mostly used for conventional energy [1, 2] throughout the globe. However, the storage capacity of the traditional sources is limited, and the fossil fuels release much amount of greenhouse effects to the environment. Due to deployment of conventional sources and to encounter the demand for energy, now the trend is changed to the sources of renewable energy [3]. The main renewable sources (RES) are solar, wind and tidal, etc. Among all the RES, solar energy [4–6] is most popular because it is easily extracted from the sunlight. The output from the solar energy is DC. To meet the application demand, DC–DC converters need to incorporate in the system [7–9]. The applications are mainly used in equipment like laptops, mobile phones, etc., which are supplied power from batteries [10, 11]. So, these batteries will give fixed voltage. However, in portable devices having a greater number of sub circuits. They require different voltage levels. So, DC–DC converters generate these required voltage levels from the fixed battery sources and renewable energy sources [12, 13]. Next, the incorporation of batteries in equipment and other systems generates difficulties due to their short life and limited power density [14, 15]. Therefore, regular replacement and maintenance are required [16]. Now, the trend is changed to implement DC–DC converters, which gives required voltage, less maintenance and increased efficiency.

Researchers and technologist put their extensive effort toward the development of LUO converters [7] which provide the boosting of the output and enhance the system performance. These are the converters of type, which give high gain with

P. S. V. Kishore · J. Bogineni (✉) · J. Rajesh · S. Halder · N. Jayaram
NIT, Tadepalligudem, Andhra Pradesh, India
e-mail: jamirajesh.sclr@nitandhra.ac.in

N. Jayaram
e-mail: jayaram@nitandhra.ac.in

N. Priyadarshi et al. (eds.), *DC–DC Converters for Future Renewable Energy Systems*, Energy Systems in Electrical Engineering,
https://doi.org/10.1007/978-981-16-4388-0_7

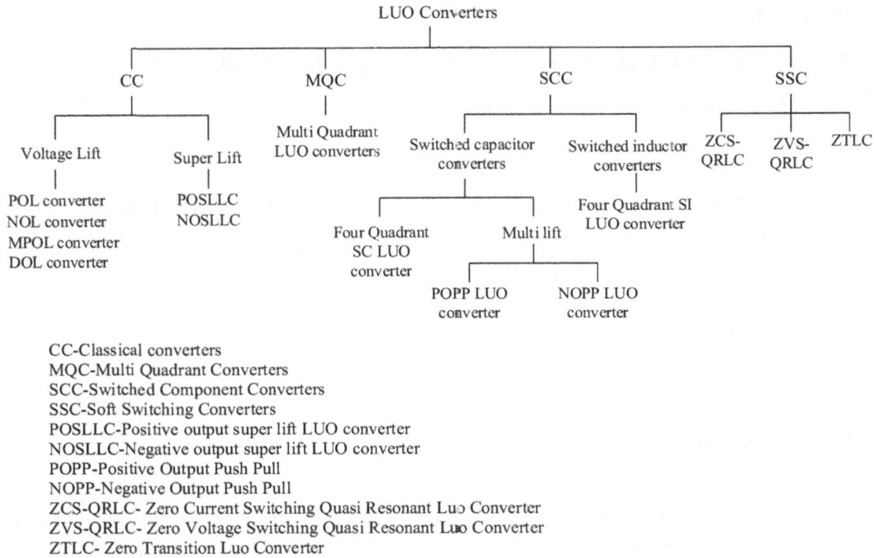

Fig. 7.1 LUO converters classification

minimum number of components. The gain will be increased in arithmetic progression from stage by stage when they are cascaded.

In this chapter, different types of LUO converters are classified and explained with the help of MATLAB simulation. The classification of LUO converters is shown in Fig. 7.1.

7.2 VL LUO Converters

In the technique of VL [8], a capacitor is charged with input voltage during switch on, and the voltage across this charged capacitor can be used for enhancing the output voltage during switch off. Accordingly, this circuit is called an first lift (FL) circuit.

Another capacitor can be charged by the same input voltage, the voltage across this capacitor also enhances the output voltage, now, the output voltage is more than the FL circuit, and this is called as second lift (SL) circuit. By doing the same operation repeatedly, the output voltage will be enhanced more and more. Subsequently, the sequence of circuits is named third lift (TL), fourth lift (FoL) circuits and so on.

7.2.1 Positive Output LUO (POL) Converters

These converters are used to convert fixed DC positive voltage to variable DC positive voltage with VL technique. These converters provide output voltage as multiples of input voltage and first quadrant operation only [9].

The first five circuits have been introduced here.

- Basic circuit
- FL circuit
- SL circuit
- TL circuit
- FoL circuit.

Basic circuit

Buck and boost operation can be done by basic circuit. Other POL converter circuits perform only boost operation. Figure 7.2 shows the POL converter basic circuit [10]. During switch off period, the energy stored in the inductor L_1 is transferred to the capacitor C_1, and during switch on period, the stored energy in the capacitor C_1 is received by the load R. During switch on time, the inductor L_1 gets energized by the voltage V_1, and at the same time, the inductor current i_{L2} flows through the capacitor C_f, load R and completes its path by freewheeling diode D. Capacitor C_1 acts like a bridge between input and output, and the voltage across the capacitor gets accumulated over the output. Therefore, if the voltage across the capacitor is higher, then the output voltage will be higher. The current i_D is flowing through the freewheeling diode during the switch off period [11]. This current decreases in the switch off period $(1 - \delta)T$.

During $(1 - \delta)T$, the charge accumulated with the capacitor C_1 rises and is equal to $(1 - \delta)T\, I_{L1}$.

During δT, the charge on the capacitor C_1 declines and is equivalent to $\delta T I_{L2}$.

In a whole period, these two charges are equal, and therefore,

$$(1 - \delta)TI_{L1} = \delta TI_{L2} \tag{7.1}$$

Fig. 7.2 Basic circuit

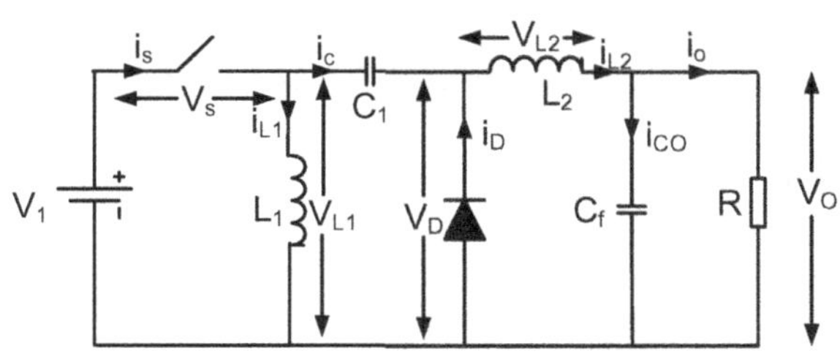

Therefore,

$$I_{L2} = \frac{1-\delta}{\delta} I_{L1} \tag{7.2}$$

Since C_f performs the operation of a low-pass filter, therefore, the output current

$$I_{L2} = I_O \tag{7.3}$$

The above Eqs. (7.2) and (7.3) are applicable for all the POL converters.

During switch on period, the current $i_S = i_{L1} + i_{L2}$, and $i_S = 0$ during switch off time. Thus, the mean current from the source (I_1) is

$$I_1 = \delta i_S = \delta(i_{L1} + i_{L2}) = \delta\left(1 + \frac{1-\delta}{\delta}\right) I_{L1} = I_{L1} \tag{7.4}$$

Therefore, from Eqs. (7.2), (7.3) and (7.4), the current flowing through the load (I_O) is

$$I_O = \frac{1-\delta}{\delta} I_1 \tag{7.5}$$

Assuming the power balance in the converters, $P_{in} = P_{out}$, i.e.,

$$V_1 I_1 = V_O I_O \tag{7.6}$$

From Eqs. (7.4) and (7.5), output voltage is

$$V_O = \frac{\delta}{1-\delta} V_1 \tag{7.7}$$

Figure 7.3 depicts the circuit diagrams of the remaining POL converters [10]. By analyzing these circuits, one can get the output voltage equations given in Table 7.1.

7.2.2 Negative Output LUO (NOL) Converters

With the VL technique, NOL converters transform fixed positive DC voltage to variable negative DC voltage. They work with high voltage magnification in the third quadrant [12]. Five circuits are added here. They are as follows:

- Basic circuit
- FL circuit
- SL circuit

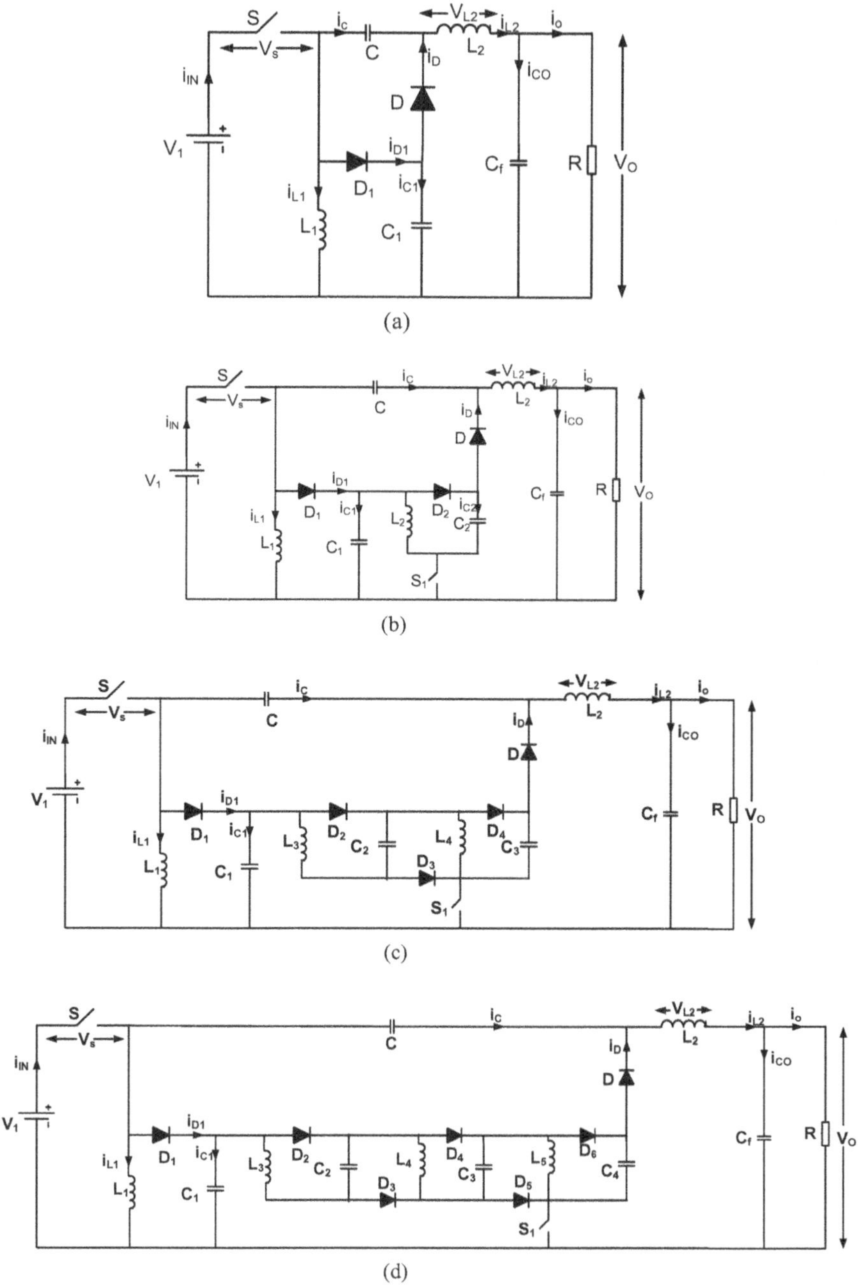

Fig. 7.3 POL converters **a** FL, **b** SL, **c** TL, **d** FoL circuits

Table 7.1 For all voltage lift circuits

Circuit name	Output voltage equation	Output voltage for V_1 = 30 V			
		$\delta = 0.3$	$\delta = 0.5$	$\delta = 0.7$	$\delta = 0.9$
Basic circuit	$V_O = \frac{\delta}{1-\delta} V_1$	12.85	30	70	210
First lift	$V_O = \frac{1}{1-\delta} V_1$	42.85	60	100	300
Second lift	$V_O = \frac{2}{1-\delta} V_1$	85.7	120	200	600
Third lift	$V_O = \frac{3}{1-\delta} V_1$	128.55	180	300	900
Fourth lift	$V_O = \frac{4}{1-\delta} V_1$	171.4	240	400	1200

- TL circuit
- FoL circuit.

Basic circuit

Buck and boost DC–DC conversion can be done by the basic circuit. The other NOL converters only perform a boost operation. The basic NOL converter circuit is depicted in Fig. 7.4. Capacitor C works as energy storing and transferring element from source to load.

The current does not flow through the diode while switch S is on, since it is reverse biased. During switch on condition, as the diode is reverse biased, the total current coming from the source flows into the inductor, and this current increases linearly. At the same time, the energy from the capacitor C flows to the load R through the inductor L_O, and therefore, $i_{C-\delta T} = i_{LO}$.

During switch off condition, the current coming from the inductor L is used to charge the capacitor C by closing the circuit through the freewheeling diode. Inductor L transfers its stored energy through inductor L_O to capacitor C and load R, and therefore, $i_L = i_{C-(1-\delta)T} + i_{LO}$. Thus, i_L current is diminished.

As the capacitor C_f does not absorb any energy in the steady state, the current through the load is $i_O = i_{LO}$. The average current output is

$$i_O = i_{LO} = i_{C-\delta T} \tag{7.8}$$

During $(1 - \delta)T$, the charge increased on the capacitor C is equal to $(1 - \delta) Ti_{C-(1-\delta)T}$.

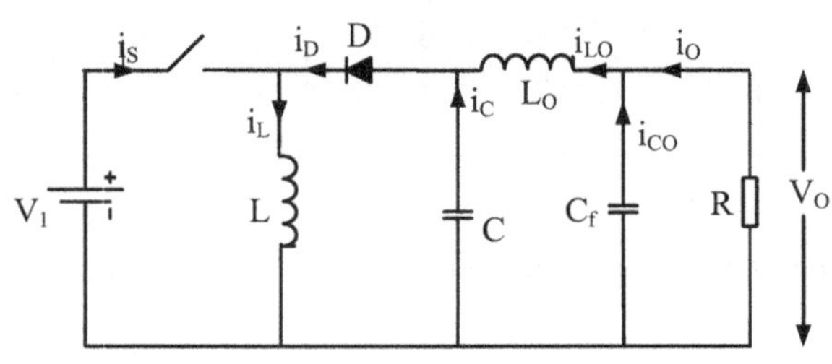

Fig. 7.4 Basic circuit

And it declines during switch on which is equal to $\delta T i_{C-\delta T}$.

In a complete period T, the above two charges are equal, and therefore

$$(1-\delta)Ti_{C-(1-\delta)T} = \delta Ti_{C-\delta T} \tag{7.9}$$

$$i_{C-(1-\delta)T} = \frac{\delta}{1-\delta} i_{C-\delta T} = \frac{\delta}{1-\delta} i_o \tag{7.10}$$

Therefore, the current flowing in the inductor, I_L, is

$$i_L = i_{C-(1-\delta)T} + i_o = \frac{i_o}{1-\delta} \tag{7.11}$$

Eqs. (7.8) and (7.9) are applicable for all the NOL converters. It is known that $i_S = i_L$ during the time δT.

Therefore, the mean current through the source, i_1, is

$$i_1 = \delta i_1 = \delta i_L = \frac{\delta}{1-\delta} i_o \tag{7.12}$$

$$i_o = \frac{1-\delta}{\delta} i_1 \tag{7.13}$$

and the output voltage is

$$V_o = \frac{\delta}{1-\delta} V_1 \tag{7.14}$$

The remaining circuits of NOL converters are depicted in Fig. 7.5 [10]. By analyzing these circuits, one can get the output voltage equations for these circuits and is shown in Table 7.1.

7.2.3 Modified Positive Output LUO (MPOL) Converters

POL converters use two switches in their circuits from second lift circuit onwards. This section describes the technique to amend POL converters [13] that only one switch is sufficient for all the circuits. There have been some circuits added here. They are as follows:

- Basic circuit
- FL circuit
- SL circuit
- TL circuit
- FoL circuit.

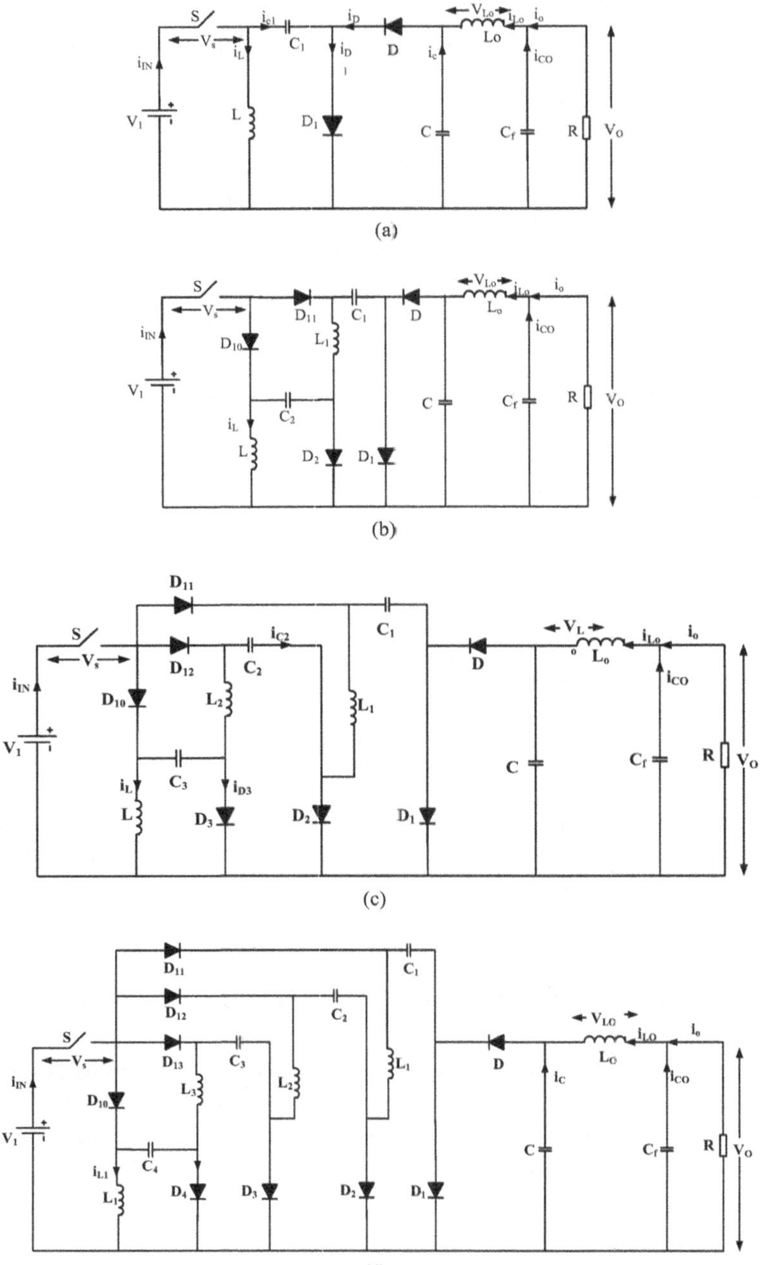

Fig. 7.5 NOL converters **a** FL, **b** SL, **c** TL, **d** FoL circuits

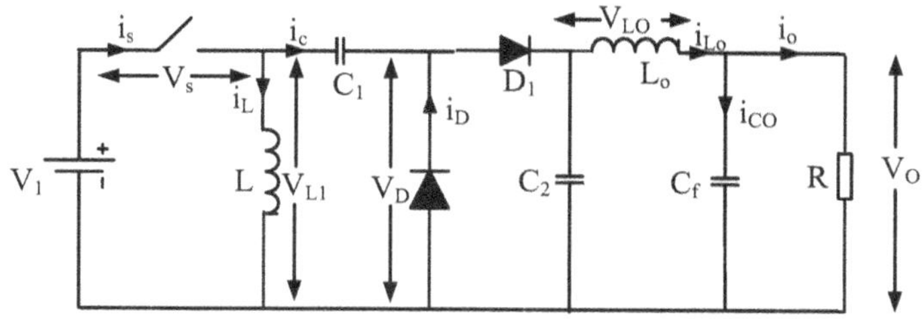

Fig. 7.6 First lift circuit

The basic circuit is same as the basic circuit of the POL converter. FL circuit is discussed here. The FL circuit of the MPOL converter is depicted in Fig. 7.6.

The mean voltage across the inductor over the complete period vanishes in the study state. Thus,

$$V_{C2} = V_{Cf} = V_O$$

During the time δT, the inducer current rises and decreases during the switch off time. V_1 and $-V_{C1}$ are the equivalent voltages across the L during switch on and switch off, respectively.

Therefore, $\delta TV_1 = (1 - \delta)TV_{C1}$ (7.15).

Hence,

$$V_{C1} = \frac{\delta}{1-\delta} V_1 \tag{7.16}$$

During switch on time,

$$V_{C2} = V_1 + V_{C1} \tag{7.17}$$

Assume that the capacitors C_1 and C_2 are large enough,
Therefore,

$$V_{C2} = V_1 + \frac{\delta}{1-\delta} V_1 = \frac{1}{1-\delta} V_1 \tag{7.18}$$

$$V_O = V_{Cf} = V_{C2} = \frac{1}{1-\delta} V_1 \tag{7.19}$$

The remaining circuits [10] and their voltage quations are shown in Fig. 7.7 and Table 7.1, respectively.

Simulations are carried out for the above circuits for the input voltage of 30 V. The duty ratio δ is varied, and the output voltages are shown in Table 7.1. Figures 7.8, 7.9 and 7.10 agree with the values in Table 7.1.

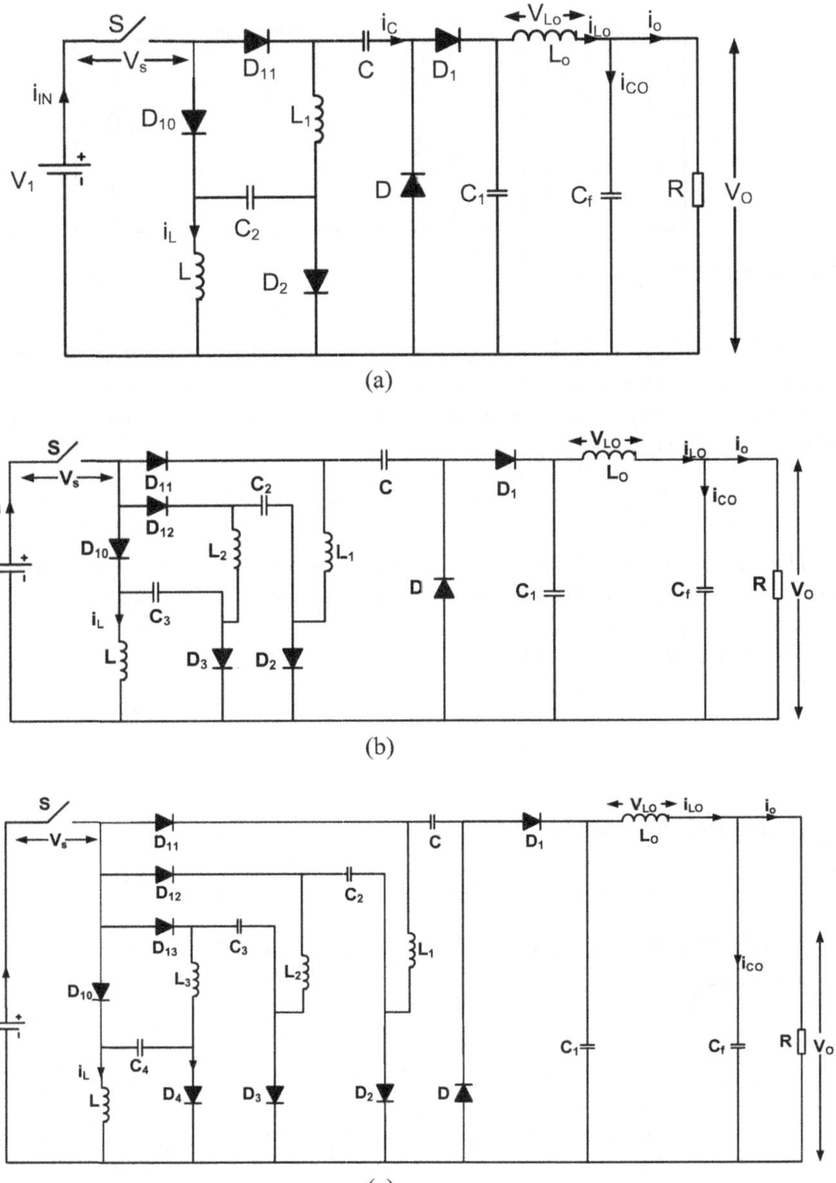

Fig. 7.7 Modified POL converters **a** SL, **b** TL, **c** FoL circuits

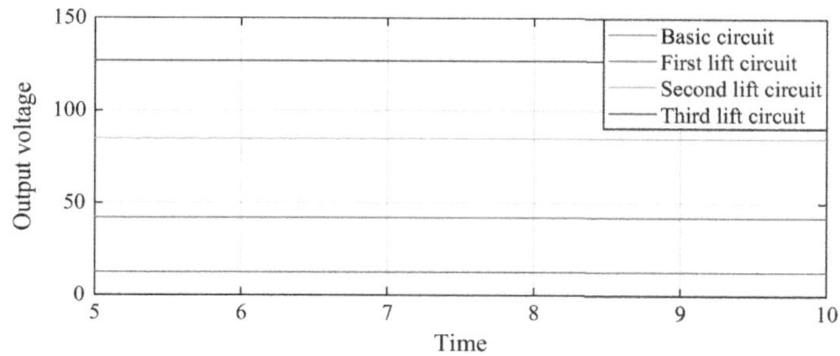

Fig. 7.8 Output voltages of different voltage lift LUO converters for $\delta = 0.3$

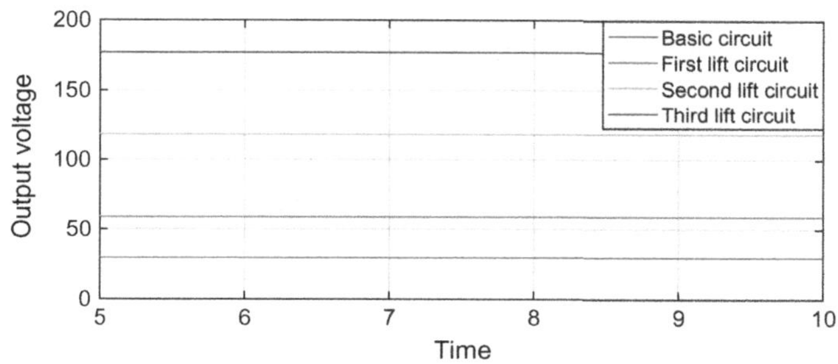

Fig. 7.9 Output voltages of different voltage lift LUO converters for $\delta = 0.5$

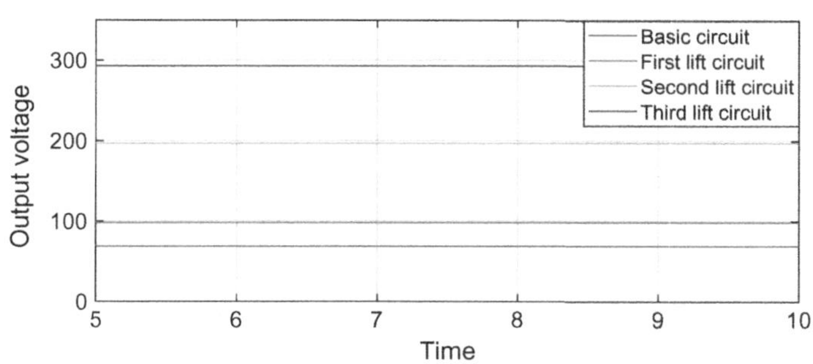

Fig. 7.10 Output voltages of different voltage lift LUO converters for $\delta = 0.7$

7.3 Two-Quadrant LUO (TQL) Converter

All the converters discussed above work only in one quadrant. In this section, two-quadrant LUO converters are discussed which work in both first and second quadrants. The TQL converter circuit diagram is depicted in Fig. 7.11. It is having

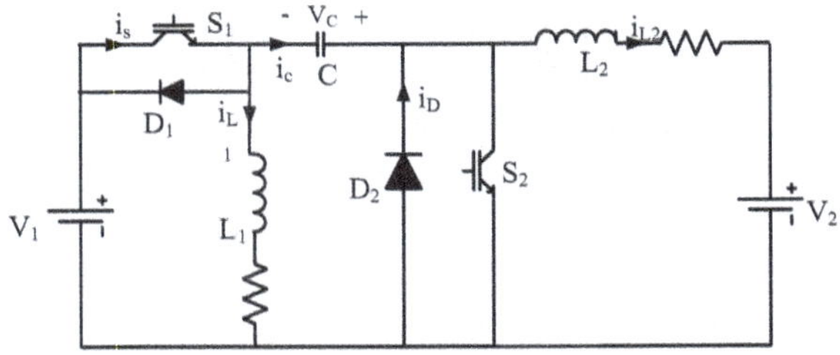

Fig. 7.11 Two-quadrant LUO converter

two operational modes. In mode 1, switch S_1 operates, and switch S_2 operates in mode 2.

Mode 1

In this mode, the converter operates in quadrant 1, and the energy is extracted from the source and transferred to the load.

Switch S_1 is ON: When the switch S_1 is on, switches S_2, D_1 and D_2 do not conduct. The equivalent circuit diagram is depicted in Fig. 7.12a. Here, L_1 stores energy from the source, and at the same time, L_2 stores energy from the capacitor and the source. Therefore, both the inductor currents will increase.

Switch S_1 is OFF: When S_1 is off, S_2, D_1 is also off, and D_2 only conducts. The equivalent circuit diagram is depicted in Fig. 7.12b. The stored energy in L_1 is transferred to the C through D_2. At the same time, the energy stored in L_2 will be

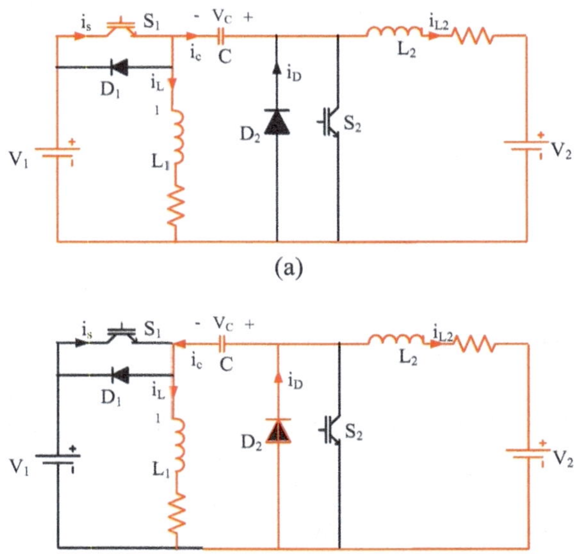

Fig. 7.12 States of mode 1 **a** S_1 is on **b** S_1 is off

released to the load V_2 and takes the path through the freewheeling diode D_2. Therefore, both inductor currents will decrease.

Mode 2

In this mode, the converter operates in second quadrant, and the energy is transferred from load to source.

Switch S_2 is ON: When S_2 is on, S_1, D_1 and D_2 do not conduct. The corresponding circuit diagram is depicted in Fig. 7.13a. The current i_{L1} of inductor L_1 flows through the switch S_2 and increases by the capacitor voltage V_c. Therefore, capacitor voltage decreases. At the same time, the current i_{L2} of inductor L_2 also increases by the load voltage V_2. Therefore, both inductors current will increase.

Switch S_2 is OFF: When S_2 is off, S_1 and D_2 are also turned off, and only D_1 conducts. The corresponding circuit is depicted in Fig. 7.13b. The i_{L1} of L_1 flows to the source V_1 via D_1. At the same time, i_{L2} of L_2 flows through the capacitor, and capacitor voltage is increased. After capacitor, it flows through the source via D_1. That means load to source energy is transferring.

Simulation results of this converter in mode A and B are shown in Figs. 7.14 and 7.15. These simulation results agree with above discussion.

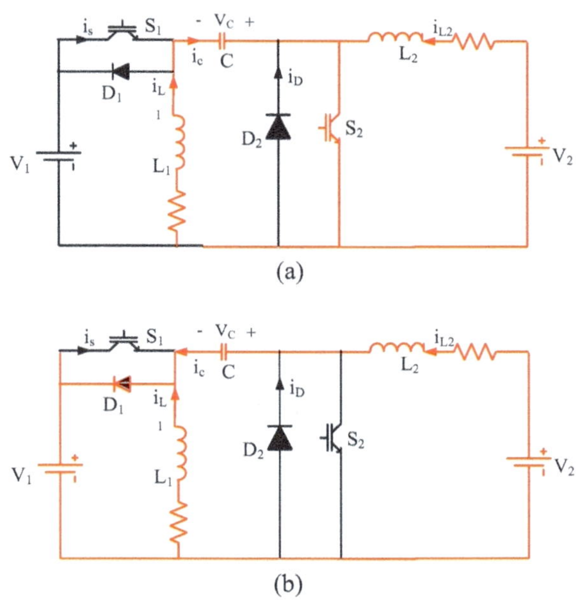

Fig. 7.13 States of mode 2 **a** S_2 is on **b** S_2 is off

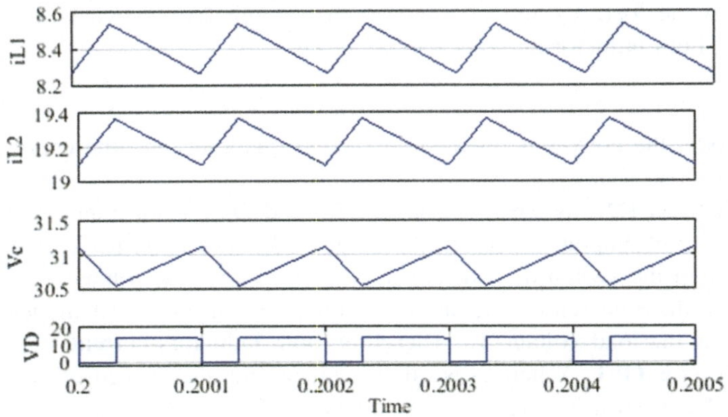

Fig. 7.14 Waveforms in mode 1 for duty ratio of 0.3

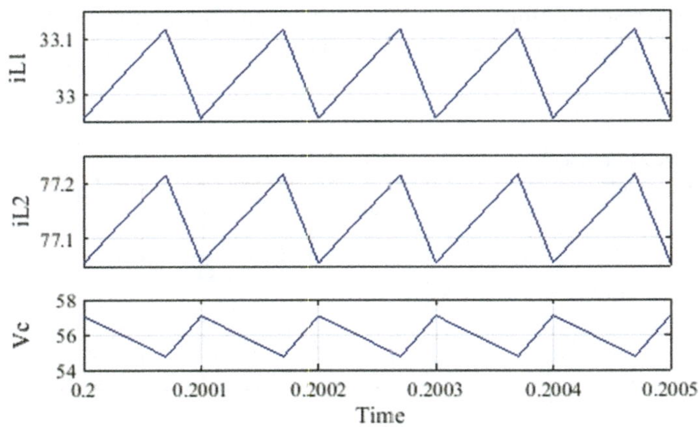

Fig. 7.15 Waveforms in mode 2 for duty ratio of 0.7

7.4 Four-Quadrant Switched Inductor LUO (FQSIL) Converter

Switched capacitor converters give greater power density but difficult in controlling and complex structure. Therefore, switched inductor DC–DC converters came into existence as it is simple circuit and has greater power density. The circuit diagram of four-quadrant switched inductor converter is depicted in Fig. 7.16.

It operates in four modes. The circuit diagram for modes 1 and 2 is depicted in Fig. 7.17a and for modes 3 and 4 is depicted in Fig. 7.17b.

Mode 1: In this mode, it operates in first quadrant that is the energy is transferred from source to positive voltage load. In this mode, all the voltages and currents are

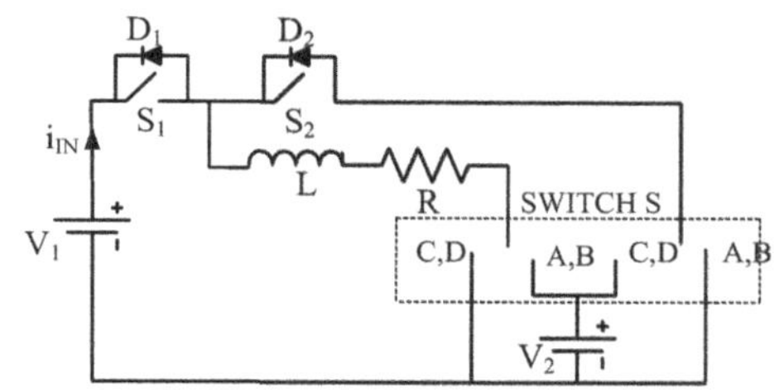

Fig. 7.16 Switched inductor LUO converter

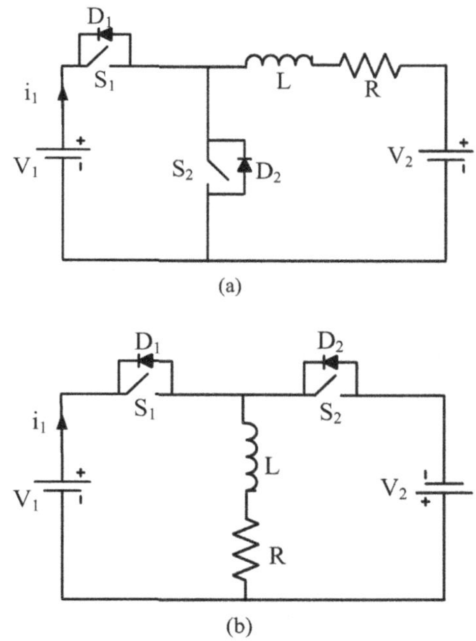

Fig. 7.17 Circuits for **a** mode 1 and mode 2 **b** mode 3 and mode 4

positive. Switch S_1 is closed during on state, and remaining all other switches are off. The source voltage V_1 supplies inductor and load, and therefore, inductor current increases. Diode D_2 is turned on during switch off state, and remaining all the switches are off. The inductor current flows to the load via D_2.

Mode 2: In this mode, it operates in second quadrant that is the energy is transferred from load with positive voltage to the source. In this mode V_1, V_2 are positive, and I_1, I_2 are negative. During switch on state, S_2 is closed, and remaining all switches are off. The load voltage V_2 supports the inductor, and inductor current enhances. Diode D_1 is turned on during switch off state, and remaining all other switches are off. The inductor current flows through V_1, V_2 through the diode D_1, and it decreases.

Mode 3: In this mode, it operates in third quadrant that is the energy is transferred from source to the load with negative voltage. In this mode, V_1 and I_1 are positive, and V_2 and I_2 are negative. Switch S_1 is closed during on state, and all other switches are off. The source V_1 supplies the inductor, and therefore, current through the inductor increases. Diode D_2 is turned on during switch off state, and remaining switches are tuned off. The inductor current flows through the load voltage V_2 and D_2, and it decreases.

Mode 4: In this mode, it operates in fourth quadrant that is the energy is transferred from load with negative voltage to source. In this mode, V_1 and I_2 are positive, and V_1 and I_2 are negative. During switch on time, S_2 is closed, and remaining switches are off. The load voltage supports the inductor, and inductor current increases. Diode D_1 is turned on during switch off state, the inductor current flows through D_1, source voltage V_1, and it decreases.

7.5 Conclusion

In this chapter, different types of LUO converters are classified and explained. The output voltages of different voltage lift circuits for the input voltage of 30 V are given in Table 7.1. The simulation studies for the above-mentioned voltage lift circuits are carried out in MATLAB/Simulink, and the corresponding results are shown. The simulation results prove that these LUO converters have high gain compared to all existing conventional converters. The LUO converters have greater power density, greater efficiency with easy to design and concurrently reduced cost and ripples in the output voltage. Some advanced LUO converters like second and fourth quadrant switched inductor LUO converters are discussed. All these LUO converters are well suited for renewable energy systems.

References

1. Khan, K.A., Hasan, M., Islam, M.A., Alim, M.A., Asma, U., Hassan, L., Ali, M.H.: A study on conventional energy sources for power production. Int. J. Adv. Res. Innov. Ideas Educ. **4** (4), 214–228 (2018)
2. Zobaa, A.F., Bansal, R.C.: Handbook of renewable energy technology. World Scientific (2011)
3. Panwar, N.L., Kaushik, S.C., Kothari, S.: Role of renewable energy sources in environmental protection: a review. Renew. Sustain. Energy Rev. **15**(3), 1513–1524 (2011)
4. Kabir, E., Kumar, P., Kumar, S., Adelodun, A.A., Kim, K.H.: Solar energy: potential and future prospects. Renew. Sustain. Energy Rev. **82**, 894–900 (2018)
5. Priyadarshi, N., Azam, F., Bhoi, A. K., Alam, S.: An artificial fuzzy logic intelligent controller based MPPT for PV grid utility. In: Proceedings of 2nd International Conference on Communication, Computing and Networking, pp. 901–909. Springer, Singapore (2019)
6. Priyadarshi, N., Sharma, A.K., Priyam, S.: Practical realization of an improved photovoltaic grid integration with MPPT. Int. J. Renew. Energy Res. (IJRER) **7**(4), 1880–1891 (2017)

7. Luo, F.L.: Luo-converters, voltage lift technique. In: PESC 98 Record. 29th Annual IEEE Power Electronics Specialists Conference (Cat. No. 98CH36196), vol. 2, pp. 1783–1789. IEEE (1998)
8. Luo, F.L.: Positive output Luo converters: voltage lift technique. IEE Proc. Electric Power Appl. **146**(4), 415–432 (1999)
9. Jiao, Y., Luo, F.L.:An improved sliding mode controller for positive output Luo converter. In: 2009 35th Annual Conference of IEEE Industrial Electronics, Porto, pp. 195–200 (2009)
10. Jiao, Y., Luo, F.L.: An improved sliding mode controller for positive output Luo converter. In: 2009 35th Annual Conference of IEEE Industrial Electronics, pp. 195–200. IEEE (2009)
11. Kayalvizhi, R., Natarajan, S.P., Pandiarajan, P.S., Vijayarajeswaran, R.: Control of paralleled negative output elementary Luo converters. In: 2005 International Conference on Power Electronics and Drives Systems, vol. 2, pp. 1234–1238. IEEE (2005)
12. Pansare, C., Sharma, S.K., Jain, C., Saxena, R.: Analysis of a modified positive output luo converter and its application to solar PV system. In: 2017 IEEE Industry Applications Society Annual Meeting, pp. 1–6. IEEE (2017)
13. Padmanaban, S., Priyadarshi, N., Holm-Nielsen, J.B., Bhaskar, M.S., Azam, F., Sharma, A. K.: A novel modified sine-cosine optimized MPPT algorithm for grid integrated PV system under real operating conditions. IEEE Access **7**, 10467–10477 (2019). https://doi.org/10.1109/ACCESS.2018.2890533
14. Padmanaban, S., Priyadarshi, N., Holm-Nielsen, J.B., Bhaskar, M.S., Hossain, E., Azam, F.: A hybrid photovoltaic-fuel cell for grid integration with Jaya-based maximum power point tracking: experimental performance evaluation. IEEE Access **7**, 82978–82990 (2019). https://doi.org/10.1109/ACCESS.2019.2924264
15. Kamalapathi, K., Priyadarshi, N., Padmanaban, S., Holm-Nielsen, J.B., Azam, F., Umayal, C., Ramachandaramurthy, V.K.: A hybrid moth-flame fuzzy logic controller based integrated cuk converter fed brushless DC motor for power factor correction. Electronics **7**, 288 (2018)
16. Berkovich, Y., Axelrod, B., Madar, R., Twina, A.: Improved Luo converter modifications with increasing voltage ratio. IET Power Electron. **8**(2), 202–212 (2014)

Chapter 8
Efficacy of Super-Lift Converter Over Fundamental Converters in Induction Motor Drive Using Solar Energy

P. Elangovan, S. Ganesh, and L. Hubert Tony Raj

8.1 Introduction

The photovoltaic (PV) power shift system has been implemented worldwide to alleviate energy crises and minimize natural contamination [1]. Nowadays for domestic uses, PV devices are commonly used [2, 3]. The key goal of PV integration in residential systems is to ensure non-discrete and highly efficient electricity for household appliances [4]. The discrete existence of solar irradiation is one of the significant shortcomings of the PV system. The SECS suffers from intermittent and stochastic sun irradiation and results in problems such as load mismatch, voltage ripple and low performance of the system. Traditionally, the SECS uses batteries to design an energy storage device. However, safe charging and discharge of energy from batteries would not be supported by a voltage ripple from the PV output terminals. The inclusion of a DC/DC converter in the center of the PV and the load for regulating the load current [5] is also necessary. With single stage power conversion (i.e.) DC–AC conversion, the conventional SECS is developed, which is unsuitable for residential applications. Complex peak power point tracking (PPPT) algorithms are also needed for the device [6, 7]. In comparison with the applications involving DC motors, residential and industrial applications built with IM motors are high. For improved efficiency, some of the applications such as rolling machines, electric vehicle (EV) and cutting machines use induction motor (IM). MPPT controllers do not need certain kinds of programs. The voltage ripple in the PV output terminals is the only thing that is unresolved. The boost converter is commonly used for SECS, among the numerous DC tied converters. The oscillation level in the voltage signal at the output terminals is a typical issue associated with boosting DC/DC converters. Apart from traditional DC/DC converters, some adapted DC/DC converters [8, 9] for SECS have been

P. Elangovan (✉) · S. Ganesh · L. Hubert Tony Raj
Department of EEE, SRM TRP Engineering College, Trichy, India

N. Priyadarshi et al. (eds.), *DC–DC Converters for Future Renewable Energy Systems*, Energy Systems in Electrical Engineering,
https://doi.org/10.1007/978-981-16-4388-0_8

added. However, their integration into SECS is constrained by the influence of parasitic elements present in the converter configuration.

Super-lift Luo converter (SLLC) [10, 11], an unorthodox DC/DC converter, finds its position in the DC/DC conversion stage of SECS by eliminating the deficiencies mentioned above. The voltage lift method is a stylish procedure, given the voltage at the yield junctures of the DC/DC converter. It is commonly used in the construction of electronic power networks. Nevertheless in number juggling action, the voltage lift converters (VLC) expands the voltage at output, stage by stage. In DC/DC transition innovation, the super-lift mechanism has achieved an amazing condition and is more widespread than the voltage lift (VL) technique. In comparison to VLC, the efficacy of super-lift converters (SLC) is superior. The SLC constructs, step by stage, the yield voltage in mathematical motion. There are several subseries of the SLLC and they are main secondary, modified, re-enhanced and multiple enhanced series. For more discussion, this research analysis chose the elementary circuit from SLLC's key sequence.

The efficacy of SLLC is checked in this manuscript over simple DC/DC converters for SECS fed IM. Identification of the optimized configuration for SECS fed IM using SLLC is the principal contribution of this research work. Next is the voltage ripple reduction to less than 1% in the DC/DC conversion point. Finally, IM's performance parameters with SLLC integrated SECS have been investigated.

The manuscript is structured in such a way that Sect. 8.2 explains the modeling of PV for the proposed method, Sect. 8.3 mathematically explains the supremacy of SLLC over simple converters, Sect. 8.4 explains the correlation of drive parameters with the SECS produced, Sect. 8.5 shows the simulation results for validating the effectiveness of SLLC over basic converters, and the conclusion of the paper has been drawn in Sect. 8.6.

8.2 Mathematical Modeling of Proposed PV

The one diode model of PV for the suggested SECS is discussed in this section. PV cells, modules, frames and arrays constitute the collection of electrical properties describing the physical characteristics of PV. The PV counterpart network of the one diode model is displayed in Fig. 8.1. It comprises of a current I_{ph} source connected to a diode D in parallel and a resistance R_{sh} in parallel. To produce a load brake, a resistance R_{se} is linked in series with the load. The load branch is connected parallel to three legs of the source end of PV.

For the proposed method, the following empirical equation leads to the PV design. The model begins with the design of the photocurrent of the module and is provided by,

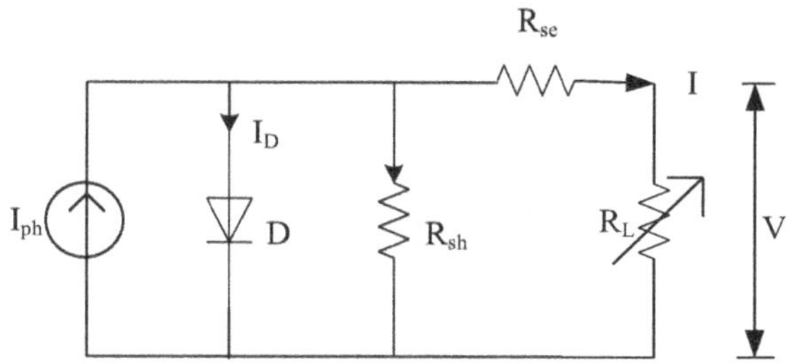

Fig. 8.1 One diode model of PV

$$I_{\text{Ph}} = [I_{\text{sc}} + K_i(T - 298)] * \frac{\lambda}{1000} \tag{8.1}$$

where K_i is temperature coefficient of current, λ is insolation, T is the module operating temperature and I_{sc} is short-circuit current of PV. The photocurrent of the module is highly dependent upon insolation and temperature.

The contrary saturation current of module (I_{rs}) is specified by,

$$I_{\text{rs}} = \frac{I_{\text{sc}}}{e^{\left(\frac{qV_{\text{oc}}}{N_s KAT}\right)} - 1} \tag{8.2}$$

where q is electron charge, I_{sc} is short-circuit current, V_{oc} is PV unit open terminal voltage, N_s is the amount of cells linked in series, K is Boltzmann value (1.3805×10^{-23} J/K) and A is the ideality factor (1.6). The reverse unit saturation current depends on the temperature and the amount of panels connected in a row.

The solar module saturation current (I_o) is specified by,

$$I_o = I_{\text{rs}} \left[\frac{T}{T_r}\right]^3 e^{\left[\frac{q*E_{\text{go}}}{BK}\left\{\frac{1}{T_r} - \frac{1}{T}\right\}\right]} \tag{8.3}$$

where T_r is the reference temperature, E_{go} is band space for silicon (1.1 eV) and A is the ideality factor (1.6).

The PV unit current (I_{PV}) is specified by,

$$I_{\text{PV}} = N_p * I_{\text{ph}} - N_p * I_o \left[e^{\left[\frac{q*(V_{\text{PV}} + I_{\text{PV}} R_{\text{se}})}{N_s AKT}\right]} - 1\right] \tag{8.4}$$

where N_p is the quantity of parallel connected panels and V_{PV} is PV yield voltage.

8.3 Superiority of SLLC Among DC/DC Converters

The initial generation DC/DC converters, where the underlying circuits of converter are primarily modified as an advanced DC/DC converter, are the focus of this research work. Figure 8.2 presents the classification of first generation DC/DC converters. The fundamentals of basic circuits of DC/DC converter like boost, buck and boost–buck converters are offered in [12]. There are several subseries converter configurations for the voltage lift converter (VLC) and the super-lift converter (SLC). From the fundamental converters, all the circuit configurations are developed. The proposed research work copes with the DC/DC converter in use in the DC connection unit for IM drives, where only positive DC connection voltage is needed. Therefore, the VLC family's negative output (N/O) Luo converter as well as the N/O cascaded boost converter together with the SLC family's N/O super-lift Luo converter are neglected.

The ascetic-lift cuk converter, ascetic-lift positive output Luo converter, inverse ascetic-lift positive output Luo converter and ascetic-lift SEPIC are among the seven self-lift positive output converters and the output voltage is same and is being expressed by,

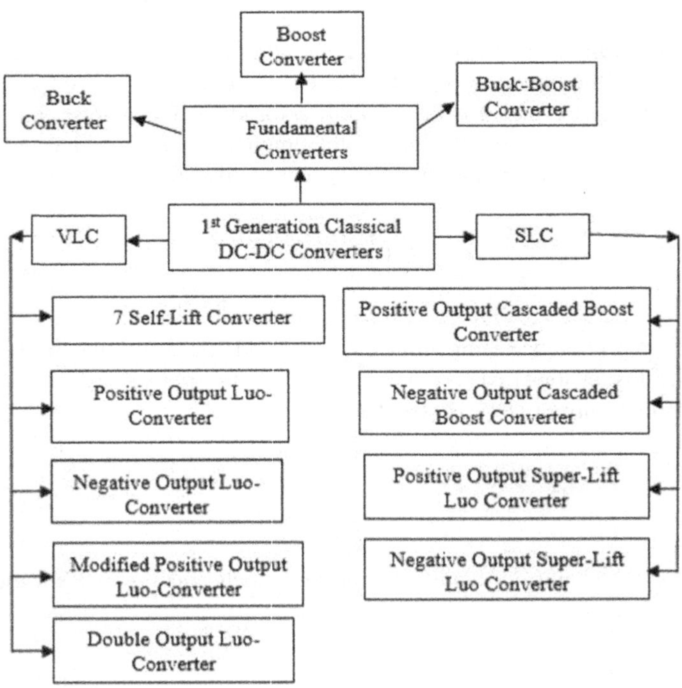

Fig. 8.2 Classification of first generation DC/DC converter

$$V_{\mathrm{odc}} = \frac{1}{1-\chi} V_{\mathrm{PV}} \tag{8.5}$$

where V_{odc} is the yield voltage from the converter. For those converters, the gain of voltage transfer can be given by,

$$T = \frac{1}{1-\chi} \tag{8.6}$$

The expressions (8.5) and (8.6) are identical to the terminal voltage and gain transfer of voltage expressions of the basic DC/DC boost converter. The lone benefit of these converters is that the yield voltage ripple quality is less than 1%. Therefore, the circuit functions of these converters are not expected to be addressed. The enhanced positive output Luo ascetic-lift converter is extracted from the P/O Luo self-lift converter, in which S and L arrangements are transacted and revealed in Fig. 8.3.

The terminal voltage of the improved ascetic-lift positive output Luo converter is specified by,

$$V_{\mathrm{odc}} = \frac{2-\chi}{1-\chi} V_{\mathrm{PV}} \tag{8.7}$$

The gain of the improved ascetic-lift positive output Luo converter is specified by,

$$T = \frac{2-\chi}{1-\chi} \tag{8.8}$$

The improved ascetic-lift positive output Luo converter achieves a terminal voltage of 1.5 times superior than other converters in the self-lift converter group as compared to Eqs. (8.5) and (8.7) for a fixed conduction duty.

Now, the operating theory of the VLC family fundamental positive output Luo converter and the SLC group fundamental positive output super-lift Luo converter

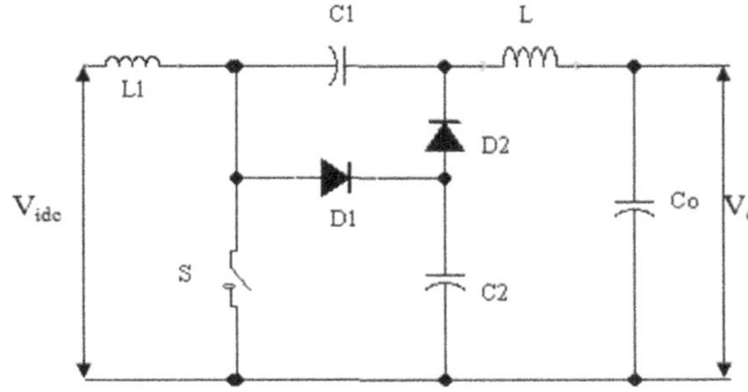

Fig. 8.3 Enhanced ascetic-lift positive output Luo converter

is explained. Figure 8.4a displays the ultimate circuit figure of the fundamental positive output Luo converter. Figure 8.4b, c shows the corresponding circuit during turn ON and OFF.

The capacitor ($C1$) behaves as a component that packs and handovers energy through the exciting inductor ($L1$) from the input terminal. The fluctuations in the voltage of the capacitor V_{C1} are ignored since the value chosen for the capacitor $C1$ is too high.

The source current is the totality of the current between inductors $L1$ and $L2$ inductors during the turn ON time. The $L1$ inductor extracts electricity directly from the input terminal, while the $L2$ inductor absorbs energy from the $C1$ capacitor from the input terminal. At the input terminals of the circuit, the current of the inductors increases with the available energy during the turn ON time. The source end current is zero during the switch OFF time and the acquired energy at switch ON period by the inductor $L1$ is utilized to excite the capacitor $C1$ through the coasting diode $D1$. In addition, the gained energy by the inductor $L2$ during the turn ON time is utilized to excite the capacitor $C0$ by the laidback diode $D1$. The steady current at the output terminals is sustained by the laidback diode $D1$.

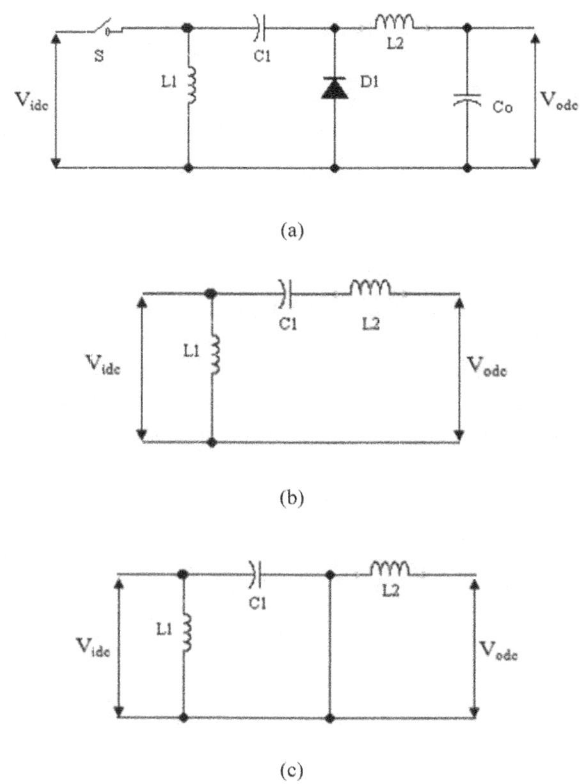

Fig. 8.4 Fundamental positive output Luo converter **a** General circuit, **b** configuration at switch ON, **c** configuration at switch OFF

The mean terminal voltage at the elementary positive Luo converter is,

$$V_{\text{odc}} = \frac{\chi}{1-\chi} V_{\text{PV}} \tag{8.9}$$

The gain of voltage transfer by the elementary positive output Luo converter is stated by,

$$T = \frac{\chi}{1-\chi} \tag{8.10}$$

Considering Eqs. (8.7) and (8.9), it is obvious that the higher output voltage level is provided by the improved ascetic-lift positive output Luo converter. But, when used for high-power applications, the improved ascetic-lift positive output Luo converter needs an extra diode and a capacitor, which induces yield voltage oscillations. Further, the elementary positive output super-lift Luo converter among the SLC group is examined for its appropriateness in IM drives.

The general circuit and the correspondent configurations in the course of ON and OFF period for the fundamental positive output super-lift Luo converter is publicized in Fig. 8.5. At ON time, the source voltage of fundamental positive output super-lift Luo converter excites C_{21} to V_{idc}. In addition, the L_{21} current lifts up with respect to V_{idc}. At OFF time, the charge in C_{21} releases and the L_{21} current decreases by means of $-(V_{\text{odc}} - 2V_{\text{idc}})$.

The mean value of yield voltage at fundamental positive output super-lift Luo converter is uttered as

$$V_{\text{odc}} = \frac{2-\chi}{1-\chi} V_{\text{PV}} \tag{8.11}$$

The transfer gain of fundamental positive output super-lift Luo converter voltage is specified by

$$T = \frac{2-\chi}{1-\chi} \tag{8.12}$$

The overhead expressions (8.11) and (8.12) are alike to the expressions (8.7) and (8.8). However, the merely cause at the back of the extended voltage lift-up in improved ascetic-lift positive output Luo converter is the existence of extra parasitic elements. Expression (8.12) clarifies that the fundamental positive output super-lift Luo converter harvests extreme level of voltage transfer with minimum parasitic elements as related to the circuits deliberated earlier. Table 8.1 displays the contrast study of boost, improved ascetic-lift positive output Luo and fundamental positive outputs per-lift Luo converter.

Fig. 8.5 Fundamental positive output super-lift Luo converter **a** General circuit, **b** configuration at switch ON, and **c** configuration at switch OFF

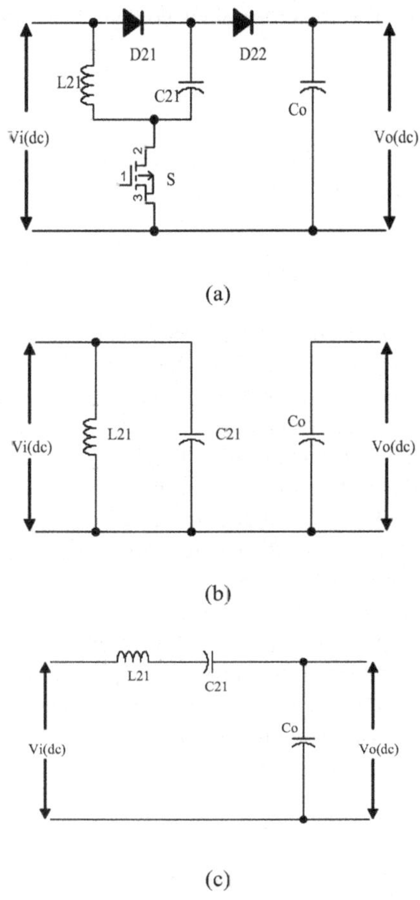

Table 8.1 Contrast study of converters

Conduction duty cycle	Gain of voltage transfer		
	Boost converter	Improved ascetic-lift positive output Luo converter	Fundamental positive output super-lift Luo converter
0.1	1.11	2.11	2.11
0.2	1.25	2.25	2.25
0.3	1.428	2.42	2.42
0.4	1.66	2.66	2.66
0.5	2	3	3
0.6	2.5	3.5	3.5
0.7	3.33	4.33	4.33
0.8	5	6	6
0.9	10	11	11

8.4 Correlation of Different Drive Parameters with SECS

In this section, via the VSI parameters, the IM parameters are correlated to the parameters of the proposed SECS. One of the fundamental principles of its action is the existence of a spinning force field (rmf) in the cavity of IM. Since the rmfs contact with the flux of cavity lets the IM to rotate. Therefore, in the stator turns of IM, which is accountable for the output of rmf, the three-phase current is given by,

$$i_a = I_p \cos(2\pi f t) \tag{8.13}$$

$$i_b = I_p \cos(2\pi f t - 120^\circ) \tag{8.14}$$

$$i_c = I_p \cos(2\pi f t + 120^\circ) \tag{8.15}$$

where i_a is phase-a current, i_b is phase-a current, i_c is phase-a current, I_p is peak current and f is the supply frequency.

The development of magneto-motive force (mmf) is responsible for each winding in the stator portion of IM. The three-phase current provided in Eqs. (8.16–8.18) is expressed at standstill (t = 0) as,

$$i_a = I_p \tag{8.16}$$

$$i_b = -\frac{I_p}{2} \tag{8.17}$$

$$i_c = -\frac{I_p}{2} \tag{8.18}$$

The prompt mmf in terms of the altitudinal angle θ can be embodied by

$$M_a(\theta) = Ni_a \cos\theta \tag{8.19}$$

$$M_b(\theta) = Ni_b \cos(\theta - 120^\circ) \tag{8.20}$$

$$M_c(\theta) = Ni_c \cos(\theta + 120^\circ) \tag{8.21}$$

The resultant mmf can be given by

$$M(\theta) = M_a(\theta) + M_b(\theta) + M_c(\theta) \tag{8.22}$$

Now, substituting the Eqs. (8.19), (8.20) and (8.21) in the Eq. (8.22), we get

$$M(\theta) = Ni_a \cos\theta + Ni_b \cos(\theta - 120^\circ) + Ni_c \cos(\theta + 120^\circ) \tag{8.23}$$

Substituting Eqs. (8.16), (8.17) and (8.18) in (8.23), we get

$$M(\theta) = NI_P[\cos 2\pi ft \cos\theta + \cos(2\pi ft - 120°)\cos(\theta - 120°) + \cos(2\pi ft + 120°)\cos(\theta + 120°)] \tag{8.24}$$

On simplifying Eq. (8.24), we get

$$M(\theta) = \frac{3}{2} NI_P \cos(2\pi ft - \theta) \tag{8.25}$$

The above expression (8.25) designates that only during the altitudinal angle is equal to *2πft*, peak mmf will be produced in the cavity of IM.

The development of IM electromagnetic torque depends on the size of the mmf air distance. The underlying cause of electromagnetic torque generation is the slip velocity, which seems to be the gap between the IM's synchronous velocity (Ne) and real velocity (*N*). The torque expression is specified by

$$T = \pi\left(\frac{P}{2}\right) lrB_P M_P \sin\varphi \tag{8.26}$$

The expression above (8.26) is the general electromagnetic torque shape. It is calculated only when the maximal level of mmf and air-gap flux are identified. But this research study concentrates about the voltage in the DC contact of the IM drive. Therefore, in terms of stator terminal voltage [13], the expression for electromagnetic torque must be considered and it is expressed by

$$T = 3\left(\frac{P}{2}\right)\left(\frac{R_r}{(S)(2\pi f)}\right)\left[\frac{V_{\text{ph}}^2}{\left(\frac{R_s + R_r}{S}\right)^2 + (2\pi f)^2 (L_s + L_r)^2}\right] \tag{8.27}$$

It is very obvious from Expression (8.27) that the inertial torque of the IM relies on the phase voltage and the frequency of input at the IM primary side. The other measurements, including the IM primary and secondary windings resistance, the inductance of the IM stator and rotor, and the quantity of IM poles, are constant. The supply's frequency and voltage are controlled by the VSI which supplies the IM.

The phase voltage equation at the primary side of IM by implying that the VSI cascading IM operating in 180° mode is given by

$$V_{\text{ph}} = \frac{V_L}{\sqrt{3}} \tag{8.28}$$

The line–line voltage equation at the primary side [14] of IM is specified by

$$V_L = \sqrt{\frac{1}{\pi} \int_0^{120^\circ} V^2{}_{\text{odc}} (2\pi ft)} = \sqrt{\frac{2}{3}} V_{\text{odc}} = 0.8165 V_{\text{odc}} \quad (8.29)$$

Equating (8.29) and (8.28), the phase voltage becomes

$$V_{\text{ph}} = 0.4714 V_{\text{odc}} \quad (8.30)$$

By replacing the Eq. (8.30) in (8.27), the expression of inertial torque in forms of DC bus voltage is engraved as

$$T = 3\left(\frac{P}{2}\right)\left(\frac{R_r}{(S)(2\pi f)}\right)\left[\frac{0.2222 V_{\text{odc}}^2}{\left(\frac{R_s + R_r}{S}\right)^2 + (2\pi f)^2 (L_s + L_r)^2}\right] \quad (8.31)$$

The relation among the measurements of VSI and IM and the recommended SECS is established by incorporating the expression of the DC-link voltage for (8.11) in the approximations (8.29), (8.30) and (8.11) (8.31). After this, the line–line and phase voltage and the IM-developed inertial torque at the IM primary side are rewritten as,

$$V_L = 0.8165 \left[\frac{2 - \delta}{1 - \delta}\right] V_{\text{PV}} \quad (8.32)$$

$$V_{\text{ph}} = 0.4714 \left[\frac{2 - \delta}{1 - \delta}\right] V_{\text{PV}} \quad (8.33)$$

$$T = 3\left(\frac{P}{2}\right)\left(\frac{R_r}{(S)(2\pi f)}\right)\left[\frac{0.2222 \left\{\frac{2-\delta}{1-\delta}\right\} V_{\text{PV}}^2}{\left(\frac{R_s + R_r}{S}\right)^2 + (2\pi f)^2 (L_s + L_r)^2}\right] \quad (8.34)$$

From the above connection, it is concluded that the line–line and phase voltage are contingent on the conduction duty and supply DC voltage of SLLC at the primary side of IM. Whereas, the inertial torque of the IM relies on the same parameters with the source frequency and slip.

8.5 Results and Discussion

Theoretically, the previous parts validated that SLC is well matched to IM drives. Using MATLAB tools, the Simulink design for boost converter incorporated IM drive, VLC incorporated IM drive and SLC incorporated IM drive is modeled to validate it effectively. In all three versions, the DC/DC converter conduction duty is set and the range is 0.5. In addition, in all three situations, the VSI works in 180°

Table 8.2 Simulation circuit specifications

Parameter	Boost converter-based IM drive	VLC-based IM drive	SLLC-based IM drive
PV output voltage	36 V		
Irradiance	1000 W/m^2		
PV module	36 cells		
Induction machine	1/2 HP, three phase, 50 Hz		
Electromagnetic inductors	$L = 3$ (mH)	$L_1 = 3$ (mH), $L_2 = 2.56$ (mH)	$L_{11} = 3$ (mH), $L_{21} = 2.56$ (mH)
Electrostatic capacitors	$C = 2200$ (μF)	$C_1 = 2000$ (μF), $C_O = 2200$ (μF)	$C_{21} = 2000$ (μF), $C_O = 2200$ (μF)

mode with fixed conduction duty. For both of those arrangements, the simulation circuit specifications are itemized in Table 8.2.

In this segment, the optimum temperature and irradiance of the built PV model for highly efficient operation are initially established. Next the temperature is set as 25 °C and 1000 W/m^2 is assigned to the irradiance. In Fig. 8.6, the plots for the V-I and V-P features of the proposed PV model are presented.

When the above properties are observed, it is clear that the highest power is obtained at a PV output voltage of 30 V. The output voltage at the junctures of the proposed PV is calculated and published in Fig. 8.7.

The PV output voltage result shows that the presence of passive elements in the PV architecture circuit lets 36 V saturate the output voltage.

Along with the PV output voltage, the voltage at DC connection is observed for all the versions and is shown in Fig. 8.8.

The mean DC bus voltage gotten for the boost converter cascaded IM drive, which is 72 V, is seen in Fig. 8.8a, which is merely double that of the DC source voltage. The voltage of the DC/DC connection fluctuates between 36 and 72 V, which shows a 200% voltage ripple. Figure 8.8b indicates the DC connection voltage for the VLC-based IM drive. It means that the voltage is not increased, but during the steady state, the ripple content in the DC connection voltage is less than 1%. The average DC connection voltage achieved for SLLC-based IM drives is shown in Fig. 8.8c. The SLLC-based SECS fed IM drive has been tested to maintain the ripple-free DC bus voltage when contrasting this finding with the findings displayed in Fig. 8.8a, b. Mostly, the DC connection voltage is elevated three times the source voltage.

The IM's velocity response is shown in Fig. 8.9 for all three simulated versions.

Because of the depleted stator voltage in the boost converter and VLC-based IM motor, the drive is unable to reach the rated velocity and is seen in Fig. 8.9a, b. Seen in Fig. 8.9c, the IM velocity response in the SLLC-based IM drive shows that the velocity response is very fast and that the velocity curve setting time is also limited to 0.18 s.

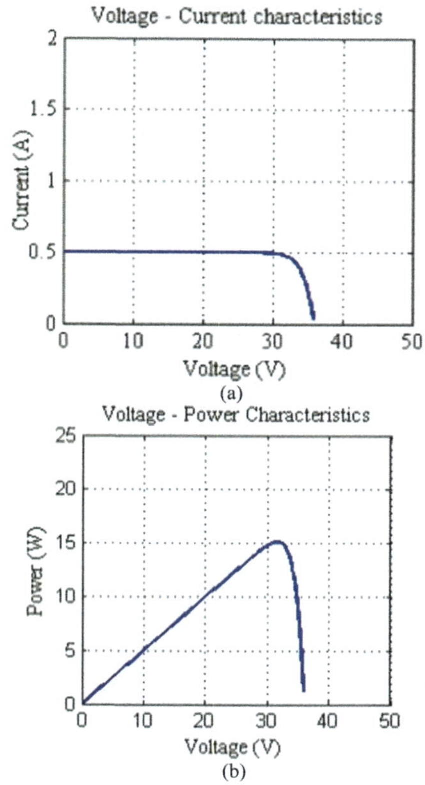

Fig. 8.6 PV module characteristics **a** V-I, **b** V-P

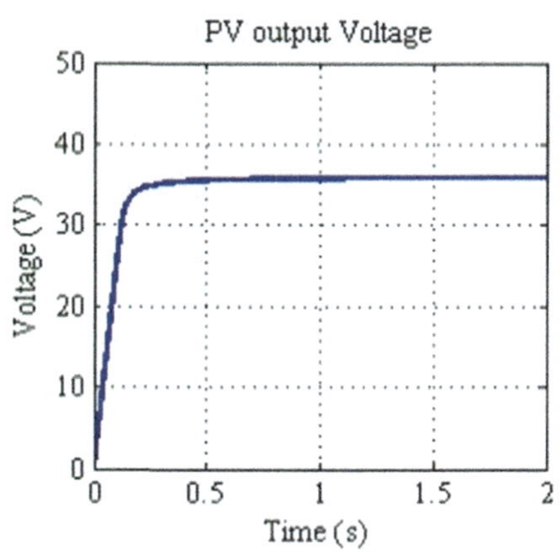

Fig. 8.7 PV output voltage

Fig. 8.8 DC connection voltage **a** cascading boost converter, **b** cascading VLC, **c** cascading SLLC

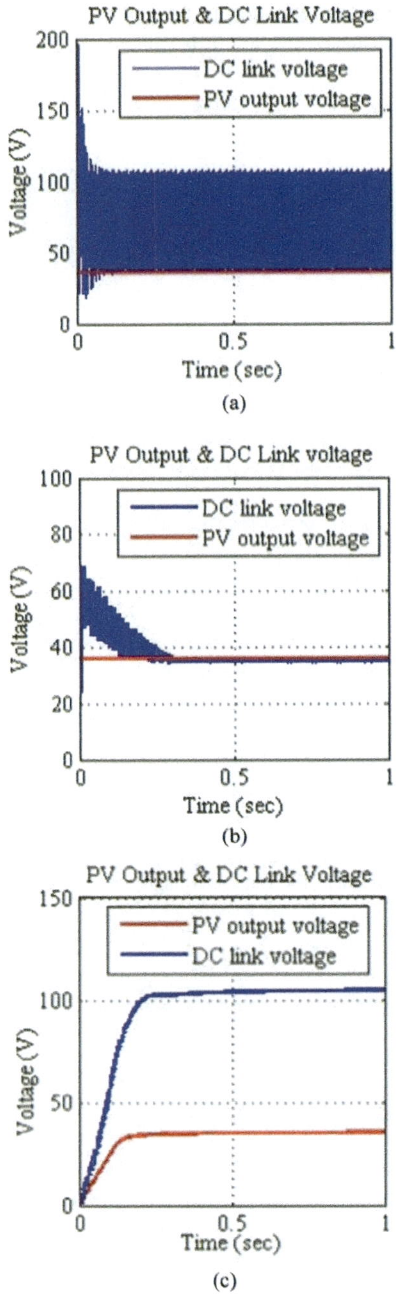

Fig. 8.9 IM's velocity response **a** cascading boost converter, **b** cascading VLC, **c** cascading SLLC

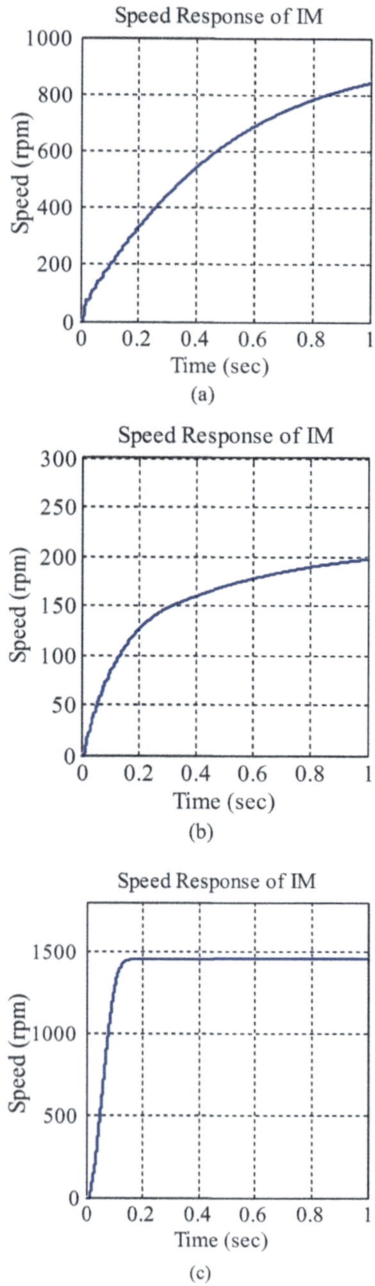

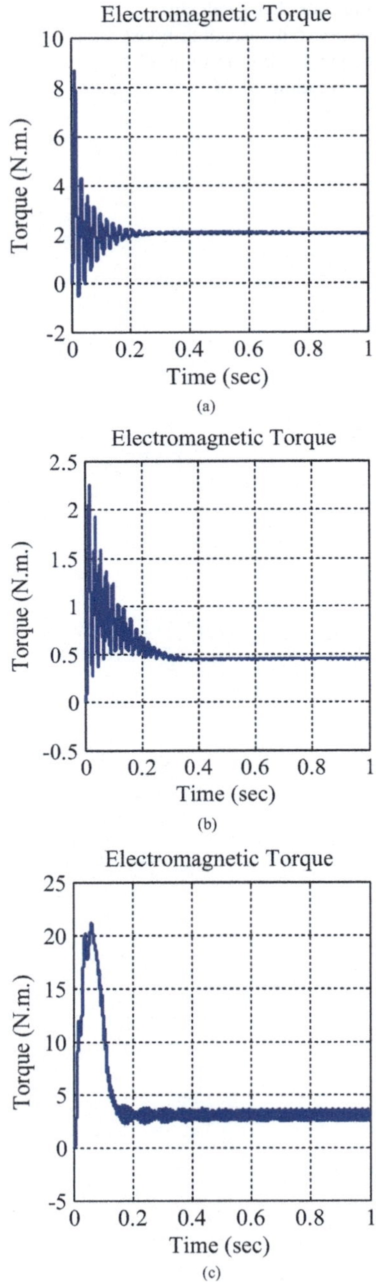

Fig. 8.10 Inertial torque of IM

Figure 8.10 displays the inertial torque produced by the IM in all the tested versions. In all three cases, the torque ripple is not removed due to the lack of closed-loop power at the end of the drive.

The inertial torque generated by the cascaded IM boost converter and the cascaded IM drive VLC, displayed Fig. 8.10a, b, stipulates that at the start of the IM, the torque oscillations are strong and allow the attached load to oscillate. On the other hand, shown in Fig. 8.10c, the inertial torque produced by IM in the SLLC-based IM drive reveals that the torque trickle is limited to a sensitivity range of 0.2 Nm during the process.

8.6 Conclusion

In this manuscript, the usefulness of SLLC over other outdated DC/DC converters such as boost converter and VLC in solar energy conversion systems (SECS) is investigated. The mathematical analysis of a single PV diode model was initially developed. Then, based on voltage transfer capacity, the VLC and SLLC were evaluated. The conceptual analysis indicated that, the SLLC was capable of producing ripple-free DC voltage in geometric evolution with a limited quantity of power storage components. The IM expression was obtained for line–line voltage, phase voltage and inertial torque and these phrases were compared via the DC/DC converter voltage expression with the PV voltage. The comparative simulation study between the SLLC, VLC and boost converter fed IM drives was agreed to verify the suitability of SLLC in IM drives. The simulation study showed that in the IM drive DC connection, the SLLC is capable of converting the PV input voltage to a threefold elevated voltage. Even the DC connection voltage produced has less than 1% trickle rate. IM's speed response adds more benefit to SLLC integrated SECS over SECS integrated basic DC/DC converter. By providing closed-loop control for the DC-link portion and motor end as well the proposed SLLC integrated SECS can be further developed.

References

1. Van Der Meer, D., Mouli, G.R.C., Mouli, G.M.E., Elizondo, L.R., Bauer, P.: Energy management system with PV power forecast to optimally charge EVs at the workplace. IEEE Trans. Ind. Inform. **14**(1), 311–320 (2018). https://doi.org/10.1109/TII.2016.2634624
2. Khan, M.T.A., Norris, G., Chattopadhyay, R., Husain, I., Bhattacharya, S.: Autoinspection and permitting With a PV utility interface (PUI) for residential plug-and-play solar photovoltaic unit. IEEE Trans. Ind. Appl. **53**(2), 1337–1346 (2017). https://doi.org/10.1109/TIA.2016.2631135
3. Farswan, R.S., Datta, A.: Low-leakage current two-stage transformerless grid-connected converter for residential PV application. IET Power Electron. **12**(5), 1043–1052 (2019), https://doi.org/10.1049/iet-pel.2018.5363

4. Awad, H., Fouad, M., Mourad, D., Rady, W.: Evaluation of various layouts of PV-based houses. 2018 20th Int. Middle East Power Syst. Conf. MEPCON 2018 - Proc., pp. 162–167 (2019). https://doi.org/10.1109/MEPCON.2018.8635255
5. Qian, T.: A converter combination scheme for efficiency improvement of PV systems. IEEE Trans. Circuits Syst. II Express Briefs. **65**(11), 1668–1672 (2018). https://doi.org/10.1109/TCSII.2017.2764027
6. Casadei, D., Grandi, G., Rossi, C.: Single-phase single-stage photovoltaic generation system based on a ripple correlation control maximum power point tracking. IEEE Trans. Energy Convers. **21**(2), 562–568 (2006). https://doi.org/10.1109/TEC.2005.853784
7. Shimizu, T., Hashimoto, O., Kimura, G.: A novel high-performance utility-interactive photovoltaic inverter system. IEEE Trans. Power Electron. **18**(2), 704–711 (2003). https://doi.org/10.1109/TPEL.2003.809375
8. Li, W., He, X.: Review of nonisolated high-step-up DC/DC converters in photovoltaic grid-connected applications. IEEE Trans. Ind. Electron. **58**(4), 1239–1250 (2011). https://doi.org/10.1109/TIE.2010.2049715
9. Das, M., Agarwal, V.: Novel high-performance stand-alone solar PV system with high-gain high-efficiency DC/DC converter power stages. IEEE Trans. Ind. Appl. **51**(6), 4718–4728 (2015). https://doi.org/10.1109/TIA.2015.2454488
10. Kushwaha, R., Singh, B.: A modified luo converter-based electric vehicle battery charger with power quality improvement. IEEE Trans. Transp. Electrification. **5**(4), 1087–1096 (2019).
11. Chandrasekar, B., et al.: Non-isolated high-gain triple port DC/DC buck-boost converter with positive output voltage for photovoltaic applications, IEEE Access. **8**, 113649–113666 (2020). https://doi.org/10.1109/ACCESS.2020.3003192
12. Zhang, B., Qiu, D.: Sneak circuits of power electronic converters. Sneak Circuits Power Electron. Convert. vol. 9781118379, 1–283 (2015). https://doi.org/10.1002/9781118379950
13. Bose, B.K.: AC Machines for Drives. In: Bernard GOODWIN, Modern Power Electronics and AC Drives. New Jersey: Prentice Hall, pp. 29–97 (2002)
14. Bimbhra, P.S.: Inverters. In: Romesh Chander KHANNA, (ed). Power Electronics. New Delhi: Khanna, pp. 414–503 (2008)

Chapter 9
Quadratic Boost Converter for Green Energy Applications

Sankar Peddapati and SVK Naresh

9.1 Introduction

High gain converters are widely adapted in applications like portable electronics, aeronautics, AC–DC integration, renewable energy systems, electric vehicles, and telecommunications. The conventional PWM boost, buck–boost converters have practical limitations on voltage gain due to the presence of parasitic nature in circuit elements of the converter and switching times of the switch. In the literature, the solutions to achieve high gain conversion are cascaded converters, magnetically coupled converters, switched-capacitor and switched-inductor structures. Based on the nature of the input current, these converters are categorized into continuous and discontinuous; to minimize electrode losses in fuel cells and maximize energy extraction in PV systems, continuous source current is desirable.

This chapter presents a detailed analysis of continuous input non-magnetic coupled cascaded DC–DC converters whose gain is quadratic in nature.

The high gain quadratic converters are classified into

(a) Quadratic boost converters [1–10]
(b) Quadratic buck–boost converters [11–17]
(c) Hybrid quadratic boost converters [18–24].

9.2 Quadratic Boost Converters

The quadratic boost converters having gain $M = \frac{1}{(1-D)^2}$ are categorized into two-switch and single-switch topologies. The two-switch topologies given in

S. Peddapati (✉) · SVK Naresh
National Institute of Technology, Andhra Pradesh, Tadepalligudem 534101, India

N. Priyadarshi et al. (eds.), *DC–DC Converters for Future Renewable Energy Systems*, Energy Systems in Electrical Engineering,
https://doi.org/10.1007/978-981-16-4388-0_9

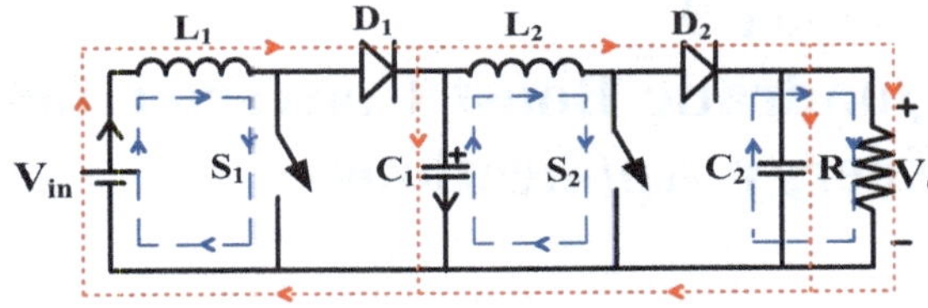

Fig. 9.1 Cascaded boost converter (Conv-1) [1]

Figs. 9.1 and 9.2 are of fourth-order circuits containing two inductors (L_1 and L_2) and two capacitors (C_1 and C_2).

9.2.1 Operation and Derivation of Steady-State Equations

In this section, the operation of the quadratic boost converters is demonstrated assuming the capacitor voltage to be constant and switching devices as ideal elements. All the converters shown in Figs. 9.1 and 9.2 have two active switches (S_1 and S_2) operated synchronously and conducts for DT time, where D is duty ratio and T is switching period. The diodes (D_1 and D_2) conduct during turn-off times $(1 - D)\ T$ of the two switches.

Cascaded boost converter [1] given in Fig. 9.1 is taken into consideration for detailing the procedure to synthesize the circuit operation and steady-state performance. In Fig. 9.1, the ON and OFF state modes of the converters are differentiated by representing the current paths in blue and red colors, respectively.

Mode 1: During the interval DT, both switches S_1 and S_2 will be turned ON, and diodes D_1 and D_2will be reverse biased, the current paths are indicated with blue color as shown in Fig. 9.1. In this mode, both inductors (L_1, L_2) charge simultaneously, and capacitors (C_1, C_2) discharge to the cascaded circuit and load, respectively. Thus, the inductor voltage and capacitor current equations are

$$v_{L1}(t) = V_{\text{in}};\ v_{L2}(t) = V_{C1};\ i_{C1}(t) = -I_{L2};\ i_{C2}(t) = -I_{O} \tag{9.1}$$

As the inductor voltage is positive, the current increases linearly as shown in Fig. 9.3.

Mode 2: In the interval $(1 - DT)$, the active switches S_1 and S_2 are closed, the inductors force diodes D_1 and D_2 to conduct. In this mode, the current paths indicated in red color shows discharge and charge paths of inductors and capacitors, respectively. The voltage and current equations are as follows:

$$v_{L1}(t) = V_{\text{in}} - V_{C1};\ v_{L2}(t) = V_{C1} - V_{O};\ i_{C1}(t) = I_{L1} - I_{L2};\ i_{C2}(t) = I_{L2} - I_{O} \tag{9.2}$$

In this mode, both inductor voltages become negative and the current decreases linearly as shown in Fig. 9.3.

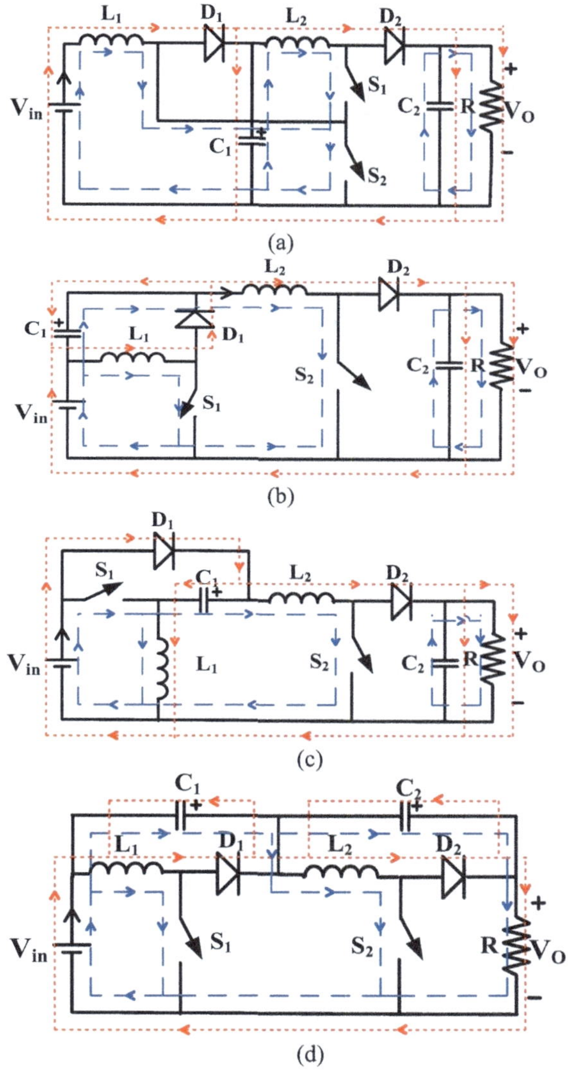

Fig. 9.2 Two-switch quadratic boost converters having gain $M = \frac{1}{(1-D)^2}$ **a** Conv-2 [2], **b** Conv-3 [3], **c** Conv-4 [4] and **d** Conv-5 [5]

The average voltage across inductor and current flowing through the capacitor over a switching period T is obtained using the following equations.

$$V_L = \frac{1}{T}\int_0^T v_L(t)\,\mathrm{d}t \tag{9.3}$$

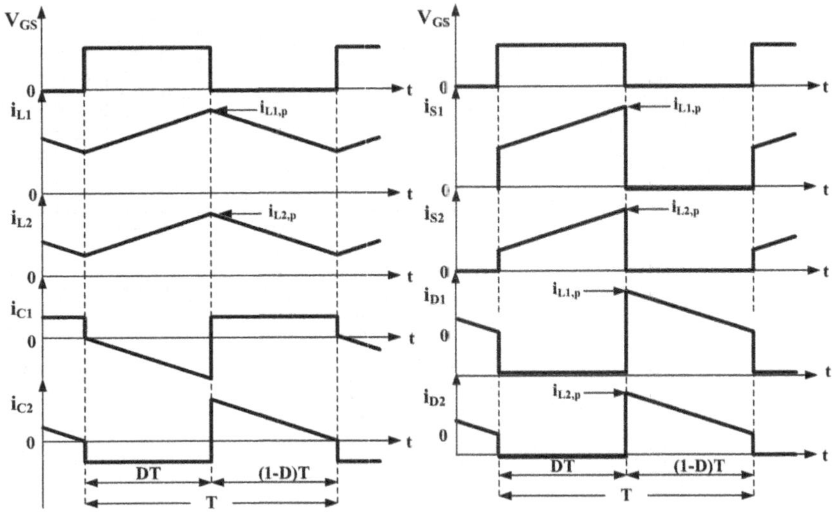

Fig. 9.3 Steady-state current waveforms of the cascaded boost converter

$$I_C = \frac{1}{T}\int_0^T i_C(\mathrm{t})\mathrm{d}t \tag{9.4}$$

The dynamic equations obtained using Eqs. (9.3) and (9.4) applied for the converter operated under different modes are

$$V_{L1} = L_1\frac{\mathrm{d}i_{L1}}{\mathrm{d}t} = V_{\mathrm{in}} - V_{C1}(1-D) \tag{9.5}$$

$$V_{L2} = L_2\frac{\mathrm{d}i_{L2}}{\mathrm{d}t} = V_{C1} - V_O(1-D) \tag{9.6}$$

$$I_{c1} = C_1\frac{\mathrm{d}v_{C1}}{\mathrm{d}t} = -I_{L2} + I_{L1}(1-D) \tag{9.7}$$

$$I_{c2} = C_2\frac{\mathrm{d}v_{C2}}{\mathrm{d}t} = I_{L2}(1-D) - \frac{V_O}{R} \tag{9.8}$$

where I_{L1} and I_{L2} are the inductor currents, D is the duty ratio of both the switches, V_{in} is the input voltage, V_{C1} and V_{C2} are the capacitor voltages, R is the load resistance, V_O and I_O are output voltage and output current, respectively. The lower-case letters represent instantaneous parameters.

Similarly, the voltage gain (M) and current gain shown in Eq. (9.11) can be computed using the principle of volt-second balance of an inductor and ampere-second balance of the capacitor over a switching cycle (T) given in Eqs. (9.9) and (9.10), respectively.

$$V_L = \frac{1}{T}\int_0^T v_L(t)\mathrm{d}t = 0 \tag{9.9}$$

$$I_C = \frac{1}{T}\int_0^T i_C(t)\mathrm{d}t = 0 \tag{9.10}$$

$$M = \frac{V_O}{V_{\text{in}}} = \frac{1}{(1-D)^2}; \frac{I_{\text{in}}}{I_O} = \frac{1}{(1-D)^2} \tag{9.11}$$

From the voltage gains and steady current waveforms shown in Fig. 9.3, the relation between capacitor voltage and input voltage and inductor current and output current is given as

$$\frac{V_{C1}}{V_{\text{in}}} = \frac{1}{(1-D)}; \frac{V_{C2}}{V_{\text{in}}} = \frac{1}{(1-D)^2}; \frac{I_{L1}}{I_O} = \frac{1}{(1-D)^2}; \frac{I_{L2}}{I_O} = \frac{1}{(1-D)} \tag{9.12}$$

9.2.2 Design of Passive Elements

The inductors (L_1 and L_2) and capacitors (C_1 and C_2) as energy storage elements are appropriately switched to get the increased voltage gain. The inductor's current ripple factor ($\Delta i_L/i_L$) and capacitor (or load) voltage ripple ($\Delta V_C/V_C$) are important design specifications for the choice of these elements.

From Eqs. (9.1) and (9.12), the ripple current in inductor (Δi_{L1}) and the steady-state inductor current (I_{L1}) is given below

$$\Delta i_{L1} = \frac{V_{L1}}{L}DT = \frac{V_{\text{in}}}{L_1}DT = \frac{V_o(1-D)^2}{L}DT \tag{9.13}$$

$$I_{L1} = \frac{I_O}{(1-D)^2} = \frac{V_O}{R(1-D)^2} \tag{9.14}$$

$$\frac{\Delta i_{L1}}{i_{L1}} = \frac{RD(1-D)^4}{fL_1} \tag{9.15}$$

where f is the switching frequency.

For the given current ripple, from Eq. (9.15) the minimum inductor required is

$$L_{1\,\text{min}} = \frac{RD(1-D)^4}{f\frac{\Delta i_{L1}}{i_{L1}}} \tag{9.16}$$

Similarly, the minimum inductor required $L_{2\min}$ is given by

$$L_{2\min} = \frac{RD(1-D)^2}{f\frac{\Delta i_{L2}}{i_{L2}}} \tag{9.17}$$

The relation between the capacitor C_1 and the ripple factor $\left(\frac{\Delta V_{C1}}{V_{C1}}\right)$ is given below:

$$\Delta V_{C1} = \frac{I_{L2}}{C_1}DT = \frac{I_o}{(1-D)C_1}DT = \frac{V_o}{R(1-D)C_1}DT \tag{9.18}$$

$$V_{C1} = \frac{V_{\text{in}}}{(1-D)} = V_O(1-D) \tag{9.19}$$

$$\frac{\Delta V_{C1}}{V_{C1}} = \frac{D}{(1-D)^2 RfC_1} \tag{9.20}$$

The capacitor C_1 is selected using the following equations based on the given ripple factor $\left(\frac{\Delta V_{C1}}{V_{C1}}\right)$

$$C_{1\min} = \frac{D}{\left(\frac{\Delta V_{C1}}{V_{C1}}\right)(1-D)^2 Rf} \tag{9.21}$$

The procedure of finding the relation between C_2 and the ripple factor $\left(\frac{\Delta V_{C2}}{V_{C2}}\right)$ is the same as the above; the capacitor C_2 can be selected using the following equation

$$C_{2\min} = \frac{D}{Rf\left(\frac{\Delta V_{C2}}{V_{C2}}\right)} \tag{9.22}$$

Equations (9.11)–(9.22) are used for the selection of power semiconductor switches, inductors and capacitors of the converter. It is important to analyze the loss component for the selected components to develop a good heat management system.

9.2.3 Power Loss Calculations

For power loss analysis, circuit non-idealities of various converter components from the corresponding datasheets are to be used. Here, the loss components are computed by assuming the current flowing through inductor and voltage across capacitor to be ripple-free.

The power losses of a converter include the losses due to switches, diodes, inductors and capacitors as given below

$$P_{L,\text{total}} = P_{LS1} + P_{LS2} + P_{LD1} + P_{LD2} + P_{LL1} + P_{LL2} + P_{LC1} + P_{LC2} \quad (9.23)$$

In switch, the conduction losses are due to the on-state resistance ($R_{ds(\text{on})}$), and the switching losses are due to turn-on $\left(t_{d(\text{on})}\right)$ and turn-off time $\left(t_{d(\text{off})}\right)$ delays. The switch losses are obtained as

$$P_{LSk} = R_{ds(\text{on})} * I^2_{sk(\text{RMS})} + \frac{1}{2T}\left[I_{sk(\text{peak})} * V_{dsk} * \left(t_{d(\text{on})} + t_{d(\text{off})}\right)\right] \quad (9.24)$$

where k represents kth element parameters (Here, $k = 1$ or $k = 2$).

In a diode, the loss component can be calculated for the given forward diode resistance ($R_{d(\text{on})}$) and forward voltage drop (V_F).

$$P_{LDk} = R_{d(\text{on})} * I^2_{Dk(\text{RMS})} + V_F * I_{Dk(\text{AVG})} \quad (9.25)$$

The inductor and capacitor loss component can be calculated as

$$P_{LLk} = R_L * I^2_{Lk(\text{RMS})} \quad (9.26)$$

$$P_{LCk} = R_{\text{esr}} * I^2_{Ck(\text{RMS})} \quad (9.27)$$

where R_L and R_{esr} are inductor resistance and ESR of capacitance; I_S, I_D, I_L and I_C are switch, diode, inductor and capacitor currents, respectively.

From Fig. 9.1 and steady-state equations given in (9.11) and (9.12), currents and voltages used to compute the converter power loss are given below:

$$I_{S1} = DI_{L1}$$

$$I_{L1,\text{peak}} = I_{S1,\text{peak}} = \frac{V_O}{R(1-D)^2}; I_{L1,\text{RMS}} = \frac{V_O}{R(1-D)^2} \text{ and } I_{S1,\text{RMS}} = \frac{V_O}{R(1-D)^2}\sqrt{D} \quad (9.28)$$

$$I_{S2} = DI_{L2}$$

$$I_{L2,\text{peak}} = I_{S2,\text{peak}} = \frac{V_O}{R(1-D)}; I_{L2,\text{RMS}} = \frac{V_O}{R(1-D)} \text{ and } I_{S2,\text{RMS}} = \frac{V_O}{R(1-D)}\sqrt{D} \quad (9.29)$$

$$V_{ds1} = V_{C1} = V_O(1-D) \quad (9.30)$$

$$V_{ds2} = V_{C2} = V_O(1-D)^2 \tag{9.31}$$

$$I_{D1} = I_{L1}(1-D); I_{D1,\text{AVG}} = \frac{V_O}{R(1-D)} \text{ and } I_{D1,\text{RMS}} = \frac{V_O}{R(1-D)^2}\sqrt{1-D} \tag{9.32}$$

$$I_{D2} = I_{L2}(1-D); I_{D2,\text{AVG}} = \frac{V_O}{R} \text{ and } I_{D2,\text{RMS}} = \frac{V_O}{R(1-D)}\sqrt{1-D} \tag{9.33}$$

$$I^2_{C1,\text{RMS}} = I^2_{L2}D + (I_{L1} - I_{L2})^2(1-D) = \frac{V_O D}{R^2(1-D)^3} \tag{9.34}$$

$$I^2_{C2,\text{RMS}} = I^2_O(D) + (I_{L2} - I_O)^2(1-D) = \frac{V_O D}{R^2(1-D)} \tag{9.35}$$

The power loss computation helps in determining the converter's efficiency using the following equation

$$\text{Efficiency}(\eta) = \frac{P_O}{P_O + P_{L,\text{total}}} \times 100\% \tag{9.36}$$

9.2.4 *Impact of Inductor Resistance on Voltage Gain*

The voltage gain in the previous section is computed by neglecting the non-idealities of the converter. However, the impact of inductor resistance on voltage gain is significant among various non-idealities of the converter. By considering the inductors resistance drop, the ON state and OFF state Eqs. (9.1) and (9.2) are modified as below.

Mode 1:

During DT time,

$$v_{L1}(t) = V_{\text{in}} - I_{L1}R_{L1} \tag{9.37}$$

$$v_{L2}(t) = V_{C1} - I_{L2}R_{L2} \tag{9.38}$$

Mode 2:

In the interval $(1 - D)T$,

$$v_{L1} = V_{\text{in}} - V_{C1} - I_{L1}R_{L1} \tag{9.39}$$

$$v_{L2} = V_{C1} - V_O - I_{L2}R_{L2} \tag{9.40}$$

Applying the volt-second balance principle to the inductor voltage, the following expressions are derived.

$$v_{L1} = V_{in} - V_{C1}(1 - D) - I_{L1}R_{L1} \tag{9.41}$$

$$v_{L2} = V_{C1} - V_O(1 - D) - I_{L2}R_{L2} \tag{9.42}$$

Solving Eqs. (9.41) and (9.42),

$$V_{C1}(1 - D) = V_{in} - I_{L1}R_{L1} \tag{9.43}$$

$$V_O(1 - D) = V_{C1} - I_{L2}R_{L2} \tag{9.44}$$

The expression for capacitor voltage from Eqs. (9.43)–(9.44) using steady-state inductor current given in Eq. (9.12) is given below:

$$V_{C1} = \frac{V_{in}}{1 - D} - \frac{V_O}{(1 - D)^2}\frac{R_{L1}}{R} \tag{9.45}$$

Gain derived in terms of per-unit inductors resistance $\left(\frac{R_{L1}}{R}, \frac{R_{L2}}{R}\right)$ from Eqs. (9.37)–(9.45) is given below

$$\frac{V_O}{V_{in}} = \frac{1}{(1 - D)^2 + \frac{1}{(1-D)^2}\frac{R_{L1}}{R} + \frac{R_{L2}}{R}} \tag{9.46}$$

The modified voltage gain is calculated for different duty ratios by varying per-unit inductors resistance of the converter and plotted in Fig. 9.4. It is clear from the figure that, the maximum gain of the converter decreases with increase in inductors resistance. Further, with increase in duty ratio beyond 0.8, the gain reduces.

9.2.5 *Conversion to Single-Switch Topology Using Graft Scheme*

The converters given in Fig. 9.1 have two active switches; it is important to verify whether the Graft scheme can be applied for the converters to minimize the number of active switches. In the Graft scheme, the two active synchronously operating switches sharing a common node can be replaced with two diodes and one switch configuration as shown in Fig. 9.5 [25]. The type of common node between the two active switches may be D–D (drain–drain), S–S (source–source), D–S (drain–source) and S–D (source–drain). The advantage of two switches to single-switch transformation helps to improve the power density, reliability and to reduce the converter's size, cost and driving circuit complexity.

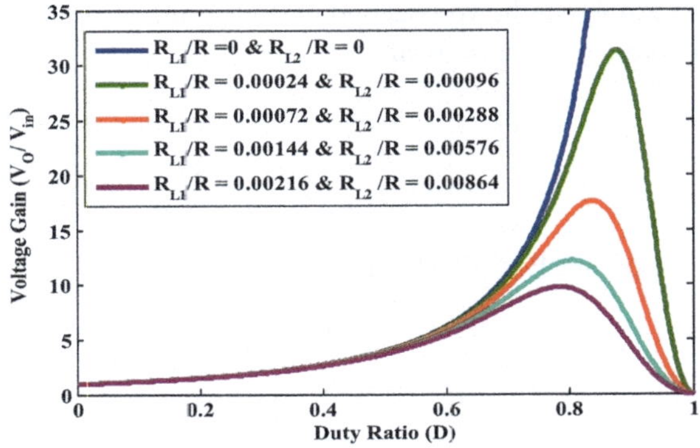

Fig. 9.4 Effect of inductor resistance on voltage gain

The converter given in Fig. 9.1 has a common node for two switches S_1 and S_2; here, source–source (S–S) connected switches can be replaced by the Graft switch as given in Fig. 9.5a. The modified circuit with the incorporation of the S–S Graft switch is given in Fig. 9.6. In this circuit, the presence of a cross-marked diode is insignificant in the circuit operation; hence, the diode can be eliminated. Further, the Graft scheme can be applied for the converters given in Fig. 9.2, whose

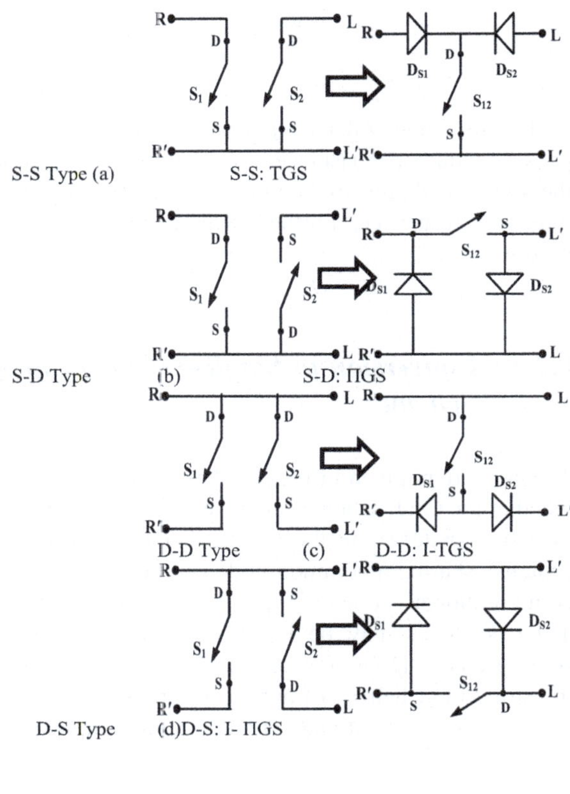

Fig. 9.5 Types of common nodes in between the two switches and corresponding grafted switch

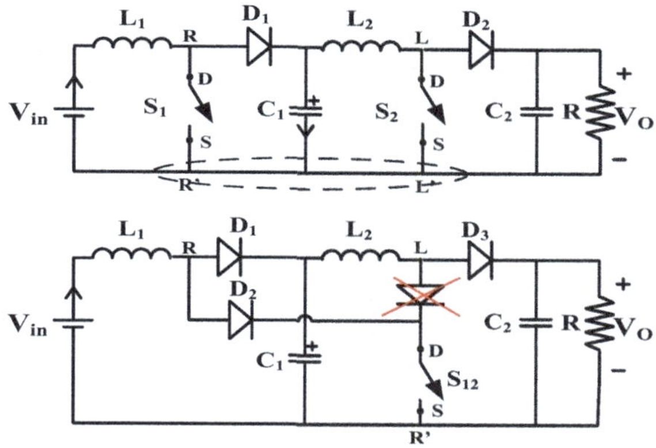

Fig. 9.6 Application of Graft scheme for reduced switch conversion

switches share the common node-to-device single-switch topologies. Coincidently, many single-switch quadratic boost converter topologies [6–10] shown in Fig. 9.7 are equivalent to the Graft scheme applied two-switch topologies.

9.2.6 *Features of Quadratic Boost Converters*

The two-switch converters given in Fig. 9.2 and single-switch topologies given in Fig. 9.7 have a gain $\left(M = \frac{1}{(1-D)^2}\right)$. Both single-switch and two-switch converters are of fourth-order circuits and have ON state and OFF state modes of operation. The dynamic equations of the converters can be found from averaging inductor currents and capacitor voltages using Eqs. (9.3) and (9.4) during the ON state and OFF state. Similar to the cascaded boost converter, the dynamic and steady-state equations of the quadratic boost converters in Figs. 9.2 and 9.7 are derived and are shown in Table. 9.1.

- From Figs. 9.2 and 9.7, the converters Conv-1, Conv-2, Conv-6 and Conv-7 are identical in both ON and OFF states; therefore, these converters get the same dynamic and steady-state equations as shown in Table 9.1. Similarly, Conv-3 and Conv-10 are identical.
- Conv-2 has the lowest switching device power rating (VA-requirements) among others, as the required blocking voltage of each switch is equal to half of the output voltage when the two switches are switching simultaneously. The other feature of the converter is by the inclusion of a new mode (i.e., S_1-ON, S_2-OFF —highlighted part of the switching pulse) as given in Fig. 9.8; the converter offers a new gain $M = \frac{(D(a-1)+1}{(D.a-1)(D-1)}$. The current paths while S_1 and S_2 operate synchronously ON and OFF are given in Fig. 9.2a while S_1-ON and S_2-OFF are given in Fig. 9.8b.

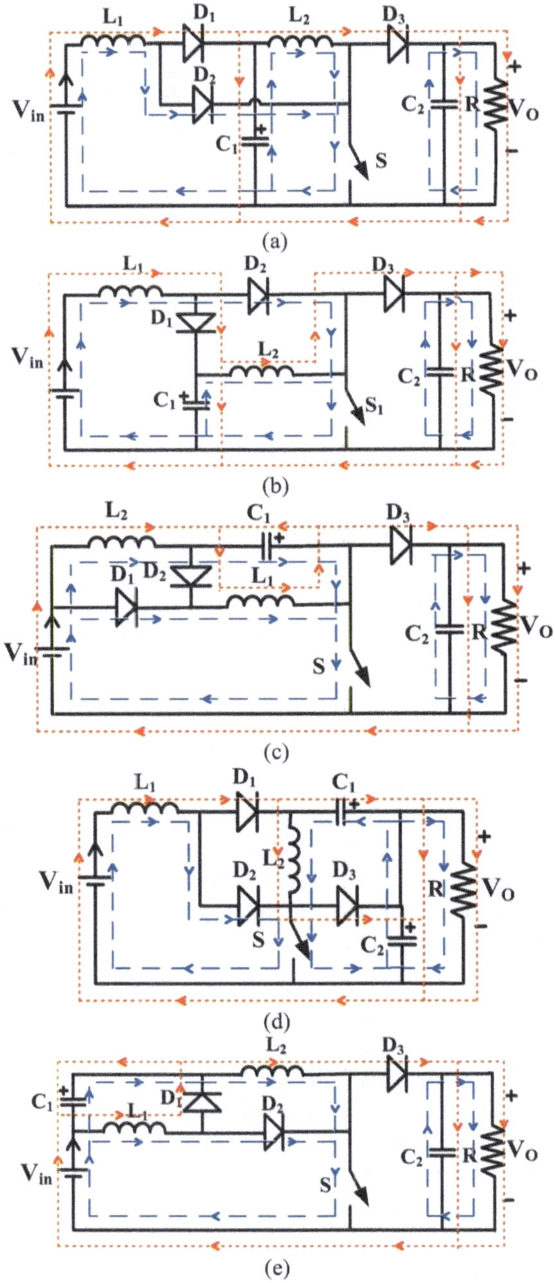

Fig. 9.7 Single-switch topologies **a** Conv-6 [6], **b** Conv-7 [7], **c** Conv-8 [8], **d** Conv-9 [9] and Conv-10 [10]

Table 9.1 Modeling of quadratic boost converters

No.	Dynamic equations	Steady-state eq.	Boundary conditions
1,2,6,7	$L_1\frac{di_{L1}}{dt}=v_{in}-v_{C1}(1-D)$ $L_2\frac{di_{L2}}{dt}=v_{C1}-v_O(1-D)$ $C_1\frac{dv_{C1}}{dt}=-i_{L2}D+i_{L1}(1-D)$ $C_2\frac{dv_{C2}}{dt}=i_{L2}(1-D)-\frac{v_O}{R}$	$\frac{V_O}{V_{in}}=\frac{1}{(1-D)^2}\frac{I_{in}}{I_O}=\frac{1}{(1-D)^2}$ $\frac{V_{C1}}{V_{in}}=\frac{1}{(1-D)}\frac{V_{C2}}{V_{in}}=\frac{1}{(1-D)^2}$ $\frac{I_{L1}}{I_O}=\frac{1}{(1-D)^2}\frac{I_{L2}}{I_O}=\frac{1}{(1-D)}$	$L_{1\min}=\frac{RD(1-D)^4}{f\left(\frac{\Delta i_{L1}}{I_{L1}}\right)}L_{2\min}=\frac{RD(1-D)^2}{f\left(\frac{\Delta i_{L2}}{I_{L2}}\right)}$ $C_{1\min}=\frac{D}{\left(\frac{\Delta v_{C1}}{V_{C1}}\right)(1-D)^2Rf}C_{2\min}=\frac{D}{Rf\left(\frac{\Delta v_{C2}}{V_{C2}}\right)}$
3,10	$L_1\frac{di_{L1}}{dt}=v_{in}-v_{C1}(1-D)$ $L_2\frac{di_{L2}}{dt}=v_{C1}D+(v_{in}+v_O)(1-D)$ $C_1\frac{dv_{C1}}{dt}=-i_{L2}D+i_{L1}(1-D)$ $C_2\frac{dv_{C2}}{dt}=i_{L2}(1-D)-\frac{v_O}{R}$		
4	$L_1\frac{di_{L1}}{dt}=v_{in}-v_{C1}(1-D)$ $L_2\frac{di_{L2}}{dt}=v_{in}-v_{C1}D-v_O(1-D)$ $C_1\frac{dv_{C1}}{dt}=(i_{in}-i_{L1})D-(i_{in}-i_{L2})(1-D)$ $C_2\frac{dv_{C2}}{dt}=i_{L2}(1-D)-\frac{v_O}{R}$		
5	$L_1\frac{di_{L1}}{dt}=v_{in}D-v_{C1}(1-D)$ $L_2\frac{di_{L2}}{dt}=(v_{in}+v_{C1})D-v_O(1-D)$ $C_1\frac{dv_{C1}}{dt}=-i_{L2}D-\frac{v_O}{R}+i_{L1}(1-D)$ $C_2\frac{dv_{C2}}{dt}=i_{L2}(1-D)-\frac{v_O}{R}$	$\frac{V_O}{V_{in}}=\frac{1}{(1-D)^2}\frac{I_{in}}{I_O}=\frac{1}{(1-D)^2}\frac{V_{C1}}{V_{in}}=\frac{D}{(1-D)}$ $\frac{V_{C2}}{V_{in}}=\frac{D}{(1-D)^2}$ $\frac{I_{L1}}{I_O}=\frac{1}{(1-D)^2}\frac{I_{L2}}{I_O}=\frac{1}{(1-D)}$	$L_{1\min}=\frac{RD(1-D)^4}{f\left(\frac{\Delta i_{L1}}{I_{L1}}\right)}L_{2\min}=\frac{R(1-D)^2}{f\left(\frac{\Delta i_{L2}}{I_{L2}}\right)}$ $C_{1\min}=\frac{D}{\left(\frac{\Delta v_{C1}}{V_{C1}}\right)(1-D)^2Rf}C_{2\min}=\frac{D}{Rf\left(\frac{\Delta v_{C2}}{V_{C2}}\right)}$
8	$L_1\frac{di_{L1}}{dt}=v_{in}D-v_{C1}(1-D)$ $L_2\frac{di_{L2}}{dt}=(v_{in}+v_{C1})D-v_O(1-D)$ $C_1\frac{dv_{C1}}{dt}=i_{L2}D-i_{L1}(1-D)$ $C_2\frac{dv_{C2}}{dt}=i_{L2}(1-D)-\frac{v_O}{R}$	$\frac{V_O}{V_{in}}=\frac{1}{(1-D)^2}\frac{I_{in}}{I_O}=\frac{1}{(1-D)^2}\frac{V_{C1}}{V_{in}}=\frac{D}{(1-D)}$ $\frac{V_{C2}}{V_{in}}=\frac{1}{(1-D)^2}$ $\frac{I_{L1}}{I_O}=\frac{1}{(1-D)^2}\frac{I_{L2}}{I_O}=\frac{1}{(1-D)}$	$L_{1\min}=\frac{RD(1-D)^4}{f\left(\frac{\Delta i_{L1}}{I_{L1}}\right)}L_{2\min}=\frac{RD(1-D)^2}{f\left(\frac{\Delta i_{L2}}{I_{L2}}\right)}$ $C_{1\min}=\frac{D}{\left(\frac{\Delta v_{C1}}{V_{C1}}\right)(1-D)^2Rf}C_{2\min}=\frac{D}{Rf\left(\frac{\Delta v_{C2}}{V_{C2}}\right)}$
9	$L_1\frac{di_{L1}}{dt}=v_{in}-(v_O-v_{C1})(1-D)$ $L_2\frac{di_{L2}}{dt}=-v_{C1}+v_OD$ $C_1\frac{dv_{C1}}{dt}=i_{L2}D-i_{L1}(1-D)$ $C_2\frac{dv_{C2}}{dt}=i_{L2}(1-D)-\frac{v_O}{R}$		

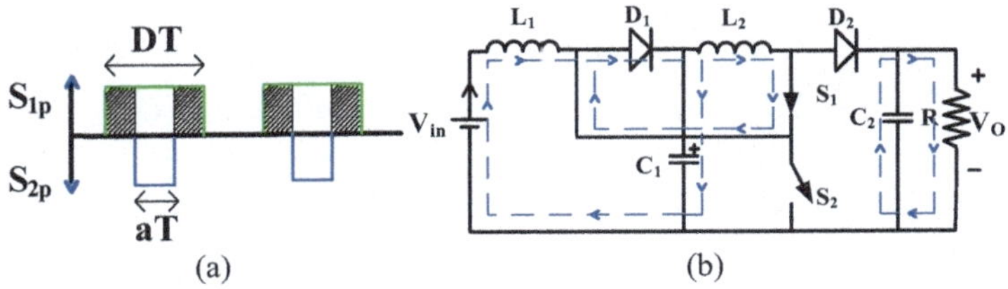

Fig. 9.8 Conv-2 **a** Switch pulse, **b** current paths in S_1-ON, S_2-OFF Mode

Fig. 9.9 Quadratic buck–boost converter Conv-13 [13]

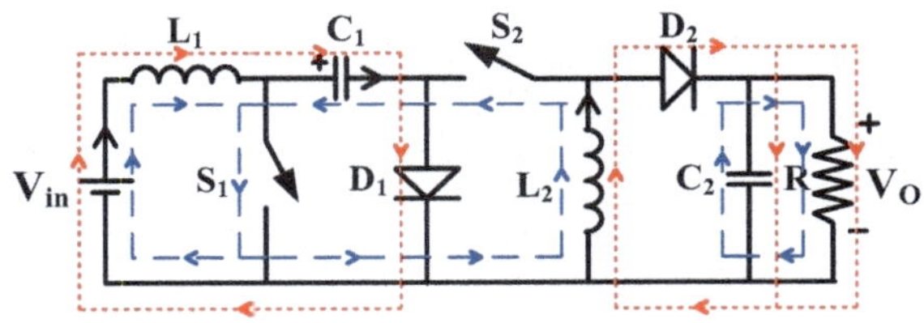

- From Fig. 9.2, converters Conv-3 and Conv-5 are efficient as it has a feature of reduced redundant power processing capacity.
- Conv-4 has the advantage of reduced ripple current at input and is a good choice for fuel cell applications.
- Conv-5 can be operated either in simultaneous mode or interleaved mode for the same voltage gain. If the converter operates in interleaved mode, the output voltage ripple can be reduced.
- The single-switch converter topologies given in Fig. 9.7 have the advantage of reduced driver circuit requirement, and their reliability is superior as compared to two-switch topologies.
- From Table 9.1, the voltage stress across the capacitor C_1 is minimum for converters Conv-5, Conv-8 and Conv-9.

9.3 Quadratic Buck–Boost Converters

The quadratic buck–boost converters shown in Figs. 9.9 and 9.10 have an output gain of $M = \frac{D}{(1-D)^2}$ and is named as Conv-11 to Conv-14. These converters are of fourth-order constituting either two synchronously operating switches or its single-switch equivalents. Here too, the ON and OFF state modes are differentiated by blue- and red-colored current paths, respectively.

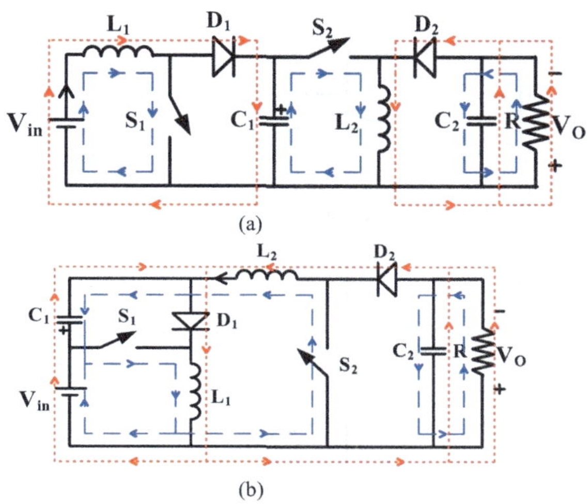

Fig. 9.10 Two-switch quadratic buck–boost converters having gain $M = \frac{D}{(1-D)^2}$ **a** Conv-11 [11] and **b** Conv-12 [12]

9.3.1 Operation and Derivation of Steady-State Equations

In this section, the operation of the recently proposed quadratic buck–boost converter Conv-13 is demonstrated assuming the capacitor voltage to be constant and switching devices as idle elements for the converter.

Mode 1: In the interval DT, both switches S_1 and S_2 will be turned ON, and diodes D_1 and D_2 will be reverse biased. In this mode, inductor L_1 and L_2 have energized from source V_{in} and capacitor C_1, and therefore, the inductor currents raise linearly as shown in Fig. 9.11. The voltage and current equations of inductor and capacitor are

$$v_{L1}(t) = V_{\text{in}}; v_{L2}(t) = V_{C1}; i_{C1}(t) = -I_{L2}; i_{C2}(t) = -I_O \quad (9.47)$$

Mode 2: In the interval $(1 - DT)$, diodes D_1 and D_2 conduct to provide the path for the inductor currents when the switch is at OFF state. In this mode, inductor voltages become negative and the current decreases linearly as shown in Fig. 9.11. Thus, the inductor voltage and capacitor current equations are

$$v_{L1}(t) = V_{\text{in}} - V_{C1}; v_{L2}(t) = -V_O; i_{C1}(t) = I_{L1} - I_{L2}; i_{C2}(t) = I_{L2} - I_O \quad (9.48)$$

The dynamic equations obtained using Eqs. (9.3) and (9.4) applied for the quadratic buck–boost converter operated under different modes are

$$L_1 \frac{\mathrm{d}i_{L1}}{\mathrm{d}t} = v_{\text{in}}D - (v_{\text{in}} - v_{C1})(1 - D) \quad (9.49)$$

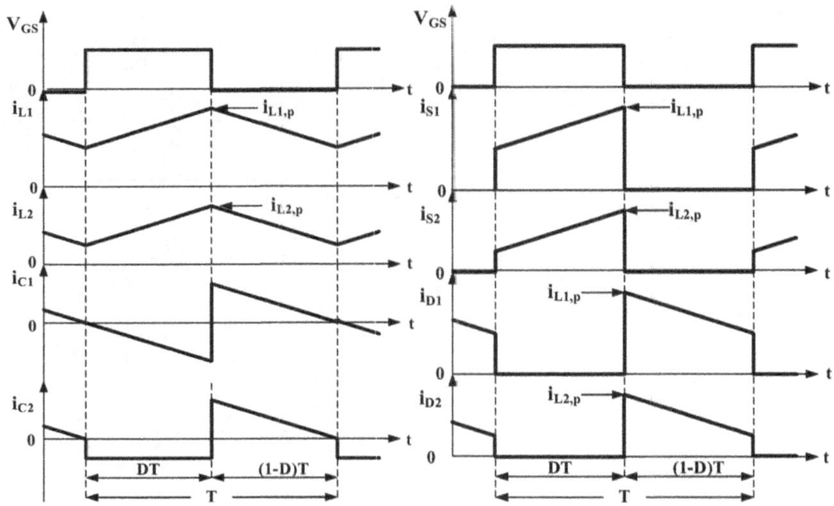

Fig. 9.11 Steady-state current waveforms of the quadratic buck–boost converter (Conv-13)

$$L_2 \frac{di_{L2}}{dt} = v_{C1}D - v_O(1 - D) \tag{9.50}$$

$$C_1 \frac{dv_{C1}}{dt} = -i_{L2}D + i_{L1}(1 - D) \tag{9.51}$$

$$C_2 \frac{dv_{C2}}{dt} = i_{L2}(1 - D) - \frac{v_O}{R} \tag{9.52}$$

Similarly, the voltage gain (M) and current gain of the converter by applying the principle of volt-second balance of an inductor and ampere-second balance of the capacitor over a switching cycle (T) are shown below:

$$M = \frac{V_O}{V_{in}} = \frac{D}{(1 - D)^2}; \frac{I_{in}}{I_O} = \frac{D}{(1 - D)^2} \tag{9.53}$$

From the voltage gains and steady current waveforms shown in Fig. 9.11, the relation between capacitor voltage and input voltage and inductor current and output current is given as

$$\frac{V_{C1}}{V_{in}} = \frac{1}{(1 - D)}; \frac{V_{C2}}{V_{in}} = \frac{D}{(1 - D)^2}; \frac{I_{L1}}{I_O} = \frac{D}{(1 - D)^2}; \frac{I_{L2}}{I_O} = \frac{1}{(1 - D)} \tag{9.54}$$

9.3.2 Design of Passive Elements

For the given design specifications of inductor's current ripple factor ($\Delta i_L/i_L$) and capacitor (or load) voltage ripple factor ($\Delta V_C/V_C$), the empirical formula for minimum allowable inductor and capacitors is presented in this section.

The relation between the capacitor L_1 and the ripple factor $\left(\frac{\Delta I_{L1}}{I_{L1}}\right)$ is given below:

$$\Delta i_{L1} = \frac{V_{L1}}{L} DT = \frac{V_{\text{in}}}{L_1} DT = \frac{V_o(1-D)^2}{L} T \tag{9.55}$$

$$I_{L1} = \frac{I_O D}{(1-D)^2} = \frac{V_O D}{R(1-D)^2} \tag{9.56}$$

$$\frac{\Delta i_{L1}}{i_{L1}} = \frac{R(1-D)^4}{DfL_1} \tag{9.57}$$

The minimum inductor required derived from Eq. (9.57) is

$$L_{1\,\min} = \frac{R(1-D)^4}{Df\left(\frac{\Delta i_{L1}}{i_{L1}}\right)}; \tag{9.58}$$

Second inductor, $L_{2\min}$, is given by

$$L_{2\,\min} = \frac{R(1-D)^2}{f\frac{\Delta i_{L2}}{i_{L2}}} \tag{9.59}$$

From dynamic and steady-state equations of the quadratic buck–boost converter,

$$\Delta V_{C1} = \frac{I_{L2}}{C_1} DT = \frac{I_O}{(1-D)C_1} DT = \frac{V_o}{R(1-D)C_1} DT \tag{9.60}$$

$$V_{C1} = \frac{V_{\text{in}}}{(1-D)} = \frac{V_O(1-D)}{D} \tag{9.61}$$

$$\frac{\Delta V_{C1}}{V_{C1}} = \frac{D^2}{(1-D)^2 RfC_1} \tag{9.62}$$

The capacitors C_1 and C_2 are selected using the following equations based on the given ripple factor $\left(\frac{\Delta V_{C1}}{V_{C1}}\right)$ and $\left(\frac{\Delta V_{C2}}{2}\right)$

$$C_{1\,\min} = \frac{D^2}{\left(\frac{\Delta V_{C1}}{V_{C1}}\right)(1-D)^2 Rf}; C_{2\,\min} = \frac{D}{Rf\left(\frac{\Delta V_{C2}}{V_{C2}}\right)} \tag{9.63}$$

9.3.3 Power Loss Calculations

Similar to power loss calculations for the quadratic boost converter given in Sect. 9.2.3, the loss components in the quadratic buck–boost converter are computed using Eqs. (9.23)–(9.27) assuming the current flowing through inductor and voltage across capacitor is ripple-free.

From Fig. 9.11 and steady-state equations are given in (9.53) and (9.54), currents and voltages expressions used to compute the converter power loss are given below:

$$I_{L1,\mathrm{peak}} = I_{S1,\mathrm{peak}} = \frac{V_O}{R(1-D)^2}; I_{L1,\mathrm{RMS}} = \frac{V_O}{R(1-D)^2} \text{ and } I_{S1,\mathrm{RMS}} = \frac{V_O}{R(1-D)^2}\sqrt{D} \tag{9.64}$$

$$I_{L2,\mathrm{peak}} = I_{S2,\mathrm{peak}} = \frac{V_O}{R(1-D)}; I_{L2,\mathrm{RMS}} = \frac{V_O}{R(1-D)} \text{ and } I_{S2,\mathrm{RMS}} = \frac{V_O}{R(1-D)}\sqrt{D} \tag{9.65}$$

$$V_{ds1} = \frac{V_O(1-D)}{D}; V_{ds2} = V_O \tag{9.66}$$

$$I_{D1,\mathrm{AVG}} = \frac{V_O D}{R(1-D)} \text{ and } I_{D1,\mathrm{RMS}} = \frac{V_O}{R(1-D)^2}\sqrt{1-D} \tag{9.67}$$

$$I_{D2,\mathrm{AVG}} = \frac{V_O}{R} \text{ and } I_{D2,\mathrm{RMS}} = \frac{V_O}{R(1-D)}\sqrt{1-D} \tag{9.68}$$

$$I^2_{C1,\mathrm{RMS}} = I^2_{L2}D + (I_{L1})^2(1-D) = \frac{P_O D}{R(1-D)^3} \tag{9.69}$$

$$I^2_{C2,\mathrm{RMS}} = I^2_O(D) + (I_{L2} - I_O)^2(1-D) = \frac{P_O D}{R^2(1-D)} \tag{9.70}$$

For the given duty ratio, Eqs. (9.64)–(9.70) are calculated and substituted in Eqs. (9.23) and (9.27) to compute total power losses of the Conv-13, in turn, helps in determining the efficiency of converter using Eq. (9.36).

9.3.4 Impact of Inductor Resistance on Voltage Gain

The steady-state models developed in Sect. 9.3.2 have not considered the non-idealities; however, inductor resistance has a significant impact on the actual voltage gain.

The relation between the voltage gain and per-unit resistance $\left(\frac{R_{L1}}{R}, \frac{R_{L2}}{R}\right)$ for the quadratic buck–boost converter given in Fig. 9.9 is derived by including the inductor resistance drop in Eqs. (9.47) and (9.48). The modified gain is given by:

$$\frac{V_O}{V_{\text{in}}} = \frac{1}{\left(\frac{D}{(1-D)^2}\right)\left(\frac{R_{L1}}{R}\right) + \frac{1}{D}\left(\frac{R_{L2}}{R}\right) + \frac{1-D}{D}} \tag{9.71}$$

For different values of per-unit resistance $\left(\frac{R_{L1}}{R}, \frac{R_{L2}}{R}\right)$, voltage gain is plotted against the duty ratio and is shown in Fig. 9.12. From the figure, if duty ratio is increased beyond 0.8, the voltage gain starts to decrease for most of the per-unit resistance values; it reflects the practical voltage gain limitations of the quadratic buck–boost converter.

9.3.5 Conversion to Single-Switch Topology Using Graft Scheme

The graft scheme can be applied for the converters where two switches share a common node to minimize the number of active switches as given in Fig. 9.5. Similarly, Fig. 9.9 can be reconfigured by shifting the switch S_2 toward the negative polarity of the source to enable the graft scheme. The single-switch topology equivalent is shown in Fig. 9.13. In the figure, the cross-marked diode can be eliminated due to its insignificance in circuit operation. Two-switch equivalent and reported single-switch topologies are shown in Fig. 9.14.

Fig. 9.12 Inductor's resistance effect on voltage gain for Conv-13

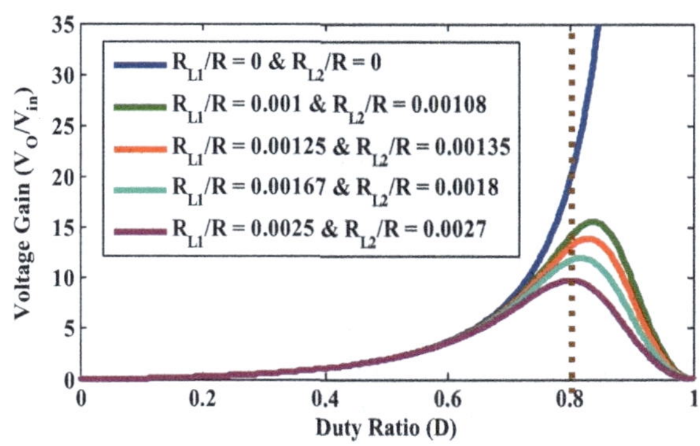

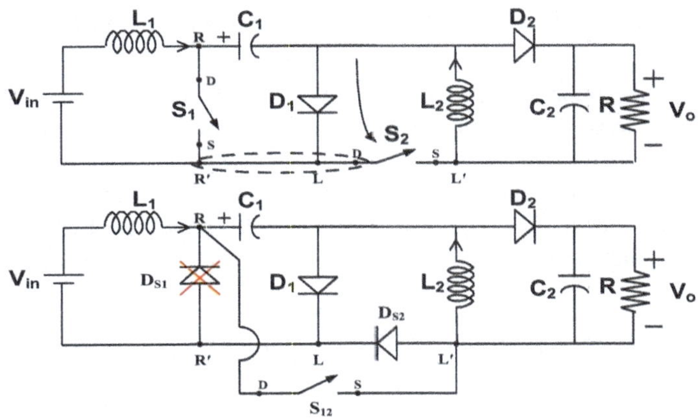

Fig. 9.13 Single-switch equivalent topology of Conv-13 (SS Conv-13)

Fig. 9.14 Single-switch topologies **a** Conv-14 [14], **b** SS Conv-13

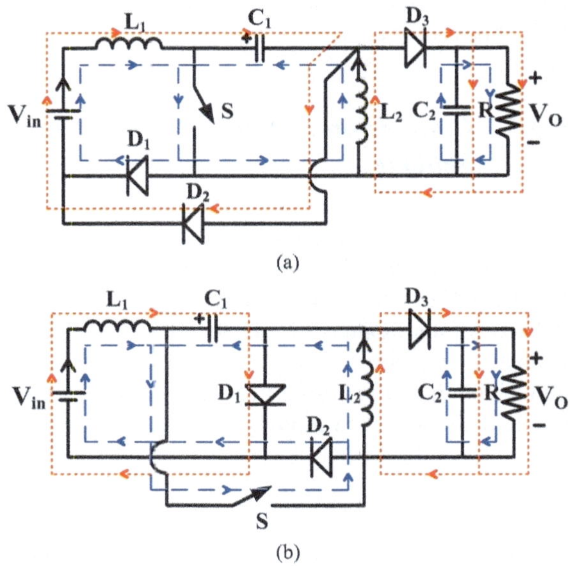

9.3.6 Quadratic Buck–Boost Converters with Gain $M = \frac{D^2}{(1-D)^2}$

The other class of quadratic buck–boost converters having a gain $M = \frac{D^2}{(1-D)^2}$ is shown in Fig. 9.15. From this figure, the ON state and OFF state equations are used to develop dynamic and steady-state equations and are listed in Tables 9.2 and 9.3.

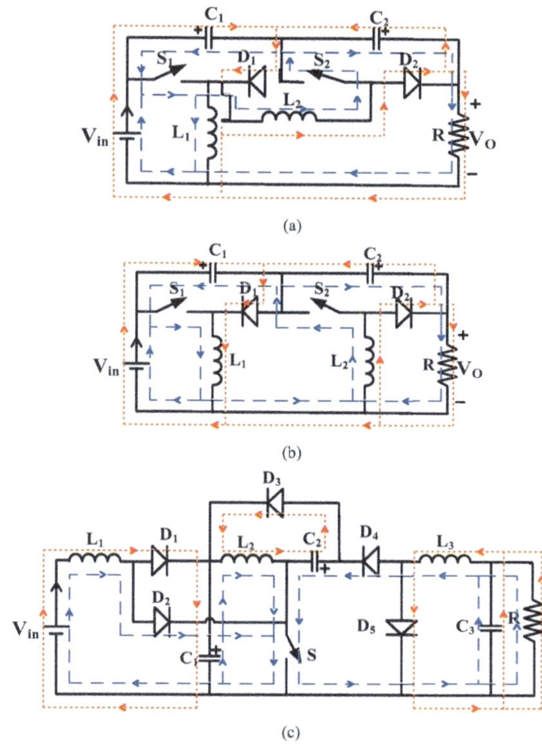

Fig. 9.15 Quadratic buck–boost converters having gain $M = \frac{D^2}{(1-D)^2}$ **a** Conv-15 [15], **b** Conv-16 [16] and **c** Conv-17 [17]

These converters are effective at applications demanding a high step-down ratio at lower duty ratios. From Fig. 9.15, Conv-15 and Conv-16 are of fourth-order and have positive output polarity, whereas Conv-17 is a sixth-order circuit with negative polarity at its output terminals. The Conv-17 is derived to a single switch by cascading boost, buck–boost and buck converter for enabling continuous input–continuous output.

9.4 Hybrid Quadratic Boost Converters

In this section, the boost converters with different voltage gain other than the quadratic boost converters of gain $M = \frac{1}{(1-D)^2}$ are presented.

The converters are classified into two-switch and single-switch topologies given in Figs. 9.16 and 9.17, respectively. The converter's gains can be computed by applying the principle of volt-second balance and ampere-second balance using ON and OFF state equations. The dynamic equations and corresponding steady-state equations are listed in Table 9.4. The Conv-18 and Conv-19 shown in Fig. 9.16a, b are reduced redundant power processing topologies having a gain $\frac{D+(1-D)^2}{(1-D)^2}$.

Table 9.2 Modeling of quadratic buck–boost converters with gain $M = \frac{D}{(1-D)^2}$

No.	Dynamic equations	Steady-state eq.	Boundary conditions
11,14	$L_1 \frac{di_{L1}}{dt} = v_{in} - v_{C1}(1-D)$ $L_2 \frac{di_{L2}}{dt} = v_{C1}D - v_O(1-D)$ $C_1 \frac{dv_{C1}}{dt} = -i_{L2}D + i_{L1}(1-D)$ $C_2 \frac{dv_{C2}}{dt} = i_{L2}(1-D) - \frac{v_O}{R}$	$\frac{V_O}{V_{in}} = \frac{D}{(1-D)^2}$ $\frac{I_{in}}{I_O} = \frac{D}{(1-D)^2}$ $\frac{V_{C1}}{V_{in}} = \frac{1}{(1-D)}$ $\frac{V_{C2}}{V_{in}} = \frac{D}{(1-D)^2}$ $\frac{I_{L1}}{I_O} = \frac{D}{(1-D)^2}$ $\frac{I_{L2}}{I_O} = \frac{1}{(1-D)}$	$L_{1\,min} = \frac{R(1-D)^4}{Df\left(\frac{\Delta i_{L1}}{I_{L1}}\right)}$ $L_{2\,min} = \frac{R(1-D)^2}{f\left(\frac{\Delta i_{L2}}{I_{L2}}\right)}$ $C_{1\,min} = \frac{D^2}{\left(\frac{\Delta V_{C1}}{V_{C1}}\right)(1-D)^2 Rf}$ $C_{2\,min} = \frac{D}{Rf\left(\frac{\Delta V_{C2}}{V_{C2}}\right)}$
13,SS-13	$L_1 \frac{di_{L1}}{dt} = v_{in}D - (v_{in} - v_{C1})(1-D)$ $L_2 \frac{di_{L2}}{dt} = v_{C1}D - v_O(1-D)$ $C_1 \frac{dv_{C1}}{dt} = -i_{L2}D + i_{L1}(1-D)$ $C_2 \frac{dv_{C2}}{dt} = i_{L2}(1-D) - \frac{v_O}{R}$		
12	$L_1 \frac{di_{L1}}{dt} = v_{in}D - (v_{in} - v_{C1})(1-D)$ $L_2 \frac{di_{L2}}{dt} = v_{C1} - v_{in} - v_O(1-D)$ $C_1 \frac{dv_{C1}}{dt} = -i_{L2}D + (i_{L1} - i_{L2})(1-D)$ $C_2 \frac{dv_{C2}}{dt} = i_{L2}(1-D) - \frac{v_O}{R}$	$\frac{V_O}{V_{in}} = \frac{D}{(1-D)^2}$ $\frac{I_{in}}{I_O} = \frac{D}{(1-D)^2}$ $\frac{V_{C1}}{V_{in}} = \frac{1}{(1-D)}$ $\frac{V_{C2}}{V_{in}} = \frac{D}{(1-D)^2}$ $\frac{I_{L1}}{I_O} = \frac{1}{(1-D)^2}$ $\frac{I_{L2}}{I_O} = \frac{1}{(1-D)}$	$L_{1\,min} = \frac{R(1-D)^4}{f\left(\frac{\Delta i_{L1}}{I_{L1}}\right)}$ $L_{2\,min} = \frac{RD(1-D)}{f\left(\frac{\Delta i_{L2}}{I_{L2}}\right)}$ $C_{1\,min} = \frac{D^2}{\left(\frac{\Delta V_{C1}}{V_{C1}}\right)(1-D)^2 Rf}$ $C_{2\,min} = \frac{D}{Rf\left(\frac{\Delta V_{C2}}{V_{C2}}\right)}$

Table 9.3 Modeling of quadratic buck–boost converters with gain $M = \frac{D^2}{(1-D)^2}$

No.	Dynamic equations	Steady-state eq.	Boundary conditions
15	$L_1 \frac{di_{L1}}{dt} = v_{in} - v_{C1}(1-D)$ $L_2 \frac{di_{L2}}{dt} = v_{C1}D - v_{C2}(1-D)$ $C_1 \frac{dv_{C1}}{dt} = (-i_{L2} + i_O)D + (i_{L1} + i_O)(1-D)$ $C_2 \frac{dv_{C2}}{dt} = i_{L2}(1-D) - \frac{v_O}{R}$	$\frac{V_O}{V_{in}} = \frac{D^2}{(1-D)^2}$ $\frac{I_{in}}{I_O} = \frac{D^2}{(1-D)^2}$ $\frac{V_{C1}}{V_{in}} = \frac{1}{(1-D)}$ $\frac{V_{C2}}{V_{in}} = \frac{D}{(1-D)^2}$ $\frac{V_{C3}}{V_{in}} = \frac{D^2}{(1-D)^2}$ $\frac{I_{L1}}{I_O} = \frac{2D-1}{(1-D)^2}$ $\frac{I_{L2}}{I_O} = \frac{1}{(1-D)}$	$L_{1\,min} = \frac{R(1-D)^4}{f\left(\frac{\Delta i_{L1}}{I_{L1}}\right)D(2D-1)}$ $L_{2\,min} = \frac{R(1-D)^2}{f\left(\frac{\Delta i_{L1}}{I_{L1}}\right)D}$ $C_{1\,min} = \frac{D^4}{\left(\frac{\Delta V_{C1}}{V_{C1}}\right)(1-D)^2 Rf}$ $C_{2\,min} = \frac{D^2}{Rf\left(\frac{\Delta V_{C2}}{V_{C2}}\right)}$
16	$L_1 \frac{di_{L1}}{dt} = v_{in} - v_{C1}(1-D)$ $L_2 \frac{di_{L2}}{dt} = v_{C1} - v_{in} - v_{C2}(1-D)$ $C_1 \frac{dv_{C1}}{dt} = -i_{L2} + i_O + i_{L1}(1-D)$ $C_2 \frac{dv_{C2}}{dt} = i_{L2}(1-D) - \frac{v_O}{R}$	$\frac{V_O}{V_{in}} = \frac{D^2}{(1-D)^2}$ $\frac{I_{in}}{I_O} = \frac{D^2}{(1-D)^2}$ $\frac{V_{C1}}{V_{in}} = \frac{1}{(1-D)}$ $\frac{V_{C2}}{V_{in}} = \frac{D}{(1-D)^2}$ $\frac{V_{C3}}{V_{in}} = \frac{D^2}{(1-D)^2}$ $\frac{I_{L1}}{I_O} = \frac{D}{(1-D)^2}$ $\frac{I_{L2}}{I_O} = \frac{1}{(1-D)}$	$L_{1\,min} = \frac{R(1-D)^4}{f\left(\frac{\Delta i_{L1}}{I_{L1}}\right)D^2}$ $L_{2\,min} = \frac{R(1-D)^2}{f\left(\frac{\Delta i_{L1}}{I_{L1}}\right)D^2}$ $C_{1\,min} = \frac{D^3}{\left(\frac{\Delta V_{C1}}{V_{C1}}\right)(1-D)^2 Rf}$ $C_{2\,min} = \frac{D^2}{Rf\left(\frac{\Delta V_{C2}}{V_{C2}}\right)}$

(continued)

Table 9.3 (continued)

No.	Dynamic equations	Steady-state eq.	Boundary conditions
17	$L_1\frac{di_{L1}}{d_t}=v_{in}-v_{C1}(1-D)$ $L_2\frac{di_{L2}}{d_t}=v_{C1}D-v_{C2}(1-D)$ $L_3\frac{di_{L3}}{d_t}=v_{C2}D-v_O$ $C_1\frac{dv_{C1}}{d_t}=i_{L2}D-i_{L1}(1-D)$ $C_2\frac{dv_{C2}}{d_t}=i_{L3}D-i_{L2}(1-D)$ $C_3\frac{dv_{C3}}{d_t}=i_{L3}-\frac{v_O}{R}$	$\frac{V_O}{V_{in}}=\frac{D^2}{(1-D)^2}$ $\frac{I_{in}}{I_O}=\frac{D^2}{(1-D)^2}$ $\frac{V_{C1}}{V_{in}}=\frac{1}{(1-D)}$ $\frac{V_{C2}}{V_{in}}=\frac{D}{(1-D)^2}$ $\frac{V_{C3}}{V_{in}}=\frac{D^2}{(1-D)^2}$ $\frac{I_{L1}}{I_O}=\frac{D^2}{(1-D)^2}$ $\frac{I_{L2}}{I_O}=\frac{D}{(1-D)}$ $\frac{I_{L3}}{I_O}=1$	$L_{1\,\min}=\frac{R(1-D)^4}{f\left(\frac{\Delta i_{L1}}{I_{L1}}\right)D^2}$ $L_{2\,\min}=\frac{R(1-D)^2}{f\left(\frac{\Delta i_{L1}}{I_{L1}}\right)D^2}$ $L_{3\,\min}=\frac{R(1-D)}{f\left(\frac{\Delta i_{L1}}{I_{L1}}\right)}$ $C_{1\,\min}=\frac{D^4}{\left(\frac{\Delta V_{C1}}{V_{C1}}\right)(1-D)^2Rf}$ $C_{2\,\min}=\frac{D^2}{Rf\left(\frac{\Delta V_{C2}}{V_{C2}}\right)}$ $C_{3\,\min}=\frac{D}{Rf\left(\frac{\Delta V_{C2}}{V_{C2}}\right)}$

Conv-20 given in Fig. 9.16c is a conventional cascaded boost converter where the two switches are operated in complementary nature and results in voltage gain $M=\frac{1}{D(1-D)}$. This converter has higher voltage gains at lower duty ratios as compared to quadratic boost converters. Conv-21, Conv-22 and Conv-23 given in Fig. 9.16d–f give same voltage gain, i.e., $M=\frac{1}{D(1-D)}$. The structure of Conv-21 is the same as that of Conv-4, but the two switches of the converter are operating in complementary nature and result in less input current ripples than Conv-4. Conv-22 is derived by the additive connection of SEPIC and boost converters operated in a complementary switching scheme. Conv-23 is a generalized converter topology with a cell containing an inductor, capacitor and two diodes which are shown in Fig. 9.16f.

Conv-24 given in Fig. 9.16g is developed by adding voltage lift cell to the Conv-20 and results in higher voltage gain $\left(M=\frac{1+D}{D(1-D)}\right)$. In Conv-25, the input current ripple is reduced while operating in a complementary switching scheme because the structure enables the input current to bear function of the difference of two inductor currents as shown in Fig. 9.16h.

The single circuit topology Conv-26 given in Fig. 9.17a is derived from quadratic boost converter by adding double enhance circuit; the gain of the circuit is $M=\frac{2}{(1-D)^2}$. This is one among many high step-up Luo converter reported in [23]. Conv-27 is devised by the inclusion of the capacitor–inductor–diode (CLD) cell to a conventional single-switch quadratic boost converter [24] and is given in Fig. 9.17b. This topology has less voltage stresses on power semiconducting devices as compared to the conventional quadratic boost converter. Figure 9.17c is a single-switch topology by applying the Graft scheme to the Conv-18 shown in Fig. 9.16a.

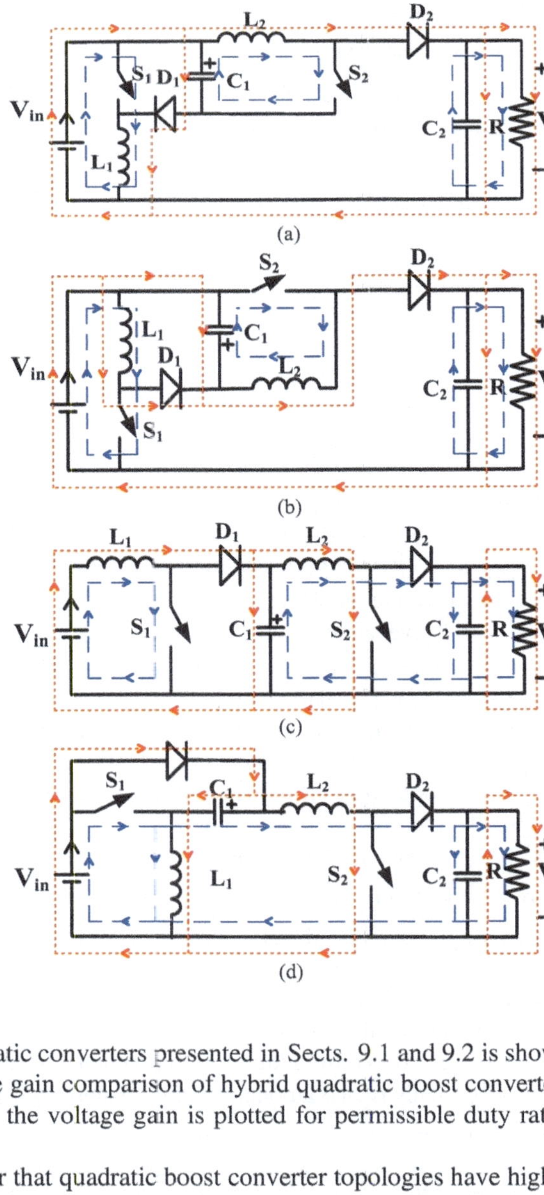

Fig. 9.16 Two-switch hybrid quadratic boost converters **a** Conv-18 [3], **b** Conv-19 [3], **c** Conv-20 [18], **d** Conv-21 [4], **e** Conv-22 [19], **f** Conv-23 [20], **g** Conv-24 [21] and **h** Conv-25 [22]

The voltage gain of quadratic converters presented in Sects. 9.1 and 9.2 is shown in Fig. 9.18a, and the voltage gain comparison of hybrid quadratic boost converters is given in Fig. 9.18b. Here, the voltage gain is plotted for permissible duty ratio, i.e., in between 0.2 and 0.8.

From Fig. 9.18a, it is clear that quadratic boost converter topologies have higher voltage gain for wide range of duty ratio as compared to conventional boost and quadratic buck–boost converter topologies. From Fig. 9.18b, the Luo double quadratic boost converter [26] has the highest voltage gain at higher values of duty ratio; in contrast, Conv-24 is best among other converters at lower duty ratios.

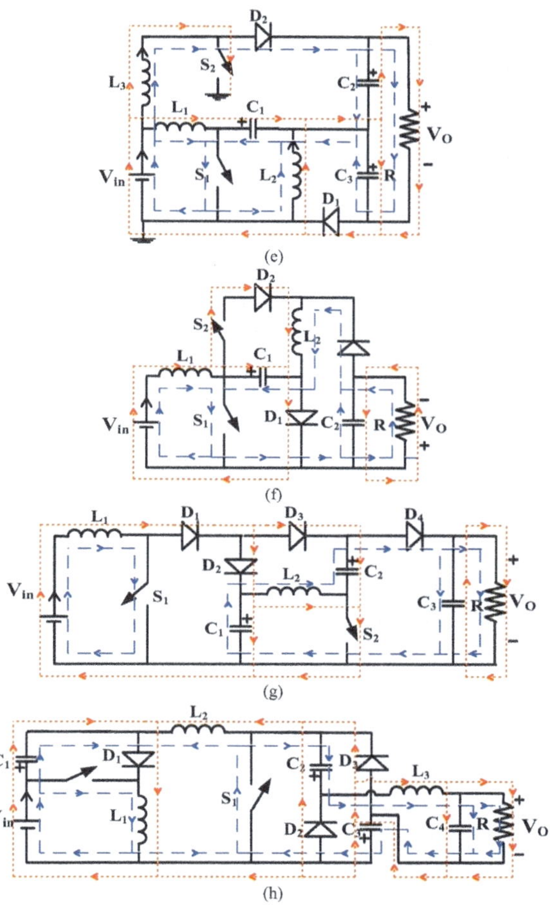

Fig. 9.16 (continued)

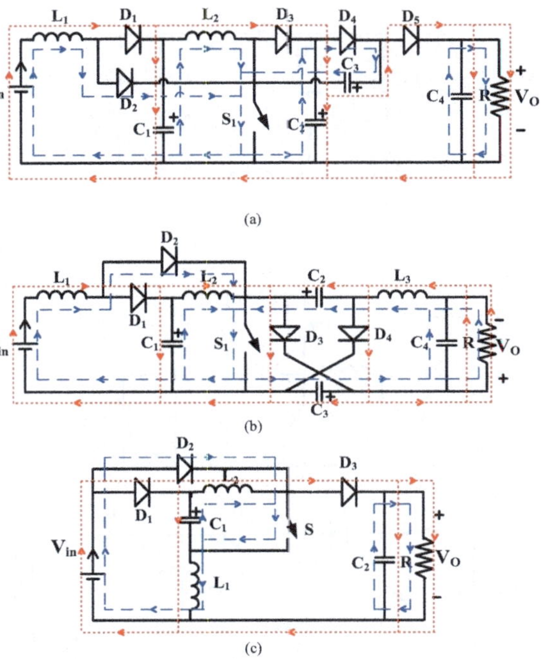

Fig. 9.17 Hybrid quadratic boost single-switch topologies **a** Conv-26 [23], **b** Conv-27 [24] and **c** SS Conv-18 [18]

Table 9.4 Modeling of hybrid quadratic boost converters

No.	Dynamic equations	Steady-state equation	Boundary conditions
20	$L_1 \frac{di_{L1}}{d_t} = v_{in} - v_{C1}(1-D)$ $L_2 \frac{di_{L2}}{d_t} = v_{C1}D - v_O(1-D)$ $C_1 \frac{dv_{C1}}{d_t} = -i_{L2}D + i_{L1}(1-D)$ $C_2 \frac{dv_{C2}}{d_t} = i_{L2}(1-D) - \frac{v_O}{R}$	$\frac{V_O}{V_{in}} = \frac{1}{D(1-D)}$ $\frac{I_{in}}{I_O} = \frac{1}{D(1-D)}$ $\frac{V_{C1}}{V_{in}} = \frac{1}{(1-D)}$ $\frac{V_{C2}}{V_{in}} = \frac{1}{D(1-D)}$ $\frac{I_{L1}}{I_O} = \frac{1}{D(1-D)}$ $\frac{I_{L2}}{I_O} = \frac{1}{D}$	$L_{1\,min} = \frac{RD^3(1-D)^2}{f\left(\frac{\Delta i_{L1}}{I_{L1}}\right)}$ $L_{2\,min} = \frac{RD^3(1-D)}{f\left(\frac{\Delta i_{L2}}{I_{L2}}\right)}$
21	$L_1 \frac{di_{L1}}{d_t} = v_{in} - v_{C1}(1-D)$ $L_2 \frac{di_{L2}}{d_t} = v_{in} - (v_O - v_{C1})(1-D)$ $C_1 \frac{dv_{C1}}{d_t} = i_{L2}D - i_{L1}(1-D)$ $C_2 \frac{dv_{C2}}{d_t} = i_{L2}D - \frac{v_O}{R}$	$\frac{V_O}{V_{in}} = \frac{1}{D(1-D)}$ $\frac{I_{in}}{I_O} = \frac{1}{D(1-D)}$ $\frac{V_{C1}}{V_{in}} = \frac{1}{(1-D)}$ $\frac{V_{C2}}{V_{in}} = \frac{1}{D(1-D)}$ $\frac{I_{L1}}{I_O} = \frac{1}{(1-D)}$ $\frac{I_{L2}}{I_O} = \frac{1}{D}$	$L_{1\,min} = \frac{RD^2(1-D)^2}{f\left(\frac{\Delta i_{L2}}{I_{L2}}\right)}$ $L_{2\,min} = \frac{RD^3(1-D)}{f\left(\frac{\Delta i_{L2}}{I_{L2}}\right)}$
22	$L_1 \frac{di_{L1}}{d_t} = v_{in} + (v_{C1} - v_{C2})(1-D)$ $L_2 \frac{di_{L2}}{d_t} = v_{in} - (v_{C1} + v_{C3})(1-D)$ $L_3 \frac{di_{L3}}{d_t} = v_{C1}D - v_{C3}(1-D)$	$\frac{V_O}{V_{in}} = \frac{1}{D(1-D)}$ $\frac{I_{in}}{I_O} = \frac{1}{D(1-D)}$ $\frac{V_{C1}}{V_{in}} = 1$ $\frac{V_{C2}}{V_{in}} = \frac{1+D}{D}$	$L_{1\,min} = \frac{RD^2(1-D)^2}{f\left(\frac{\Delta i_{L1}}{I_{L1}}\right)}$ $L_{2\,min} = \frac{RD^2(1-D)^2}{f\left(\frac{\Delta i_{L2}}{I_{L2}}\right)}$

(continued)

Table 9.4 (continued)

No.	Dynamic equations	Steady-state equation	Boundary conditions
	$C_1 \frac{\mathrm{d}v_{C1}}{\mathrm{d}t} = i_{L3}D - \frac{v_O}{R}$ $C_2 \frac{\mathrm{d}v_{C2}}{\mathrm{d}t} = \frac{v_O}{R}D + \left(i_{L1} + i_{L2} - \frac{v_O}{R}\right)(1-D)$ $C_3 \frac{\mathrm{d}v_{C3}}{\mathrm{d}t} = (i_{L2} + i_{L3}) + i_{L1}(1-D)$	$\frac{V_{C3}}{V_{\text{in}}} = \frac{D}{(1-D)}$ $\frac{I_{L1}}{I_O} = \frac{1}{(1-D)}$ $\frac{I_{L2}}{I_O} = 0$ $\frac{I_{L3}}{I_O} = \frac{1}{D}$	$L_{3\,\min} = \frac{RD^3(1-D)}{f\left(\frac{\Delta i_{L3}}{I_{L3}}\right)}$ $C_{1\,\min} = \frac{1-D}{\left(\frac{\Delta V_{C1}}{V_{C1}}\right)Rf}$ $C_{2\,\min} = \frac{D}{\left(\frac{\Delta V_{C2}}{V_{C2}}\right)Rf}$ $C_{3\,\min} = \frac{1}{\left(\frac{\Delta V_{C3}}{V_{C3}}\right)Rf}$
23	$L_1 \frac{\mathrm{d}i_{L1}}{\mathrm{d}t} = v_{\text{in}} - v_{C1}(1-D)$ $L_2 \frac{\mathrm{d}i_{L2}}{\mathrm{d}t} = (v_{C1} - v_O)D + v_{C1}(1-D)$ $C_1 \frac{\mathrm{d}v_{C1}}{\mathrm{d}t} = i_{L2} + (i_{L1} - i_{L2})(1-D)$ $C_2 \frac{\mathrm{d}v_{C2}}{\mathrm{d}t} = i_{L2}D - \frac{v_O}{R}$	$\frac{V_O}{V_{\text{in}}} = \frac{1}{D(1-D)}$ $\frac{I_{\text{in}}}{I_O} = \frac{1}{D(1-D)}$ $\frac{V_{C1}}{V_{\text{in}}} = \frac{1}{(1-D)}$ $\frac{V_{C2}}{V_{\text{in}}} = \frac{1}{D(1-D)}$	$L_{1\,\min} = \frac{RD^3(1-D)^2}{f\left(\frac{\Delta i_{L1}}{I_{L1}}\right)}$ $L_{2\,\min} = \frac{RD^3(1-D)}{f\left(\frac{\Delta i_{L2}}{I_{L2}}\right)}$
25	$L_1 \frac{\mathrm{d}i_{L1}}{\mathrm{d}t} = v_{\text{in}} - v_{C1}(1-D)$ $L_2 \frac{\mathrm{d}i_{L2}}{\mathrm{d}t} = v_{C1} - v_{\text{in}} - v_{C2}(1-D)$ $L_3 \frac{\mathrm{d}i_{L3}}{\mathrm{d}t} = 2v_{C2}D - v_{C2}(1-D) - v_O$ $C_1 \frac{\mathrm{d}v_{C1}}{\mathrm{d}t} = i_{L1}(1-D) - i_{L2}$ $C_2 \frac{\mathrm{d}v_{C2}}{\mathrm{d}t} = i_{L2}(1-D) - i_{L3}$ $C_3 \frac{\mathrm{d}v_{C3}}{\mathrm{d}t} = i_{L3} - \frac{v_O}{R}$	$\frac{V_O}{V_{\text{in}}} = \frac{D(1+D)}{(1-D)^2}$ $\frac{I_{\text{in}}}{I_O} = \frac{D(1+D)}{(1-D)^2}$ $\frac{V_{C1}}{V_{\text{in}}} = \frac{1}{1-D}$ $\frac{V_{C2}}{V_{\text{in}}} = \frac{D}{(1-D)^2}$ $\frac{V_{C3}}{V_{\text{in}}} = \frac{D(1+D)}{(1-D)^2}$ $\frac{I_{L1}}{I_O} = \frac{1+D}{(1-D)^2}$ $\frac{I_{L2}}{I_O} = \frac{1+D}{(1-D)}$ $\frac{I_{L3}}{I_O} = 1$	$L_{1\,\min} = \frac{R(1-D)^4}{f\left(\frac{\Delta i_{L1}}{I_{L1}}\right)(1+D)^2}$ $L_{2\,\min} = \frac{R(1-D)^2}{f\left(\frac{\Delta i_{L2}}{I_{L2}}\right)(1+D)^2}$ $L_{3\,\min} = \frac{R\left(2D+D^2\right)}{f\left(\frac{\Delta i_{L3}}{I_{L3}}\right)(1-D)}$ $C_{1\,\min} = \frac{D(1+D)}{\left(\frac{\Delta V_{C1}}{V_{C1}}\right)(1-D)Rf}$ $C_{2\,\min} = \frac{D}{Rf\left(\frac{\Delta V_{C2}}{V_{C2}}\right)}$
18 SS-18	$L_1 \frac{\mathrm{d}i_{L1}}{\mathrm{d}t} = v_{\text{in}} - v_{C1}(1-D)$ $L_2 \frac{\mathrm{d}i_{L2}}{\mathrm{d}t} = (v_{\text{in}} - v_O)(1-D) - v_{C1}D$ $C_1 \frac{\mathrm{d}v_{C1}}{\mathrm{d}t} = -i_{L2}D + i_{L1}(1-D)$ $C_2 \frac{\mathrm{d}v_{C2}}{\mathrm{d}t} = i_{L2}(1-D) - \frac{v_O}{R}$	$\frac{V_O}{V_{\text{in}}} = \frac{D+(1-D)^2}{(1-D)^2}$ $\frac{I_{\text{in}}}{I_O} = \frac{D+(1-D)^2}{(1-D)^2}$ $\frac{V_{C1}}{V_{\text{in}}} = \frac{1}{(1-D)}$ $\frac{V_{C2}}{V_{\text{in}}} = \frac{D+(1-D)^2}{(1-D)^2}$ $\frac{I_{L1}}{I_O} = \frac{D}{(1-D)^2}$ $\frac{I_{L2}}{I_O} = \frac{1}{1-D}$	$L_{1\,\min} = \frac{R(1-D)^4}{f\left(\frac{\Delta i_{L1}}{I_{L1}}\right)\left(D+(1-D)^2\right)}$ $L_{2\,\min} = \frac{RD(1-D)^2}{f\left(\frac{\Delta i_{L2}}{I_{L2}}\right)\left(D+(1-D)^2\right)}$
19	$L_1 \frac{\mathrm{d}i_{L1}}{\mathrm{d}t} = v_{\text{in}}D - v_{C1}(1-D)$ $L_2 \frac{\mathrm{d}i_{L2}}{\mathrm{d}t} = (v_{\text{in}} - v_O)(1-D) + v_{C1}$ $C_1 \frac{\mathrm{d}v_{C1}}{\mathrm{d}t} = -i_{L2}D + i_{L1}(1-D)$ $C_2 \frac{\mathrm{d}v_{C2}}{\mathrm{d}t} = i_{L2}(1-D) - \frac{v_O}{R}$	$\frac{V_O}{V_{\text{in}}} = \frac{D+(1-D)^2}{(1-D)^2}$ $\frac{I_{\text{in}}}{I_O} = \frac{D+(1-D)^2}{(1-D)^2}$ $\frac{V_{C1}}{V_{\text{in}}} = \frac{D}{(1-D)}$ $\frac{V_{C2}}{V_{\text{in}}} = \frac{D+(1-D)^2}{(1-D)^2}$ $\frac{I_{L1}}{I_O} = \frac{1}{(1-D)^2}$ $\frac{I_{L2}}{I_O} = \frac{1}{1-D}$	$L_{1\,\min} = \frac{RD(1-D)^4}{f\left(\frac{\Delta i_{L1}}{I_{L1}}\right)\left(D+(1-D)^2\right)}$ $L_{2\,\min} = \frac{RD^2(1-D)^2}{f\left(\frac{\Delta i_{L2}}{I_{L2}}\right)\left(D+(1-D)^2\right)}$
27	$L_1 \frac{\mathrm{d}i_{L1}}{\mathrm{d}t} = v_{\text{in}} - v_{C1}(1-D)$ $L_2 \frac{\mathrm{d}i_{L2}}{\mathrm{d}t} = v_{C1} - v_{C2}(1-D)$	$\frac{V_O}{V_{\text{in}}} = \frac{1+D}{(1-D)^2}$ $\frac{I_{\text{in}}}{I_O} = \frac{1+D}{(1-D)^2}$	$L_{1\,\min} = \frac{RD(1-D)^4}{f\left(\frac{\Delta i_{L2}}{I_{L2}}\right)(1+D)^2}$

(continued)

Table 9.4 (continued)

No.	Dynamic equations	Steady-state equation	Boundary conditions
	$L_3 \frac{di_{L3}}{dt} = 2v_{C2}D + v_{C2}(1-D) - v_O$	$\frac{V_{C1}}{V_{in}} = \frac{1}{(1-D)}$ $\frac{V_{C2}}{V_{in}} = \frac{1}{(1-D)^2}$ $\frac{V_{C3}}{V_{in}} = \frac{(1+D)}{(1-D)^2}$	$L_{2\,min} = \frac{RD(1-D)^2}{f\left(\frac{\Delta i_{L2}}{I_{L2}}\right)(1+D)^2}$
24	$L_1 \frac{di_{L1}}{dt} = v_{in} - v_{C1}(1-D)$ $L_2 \frac{di_{L2}}{dt} = v_{C1} + (v_{C2} - v_O)D$ $C_1 \frac{dv_{C1}}{dt} = i_{L2}D + (i_{L1} - i_{L2} - i_{C2})(1-D)$ $C_2 \frac{dv_{C2}}{dt} = i_{L2}D + (i_{L1} - i_{L2} - i_{C1})(1-D)$ $C_3 \frac{dv_{C3}}{dt} = i_{L2}D - \frac{v_O}{R}$	$\frac{V_O}{V_{in}} = \frac{1+D}{D(1-D)}$ $\frac{I_{in}}{I_O} = \frac{1+D}{D(1-D)}$ $\frac{V_{C1}}{V_{in}} = \frac{1}{(1-D)}$ $\frac{V_{C2}}{V_{in}} = \frac{1}{(1-D)}$ $\frac{V_{C3}}{V_{in}} = \frac{1+D}{D(1-D)}$	$L_{1\,min} = \frac{RD^3(1-D)^2}{f\left(\frac{\Delta i_{L1}}{I_{L1}}\right)(1+D)^2}$ $L_{2\,min} = \frac{RD^2(1-D)}{f\left(\frac{\Delta i_{L2}}{I_{L2}}\right)(1+D)}$
26	$L_1 \frac{di_{L1}}{dt} = v_{in} - v_{C1}(1-D)$ $L_2 \frac{di_{L2}}{dt} = v_{C1} - v_{C2}(1-D)$	$\frac{V_O}{V_{in}} = \frac{2}{(1-D)^2}$ $\frac{I_n}{I_O} = \frac{2}{(1-D)^2}$ $\frac{V_{C1}}{V_{in}} = \frac{1}{(1-D)}$ $\frac{V_{C2}}{V_{in}} = \frac{1}{(1-D)^2}$	$L_{1\,min} = \frac{RD(1-D)^4}{4f\left(\frac{\Delta i_{L1}}{I_{L1}}\right)}$ $L_{1\,min} = \frac{RD(1-D)^3}{4f\left(\frac{\Delta i_{L2}}{I_{L2}}\right)}$

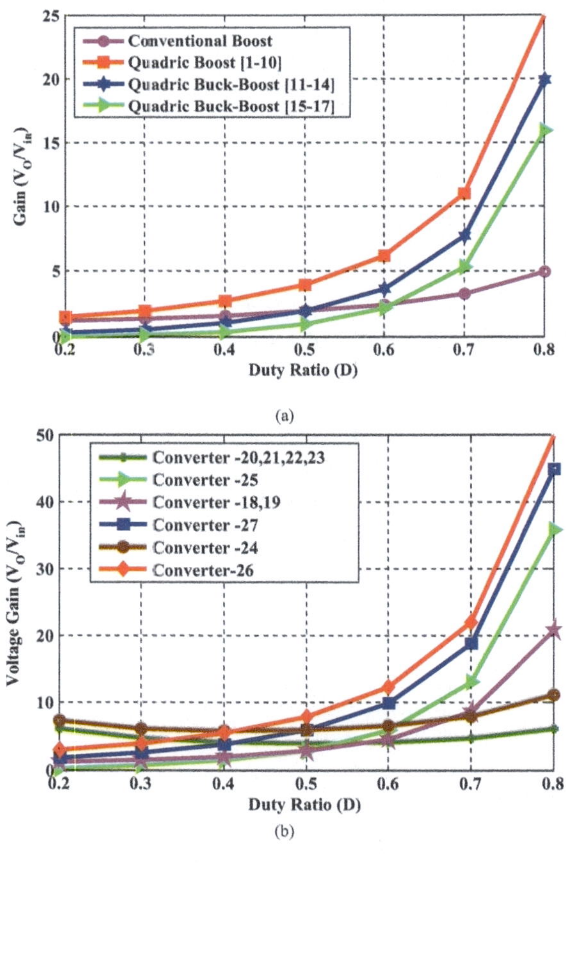

Fig. 9.18 Voltage gain comparison of quadratic boost converters

9.5 Conclusion

High-voltage gain converters with continuous input are required to enter the sustainable power market filling in as a connection between low-voltage sources and inverters. This chapter presented a comprehensive procedure to develop mathematical modeling of two-switch quadratic boost and buck–boost converter circuits for renewable energy applications; followed by devising single-switch equivalent circuits using the Graft scheme. The steady-state and dynamic equations are tabulated for three group of converters, namely quadratic boost $\left(M = \frac{1}{(1-D)^2}\right)$, quadratic buck–boost $\left(M = \frac{D}{(1-D)^2}, \frac{D^2}{(1-D)^2}\right)$ and hybrid quadratic boost converters $\left(M = \frac{1}{D(1-D)}, \frac{1+D}{D(1-D)}, \frac{D(1+D)}{(1-D)^2}, \frac{1+D}{(1-D)^2}, \frac{2}{(1-D)^2}\right)$. The chapter also addressed the design of converters, the impact of inductor's resistance on voltage gain and converter's power loss calculation. Further, key application-specific aspects of selected quadratic boost DC–DC converters are included to help practice engineers.

References

1. Morales-Saldana, J.A., Gutierrez, E.E.C., Leyva-Ramos, J.: Modeling of switch-mode dc-dc cascade converters. IEEE Trans. Aerosp. Electron. Syst. **38**(1), 295–299 (2002). https://doi.org/10.1109/7.993249
2. de Novaes, Y.R., Rufer, A., Barbi, I.: A new quadratic, three-level, DC/DC converter suitable for fuel cell applications. In: 2007 Power Conversion Conference—Nagoya, Nagoya, pp. 601–607 (2007). https://doi.org/10.1109/PCCON.2007.373028
3. Loera-Palomo, R., Morales-Saldaña, J.A.: Family of quadratic step-up dc–dc converters based on non-cascading structures, In: IET Power Electronics, vol. 8, no. 5, pp. 793–801 (2015). https://doi.org/10.1049/iet-pel.2013.0879
4. Naderi, A., Abbaszadeh, K.: High step-up DC–DC converter with input current ripple cancellation. In: IET Power Electronics, vol. 9, no. 12, pp. 2394–2403 (2016). https://doi.org/10.1049/iet-pel.2015.0723
5. Lopez-Santos, O., Mayo-Maldonado, J.C., Rosas-Caro, J.C., Valdez-Resendiz, J.E., Zambrano-Prada, D.A., Ruiz-Martinez, O.F.: Quadratic boost converter with low-output-voltage ripple. In: IET Power Electronics, vol. 13, no. 8, pp. 1605–1612 (2020). https://doi.org/10.1049/iet-pel.2019.0472
6. Leyva-Ramos, J., Ortiz-Lopez, M.G., Diaz-Saldierna, L.H., Morales-Saldana, J.A.: Switching regulator using a quadratic boost converter for wide DC conversion ratios. IET Power Electron. **2**(5), 605–613 (2009). https://doi.org/10.1049/iet-pel.2008.0169
7. Lin, B., Chen, J.: Analysis and implementation of a soft switching converter with high-voltage conversion ratio. In: IET Power Electronics, vol. 1, no. 3, pp. 386–394 (2008). https://doi.org/10.1049/iet-pel:20070315
8. Gaubert, J., Chanedeau, G.: Evaluation of DC-to-DC converters topologies with quadratic conversion ratios for photovoltaic power systems. In: 2009 13th European Conference on Power Electronics and Applications, Barcelona, pp. 1–10 (2009)
9. Lopez-Santos, O., Martinez-Salamero, L., Garcia, G., Valderrama-Blavi, H., Sierra-Polanco, T.: Comparison of quadratic boost topologies operating under sliding-mode control. In: 2013 Brazilian Power Electronics Conference, Gramado, pp. 66–71 (2013). https://doi.org/10.1109/COBEP.2013.6785096

10. Ye, Y., Cheng, K.W.E.: Quadratic boost converter with low buffer capacitor stress. IET Power Electron. **7**(5), 1162–1170 (2014). https://doi.org/10.1049/iet-pel.2013.0205
11. Tymerski, R., Vorperian, V.: Generation, classification and analysis of switched-mode DC-to-DC converters by the use of converter cells. In: INTELEC '86—International Telecommunications Energy Conference, Toronto, Canada, pp. 181–195 (1986). https://doi.org/10.1109/INTLEC.1986.4794425
12. García-Vite, P.M., Soriano-Rangel, C.A., Rosas-Caro, J.C., Mancilla-David, F.: A DC_DC converter with quadratic gain and input current ripple cancelation at a selectable duty cycle. Renew. Energy **101**, 431–436 (2017)
13. Hasanpour, S., Mostaan, A., Baghramian, A., Mojallali, H.: Analysis, modeling, and implementation of a new transformerless semi-quadratic Buck–boost DC/DC converter. Int. J. Circ. Theor. Appl. **47**, 862–883 (2019). https://doi.org/10.1002/cta.2620
14. Jian, F., Zhang, B., Qiu, D., Xiao, W.: A novel single-switch cascaded DC-DC converter of Boost and Buck-boost converters. In: 2014 16th European Conference on Power Electronics and Applications, Lappeenranta, pp. 1–9 (2014). https://doi.org/10.1109/EPE.2014.6910723
15. Azam, F., Priyadarshi, N., Nagar, H., Kumar, S., Bhoi, A.K.: An overview of solar-powered electric vehicle charging in vehicular adhoc network. In: Electric Vehicles. Green Energy and Technology. Springer, Singapore (2021)
16. Rosas-Caro, J.C., Sanchez, V.M., Valdez-Resendiz, J.E., Mayo-Maldonado, J.C., Beltran-Carbajal, F., Valderrabano-Gonzalez, A.: Quadratic buck–boost converter with positive output voltage and continuous input current for PEMFC systems. Int. J. Hydrogen Energy **42**(51), 30400–30406 (2017)
17. Rosas-Caro, J.C., Valdez-Resendiz, J.E., Mayo-Maldonado, J.C., Alejo-Reyes, A., Valderrabano-Gonzalez, A.: Quadratic buck–boost converter with positive output voltage and minimum ripple point design. In: IET Power Electronics, vol. 11, no. 7, pp. 1306–1313 (2018). https://doi.org/10.1049/iet-pel.2017.0090
18. Zhang, N., Zhang, G., See, K.W., Zhang, B.: A single-switch quadratic buck-boost converter with continuous input port current and continuous output port current. IEEE Trans. Power Electron. **33**(5), 4157–4166 (2018). https://doi.org/10.1109/TPEL.2017.2717462
19. Shahir, F.M., Babaei, E., Farsadi, M.: A new structure for nonisolated boost dc–dc converter. J. Circuits Syst. Comput. (JCSC) **26**(1), 1–26 (2016)
20. García-Vite, P.M., Rivera-Espinosa, M.R., Alejandre-López, A.Y., Castillo-Gutiérrez, R., González-Rodríguez, A., Hernández-Angel, F.: Analysis and implementation of a step-up power converter with input current ripple cancelation. Int. J. Circ. Theor. Appl. **46**, 1338–1357 (2018). https://doi.org/10.1002/cta.2478
21. Shahir, F.M., Babaei, E., Farsadi, M.: Extended topology for a boost DC–DC converter. IEEE Trans. Power Electron. **34**(3), 2375–2384 (2019). https://doi.org/10.1109/TPEL.2018.2840683
22. Mohammadzadeh Shahir, F., Babaei, E., Farsadi, M.: Analysis and design of voltage-lift technique-based non-isolated boost dc–dc converter. In: IET Power Electronics, vol. 11, no. 6, pp. 1083–1091 (2018). https://doi.org/10.1049/iet-pel.2017.0259
23. García-Vite, P.M., Rosas-Caro, J.C., Martínez-Salazar, A.L., Chavez, J.d.J., Valderrábano-González, A., Sánchez-Huerta, V.M.: Quadratic buck–boost converter with reduced input current ripple and wide conversion range. In: IET Power Electronics, vol. 12, no. 15, pp. 3977–3986 (2019). https://doi.org/10.1049/iet-pel.2018.5616
24. Luo, F.L., Ye, H.: Positive output cascade boost converters. In: IEE Proceedings—Electric Power Applications, vol. 151, no. 5, pp. 590–606 (2004). https://doi.org/10.1049/ip-epa:20040511
25. Yang, P., Xu, J., Zhou, G., Zhang, S.: A new quadratic boost converter with high voltage step-up ratio and reduced voltage stress. Proceedings of The 7th International Power Electronics and Motion Control Conference, Harbin, pp. 1164–1168 (2012). https://doi.org/10.1109/IPEMC.2012.6258989.
26. Tsai-Fu, Wu., Chen, Y.-K.: A systematic and unified approach to modeling PWM DC/DC converters based on the graft scheme. IEEE Trans. Industr. Electron. **45**(1), 88–98 (1998). https://doi.org/10.1109/41.661309.

Chapter 10
Design and Implementation of Low-Cost Solar Photovoltaic Emulator Utilizing Arduino Controller and DC–DC Buck Converter Topology

Jordan S. Z. Lee, Rodney H. G. Tan, and T. Sudhakar Babu

10.1 Introduction

The energy demand of the world has increased tremendously because of the advancement in technology and the rising of population growth. Non-renewable energy sources will run out at an alarming rate due to the high demands of energy and cannot be replenished. Moreover, non-renewable energy sources have caused many environmental issues, such as CO_2 emission and global warming. Therefore, renewable energy sources have become an important prospect in energy development. Solar energy is the most popular renewable energy because of its abundance and environment friendly. Besides, the cost of solar panels decreased significantly, and the solar panel's efficiency has improved because of the advancement in solar technology in recent years. However, the solar energy source is highly dependent on meteorological factors such as solar irradiance, temperature and cloud.

Many researchers have worked on the improvement of the efficiency and the yield of a solar panel. Maximum power point tracking (MPPT) [1, 2] and partial shading study [3, 4] are important research to improve the yield of a solar panel. MPPT technology is implemented in almost all the solar inverter which is required in a solar plant to match the electrical characteristics of the solar source to the load [5]. A solar panel is required to assess the MPPT algorithm. However, if many solar panels with different power ratings are required to validate the MPPT algorithm, the cost of research will increase tremendously. Besides, the solar panel is affected by the meteorological factor, which varied throughout the day and will cause many inconveniences during the test. Therefore, this chapter will introduce the design of a low-cost PV emulator which has the function of emulating the electrical characteristics of a solar panel. Many researchers have introduced different designs of photovoltaic (PV) emulator by using different modeling approaches, power

J. S. Z. Lee · R. H. G. Tan (✉) · T. S. Babu
UCSI University, Kuala Lumpur, Malaysia
e-mail: rodneytan@ucsiuniversity.edu.my

N. Priyadarshi et al. (eds.), *DC–DC Converters for Future Renewable Energy Systems*, Energy Systems in Electrical Engineering,
https://doi.org/10.1007/978-981-16-4388-0_10

converter topology and controller. MATLAB/Simulink is chosen as the design and simulation platform, and the Arduino Uno is used as the controller for hardware implementation.

10.2 Literature Review

PV emulator has been around for a few decades. In 1985, Easwarakhanthan has proposed a simulator of a PV panel using a programmable voltage generator [6]. Khouzam has proposed a PV emulator using three types of the mathematical model, which are parametric, interpolation model and radiation damage model [7]. In the 2010s, the research value of the PV emulator is growing due to the development of renewable energy. The main objective of the research of the PV emulator is to improve its accuracy. The application of buck converter in a PV emulator is one of the most popular methods in emulating the electrical characteristics of a PV panel. The lookup table method is avoided by most of the researchers because of the large memory requirement. In 2011, Yusivar has proposed a buck converter-based PV emulator using a bode diagram as the analysis method. A state space simulator model is derived by the researchers to model a detailed nonlinear model [8]. In 2013, Rana and Patel have proposed a PV emulator using a current-controlled buck converter. A single-diode model is adopted by the researchers, but the shunt resistance is neglected. Linear and nonlinear loads are used to test the PV emulator's effectiveness, but the IV curve produced has ripple [9]. Balakishan and Sandeep have introduced a PV emulator using a buck converter as well. Various cases such as load change, varied irradiance and temperature changes are simulated by the researchers. The researchers have concluded that when the load resistance and temperature increase, the output power increases, and when the irradiance increases, the output power increases [10]. A few buck converter-based PV emulator using the dSPACE controller for hardware implementation was proposed by a few researchers [11]. The PV emulator's results of Rachid required further improvement as mentioned by the researcher. Azharuddin has included a real-time control feature in his proposed PV emulator [12]. The researcher has further improved the PV emulator by adding MPPT capability using Perturb and Observe algorithm [13] and the hill-climbing method [14]. Mishra has introduced an economical multiple panels PV emulator for research and education. 'Workbench' software is used by the researcher to generate C code to be implemented in hardware [15]. In 2019, Mallal has proposed a PV emulator by using a buck converter as well. Environmental factor such as solar incidence angle is taken into consideration by the researcher [16].

There are other power converters adopted by the researcher in modeling the PV emulator as well. The boost converter is adopted by Koutoulis in modeling the PV emulator with MPPT capability [17]. Lu and Nguyen have proposed a PV emulator employing a buck–boost converter. Piecewise linear method is adopted by the researchers to model the PV emulator [18]. Carralero has introduced a PV emulator using a four-switch buck–boost converter [19]. A DC chopper is used by Zeng to simulate the electrical characteristics of a PV panel [20]. Martin-Segura has proposed a PV emulator with an AC–DC power converter based on a DC–DC full-bridge structure and high-frequency transformer [21]. Castillo has proposed a PV simulator system using the ARM processor. A double-diode model is adopted by the researcher, and a human–system interface is added into the system [22].

Most of the literature above does not have hardware implementation to validate the simulation results. The PV emulator with hardware implementation proposed required high-costs controllers such as dSPACE controller and ARM processor. Some did not elaborate on the methodology clearly, making it difficult to reproduce. In this chapter, the methodology of modeling a PV emulator using buck converter topology is clearly presented, and a low-cost Arduino controller is used for hardware implementation.

10.3 Methodology

This section will present the development of PV emulator. It is divided into four subsections, including PV panel modeling, DC–DC buck converter, PI controller tuning and hardware controller implementation. The overview of PV emulator model developed in MATLAB/Simulink is shown in Fig. 10.1.

It is consists of a current-controlled DC–DC buck converter, PV model and PI controller. For the buck converter to work as a PV emulator, the PV model is integrated into the feedback control path of the buck converter. The PV model will receive the output voltage and current as feedback from the buck converter to compute the reference current. The difference between the reference and output

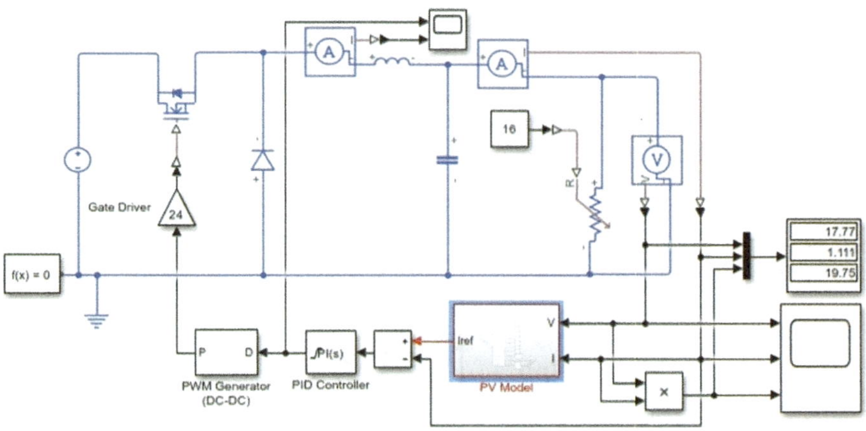

Fig. 10.1 Overview of PV emulator model in MATLAB/Simulink

current of the buck converter will feed to the PI controller to determine the duty cycle which will control the buck converter. MATLAB/Simulink is used to model and develop PV emulator. Arduino Uno is selected for the hardware controller implementation, because it is low cost and has simplicity of code generation from Simulink.

10.3.1 PV Panel Model

A single-diode equivalent circuit is used to model the PV panel [23, 24]. Mathematical equations of the model are derived based on Kirchhoff's current law. [25] Single-diode model is chosen because of its simplicity with good accuracy. Single-diode model equivalent circuit consists of a current source representing the photocurrent generated by the solar irradiance, a diode representing the semiconductor material characteristic and two resistors which are series resistor R_s representing the current losses and shunt resistor R_{sh} representing the leakage current for the diode. Figure 10.2 shows the single-diode model equivalent circuit.

The output current of the single-diode equivalent circuit can be determined using Eq. (10.1).

$$I = I_{ph} - \left(I_s \times e^{\left(\frac{V_o + (I_o \times R_s)}{V_t}\right) - 1} - \right) - \left(\frac{V_o + (I_o \times R_s)}{R_{sh}}\right) \tag{10.1}$$

where I_{ph} is the photocurrent, I_s is the diode reverse saturation current, V_o is the output voltage, I_o is the output current, v_t is the thermal voltage, R_s is the series resistance, and R_{sh} is the shunt resistance. The thermal voltage v_t is computed as shown in Eq. (10.2).

$$v_t = \frac{A \times k \times T_{cell} \times N_s}{q} \tag{10.2}$$

where A is the ideality factor of a diode, k is the Boltzmann constant (1.38×10^{-23}), T_{cell} is the solar cell temperature in Kelvin, N_s is the number of solar cells in series, and q is the electron charge (1.6×10^{-19}). The open-circuit voltage (V_{oc}) and

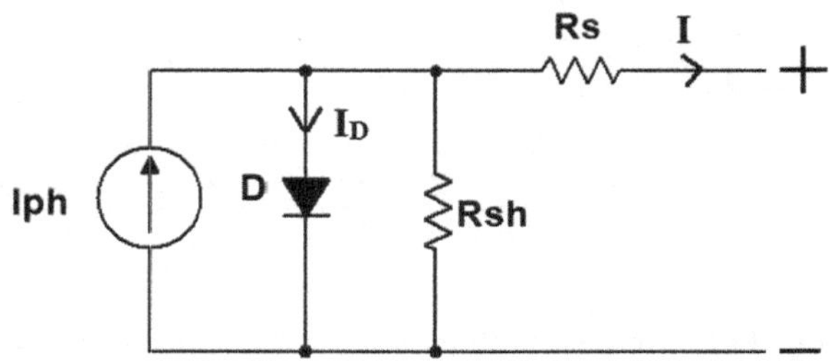

Fig. 10.2 Single-diode equivalent circuit PV model

short-circuit current (I_{sc}) of a PV panel are affected by the temperature of the solar cell. Thus, every PV panel has a temperature coefficient for current and voltage. Most of PV panel models have a similar value of temperature coefficient for current and voltage. The short-circuit current with temperature coefficient (I_{sc_T}) and open-circuit voltage with temperature coefficient (V_{oc_T}) are calculated using Eqs. (10.3) and (10.4), respectively.

$$I_{sc\_T} = I_{sc} + \left(\left(I_{sc} \times \left(\frac{K_I}{100} \right) \right) \times (T - T_{stc}) \right) \quad (10.3)$$

$$V_{oc\_T} = V_{oc} + \left(\left(V_{oc} \times \left(\frac{K_V}{100} \right) \right) \times (T - T_{stc}) \right) \quad (10.4)$$

where K_I is the current temperature coefficient, K_V is the voltage temperature coefficient, T is the temperature of the solar cell in degree Celsius, and T_{stc} is the temperature of the standard testing condition in degree Celsius which is 25 °C. The diode reverse saturation current is calculated using I_{sc_T} and V_{oc_T}, as shown in Eq. (10.5).

$$I_s = \frac{I_{sc\_T}}{e^{\frac{V_{oc\_T}}{v_t}} - 1} \quad (10.5)$$

According to the equation of computing the photocurrent, I_{ph}, as shown in Eq. (10.6), the photocurrent is directly proportional to the irradiance.

$$I_{ph} = I_{sc\_T} \times \left(\frac{G}{G_{stc}} \right) \quad (10.6)$$

where G is the irradiance in W/m^2, and G_{stc} is the irradiance of standard testing condition, which is usually 1000 W/m^2.

The PV model subsystem is implemented using MATLAB function block as shown in Fig. 10.4. There are two inputs and one output for the PV model subsystem. The two inputs are output voltage and output current from buck converter, while the output is reference current. The rest of the inputs are the required PV model parameters. Based on Eq. (10.4), the parameters required to model a PV panel are V_{oc}, I_{sc}, A, R_s, R_{sh}, N_s, K_I and K_V. However, parameters such as R_s, R_{sh} and A are not obtainable from the product datasheet. These parameters can be obtained by using DeSoto algorithm from NREL. In this chapter, a 20 W PV module is used. The PV model inputs parameters are clearly illustrated in Fig. 10.3.

The PV model Eqs. (10.4)–(10.9) are coded using MATLAB script in MATLAB function block as shown in Fig. 10.4.

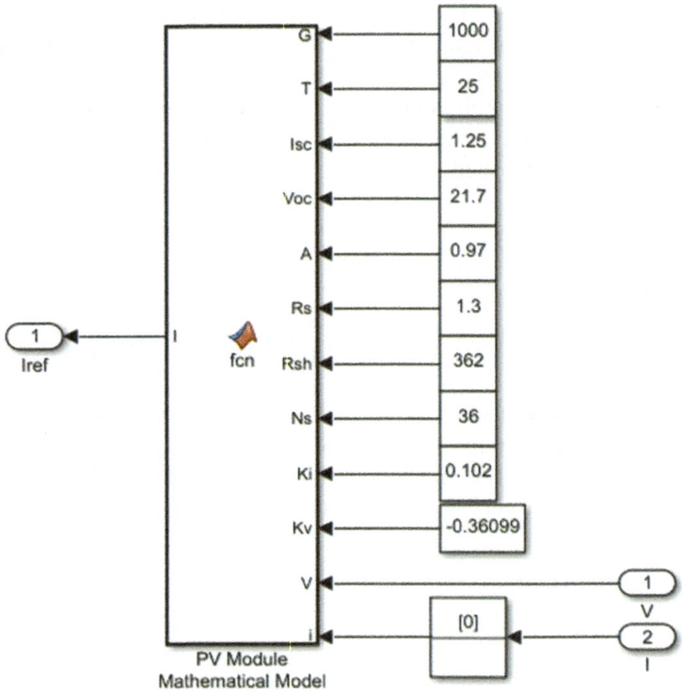

Fig. 10.3 PV model subsystem

```
function I = fcn(G,T,Isc,Voc,A,Rs,Rsh,Ns,Ki,Kv,V,i)

    q = 1.6e-19;                              % Electron charge
    k = 1.38e-23;                             % Boltzmann constant
    Tcell = 273+T;                            % Cell temperature in Kelvin
    vt=(A*k*Tcell*Ns)/q;                      % Thermal voltage Equation(5)

    Isc_T = Isc+((Isc*(Ki/100))*(T-25));      % Isc temperature coef Equation(6)
    Voc_T = Voc+((Voc*(Kv/100))*(T-25));      % Voc temperature coef Equation(7)
    Is = Isc_T/(exp(Voc_T/vt)-1);             % Reverse saturation current Equation(8)
    Iph = Isc_T*(G/1000);                     % Photon current Equation(9)

    % Single diode model output current Equation(4)
    I = Iph-(Is*(exp((V+(i*Rs))/vt)-1))-((V+(i*Rs))/Rsh);
end
```

Fig. 10.4 PV model code in MATLAB function block

10.3.2 DC–DC Buck Converter

The primary function of a DC–DC buck converter is to step down the input voltage to a lower output voltage to the load while maintaining its power. This function is achieved by controlling the switching frequency and duty cycle to the power transistor. Since solar PV panel is a current source device, a current-controlled DC–DC buck converter will be used. This section focuses on the operation principle and design aspect of the DC–DC buck converter. The design of buck converter and PI controller tuning will be discussed. Referring to Fig. 10.1, the DC source is connected to the inductor through the MOSFET power transistor, which acts as a switch. The operation of the MOSFET can be described in two states, namely on state and off state. During on state, the DC source will supply current to the load and charging the inductor simultaneously. The diode works in reverse biased in this state. During off state, the DC source is disconnected to the circuit, and the inductor will discharge and act as the source to continue supply current in the circuit. The diode will work in a forward-biased mode to complete the circuit. To ensure the buck converter is operating in continuous current mode (CCM), the switch needed to close again before the inductor is fully discharged. The duty cycle that controls the MOSFET on and off state switching can be determined by Eq. (10.7), where V_{out} is the output voltage, V_{in} is the input voltage, I_{out} is output current, and I_{in} is input current.

$$D = \frac{V_{out}}{V_{in}} \text{or} \frac{I_{out}}{I_{in}} \tag{10.7}$$

A buck converter can operate in both voltage-controlled and current-controlled mode. In voltage-controlled mode, a reference voltage is used to determine the duty cycle. In current-controlled mode, the reference current computes by the PV model is used to determine the duty cycle as shown in Fig. 10.1. It is important to reduce the ripple magnitude at the output to as low as possible during the design stage. The parameters that will affect the ripple current and voltage at the output are inductor value, capacitor value and the switching frequency. The inductor and capacitor value can be determined using Eqs. (10.8) and (10.9), respectively.

$$L = \frac{V_{in}(1-D)D}{\Delta I \times f} \tag{10.8}$$

$$C = \frac{V_{in}(1-D)D}{8 \times \Delta V \times L \times f^2} \tag{10.9}$$

where L is the inductor value, C is the capacitor value, V_{in} is the input voltage, ΔI is the ripple currentallowable magnitude, ΔV is the ripple voltage allowable magnitude, D is the duty cycle, and f is the switching frequency. According to Eq. (10.8), the ripple current is greatly affected by the inductor, because the inductor is inversely proportional to the ripple current. For ripple voltage, it is affected by the

Table 10.1 Components value and rating

Component	Value/model	Voltage rating	Current rating	Power rating
Inductor	1 mH	–	10 A	–
Capacitor	4700 μF	50 V	–	–
Variable resistor /electronic load	0 Ω to 100 Ω	–	–	300 W
Schottky diode	1N5825	40 V	5 A	–
MOSFET	IRF740	400 V	10 A	–
MOSFET Driver	IR2110	525 V	–	–
Current sensor	ACS712	–	30 A	–
Voltage sensor Resistor network	R1 = 3.9 kΩ R2 = 1 kΩ	–	–	0.25 W

capacitor value according to Eq. (10.9). The current rating for inductor and voltage rating for the capacitor should be also taken into considerations for hardware implementation. This is to ensure the component can withstand the operating power of the buck converter. According to Eqs. (10.8) and (10.9), the frequency will affect both the ripple current and voltage. It is a common practice to increase the switching frequency to reduce the size and cost of the inductor and capacitor. However, the higher switching frequency will contribute to more switching loss, leading to lower conversion efficiency. Hence, this factor should be taken into consideration when determining the switching frequency. The components value set is based on actual parts value that is commercially available for hardware implementation as shown in Table 10.1.

10.3.3 PI Controller Tuning

The feedback control system consists of a PV model to generate the reference current, a PI controller and a PWM generator. The PV model is located at the feedback control path of the buck converter to form a PV emulator model as shown in Fig. 10.1. The reference current is now computed by the PV model based on the output voltage and current of the buck converter. When the load resistance changes, the output voltage and current change, and new corresponding reference current is computed. The PI controller will determine the duty cycle based on the difference between the reference current and output current to achieve the desired output current. The PWM generator will send a pulse width-modulated duty cycle to the gate of MOSFET based on the PI controller output. The PWM signal should pass through a gain block which acts as the gate driver to drive the gate of MOSFET. The tuning of proportional K_p and integral K_i coefficients in the PI controller block are essential to ensure the controller has a fast and smooth response without any

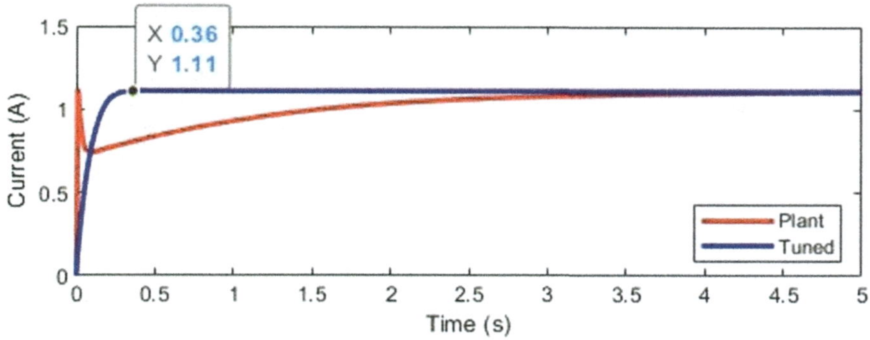

Fig. 10.5 PI tuning based on PV emulator model plant response

overshooting or oscillation during steady state. The PI controller can be tuned by first setting the K_p and K_i to 1, PV model solar irradiance to 1000 and load resistor to 16 Ω which is at the maximum power point. This will produce the step response of the plant as shown in Fig. 10.5.

It can be observed that the plant has a sharp rise and drop at the beginning, and it took up to 5 s to reach the steady state. By slowly decreasing the K_i until the sharp spike reduces below 0.5 A, then slowly increasing the K_i until the transient response time is less than 0.5 s. After some tuning, $K_p = 0.03$ and $K_i = 7$ were decided. The tuned PI controller can reach steady-state smoothly without any spike or oscillation in about 0.36 s, which is about 13.9 times faster than the untune plant model. This transient response time is good enough to emulate real PV panel. Please note that this PI coefficient is only applicable to 20 W PV panel model. Change of PV panel model requires to retune the PI coefficients again.

10.3.4 PV Emulator Controller Hardware Implementation

The overview of the PV emulator hardware implementation is shown in Fig. 10.6. The PV emulator hardware implementation is divided into two parts. There are current-controlled buck converter circuit and PV emulator controller. For the buck converter circuit construction, the components rating selection is very important. The voltage rating of the components required will need to be higher than the DC source, the open-circuit voltage of the PV panel, and the current rating of components required will be higher than the short-circuit current of the PV panel.

To drive the gate of the MOSFET from the Arduino controller, a MOSFET driver is required. A current sensor and voltage divider sensing resistor network is required to measure the output current and voltage, respectively, for the PV emulator controller. The load resistor power rating must be able to withstand the power delivered from the PV emulator. Therefore, a wire-wound-type resistor or electronic

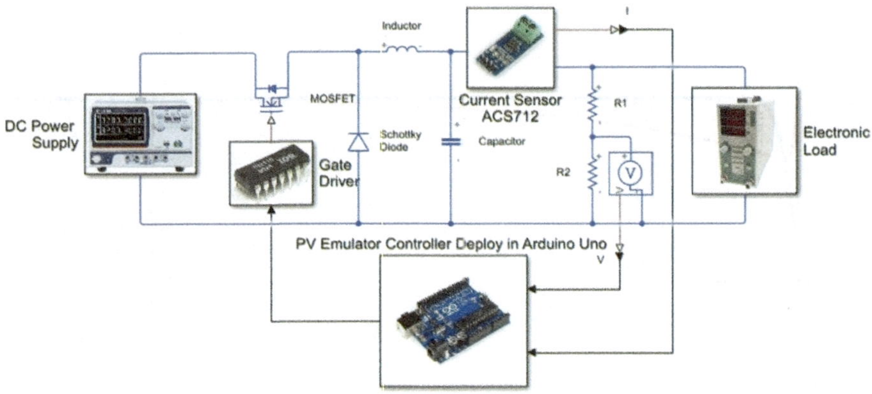

Fig. 10.6 Overview of PV emulator hardware implementation

load is used. Table 10.1 shows the components value and rating used for the construction of the buck converter circuit.

The PV emulator controller hardware part is implemented using the Arduino Uno board. It has the following advantages: There are low cost, availability of MATLAB/Simulink Arduino library block, MATLAB S-function user-defined block and automatic code generation. A PV emulator controller model to be deployed to the Arduino Uno board is developed in MATLAB/Simulink as shown in Fig. 10.7.

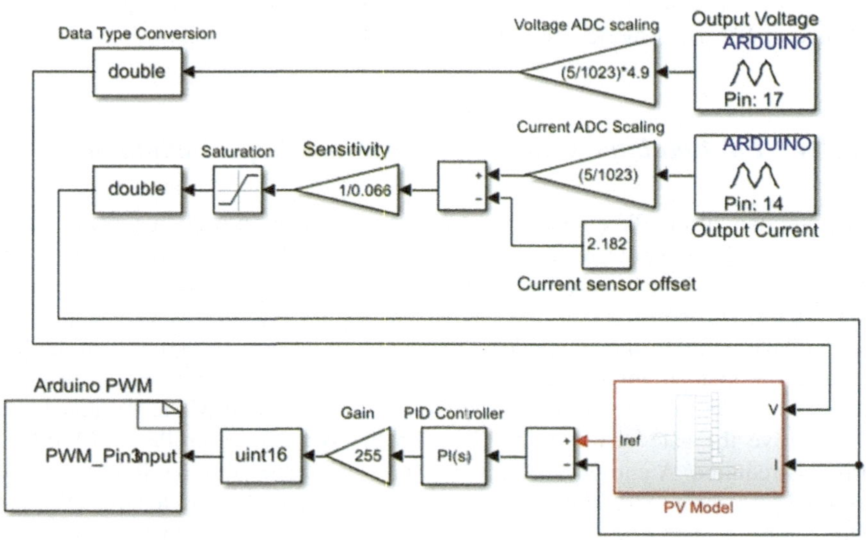

Fig. 10.7 PV emulator controller model to be deployed to Arduino Uno board

The Arduino analog input pin assigned to pin 17 and 14 to receive the output voltage and current from the buck converter circuit, respectively. The output voltage and output current to Arduino Uno are multiplied by a factor of 5/1023 for ADC scaling, because the maximum voltage can be received by Arduino Uno analog input pin is 5 V and 1023 is the maximum digital numerical value for an Arduino 10-bit analog to digital converter (ADC). A voltage divider resistor network form by two resistors is to reduce the buck converter output voltage to under 5 V to be within the readable range by the analog input. Resistor R1 is set as 3.9 kΩ, and R2 is set as 1 kΩ at the output of the buck converter circuit. Thus, the voltage input to Arduino analog input is required to be multiplied by 4.9 based on the voltage divider rule.

Current sensor with different ratings will have different offset voltage and sensitivity. Thus, the input current to Arduino will be compared with the offset voltage and multiplied by 1/sensitivity. The offset voltage 2.182 is obtained by testing the output of the current sensor is zero when there is no input current. The sensitivity of a 30 A current sensor is 66 mA. A saturation block is used to prevent any negative output value. The saturation block parameter value is set to range from 0 to 30. The data-type conversion block is required to change the signal to double data type so that the PV model can read the correct data type of output voltage and current. The PV model reference current is then compared with the buck converter output current, and the difference is sent to the PI controller block to determine the duty cycle. To convert the duty cycle output range from 0 to 1 to the range of 0 to 255, the value of the duty cycle is multiplied by 255.

The Arduino PWM block is implemented using the S-function builder block. The reason for using the S-function builder block is to access the Arduino AT mega328 controller PWM control register directly to set its native 3921.16 Hz PWM frequency. Since the S-function block only accepts a uint16 data type as input, a data type conversion block is used to change the data type to uint16. To set the frequency of 3921.16 Hz, the code to access the PWM control register in the S-function builder block is shown in Fig. 10.8.

The PV emulator hardware setup is shown in Fig. 10.9. It includes a DC–DC buck converter circuit with current sensor and voltage-sensing resistor network, an Arduino UNO board deploy using PV emulator controller model as shown in Fig. 10.6, A laboratory power supply is used as DC source, and a 300 W electronic load or wire wound variable resistor is used as a load. The computer with MATLAB/Simulink is used to deploy the PV emulator controller model to the Arduino UNO board as well as monitoring the PV emulator output voltage and current for a given test load for PV characteristic curve plotting.

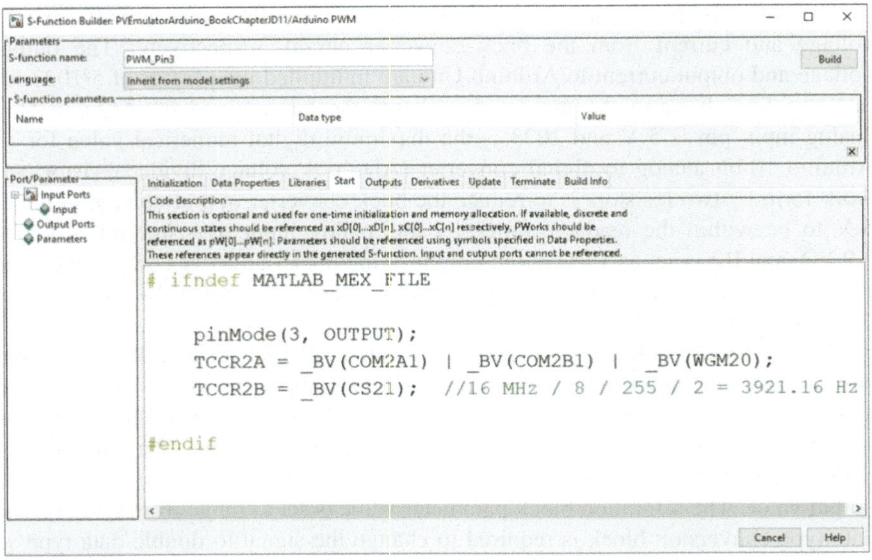

Fig. 10.8 S-function builder block to set the PWM frequency

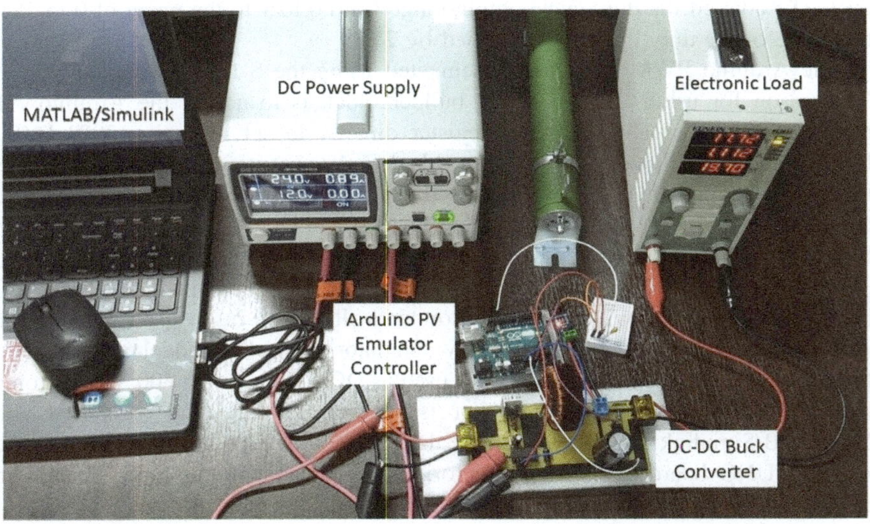

Fig. 10.9 PV emulator hardware prototype setup

10.4 Results and Discussions

The PV emulator simulation model and its hardware prototype performance are evaluated in this section. A reference PV characteristic curve is generated from the PV panel model for evaluation. Mean absolute percentage error (MAPE) is used as a performance indicator.

10.4.1 Reference PV Model Characteristic Curve

A reference PV model characteristic curve is needed to evaluate the performance of the proposed PV emulator. A simple circuit consists of a controlled current source, a variable resistor and the PV panel model as shown in Sect. 4.1, is connected in the feedback path as shown in Fig. 10.10 and is developed to generate the reference PV characteristic curve. The ramp slope block is set to 200 and run for 0.3 s, and the scopes are sampled at 0.01 s. This will sweep the variable resistor value from 2 Ω to 60 Ω in the interval of 2 Ω producing up to 30 measurement points. The voltage and current across the resistor are then fedback to the PV panel model to compute the PV current to the current source.

The voltage and current from the measurement scope are sampled at 0.01 s to plot the PV characteristic curve as shown in Fig. 10.11.

It can be observed from the reference PV characteristic curve that the maximum power of the PV panel is 19.75 W at 17.78 V and 1.111 A which is close to the rated power of 20 W. This PV characteristics curve will be used as a reference to evaluate the performance of the PV emulator simulation model and hardware implementation.

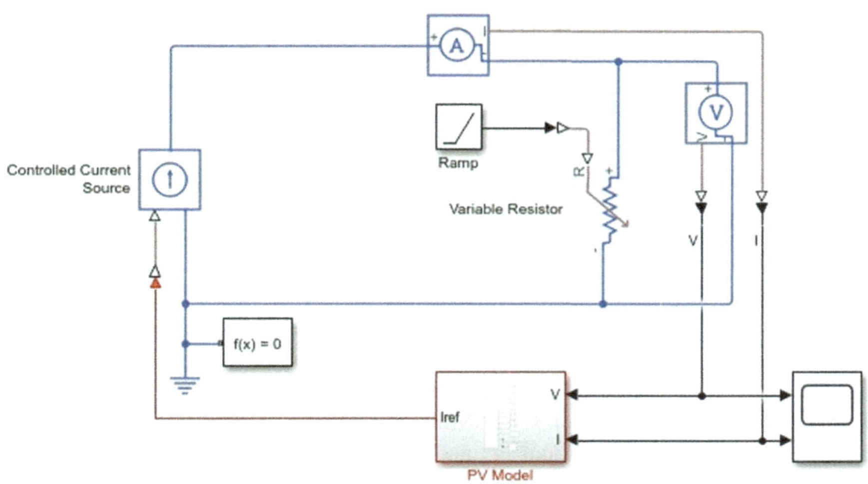

Fig. 10.10 Circuit model to plot reference PV characteristic curve

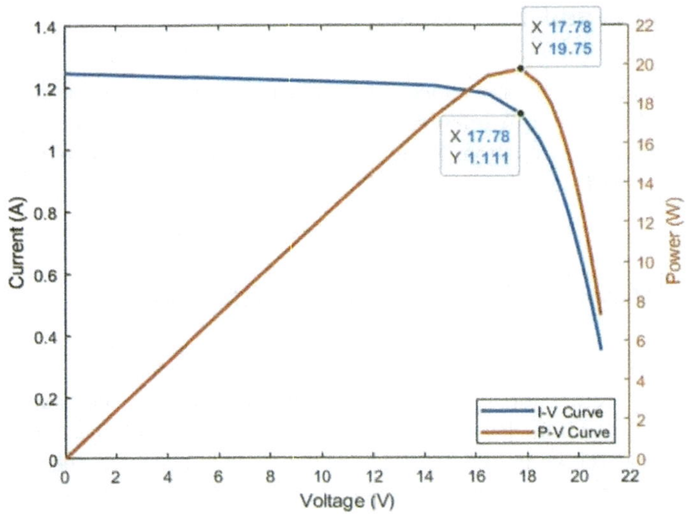

Fig. 10.11 Reference PV current and power characteristic curve

10.4.2 Mean Absolute Percentage Error

Mean absolute percentage error (MAPE) is used as a performance indicator to evaluate how accurate the PV emulator output PV characteristic curve compares with the reference PV characteristic curve. The lower the MAPE, the higher the accuracy of the PV emulator performs. The MAPE is defined in Eq. (10.10), where $P_{\mathrm{PVM}}(V_i)$ is the power of the PV panel model at an output voltage, $P_{\mathrm{PVE}}(V_i)$ is the power of the PV emulator at an output voltage, and n is the number of the output voltage point.

$$\mathrm{MAPE} = \frac{1}{n}\left(\sum_{i=1}^{n}\left|\frac{P_{\mathrm{PVM}}(V_i) - P_{\mathrm{PVE}}(V_i)}{P_{\mathrm{PVM}}(V_i)}\right|\right) \times 100\% \tag{10.10}$$

10.4.3 PV Emulator Simulation Model Performance Evaluation

The evaluation looks into the performance of the PV emulator simulation model accuracy in terms of the PV characteristic curve. The evaluation was carried out with the load resistance running from 2 to 60 Ω in the interval of 2 Ω. This resistance range covers the 20 W PV panel operating voltage and current. Each load resistance simulation runs for 1 s. The voltage, current and power are then recorded for PV curve plotting. The PV current and power characteristic curves are shown in Figs. 10.12 and 10.13, respectively.

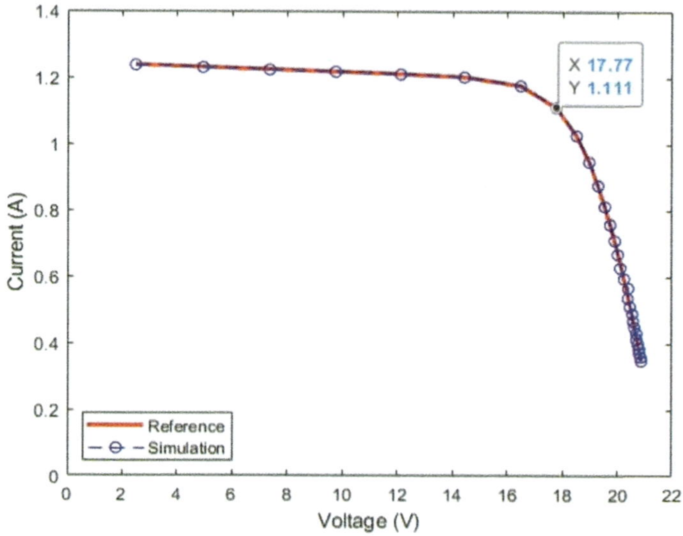

Fig. 10.12 PV emulator simulation *I–V* curve performance

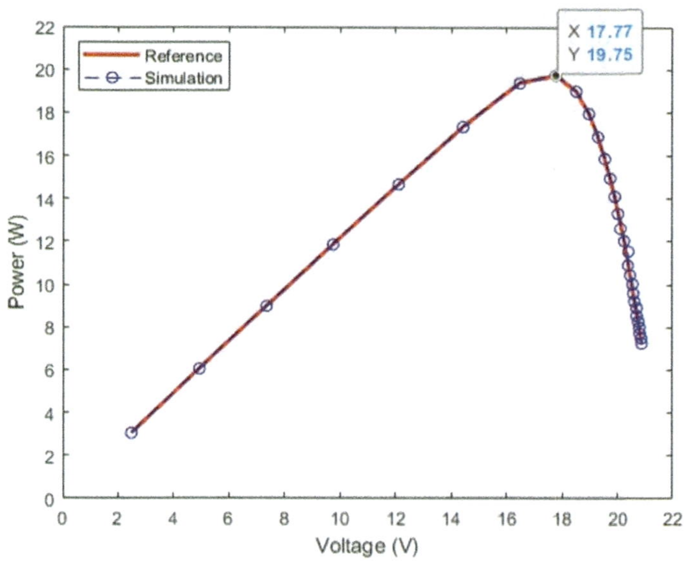

Fig. 10.13 PV emulator simulation *P–V* curve performance

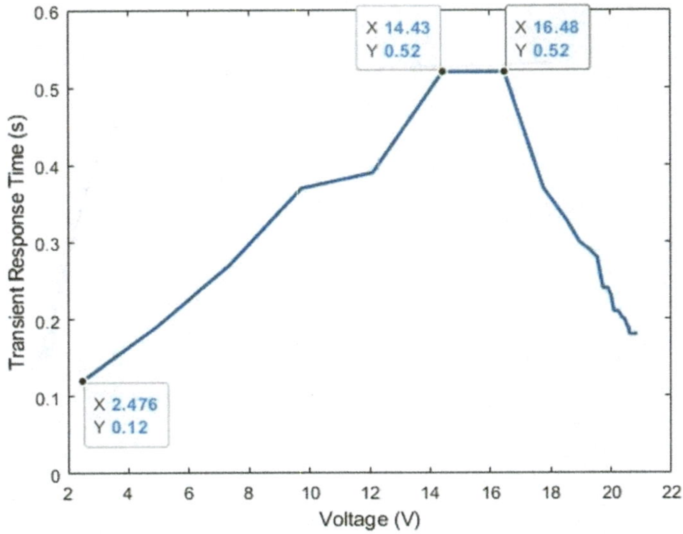

Fig. 10.14 PV emulator model transient response time performance

It is clearly observed that the PV emulator output characteristic curves overlapped on the reference PV model. The maximum power is located at 17.77 V, 1.111 A at 19.75 W with MAPE of 0.12%. Since the simulation model is theoretical, the MAPE error is so low and can be neglected. The PV emulator model transient response time performance is shown in Fig. 10.14.

It can be observed that the lowest tranisent response time is 0.12 s, and the highest transient response time to reach steady state is 0.52 s located between 14.4 and 16.5 V. This is because at this point where the voltage and current are almost at equal high, therefore it took longer time for the voltage and current to reach the steady state. In summary, the PV emulator simulation model performance closely matches the theoretical reference PV model. The average transient response time is about 0.25 s which is good enough to emulate the real PV panel for practical application.

10.4.4 PV Emulator Hardware Prototype Test

This test is to evaluate the PV emulator hardware performance against the reference PV model. Similarly, the hardware PI coefficients $K_p = 0.03$ and $K_i = 7$ are used for this evaluation. The load resistance range use for this test is also from 2 to 60 Ω with the interval of 2 Ω. Electronic load is used to achieve these resistance values with higher precision. The performance of the PV emulator hardware and the reference PV model current and power characteristic curve are shown in Figs. 10.15 and 10.16, respectively. It is be clearly observed that the PV emulator

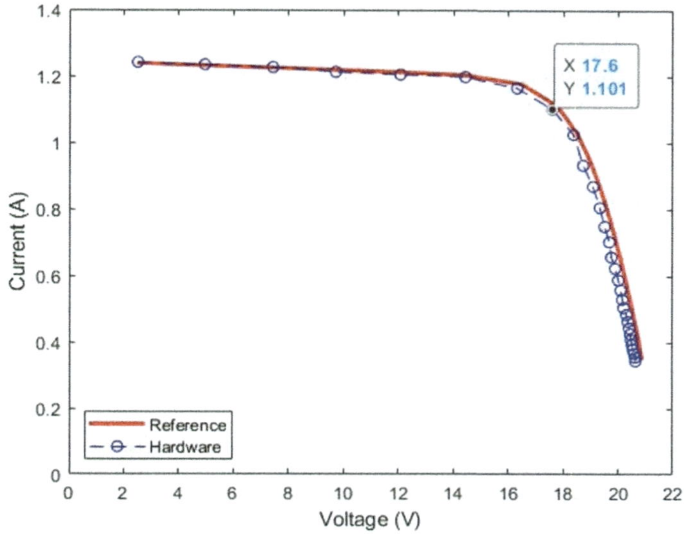

Fig. 10.15 PV emulator hardware and reference *I*–*V* curve performance

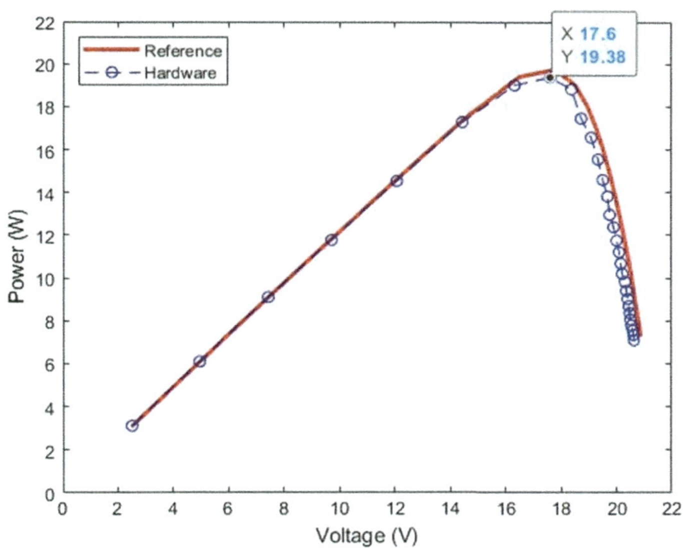

Fig. 10.16 PV emulator hardware and reference *P*–*V* curve performance

hardware PV power characteristic curve closely matches the PV reference model power characteristic curve. The maximum power is located at 17.6 V, 1.101 A at 19.38 W with MAPE of 2.02%. The error is mainly when the voltage approaches the open-circuit voltage, the current decreases as the load resistance increases, and the buck converter will begin to operate in discontinuous current mode (DCM). In DCM, the output voltage and current ripple will increase, hence causing the voltage error. This issue can be resolved by either increasing the inductor size, switching frequency or integral coefficient K_i. The other error factors were mainly contributed by the tolerance of the buck converter hardware components and the feedback current sensor sensitivity and offset stability.

This evaluation proves that the proposed PV emulator simulation model developed in MATLAB/Simulink can be practically implemented using low-cost Arduino controller to achieve a relatively low error performance.

10.5 Conclusion

The design and implementation of PV emulator using low-cost Arduino controller are presented. The design begins with current-controlled DC–DC buck converter modeling, PV panel modeling based on first principle single-diode equivalent circuit and PI controller in the MATLAB/Simulink environment. Followed by developing the PV emulator controller model for Arduino controller, deployment and the selection of the hardware components are discussed. The performance evaluation looks into how the PI controller coefficients affect the accuracy of the PV emulator in terms of PV characteristic power curve in the simulation environment. Lastly, the performance of the PV emulator hardware prototype is validated with the PV emulator simulation result. The performance of the PV emulator hardware achieved a MAPE of 2.02%. The presented PV emulator is a useful tool to emulate solar PV panel output with any weather conditions in the laboratory. The MATLAB/Simulink model and hardware prototype of the PV emulator contributed to the development of solar PV energy research as well as education learning curriculum. The PV emulator model presented in this chapter is made available for the reader to download from Mathworks official MATLAB central file exchange link below for education purpose.

https://www.mathworks.com/matlabcentral/fileexchange/82853-pv-emulator-model.

References

1. Karami, N., Moubayed, N., Outbib, R.: General review and classification of different MPPT. Techniques (2017). https://doi.org/10.1016/j.rser.2016.09.132
2. Podder, A.K., Roy, N.K., Pota, H.R.: MPPT Methods for Solar PV systems: A Critical Review Based on Tracking Nature (2019). https://doi.org/10.1049/iet-rpg.2018.5946
3. Priyadarshi, N., Anand, A., Sharma, A.K., Azam, F., Singh, V.K., Sinha, R.K.: An experimental implementation and testing of GA based maximum power point tracking for PV system under varying ambient conditions using dSPACE DS 1104 controller. Int. J. Renew. Energy Res. **7**(1), 255–265 (2017)

4. Gonzalez Montoya, D., Bastidas-Rodriguez, J.D., Trejos-Grisales, L.A., Ramos-Paja, C.A., Petrone, G., Spagnuolo, G.: A procedure for modeling photovoltaic arrays under any configuration and shading conditions. Energies **11**, 767 (2018). https://doi.org/10.3390/en11040767
5. Komerath, N.M., Komerath, P.P.: Terrestrial micro renewable energy applications of space technology. Phys. Proc. (2011). https://doi.org/10.1016/j.phpro.2011.08.024
6. Easwarakhanthan, T., Bottin, J., El-Slassi, A., Ravelet, R., Ravelet, S.: Microcomputer-controlled simulator of a photovoltaic generator using a programmable voltage generator. Sol. Cells. (1986). https://doi.org/10.1016/0379-6787(86)90025-6
7. Khouzam, K., Hoffman, K.: Real-time simulation of photovoltaic modules. Sol. Energy. (1996). https://doi.org/10.1016/0038-092X(96)00008-4
8. Yusivar, F., Farabi, M.Y., Suryadiningrat, R., Ananduta, W.W., Syaifudin, Y.: Buck-converter photovoltaic simulator. Int. J. Power Electron. Drive Syst. (2011). https://doi.org/10.11591/ijpeds.v1i2.92
9. Rana, A.V., Patel, H.H.: Current controlled buck converter based photovoltaic emulator. J. Ind. Intell. Inf. (2013). https://doi.org/10.12720/jiii.1.2.91-96
10. Balakishan, C.H., Sandeep, N.: Development of a microcontroller based PV emulator with current controlled DC-DC buck converter. Int. J. Renew. Energy Res. (2014). https://doi.org/10.20508/ijrer.00586
11. Rachid, A., Kerrour, F., Chenni, R., Djeghloud, H.: PV emulator based buck converter using dSPACE controller. In: EEEIC 2016—International Conference on Environment and Electrical Engineering (2016). https://doi.org/10.1109/EEEIC.2016.7555762
12. Azharuddin, S.M., Vysakh, M., Thakur, H.V., Nishant, B., Babu, T.S., Muralidhar, K., Paul, D., Jacob, B., Balasubramanian, K., Rajasekar, N.: A near accurate solar PV emulator using dSPACE controller for real-Time control. Energy Proc. (2014). https://doi.org/10.1016/j.egypro.2014.12.266
13. Sudhakar Babu, T., Mohammed Azharuddin, S., Nishant, B., Rajasekar, N.: A dynamic photo voltaic emulator using dSPACE controller with high accuracy solar photo voltaic characteristics. J. Renew. Sustain. Energy. (2016). https://doi.org/10.1063/1.4940414
14. Priyadarshi, N., Padmanaban, S., Bhaskar, M.S., Blaabjerg, F., Sharma, A.: A Fuzzy SVPWM based inverter control realization of grid integrated PV-wind system with FPSO MPPT algorithm for a grid-connected PV/wind power generation system: hardware implementation. IET Electric Power Appl., pp. 1–12 (2018)
15. Mishra, S., Raju, S., Kshirsagar, A., Mohan, N.: An extremely low-cost multi-panel PV emulator for research and education. In: 2018 IEEE Energy Conversion Congress and Exposition, ECCE 2018 (2018). https://doi.org/10.1109/ECCE.2018.8557361
16. Mallal, Y., El Bahir, L., Hassboun, T.: High-performance emulator for fixed photovoltaic panels. Int. J. Photoenergy. (2019). https://doi.org/10.1155/2019/3951841
17. Koutroulis, E., Kalaitzakis, K., Voulgaris, N.C.: Development of a microcontroller-based, photovoltaic maximum power point tracking control system. IEEE Trans. Power Electron. (2001). https://doi.org/10.1109/63.903988
18. Lu, D.D.C., Nguyen, Q.N.: A photovoltaic panel emulator using a buck-boost DC/DC converter and a low cost micro-controller. Sol. Energy. (2012). https://doi.org/10.1016/j.solener.2012.02.008
19. Carralero, L.L.O., Barbara da, G.S., Silva, S., Costa, F.F., Tahim, A.P.N.: PV emulator based on a four-switch buck-boost DC-DC converter. In: 2019 IEEE 15th Brazilian Power Electronics Conference and 5th IEEE Southern Power Electronics Conference, COBEP/SPEC (2019). https://doi.org/10.1109/COBEP/SPEC44138.2019.9065364
20. Zeng, Q., Song, P., Chang, L.: A photovoltaic simulator based on dc chopper. Can. Conf. Electr. Comput. Eng. (2002). https://doi.org/10.1109/CCECE.2002.1015221
21. Martín-Segura, G., López-Mestre, J., Teixidó-Casas, M., Sudrià-Andreu, A.: Development of a photovoltaic array emulator system based on a full-bridge structure. In: 2007 9th International Conference on Electrical Power Quality and Utilisation, EPQU (2007). https://doi.org/10.1109/EPQU.2007.4424165

22. Castillo Atoche, A., Vázquez Castillo, J., Ortegón-Aguilar, J., Carrasco-Alvarez, R., Sandoval Gío, J., Colli-Menchi, A.: A high-accuracy photovoltaic emulator system using ARM processors. Sol. Energy. (2015). https://doi.org/10.1016/j.solener.2015.05.026
23. Padmanaban, S., Priyadarshi, N., Holm-Nielsen, J.B., Bhaskar, M.S., Azam, F., Sharma, A. K.: A novel modified sine-cosine optimized MPPT algorithm for grid integrated PV system under real operating conditions. IEEE Access **7**, 10467–10477 (2019). https://doi.org/10.1109/ACCESS.2018.2890533
24. Padmanaban, S., Priyadarshi, N., Holm-Nielsen, J.B., Bhaskar, M.S., Hossain, E., Azam, F.: A hybrid photovoltaic-fuel cell for grid integration with jaya-based maximum power point tracking: experimental performance evaluation. IEEE Access **7**, 82978–82990 (2019). https://doi.org/10.1109/ACCESS.2019.2924264
25. Priyadarshi, N., Ramachandaramurthy, V.K., Padmanaban, S., Azam, F.: An ant colony optimized MPPT for standalone hybrid PV-wind power system with single cuk converter. Energies **12**(1), 167 (2019)
26. De Soto, W., Klein, S.A., Beckman, W.A.: Improvement and validation of a model for photovoltaic array performance. Sol. Energy. (2006). https://doi.org/10.1016/j.solener.2005.06.010

Chapter 11
DC–DC Converter for RES-Based Smart Resilient LVDC Distribution System

M. Ankush Kumar and A. Jaya Laxmi

11.1 Introduction

In present world of smart cities, the utilization of the electronic gadgets like mobiles, laptops, LED-based lighting systems, BLDC-based fans, etc., is increasing day by day. On the other hand, utilization of batteries-based home inverter systems as alternate power sources is also increased. Also, the RES are replacing the conventional methods of electricity generation to keep the environment clean. These batteries and the green sources mostly provide electricity in the form of DC. The power generated from the RES is varying in nature due to the varying property of natural resources like sunlight and wind. These variations may lead to the damage of end equipment. As the natural resources which are being converted to electricity cannot be made constant at all times, the only opportunity is to make the generated power constant. This is achieved with the help of a DC–DC converter. Thus, a DC–DC converter plays a major role in utilizing the power from any green energy or renewable energy source [1].

In literature, many converter topologies are presented to convert a variable DC power to a constant DC. Most of the renewable energy applications use DC–DC converters like buck, boost, buck–boost, cuk, sepic and zeta converters [2, 3, 4, 5, 6, 7]. These converters basically track the generated voltage from RES and increase or decrease the voltage as per the reference value. Hence, these are also named as charge controllers such that these provide a constant voltage DC power to charge the batteries. The power available from batteries is further utilized for loads.

To utilize this battery power for AC loads, it needs one more stage of power conversion based on inverter or multilevel inverter, which leads to the increased

M. Ankush Kumar (✉)
Department of EEE, Lords Institute of Engineering and Technology, Hyderabad, India

A. Jaya Laxmi
Department of EEE, JNTUHCEH, Hyderabad, India

N. Priyadarshi et al. (eds.), *DC–DC Converters for Future Renewable Energy Systems*, Energy Systems in Electrical Engineering,
https://doi.org/10.1007/978-981-16-4388-0_11

penetration of harmonics and in turn leads to the increased loss of power and also reduces the life of end equipment. So as to avoid these situations, LVDC power for residential applications is preferred in this paper [8]. The voltage level of the LVDC system for Indian standards is proposed as 48 V DC [9].

In this work, a prototype LVDC distribution system of 48 V DC is designed in real time and powered from the integrated DC microgrid made up of SPV and BLDC-based WECS. In this system, a DC–DC converter for each source is utilized to obtain a constant DC power of 48 V from each source. This makes it possible for simpler integration of the RES [10]. The LVDC distribution system is also provided with automation system for continuous monitoring and protection.

11.2 DC–DC Converter Topology

As discussed in earlier section, DC–DC converter plays a major role in developing the proposed LVDC distribution system. This simple converter provides many advantages like simple integration of RES, battery storage and a constant DC power to loads. In this work, a simple DC–DC buck–boost converter topology is utilized for each source. This converter senses the output voltage of the source and compares it with the reference value, i.e., 48 V. If the voltage is less than the reference voltage, then this converter works as boost converter and increases the voltage to 48 V DC. If the voltage is greater than the reference value, then this converter works as a buck converter and decreases the output voltage. Likewise a constant DC output of 48 V DC is obtained from this converter. This constant DC voltage level helps in charging the batteries. Hence, this converter is also named as a charge controller. The topology of the buck boost converter is shown in Fig. 11.1.

Figure 11.1 depicts the topology of converter used in this work. This converter topology is generally named as four-switch converter topology. It is observed from figure that this topology uses four switches Q_1, Q_2, Q_3 and Q_4. The switches Q_1 and Q_3 work as one group, and Q_2 and Q_4 work as another group, such that when Q_1 and Q_3 are turned on, the switches Q_2 and Q_4 will be turned OFF, and vice versa. The two groups of switches help to operate the converter as boost or buck mode of

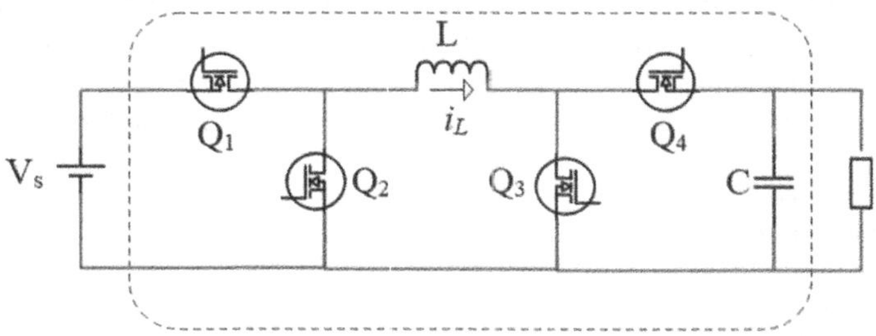

Fig. 11.1 Four-switch converter

operation, respectively. Thus, the voltage can be regulated accordingly to achieve the constant voltage output.

11.3 Integration of SPV and WECS

In this work, the RES considered are SPV system and BLDC-based WECS. The power obtained from SPV system is in DC naturally [1, 11]. The power generated by the WECS is internally AC, but the recent advancements, like BLDC-based PMSG machines, have led to the development of DC power generation from WECS. These provide many advantages like removal of gear systems, reduced losses, low maintenance and high reliability, etc. Due to these advantages, BLDC-based WECS are considered as second source such that, during sun hours, the SPV provides power to the system and during off sun hours, the WECS can provide the power to the distribution system. In this regard, a hybrid DC microgrid is designed with SPV and WECS.

To achieve this, in the first stage, with the help of DC–DC converter, a constant DC power is achieved from the SPV and BLDC-based WECS individually. The generated DC power from both the sources is stabilized at a voltage of 48 V DC with the help of DC–DC converters as discussed in earlier section. This provides an opportunity to combine the sources without the requirement of synchronization. These two sources now can be combined as two parallel sources working at same voltages. The block diagram of the integration of two sources in parallel as DC sources is shown in Fig. 11.2.

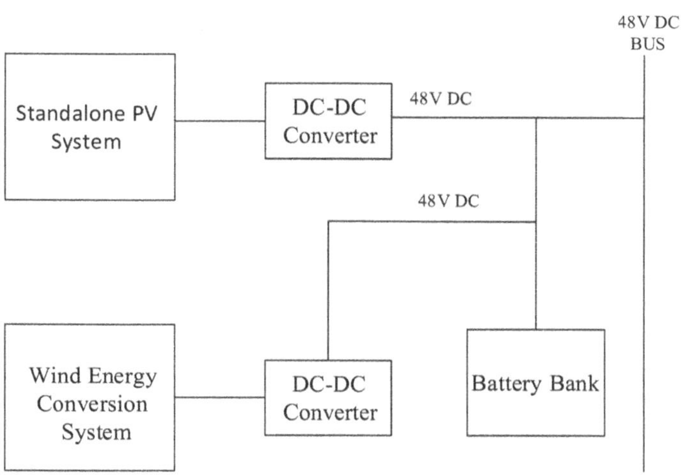

Fig. 11.2 Integration of RES in parallel as DC sources

Fig. 11.3 Real-time installation of hybrid SPV and WECS

The figure depicts the simple integration of two sources as two parallel DC sources working at same voltage. This simplest integration of two sources is possible only due to the DC–DC converter. Thus, a DC–DC converter helps in designing a hybrid DC microgrid with SPV and WECS. The DC microgrid installed in real time is shown in Fig. 11.3.

It is observed from Fig. 11.3 that the PV panels are placed at a regular angle of 17° from ground as per Indian standards such that maximum power is achieved from sunlight. The WECS system is designed using a vertical axis BLDC-based PMSG machine such that the vertical axis helps to generate power even at low wind speeds.

11.4 LVDC Distribution System

As per the present load requirements in the world, many articles show the importance of the LVDC systems while utilizing power from RES. Few articles discuss the economical analysis of the SPV-based LVDC systems. These articles show that, due to the removal of the inverter, the proposed RES-based LVDC is economical when compared to the AC system [8]. As discussed in introduction for Indian standards, a 48 V DC is proposed as base voltage for the LVDC distribution system. Thus, in this work, the obtained power from DC microgird is utilized in a LVDC distribution system such that the number of conversion stages is reduced which leads to the reduced harmonics and reduced losses in the system.

To obtain control and protection in the LVDC distribution system, an automation system is also installed in this system. The automation system installed is

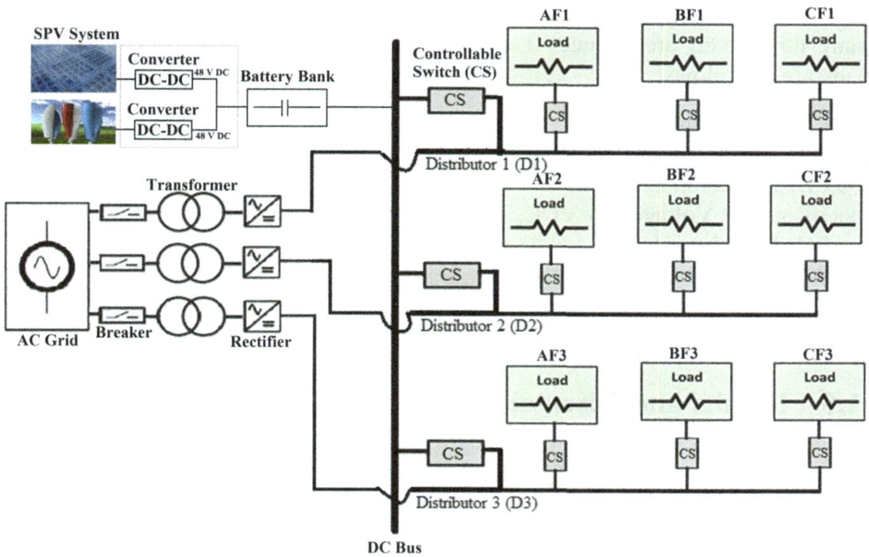

Fig. 11.4 Smart LVDC distribution system

supervisory control and data acquisition (SCADA) system. The automation system provides live data across each and every line and load to the server system at remote end. This provides an opportunity to make possible decisions from remote end. This also helps to create intentional islands of the DC loads with RES. The block diagram of the smart LVDC distribution system is shown in Fig. 11.4.

It is observed from Fig. 11.4 that the LVDC distribution system is designed with three radial DC feeders and three loads on each feeder. All loads are designed to work according to the standards, i.e., 48 V DC. The LVDC distribution system is powered from main grid through transformer and rectifier circuit such that, the three-phase AC power obtained from grid is stepped down to 48 V AC and then rectified to DC with the help of rectifier circuit. Thus, 48 V DC is obtained to the distribution system from main grid. Further, the LVDC distribution system is also powered from the hybrid DC microgird in parallel to the main grid. Figure 11.4 depicts that while connecting the microgrid to the LVDC system, controllable switches (CS) are used across each feeder. These CS help to connect or disconnect the DC microgrid to the loads from remote end using SCADA.

In real-time system, the CS shown in block diagram are replaced with the feeder monitoring units (FMU) of the SCADA hardware. Each FMU includes a CS operated based on the binary data, i.e., 0 or 1. A binary '0' turns OFF the switch and a binary '1' turns the switch ON. This control is obtained from the remote end server system. Every line (DC feeder) is connected to the DC microgird through FMU such that the connection of loads RES is controlled by SCADA. On the other hand, the same feeders are connected to grid through the rectifier. This rectifier converts a three single-phase AC lines to three DC feeders. Similar to block diagram, three loads are connected on each DC feeder. The specifications of these loads are as follows:

Load A: DC Heater Load
Specifications: Voltage: 48 VDC, Power: variable from 50 to 250 W.
Load B: DC Lamp Load
Specifications: Voltage: 48 VDC, Power: variable from 50 to 250 W.
Load C: DC Motor Load
Specifications: Voltage: 48 VDC, Power: variable from 50 to 250 W.

11.5 Results and Discussion

In earlier sections, the design of hybrid DC microgrid with two RES, i.e., SPV and WECS, is discussed. The design of the smart LVDC distribution system is also discussed in brief. Further, this section describes the obtained results in two case studies. First case study is when the LVDC distribution system is powered from main grid, and the second case study is when the distribution system is powered from hybrid DC microgrid.

11.5.1 LVDC Distribution System Powered from Main Grid

In this mode of operation, the LVDC distribution system is powered from main grid. The three phases R, Y and B are being converted to DC. It is observed that the rectifier unit is converting the R, Y and B phases to three DC feeders of 48 V each. Three DC loads, as discussed in earlier section, are connected on each DC feeder. It is observed that all the loads are powered through main grid and successfully working. In this mode of operation, the power generated from the DC microgrid is stored in batteries. The power flow to batteries through charge controller (DC–DC converter) is shown in Fig. 11.5.

It is observed from Fig. 11.5 that the power generated from DC microgrid is 260 W. This is given to batteries. There is no power delivered to loads in this mode of operation. The status of all the FMUs and the loads recorded in the remote server system are shown in Fig. 11.6.

It is observed from Fig. 11.6 that the FMUs of the three DC feeders (F1, F2 and F3) are OFF and all the loads (AF1, BF1, CF1, AF2, BF2, CF2, AF3, BF3, CF3) are supplied from main grid with respective power requirements.

Fig. 11.5 Data logger across batteries

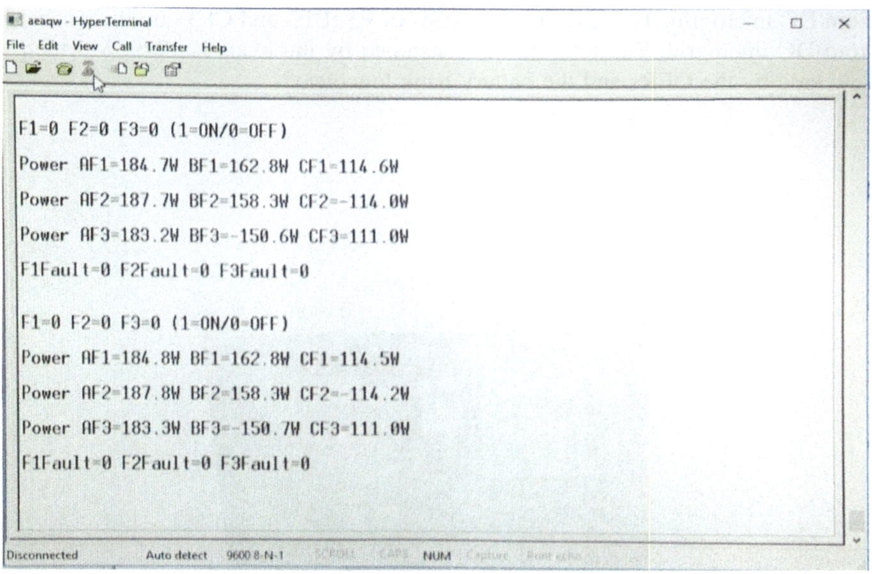

Fig. 11.6 Status of CS and loads in the remote system

11.5.2 LVDC Distribution System Powered from DC Microgrid

In this mode of operation, the LVDC distribution system is disconnected from the main grid with the help of rectifier unit. The rectifier unit is turned OFF by using FPST switch.

It is observed that the rectifier unit is OFF, and there is no power conversion from AC to DC. The LVDC loads are disconnected from the main grid. In this situation, the DC microgird is connected to the LVDC distribution system by controlling the FMU of DC feeders. The power flow across the charge controller (DC–DC converter) is shown in data logger in Fig. 11.7.

It is observed from Fig. 11.7 that the power generated from DC microgrid is 334.5 W and with the available battery power, the loads of feeder 1, 2 and 3 are powered and the least priority loads (BF3 and CF3) of F3 are kept OFF with the help of CS to satisfy the available power. The power delivered to the loads is 1212 W. The status of all FMUs and the loads observed in remote server system are shown in Fig. 11.8.

It is observed from Fig. 11.8 that the FMUs of the three DC feeders (F1, F2 and F3) are ON and the loads AF1, BF1, CF1, AF2, BF2, CF2 and AF3 are supplied from DC microgrid. The least priority loads of F3 (BF3 and CF3) are disconnected from DC microgrid. The total power consumed by the loads is 1229 W which is delivered by the DERs and the battery bank together.

Fig. 11.7 Data logger

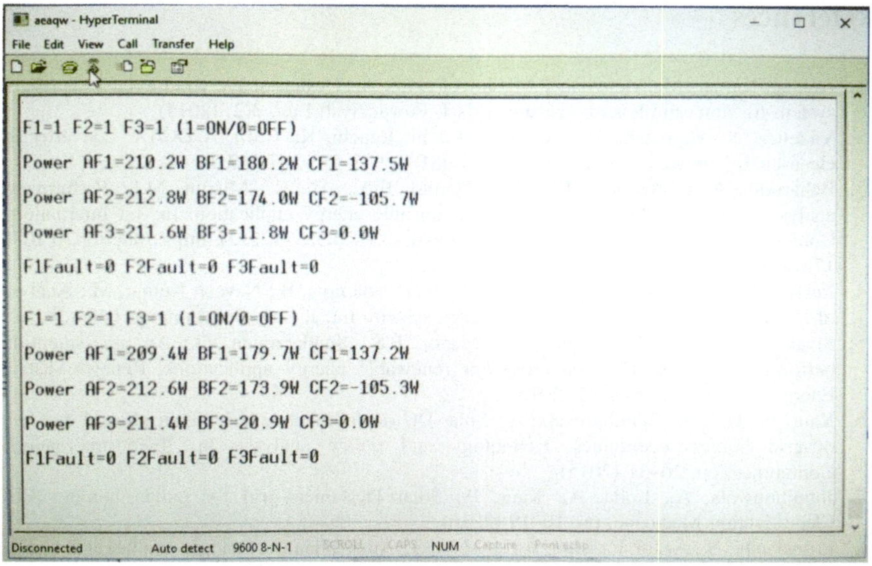

Fig. 11.8 Status of CS and loads in the remote system

11.6 Conclusion

A DC–DC converter plays a major role to obtain a constant DC power output from the varying resources like SPV system and WECS. This paper describes the effective utilization and the importance of the DC–DC converters for RES or green energy applications. The simple design of the hybrid DC microgrid in real time with SPV and WECS without the need of synchronization describes the role of the DC–DC converters. Also the working of real-time smart LVDC distribution system while powered from RES is completely dependent on DC–DC converters. Results of the system describe the successful operation of the smart LVDC loads at constant voltage of 48 V DC obtained from the converters. It is observed from the system design that the LVDC distribution system helps in reducing the number of conversion stages when powered from RES. It is concluded that the DC–DC converters play a major role in interconnecting the RES and also operating the smart LVDC distribution system in real time.

Acknowledgements Authors are thankful to the Centre of Excellence under TEQIP-II, JNTU Hyderabad, for giving us an opportunity to be a part of the prestigious project titled Disaster Management and supporting us to implement the project in real time.

References

1. Ankush Kumar, M., Menghal, P.M., Jaya Laxmi, A.: Design of PV-wind hybrid micro-grid system for domestic loading. i-manager's J. Power Syst. Eng. **3**(2) (2015)
2. Vázquez, N., Hernández, C., Vázquez, E.: In: Eguchi, K., (Ed.), A DC/DC converter for clean-energy applications. InTech (2010). ISBN: 978-953-307-147-3
3. Baharudin, N.H., Mansur, T.M.N.T., Hamid, F.A., Ali, R., Misrun, M.I.: Performance analysis of DC-DC buck converter for renewable energy application. In: 1st International Conference on Green and Sustainable Computing (ICoGeS) (2017). https://doi.org/10.1088/1742-6596/1019/1/012020.
4. Janaki, P., Prasanthi, Y., Vijaya Laxmi, S., Narayanamma, B., Naveen Kumar, M.: Analysis of DC-DC converters for renewable energy system. Int. J. Eng. Sci. Comput. **6**(4) (2016)
5. Sivakumar, S., Jagabar Sathik, M., Manoj, P.S., Sundararajan, G.: An assessment on performance of DC–DC converters for renewable energy applications. Renew. Sustain. Energy Rev. **58**, 1475–1485 (2016)
6. Kaur, P., Jain, S., Jhunjhunwala, A.: Solar-DC deployment experience in off-grid and near off-grid homes: economics, technology and policy analysis. In: IEEE International Conference, pp. 26–31 (2015)
7. Jhunjhunwala, A., Lolla, A., Kaur, P.: Solar DC micro-grid for indian homes. IEEE Electrification Magazine, pp. 10–19 (2016)
8. Priyadarshi, N., Azam, F., Bhoi, A.K., Alam, S. An artificial fuzzy logic intelligent controller based MPPT for PV grid utility. In: Lecture Notes in Networks and Systems, vol. 46. https://doi.org/10.1007/978-981-13-1217-5_88
9. Padmanaban, S., Priyadarshi, N., Holm-Nielsen, J.B., Bhaskar, M.S., Azam, F., Sharma, A. K.: A novel modified sine-cosine optimized MPPT algorithm for grid integrated PV system under real operating conditions. IEEE Access **7**, 10467–10477 (2019). https://doi.org/10.1109/ACCESS.2018.2890533
10. Padmanaban, S., Priyadarshi, N., Holm-Nielsen, J.B., Bhaskar, M.S., Hossain, E., Azam, F.: A hybrid photovoltaic-fuel cell for grid integration with jaya-based maximum power point tracking: experimental performance evaluation. IEEE Access **7**, 82978–82990 (2019). https://doi.org/10.1109/ACCESS.2019.2924264
11. Priyadarshi, N., Padmanaban, N., Holm-Nielsen, J.B., Blaabjerg, F., Bhaskar, M.S.: An experimental estimation of hybrid ANFIS–PSO-based MPPT for PV grid integration under fluctuating sun irradiance. IEEE Syst. J. **14**(1), 1218–1229 (2020). https://doi.org/10.1109/JSYST.2019.2949083

Chapter 12
Multiple Input Converter for Photovoltaic Applications

Allamsetty Hema Chander, Lalit Kumar Sahu, and Subhojit Ghosh

12.1 Introduction

Partial shading (PS) is a very regular and inevitable scenario that occurs due to shading of few cells within a module or array resulting in significant power loss of the PV systems [1]. Under PS condition, multiple power peaks are formed of which only one is global peak. Due to these multiple peaks, the standard MPPT algorithms trap in the local peak resulting in huge power loss of the system. In order to overcome the PV operation at the local suboptimal point and limit the power loss due to partial shading, several methods have been proposed. Among them, employing bypass diodes is the first and most widely used method. In this method, the diodes linked across the shadowed cells bypass the current through the diode [2]. Due to more number of diodes, this method is expensive, and also the module is underutilized during partial shading.

The other preferred solution for reducing the power loss during partial shading involves tracking and operating the array at the global peak. Few conventional algorithms are modified for tracking the global peak [3, 4]. In order to operate at the global peak, several optimization techniques have been reported such as particle swarm optimization (PSO) [5, 6], Fireworks enriched P&O [7], DEPSO [8], Cuckoo search [9], etc. However, these optimization techniques increase the computational head and memory requirements of the system. These limitations

A. H. Chander
Department of Electrical and Electronics Engineering, Gayatri VidyaParishad College of Engineering for Women, Visakhapatnam, India

L. K. Sahu (✉) · S. Ghosh
Department of Electrical Engineering, National Institute of Technology, Raipur, India
e-mail: lkumar.ele@nitrr.ac.in

S. Ghosh
e-mail: sghosh.ele@nitrr.ac.in

N. Priyadarshi et al. (eds.), *DC–DC Converters for Future Renewable Energy Systems*, Energy Systems in Electrical Engineering,
https://doi.org/10.1007/978-981-16-4388-0_12

hinder the use of GMPPT techniques for real-time applications involving rapid changes in the weather scenario.

To overcome the above-mentioned issues, reconfiguration techniques were developed. In these techniques, the modules in the array are divided into reconfigurable and fixed groups. Each module of the reconfigurable group gets associated to the fixed group employing appropriate switching for continuous relocation [10, 11]. These methods have large computational time requirements. Also finding the best optimal configuration is often a challenge. The complex computational efforts along with large memory requirements motivated the evolution of distributed MPPT (DMPPT) techniques. In DMPPT, each module has a dedicated converter for maximum power realization under its individual environmental conditions. The module-level approach allows effective and maximum power extraction from the module under all environmental circumstances. These DMPPT techniques can be classified into two methods. In the first method, each module is associated to its respective DC–DC/AC converter, and all these are series connected. In the second method, the converter connection is similar to the first method and all these are connected in parallel. The DMPPT techniques have been the recent point of focus among researchers for extracting the maximum power [12–18]. The concept of micro-inverter which falls into the category of DMPPT has been commercially well established by Enphase energy. However, dedicated converter/inverter for each module makes the system costly and unsuitable for most of the applications.

In this regard, this chapter presents a MIC based scheme for harvesting maximum power from each module under wavering atmospheric conditions in a simple, efficient and cost-effective method. The presented approach neither uses computationally intensive optimization techniques as in the case of GMPP nor uses complex algorithms as in the case of reconfiguration techniques. Moreover, the proposed approach has less component count compared to most of the existing DMPPT methods, which makes it relatively simple, cost-effective and efficient. Also, the modular scheme allows for high reliability with inherent fault tolerant and power-sharing capability. The presented approach has been extensively validated using simulation in MATLAB/Simulink environment. The superiority of the proposed approach over the existing approaches has been verified both qualitatively and quantitatively.

12.1.1 Synthesis of Multiple Input Converter

The objective of this section is to deliver a clear idea of MIC and its design aspects. In general, the MICs are synthesized by simple series–parallel combination of individual sources. However, sources having dissimilar characteristics (voltage–current) cannot be connected in this passion. In this regard, for effective integration and utilization of these sources, pulsating source cells (PSCs) are discussed in [19–21]. Depending on the input source, PSCs can be categorized as pulsating voltage source cell (PVSC) and pulsating current source cell (PCSC). The formation of PVSC and PCSC is

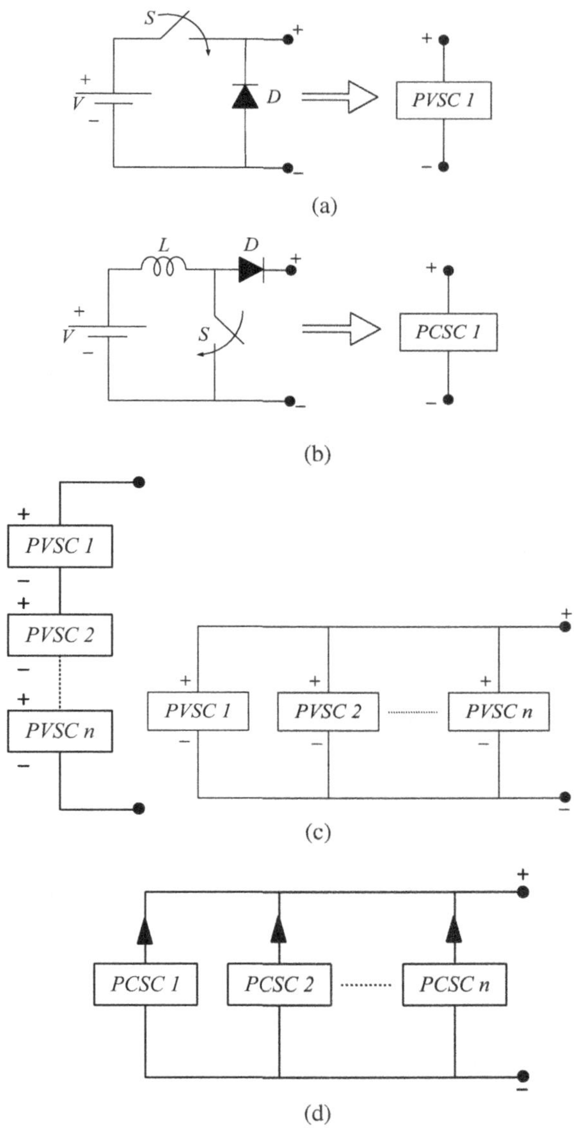

Fig. 12.1 Realization of **a** PVSC, **b** PCSC, **c** series and parallel combination of PVSCs, **d** parallel combination of PCSCs

depicted in Fig. 12.1a, b. The generalized architecture of PVSC and PCSC for multiple inputs can be the series/parallel/series–parallel combination of these cells as illustrated in Fig. 12.1c, d.

The PVSC is realized by a constant DC source connected to a switch at the output terminal by an accurately controlled power conversion method as depicted in Fig. 12.1a. Similarly, the PCSC is realized with voltage source connected to a high inductor in series (to resemble the current source) and then to a switch as shown in Fig. 12.1b. The DC source employed can be a constant voltage source, as shown in Fig. 12.2a or a constant voltage source with energy buffering unit (EBU) as in Fig. 12.2b. In the later PVSC design, the power is delivered to load after storing in the EBU.

PCSC affords a high-frequency controllable power with pulsating current. PCSC comprises of a current source(s) connected to a switch. The output terminal delivers a constant current through the source and the respective switch. The considered DC current source can be realized as shown in Fig. 12.3a, or as shown in Fig. 12.3b using an intermediate storage element.

The design rules for realization of PSCs can be summarized as.

Rule 1: PVSCs in series and PCSCs in parallel connection are allowed for simultaneous or independent power delivery.
Rule 2: Parallel connection of PVSCs is possible only by proper selection of the switching scheme. However, it is prerequisite that the source must have a series connected switch.
Rule 3: A filter is a prerequisite in MIC. It is a general practice to use LC filter for PVSC and C filter for PCSC (Fig. 12.4).

In the recent years, various MIC topologies have been realized based on the rules discussed. Besides PVSC and PCSC, few topologies have evolved combining PVSC and PCSC, and are termed as hybrid PSCs. The selection of sources and filter is the major challenge in realization of hybrid PSCs. In general, the rules 1 and 2 discussed above are used to realize the hybrid PSCs.

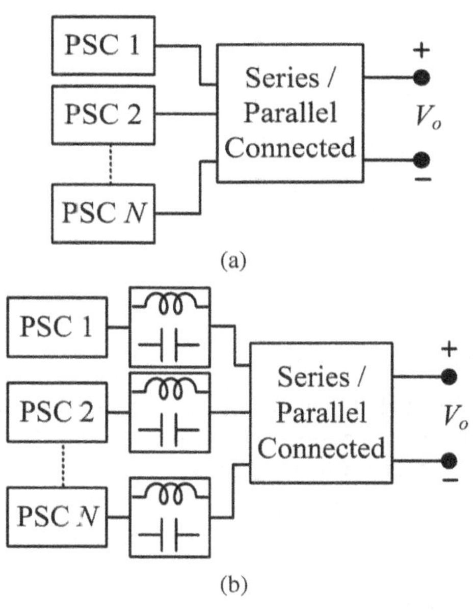

Fig. 12.2 Generalized design of PSC in **a** series/parallel, **b** series/parallel with EBU

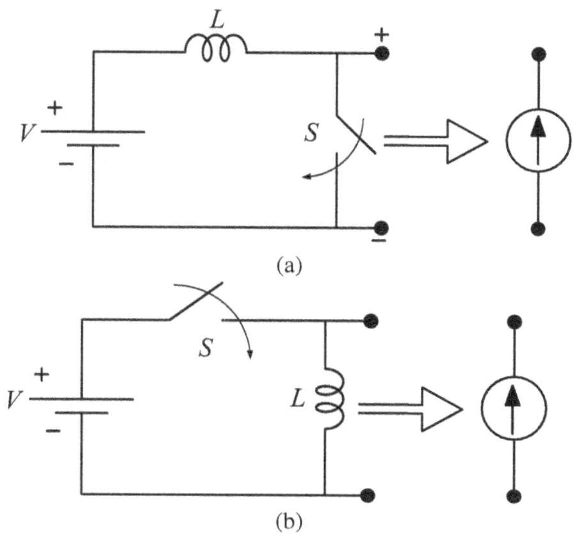

Fig. 12.3 Generalized schematic depicting realization of PCSC **a** voltage source and inductor in series, **b** voltage source, switch and inductor in series

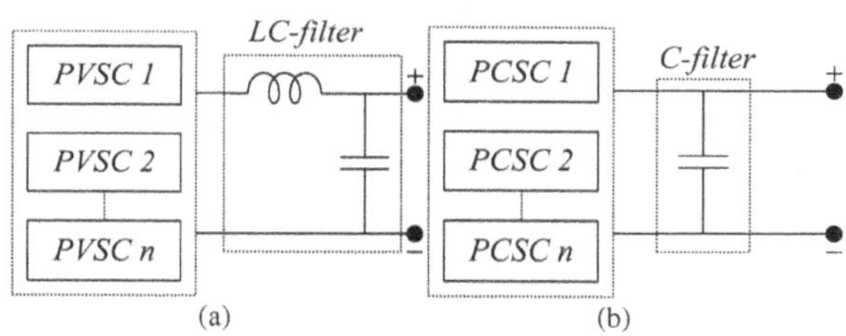

Fig. 12.4 **a** LC filter, **b** C filter

The rules discussed above can be reframed for realizing MICs with individual PVSCs and PCSCs as follows.

Rule 1: While employing PVSC, for the current to flow from positive terminal, a current buffer is connected in series with it.
Rule 2: Similarly, while employing PVSC at the converter output, for the current to flow from positive terminal, a current sink is connected in series with it.
Rule 3: PVSC should be connected in a mesh with an output sink.

Similarly, the rules [22] for synthesis of MIC with PCSC can be summarized:

Rule 1: While employing PCSC, for the outgoing terminal to be connected at the positive terminal, a voltage buffer is connected across it.
Rule 2: Similarly, while employing PCSC at the converter output, for the outgoing terminal to be connected at the positive terminal, a voltage sink is connected across it.
Rule 3: There must be a mesh formed by PCSC with output sink.

Several MICs have been synthesized and evolved in the recent past following the rules using the conventional converters. In the initial times, the series connection of two buck converters has been realized and widely accepted. This has also been considered as the basic motivation in realizing new MIC topologies. The proceeding developments include series connection of buck and Cuk converters resembling the series connection of two PVSCs. The state of the art reports several such MIC topologies.

12.1.2 System Configuration

The structure of the converter topology with dual input for PV applications is illustrated in Fig. 12.5 [23]. The switches S_{pv1} and S_{pv2} are used for power delivery, and the mode of operation is defined by the switch S_{dc}. The switches S_{pv1} and S_{pv2} aid in realizing MPPT of individual modules, and hence, their switching signals are fed from MPPT algorithm. Owing to the advantages like easy implementation and low cost, perturb and observe (P&O) method is used for realizing MPPT [24]. The presented approach effectively realizes individual MPPT in the worst-case scenario like partial shading without any computational efforts like GMPP techniques, which not only reduce the complex computational efforts involved but also save memory and time requirements. This topology also realizes MPP during module mismatch conditions. The modules are electrically separated, i.e., neither series nor parallel connected as depicted in Fig. 12.5. This electrical separation aids in the redundant and modular structure, i.e., under a faulty module condition, as the remaining module(s) are not electrically connected, they remain unaffected and realize their individual maximum power, thereby increasing the system reliability. The configuration of the topology makes monitoring and fault detection of the module easier. Further, the configuration provides an advantage of replacing the faulty module without disturbing the other modules. Moreover, employing a single inductor and capacitor results in reduced size, cost and increased reliability of the system. The switch S_{dc} is responsible for different modes of operation. The independent control of switches provides more flexibility and reduces the control complexity. Also, employing less component count (passive and semiconductors) with regard to the DMPPT techniques results in cost-effective and efficient solution for realizing individual MPP.

12.1.3 Control Strategy

The power delivery of the modules is majorly decided by the switching scheme employed. Further, the switching scheme decides the power diversification. In this regard, the schemes used for switching has been discussed in this section. The switching schemes are based on the concept of time multiplexing of the generated

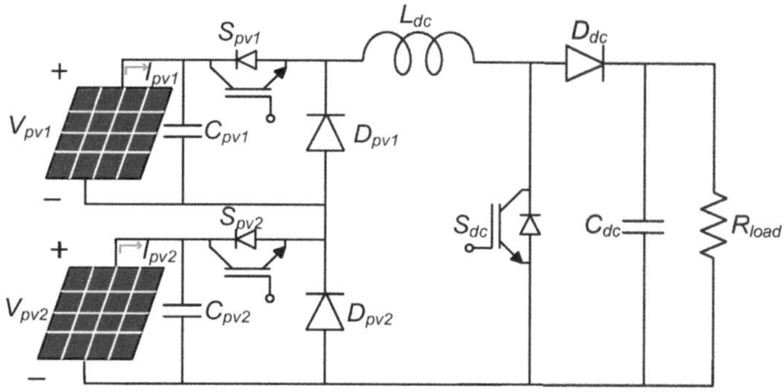

Fig. 12.5 Presented approach with two inputs

switching pulses. These switching pulses are generated using any one of the schemes,

(a) Rising edge synchronization
(b) Intermediate synchronization
(c) Falling edge synchronization.

These pulse generation schemes are illustrated in Fig. 12.6. The working states of the switching schemes in relation with the duty ratio is as follows.

For rising edge synchronization from Fig. 12.6a

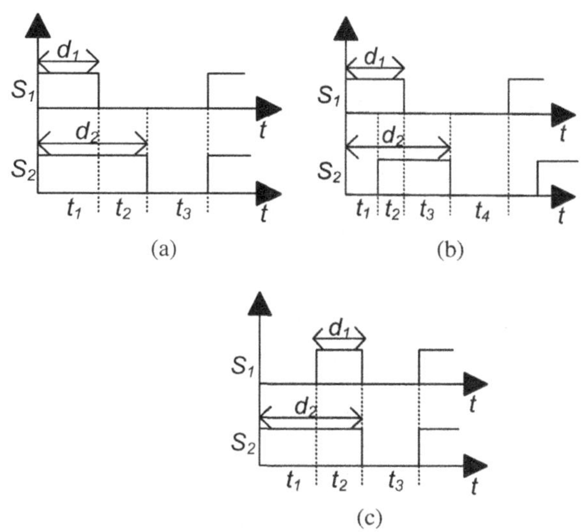

Fig. 12.6 Switching signals using different synchronization techniques. **a** Rising edge, **b** intermediate, **c** falling edge

$$\left.\begin{aligned} t_1 &= d_1 T_s \\ t_2 &= (d_2 - d_1) T_s \\ t_3 &= (1 - d_2) T_s \end{aligned}\right\} \tag{12.1}$$

For intermediate synchronization Fig. 12.6b

$$\left.\begin{aligned} t_1 &= (d_1 - d_{12}) T_s \\ t_2 &= (d_{12}) T_s \\ t_3 &= (d_2 - d_{12}) T_s \\ t_4 &= (1 - d_1 - d_2 + d_{12}) T_s \end{aligned}\right\} \tag{12.2}$$

For falling edge synchronization Fig. 12.6c

$$\left.\begin{aligned} t_1 &= (d_2 - d_1) T_s \\ t_2 &= d_1 T_s \\ t_3 &= (1 - d_2) T_s \end{aligned}\right\} \tag{12.3}$$

12.1.4 Operation and Working of Presented MIC

The MIC structure illustrated in Fig. 12.5 provides flexibility to operate in all the three basic modes of operation namely, boost, buck and buck–boost. The operation in boost configuration is detailed and can be extended for the other configurations. The basic concept of boost configuration is to charge the inductor and then discharge to load along with the source. The same has been applied to the presented structure and divided into two modes. In mode 1, inductor charges with one or both the source(s), and in mode 2, inductor discharges to the load along with one or both the source(s). In the present study, inductor is charged with module 1 and discharged with module 2 as illustrated in Fig. 12.7. This avoids the series connection of the modules which significantly aids in realizing MPP of individual modules, especially during partial shading conditions. The duty ratio of the switches S_{pv1} and S_{pv2} is decided by the MPP algorithm employed. Among the above-discussed switching schemes, the rising edge synchronization is implemented for the switches S_{pv1} and S_{dc} and the intermediate synchronization for the switches S_{dc} and S_{pv2}.

Mode 1: Inductor charging

In this mode, the inductor (L_{dc}) is charged using either of the sources. As discussed, the inductor is charged through module 1. In this mode, the capacitor (C_{dc}) supplies energy to the load as depicted in Fig. 12.7a.

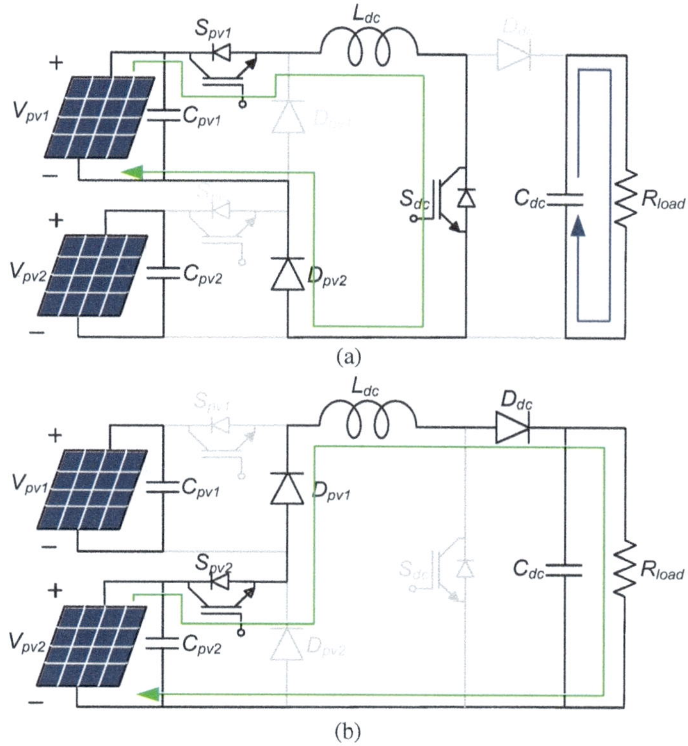

Fig. 12.7 Operational modes of the presented approach in boost configuration

Applying KVL,

$$V_{\mathrm{pv1}} - L\frac{\mathrm{d}i_L}{\mathrm{d}t} = 0 \tag{12.4}$$

$$V_C = V_o \tag{12.5}$$

Applying KCL at the capacitor

$$i_C = \frac{V_o}{R} \tag{12.6}$$

Mode 2: Inductor discharging

In this mode, inductor (L_{dc}) is discharged through the load along with module 2 as depicted in Fig. 12.7b. The capacitor (C_{dc}) gets charged in this mode. The inductor current and the capacitor voltage in this mode are given by

$$V_{pv2} + L\frac{\mathrm{d}i_L}{\mathrm{d}t} = V_o \tag{12.7}$$

Applying KCL at the capacitor

$$i_C = i_L - \frac{V_C}{R} \tag{12.8}$$

According to volt-second balance theory, the average value of inductor voltage in one switching period is zero. Hence,

$$\int_0^{T_s} v_L = 0 \tag{12.9}$$

For different modes of operation, (12.9) can be written as

$$\int_0^{T_s} v_L = L\frac{\mathrm{d}i_L}{\mathrm{d}t}\bigg|_{T_s} = 0 \tag{12.10}$$

i.e.,

$$V_{pv1}t_1 + (V_{pv2} - V_o)t_2 = 0 \tag{12.11}$$

Substituting the time period in terms of duty ratios and simplifying, we obtain

$$V_o = \frac{V_{pv1}d_{pv1} + V_{pv2}d_{pv2}}{1 - d_{dc}} \tag{12.12}$$

where d_{pv1}, d_{pv2} and d_{dc} are the respective duty cycles of the S_{pv1}, S_{pv2} and S_{dc} switches.

For an ideal system,

$$P_{input} = P_{output} \tag{12.13}$$

$$V_{pv1}I_{pv1} + V_{pv2}I_{pv2} = V_oI_o \tag{12.14}$$

$$I_o = \frac{V_{pv1}I_{pv1} + V_{pv2}I_{pv2}}{V_o} \tag{12.15}$$

$$I_o = \frac{(V_{pv1}I_{pv1} + V_{pv2}I_{pv2})(1 - d_{dc})}{V_{pv1}d_{pv1} + V_{pv2}d_{pv2}} \tag{12.16}$$

The source current I_{pv1} and I_{pv2} (average values) expressed in terms of inductor current I_L as

$$I_{\mathrm{pv1}} = d_{\mathrm{pv1}} I_L \tag{12.17}$$

$$I_{\mathrm{pv2}} = d_{\mathrm{pv2}} I_L \tag{12.18}$$

Substituting (12.17) and (12.18) in (12.16), load current is expressed as

$$I_o = (1 - d_{\mathrm{dc}}) I_L \tag{12.19}$$

Substituting I_L back into (12.17) and (12.18), the source currents are given by

$$I_{\mathrm{pv1}} = d_{\mathrm{pv1}} I_L = \frac{d_{\mathrm{pv1}} I_o}{(1 - d_{\mathrm{dc}})} \tag{12.20}$$

$$I_{\mathrm{pv2}} = d_{\mathrm{pv2}} I_L = \frac{d_{\mathrm{pv2}} I_o}{(1 - d_{\mathrm{dc}})} \tag{12.21}$$

Dividing (12.20) and (12.21)

$$\frac{I_{\mathrm{pv1}}}{I_{\mathrm{pv2}}} = \frac{d_{\mathrm{pv1}}}{d_{\mathrm{pv2}}} \tag{12.22}$$

From (12.22), it can be concluded that the average source current varies with the variation of the duty ratio of the corresponding source switch. On the same lines, the voltage and current expressions for other configurations are derived and generalized for N-inputs as given in Table 12.1.

The presented topology is capable to operate in all configurations; however, individual MPP is realized only in boost configuration. In buck and buck–boost configurations, the modules form a series connection which limits in realizing individual MPP, especially during partial shading conditions. Hence, throughout the work, the topology is operated in boost configuration only.

Table 12.1 Output voltage expressions

Configuration	Voltage expression
Boost	$V_o = \frac{\sum (V_{\mathrm{pv}i} d_{\mathrm{pv}i})}{1 - d_{\mathrm{dc}}}$
Buck	$V_o = \sum (V_{\mathrm{pv}i} d_{\mathrm{pv}i})$
Buck–boost	$V_o = \frac{\sum (V_{\mathrm{pv}i} d_{\mathrm{pv}i})}{1 - \sum_{i=1,\ i \neq j}^{N} (d_{\mathrm{pv}i} + d_{\mathrm{pv}ij})}$

Where $V_{\mathrm{pv}i}$ is the ith source voltage with duty ratio $d_{\mathrm{pv}i}$, V_o is the output voltage, $d_{\mathrm{pv}ij}$ is the common instant duty cycle in which inputs i, j operates

12.1.5 Design of Passive Elements

The passive elements have a significant role in regulating the ripples in load voltage and inductor current to permissible limits. The values of the passive elements are highly dependent on the load. With the change in load, the load voltage and current changes which highly influence the ripples in load voltage and inductor current. Hence, their design is crucial and important in converter design. The design of these passive elements for boost configuration has been presented in this section.

The inductor used decides the percentage of inductor current ripple (Δi). Similarly, the output voltage ripple (Δv) decides the value of the capacitor. The value of inductor and capacitor values are given by [25].

$$L \geq \frac{\sum \left(V_{pvi} d_{pvi}\right)}{\Delta i.f_s} \tag{12.23}$$

$$C \geq \frac{\sum \left(V_{pvi} d_{pvi}\right)}{\Delta v.f_s.R} \tag{12.24}$$

where R is the load resistance and f_s is the switching frequency.

The higher the value capacitor stiffer is DC link voltage; however, higher value of capacitor reduces the reliability; hence, trade-off should be maintained.

12.1.6 Simulation Results

The presented approach has been validated in the MATLAB/Simulink environment considering different operating scenarios. The parameters of the module at standard test conditions (STC) and the operational environment are listed in Table 12.2. The parameter values considered are listed in Table 12.3. The values of inductor and capacitor are calculated using (12.23) and (12.24) considering 5% ripple.

Case 1: Partial shading condition

Among numerous conditions that exist for partial shading, partial shading with variation of module irradiance value has been considered in this study. The output power of the module is estimated for different levels of irradiance which is given by [26].

Table 12.2 Specifications of module at STC and operating conditions of the module

V_{oc} (V)	I_{sc} (A)	V_{mp} (V)	I_{mp} (A)	P_{max} (W)	Irradiance (W/m^2)	Temperature (°C)
43.2	4.8	34.3	4.37	150	821	30

Table 12.3 Parameters and their values

L (mH)	C (μF)	f_s (kHz)
5	1200	10

$$P_{\text{estimate}} = P_{\max}\left(\frac{G}{G_o}\right) \tag{12.25}$$

where G_o is the irradiance at 1000 W/m^2, P_{max} is the rated maximum power at $G_{o,}$ and G is the irradiance at which power is to be determined.

Initially, both the modules were imposed irradiance level of 821 W/m^2, thereby each panel generating 119 W, which is near the MPP of 123 W calculated using (12.25). The partial shading condition is imposed on module 2 by reducing the irradiance level to 611 W/m^2 at t = 0.7 s. As a result, the module power has been decreased from 119 to 87 W, which is near MPP (89 W) for 611 W/m^2, as depicted in Fig. 12.8b. As observed in Fig. 12.8a, the change in irradiance level has very less impact on the first module and continues to operate near MPP. This does not occur in conventional array configuration, and also the partially shaded modules are bypassed using diodes which results in huge power loss.

Moreover, the proposed approach does not employ any complex computational algorithms for MPP realization as in GMPPT and RMPPT. Thus, it can be concluded that the presented approach realizes individual MPP during partial shading in a simple, efficient and cost-effective manner. The load parameters depicted in Fig. 12.8c shows a decrease in the total load power due to the effect of partial shading. This approach performs equally well in case of module mismatch, where individual MPP is realized as modules are independently operated at their respective irradiance levels. This results in maintaining healthier performance and efficiency of modules reducing the long run degradation issue.

Case 2: Open circuit of a module

In the configuration of conventional array, huge power loss occurs during open circuit of a module or complete failure of the module. The presented approach possesses inherent fault-tolerant capability resulting in an uninterrupted power supply with failed/disconnected module further resulting in increased reliability and redundancy. To validate this scenario, module 1 is disconnected at t = 0.7 s, as depicted in Fig. 12.9a. Due to a sudden change in the system, the module 2 undergoes transients and reaches the near MPP within a short period of time as illustrated in Fig. 12.9b. This confirms that open/disconnection of a module does not affect the entire system as in conventional configuration. As observed from Fig. 12.9a due to decrease in the power generated, there is a decrease in the load power.

From the studies and results of the selected cases, the following conclusions are drawn.

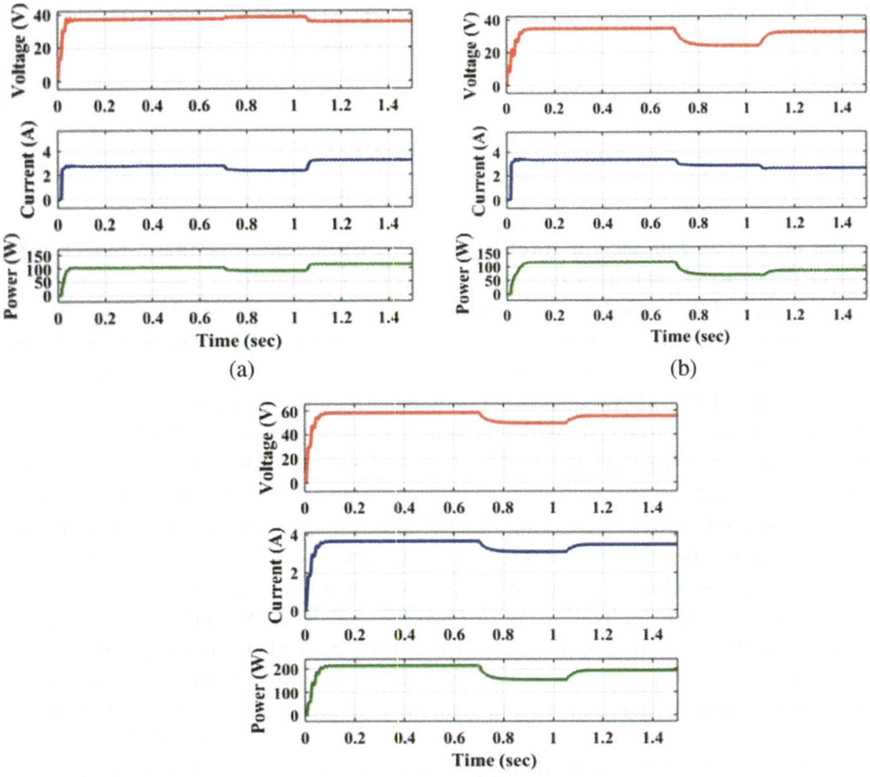

(a) (b) (c)

Fig. 12.8 Simulation results of **a** module 1, **b** module 2, **c** load under partial shading conditions

- The presented approach effectively handles the partial shading condition by realizing individual MPP which reduces the loss of power compared to the conventional approaches in a simple, efficient and cost-effective way.
- Further, this approach also reduces the power loss due to open circuit or failure of the module providing flexibility to replace it without actually disturbing the complete system. This confirms the inherent fault-tolerant capability of the approach resulting in increased reliability.
- Also, the load power has been shared equally among the modules under similar irradiance conditions confirming the inherent power-sharing ability of the presented approach.

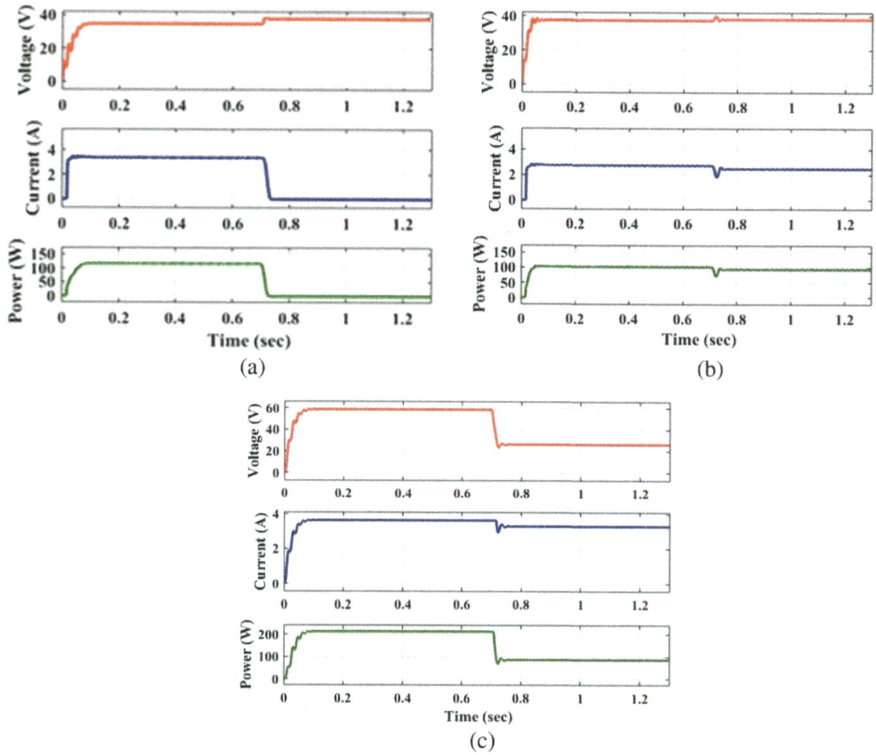

Fig. 12.9 Voltage, current and power simulation results for sudden disconnection of module condition of **a** module 1, **b** module 2, **c** load

12.1.7 MPPT Efficiency Analysis

The power realized by individual modules highly depends on environmental factors, and the maximum power at various irradiances is given by (12.25). The MPP tracking efficiency under diverse situations is quantified as [27]

$$\eta_{\text{MPP}} = \frac{P_{\text{PV}}}{P_{\text{mpp}}} \times 100 \tag{12.26}$$

The theoretical and experimental power realized under various irradiance levels has been estimated and depicted in Fig. 12.10. It has been observed that the MPP realized experimentally is in line with the calculated values using (12.25) reflecting the efficacy of the proposed approach for various levels of irradiance.

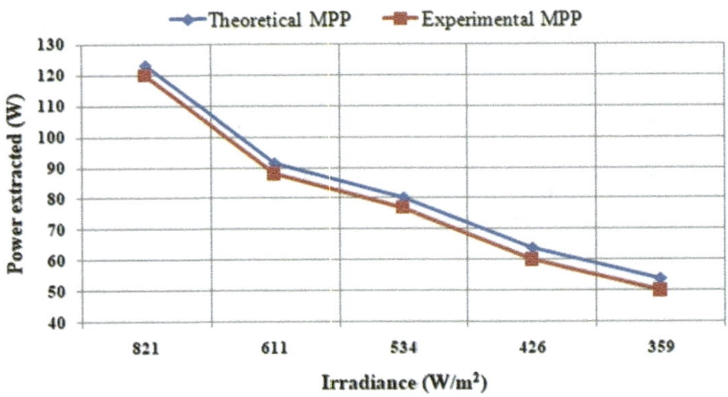

Fig. 12.10 MPPT efficiency graph

12.1.8 Comparative Analysis

A comparative analysis of the proposed approach with state-of-the-art solutions for maximum power extraction under PS conditions has been presented in this section. The presented approach has been qualitatively compared with the traditional GMPP, reconfiguration and DMPPT techniques as given in Table 12.4. Compared with GMPP techniques [6–8, 28–30], the presented approach does not involve complex computational efforts for MPP realization. Also, the memory requirement for the presented approach is very less since it does not employ any complex calculations. Moreover, in GMPP approaches, the array as a whole operate at GMPP which has a high impact on the module's lifetime and might lead to long-term degradation, while in the presented approach as each module is independent of other, the lifespan and efficiency of the module remain high comparatively. Further, the GMPP approaches in the state of art mentions no remark on open circuit or disconnection of modules. For instance, if a module having higher irradiance level is open or faulty, then the GMPP point alters which have a severe effect on the overall power generation. The presented approach efficiently handles the open-circuit condition without altering the power realization of other connected modules. On the other hand, unlike in RMPPT techniques, the presented approach need not reconfigure the modules based on the irradiation levels, thereby decreasing the complex mathematical calculations and thereby the control complexity of the system [10, 11, 31–33]. However, this technique also involves high-computational burden and memory requirements. Since no electrical connections are disturbed, the presented approach has increased reliability comparatively [34].

The presented approach successfully competes with the existing DMPPT techniques [12, 13, 15, 35–40] in terms of several qualitative parameters like tracking efficiency, memory requirements, simplicity, etc., with the added advantage of the inherent fault-tolerant capability.

Table 12.4 Comparative analysis

Parameters	DMPPT technique [12, 13, 15, 35–40]	GMPP technique [6–8, 28–30]	Array reconfiguration technique [10, 11, 31–33]	Presented technique
Memory requirements	Low	High	Medium	Very low
Computational complexity involved	Very low	High	Medium	Very low
Able to withstand open circuit	Yes	No	Yes	Yes
Ability to withstand fault	Yes	No	No	Yes
Module mismatch condition handling capability	Yes	No	No	Yes
Accuracy of MPPT tracking	Accurate	Accurate	Accurate	Accurate
Reliability	High	Moderate	Low	High
Implementation cost	High	Low	Moderate	Low

Moreover, the presented approach can be considered as a DMPPT approach since deals with module-level approach for realizing individual MPP. In this regard, the proposed approach is compared with the recent DMPPT techniques quantitatively for five inputs in Fig. 12.11. As observed, unlike other approaches, the number of inductors and capacitors do not vary with the number of inputs in the proposed approach. Further, the modular structure resulting in the reduced component count, simple design and reliable operation are the advantages of the proposed approach.

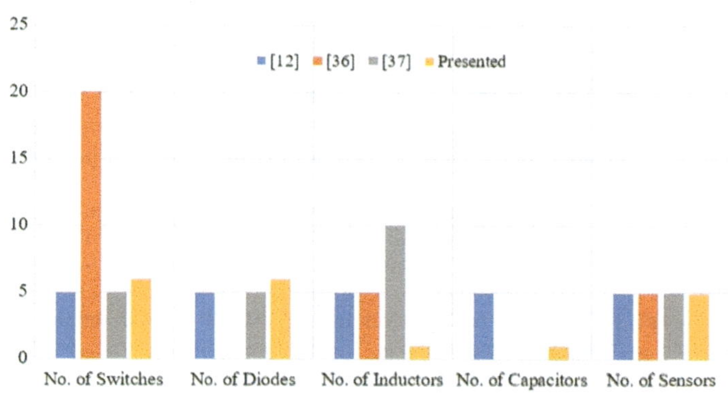

Fig. 12.11 Quantitative comparison

12.1.9 Conclusion

In this chapter, a MIC based module-level approach for realizing the MPP of individual modules of a PV array has been presented. The MIC based approach efficiently realizes the MPP of the individual modules under various operating circumstances including partial shading. The proposed scheme can handle the open-circuit condition of the module and effectively minimize the power loss compared to the conventional approaches. The approach possesses inherent fault-tolerant and power-sharing capabilities in addition to simple, modular, flexible and reliable topological configuration. Further, the comparison of the proposed scheme with the reported techniques reflects its superiority in terms of the lesser passive element count, reduced computational effort and modular structure.

References

1. Al Mamun, M.A., Hasanuzzaman, M., Selvaraj, J.: Experimental investigation of the effect of partial shading on photovoltaic performance. IET Renew. Power Gener. **11**(7), 912–921 (2017). https://doi.org/10.1049/iet-rpg.2016.0902.
2. Silvestre, S., Boronat, A., Chouder, A.: Study of bypass diodes configuration on PV modules. Appl. Energy. **86**(9), 1632–1640 (2009). https://doi.org/10.1016/j.apenergy.2009.01.020.
3. Ahmed, J., Salam, Z.: An improved perturb and observe (P&O) maximum power point tracking (MPPT) algorithm for higher efficiency. Appl. Energy. **150**, 97–108 (2015). https://doi.org/10.1016/j.apenergy.2015.04.006.
4. Priyadarshi, N., Padmanaban, S., Maroti, P.K., Sharma. A.: An extensive practical investigation of FPSO-based MPPT for grid integrated PV system under variable operating conditions with anti-islanding protection. IEEE Syst. J. pp. 1–11 (2018).
5. Priyadarshi, N., Padmanaban, S., Bhaskar, M.S., Blaabjerg, F., Sharma, A.: A fuzzy SVPWM based inverter control realization of grid integrated PV-Wind system with FPSO MPPT algorithm for a grid-connected PV/Wind power generation system: Hardware implementation. IET Electric Power Appl. pp. 1–12 (2018).
6. Priyadarshi, N., Padmanaban, N., Holm-Nielsen, J. B., Blaabjerg, F., Bhaskar, M.S.: An experimental estimation of hybrid ANFIS–PSO-based MPPT for PV grid integration under fluctuating sun irradiance. In: IEEE Systems Journal. **14**(1), 1218–1229 (2020). https://doi.org/10.1109/JSYST.2019.2949083.
7. Priyadarshi, N., Bhaskar, M.S., Padmanaban, N., Blaabjerg, F., Azam, F.: New CUK–SEPIC converter based photovoltaic power system with hybrid GSA–PSO algorithm employing MPPT for water pumping applications. IET Power Electron. 1–0 (2020). https://doi.org/10.1049/iet-pel.2019.1154.
8. Seyedmahmoudian, M., et al.: Simulation and hardware implementation of new maximum power point tracking technique for partially shaded PV system using hybrid DEPSO method. IEEE Trans. Sustain. Energy. **6**(3), 850–862 (2015). https://doi.org/10.1109/TSTE.2015.2413359.
9. Peng, B.R., Ho, K.C., Liu, Y.H.: A novel and fast MPPT method suitable for both fast changing and partially shaded conditions. IEEE Trans. Ind. Electron. **65**(4), 3240–3251 (2018). https://doi.org/10.1109/TIE.2017.2736484.
10. Padmanaban, S., Priyadarshi, N., Holm-Nielsen, J.B., Bhaskar, M.S., Azam, F., Sharma, A.K.: A novel modified sine-cosine optimized MPPT algorithm for grid integrated PV system under real operating conditions. In: IEEE Access, **7**, 10467–10477 (2019). https://doi.org/10.1109/ACCESS.2018.2890533.

11. Babu, T.S., Ram, J.P., Dragiˇ, T., Miyatake, M., Blaabjerg, F., Natarajan, R.: Particle swarm optimization based solar PV array reconfiguration of the maximum power extraction under partial shading conditions flexible array switch matrix. IEEE Trans. Sustain. Energy. **9**(1), 74–85 (2018).
12. Femia, N., Lisi, G., Petrone, G., Spagnuolo, G., Vitelli, M.: Distributed maximum power point tracking of photovoltaic arrays: Novel approach and system analysis. IEEE Trans. Ind. Electro. **55**(7), 2610–2621 (2008). https://doi.org/10.1109/TIE.2008.924035.
13. Ahmadi, D., Mansouri, S.A., Wang, J.: Circuit topology study for distributed MPPT in very large scale PV power plants. Conf. Proc. - IEEE Appl. Power Electron. Conf. Expo. - APEC, **1**(d), 786–791 (2011). https://doi.org/10.1109/APEC.2011.5744685.
14. Elasser, A., Agamy, M., Sabate, J., Steigerwald, R., Fisher, R., Harfman-Todorovic, M.: A comparative study of central and distributed MPPT architectures for megawatt utility and large scale commercial photovoltaic plants. IECON Proc. (Industrial Electron. Conf. pp. 2753–2758 (2010). https://doi.org/10.1109/IECON.2010.5675108.
15. Zhao, T., Ju, Z., Wang, H., Wei, X., Li, X., Zhang, S.: The distributed maximum power point tracking method and application in the PV grid-connected generation. 2010 Int. Conf. Intell. Syst. Des. Eng. Appl. pp. 639–642 (2010). https://doi.org/10.1109/ISDEA.2010.407.
16. Poshtkouhi, S., Palaniappan, V., Fard, M., Trescases, O.: A general approach for quantifying the benefit of distributed power electronics for fine grained MPPT in photovoltaic applications using 3-D modeling. IEEE Trans. Power Electron. **27**(11), 4656–4666 (2012). https://doi.org/10.1109/TPEL.2011.2173353.
17. Li, Q., Wolfs, P.: "A preliminary study of the distributed maximum power point tracker designs for different types of solar cells in solar and electric vehicle arrays. 2007 Australas. Univ. Power Eng. Conf. AUPEC (2007). https://doi.org/10.1109/AUPEC.2007.4548043.
18. Yao, W., Gao, M., Ren, Z., Chen, M., Qian, Z.: Improvement of performance and flexibility for photovoltaic module using individual DC/DC converter. 2009 IEEE 6th Int. Power Electron. Motion Control Conf. IPEMC '09, **3**, 441–444 (2009). https://doi.org/10.1109/IPEMC.2009.5157428.
19. Li, Y., Ruan, X., Member, S., Yang, D., Liu, F., Tse, C.K.: Synthesis of Multiple-Input DC/DC Converters. **25**(9), 2372–2385 (2010).
20. Kumar L., Jain, S.: Multiple-input DC/DC converter topology for hybrid energy system. IET Power Electron. **6**(8), 1483–1501 (2013). https://doi.org/10.1049/iet-pel.2012.0309.
21. Chander, A.H., Kumar, L., Ghosh, S., Gupta, K.K.: Comparative analysis on selection and synthesis of multiple input converters: a review. pp. 1–17 (2019). https://doi.org/10.1049/iet-pel.2019.0732.
22. Liu Y.C., Chen, Y.M.: A systematic approach to synthesizing multi-input DC-DC converters. IEEE Trans. Power Electron. **24**(1), 116–127 (2009). https://doi.org/10.1109/TPEL.2008.2009170.
23. Chander, A.H., Kumar, L.: MIC for reliable and efficient harvesting of solar energy. IET Power Electron. **12**(2), 267–275 (2019). https://doi.org/10.1049/iet-pel.2018.5079.
24. Khaled, M., Ali, H., Abd-El Sattar, M., Elbaset, A.A.: Implementation of a modified perturb and observe maximum power point tracking algorithm for photovoltaic system using an embedded microcontroller. IET Renew. Power Gener. **10**(4), 551–560 (2016). https://doi.org/10.1049/iet-rpg.2015.0309.
25. Erickson R.W., Maksimovic, D.: Fundamentals of Power Electronics, 2nd ed. Springer-Verlag, 2001.
26. Hariharan, R., Chakkarapani, M., Saravana Ilango, G., Nagamani, C.: A method to detect photovoltaic array faults and partial shading in PV systems. IEEE J. Photovoltaics. **6**(5), 1278–1285 (2016). https://doi.org/10.1109/JPHOTOV.2016.2581478.
27. Subudhi, B., Pradhan, R.: A comparative study on maximum power point tracking techniques for photovoltaic power systems. Sustain. Energy, IEEE Trans. **4**(1), 89–98 (2013). https://doi.org/10.1109/TSTE.2012.2202294.

28. Seyedmahmoudian, M., Mekhilef, S., Rahmani, R., Yusof, R., Asghar Shojaei, A.: Maximum power point tracking of partial shaded photovoltaic array using an evolutionary algorithm: A particle swarm optimization technique. J. Renew. Sustain. Energy. **6**(2), 2014. https://doi.org/10.1063/1.4868025.
29. Sundareswaran, K., Vigneshkumar, V., Sankar, P., Simon, S.P., Srinivasa Rao Nayak, P., Palani, S.: Development of an improved P&O algorithm assisted through a colony of foraging ants for MPPT in PV system. IEEE Trans. Ind. Informatics. **12**(1), 187–200 (2016). https://doi.org/10.1109/TII.2015.2502428.
30. Sundareswaran, K., Palani, S., Vigneshkumar, V.: Development of a hybrid genetic algorithm/perturb and observe algorithm for maximum power point tracking in photovoltaic systems under non-uniform insolation. IET Renew. Power Gener. **9**(7), 757–765 (2015). https://doi.org/10.1049/iet-rpg.2014.0333.
31. Velasco-Quesada, G., Guinjoan-Gispert, F., Pique-Lopez, R., Roman-Lumbreras, M., Conesa-Roca, A.: Electrical PV array reconfiguration strategy for energy extraction improvement in grid-connected PV systems. IEEE Trans. Ind. Electron. **56**(11), 4319–4331 (2009). https://doi.org/10.1109/TIE.2009.2024664.
32. Kim, H., Parkhideh, B., Bongers, T.D., Gao, H.: Reconfigurable solar converter: A single-stage power conversion PV-battery system. IEEE Trans. Power Electron. **28**(8), 3788–3797 (2013). https://doi.org/10.1109/TPEL.2012.2229393.
33. Elserougi A.A., et al.: A switched PV approach for extracted maximum power enhancement of PV arrays during partial shading. IEEE Trans. Sustain. Energy. **6**(3), 767–772 (2015).
34. Sahu, L.K., Allamsetty, H.C., Ghosh, S.: Performance analysis of multiple input converter for standalone photovoltaic system. IET Power Electron. **12**(5), 1295–1306 (2019). https://doi.org/10.1049/iet-pel.2018.6148.
35. Femia, N., Petrone, G., Spangnuolo, G., Vitelli, M.: Power Electronics and Control Techniques for Maximum Energy Harvesting in Photovoltaic Systems. CRC Press, 2013.
36. Li, Q., Wolfs, P.: A review of the single phase photovoltaic module integrated converter topologies with three different DC link configurations. IEEE Trans. Power Electron. **23**(3), 1320–1333 (2008). https://doi.org/10.1109/TPEL.2008.920883.
37. Yao, W., Gao, M., Ren, Z., Chen, M., Qian, Z.: Improvement of performance and flexibility for photovoltaic module using individual DC/DC converter. 2009 IEEE 6th Int. Power Electron. Motion Control Conf. **3**, 441–444 (2009). https://doi.org/10.1109/IPEMC.2009.5157428.
38. Wang, F., Zhu, T., Zhuo, F., Yi, H.: An improved submodule differential power processing-based PV system with flexible multi-MPPT control. IEEE J. Emerg. Sel. Top. Power Electron. **6**(1), 94–102 (2018). https://doi.org/10.1109/JESTPE.2017.2719919.
39. Biswas, J., Kamath, A.M., Gopi, A.K. and Barai, M.: Design, architecture and real time distributed coordination DMPPT algorithm for PV systems. IEEE J. Emerg. Sel. Top. Power Electron. **6777**(c), 2017. https://doi.org/10.1109/JESTPE.2017.2756698.
40. Chen, M., Gao, F., Li, R., Li, X.: A dual-input central capacitor DC/DC converter for distributed photovoltaic architectures. IEEE Trans. Ind. Appl. **53**(1), 305–318 (2017). https://doi.org/10.1109/TIA.2016.2606604.

Chapter 13
Overview of Control Strategies and Design of Isolated Bidirectional Dual Active Bridge Converter for Renewable Energy Systems

Anup Kumar Panda, Nishit Tiwary, and N. Venkataramana Naik

13.1 Introduction

The dual active bridge converter (DAB) is among the recent topologies for isolated DC–DC power converters with various advantages, namely high efficiency, ohmic isolation, soft switching, large power density, bidirectional power handling, and decreased size and weight [1–4]. The isolation provided by the DAB converter is an essential feature for renewable energy applications because it dissociates load ground from the source and provides protection to the installed system. Also, the DAB converter utilizes the transformer turns ratio for high voltage conversion. The high power density is achievable with a high-frequency transformer (HFT), making the converter suitable for high-power renewable systems. The soft switching capability ensures the reduction in switching stress thereby increasing the power conversion efficiency. Additionally, the bidirectional power flow control is easier in DAB due to the symmetric structure. Thus, it is preferable for interfacing the renewable energy sources with energy storage components such as battery bank, ultra-capacitors, etc. [5–7]. As depicted in Fig. 13.1, the DAB converter can be functional in the renewable energy systems as:

1. Front-end converter directly connected to the renewable energy source.
2. Bidirectional converter for energy storage system like battery, ultra-capacitor, etc.

A. K. Panda (✉) · N. Tiwary · N. V. Naik
Department of Electrical Engineering, National Institute of Technology, Rourkela, India
e-mail: akpanda@nitrkl.ac.in

N. Tiwary
e-mail: 517ee1002@nitrkl.ac.in

N. V. Naik
e-mail: nenavathv@nitrkl.ac.in

N. Priyadarshi et al. (eds.), *DC–DC Converters for Future Renewable Energy Systems*, Energy Systems in Electrical Engineering,
https://doi.org/10.1007/978-981-16-4388-0_13

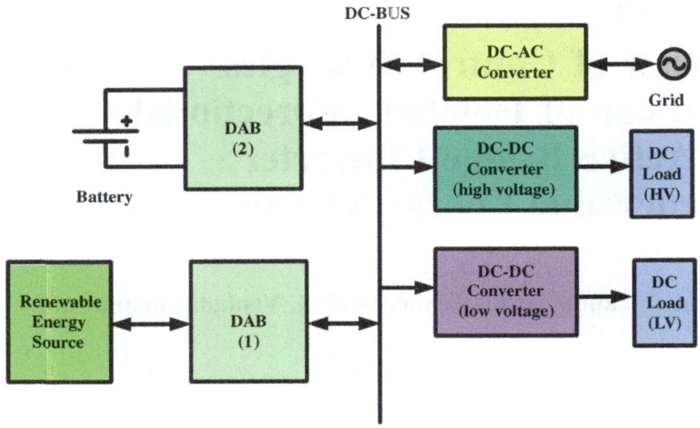

Fig. 13.1 DAB converter in renewable energy system

In recent years, significant research has been reported for design, control and other technical issues for the DAB converter and its application in renewable energy systems [8–11]. The reported work on advanced control of DAB converters is relatively scarce due to the more complexity in converter models. The control algorithm and DAB parameter design considered renewable energy sources like PV and implementing MPPT and battery management algorithms [12, 13].

This chapter presents the overview of control methods and designing the DAB converter, suitable for the renewable energy system. Subsequently, it provides efficient power handling and voltage regulation. The renewable energy systems suffer from frequent voltage change; thereby, the control method plays a vital role in delivering a satisfactory output. The control methods are broadly classified into three categories based on the strategy and implementation of the control algorithm. These control strategies for DAB are proposed with the objective to provide a fast transient response, stable steady-state response and improved voltage regulation. The controller's accurate reference tracking performance ensures a stable and fixed output terminal voltage even for the situation of sudden load change and input voltage fluctuation conditions. Additionally, a direct power control technique is suggested, which utilizes the power transfer equation and dynamic voltage equation for reference power generation and thereby controlling the output voltage. The performance of the suggested controller is evaluated in Simulink (MATLAB). The results demonstrate the suggested control method's effectiveness and the DAB converter's advantages in the renewable energy system.

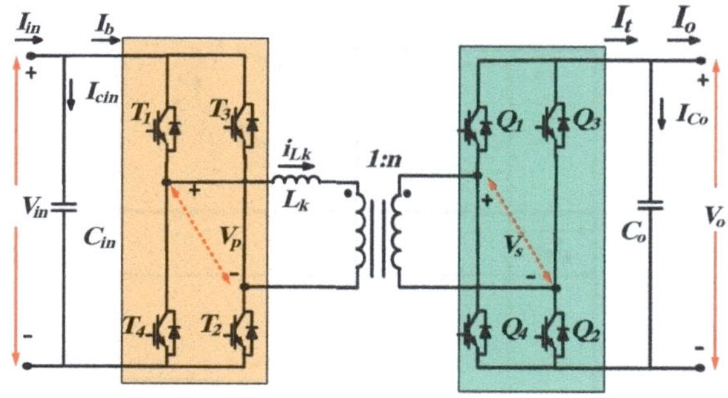

Fig. 13.2 Dual active bridge circuit

13.2 The Dual Active Bridge Converter

The DAB converter offers a bidirectional isolated power conversion technique. The circuit diagram of a DAB converter is depicted in Fig. 13.2, with 'V_{in}' and 'V_o' are the voltages in input terminal and output terminal, 'L_k' denotes the equivalent inductor, which comprise the leakage inductance of the HFT including an auxiliary inductor in series, required for the power transfer. The semiconductor switches T_{1-4} and Q_{1-4} make the two symmetric H-bridges on either side of the HFT. The DC voltage is converted to high frequency square wave AC voltage in the first stage of conversion implemented with the help of the first H-bridge, denoted as the primary bridge. Thus, the stage can be denoted as the inverter stage and 'V_p' denoting the primary bridge voltage.

The transformer transfers the power from its primary winding to secondary winding. The secondary bridge then transforms the square wave AC voltage to DC voltage, which is obtained in the output terminal of the DAB converter. The stage can be denoted as the rectifier stage and 'V_s' denoting the secondary bridge voltage. The equivalent circuit diagram is displayed in Fig. 13.3 with the two square waves and an equivalent inductance between them. The transformer turn ratio, on the requirement, can be applied for step-up and step-down of the primary to secondary side voltage. The power transfer from the primary to secondary side occurs when a phase shift is introduced in these voltage waveforms, similar to the power transfer between two voltage buses in a power system. The phase shift makes one of the bridges as leading and the other as a lagging bridge with transfer of power from the

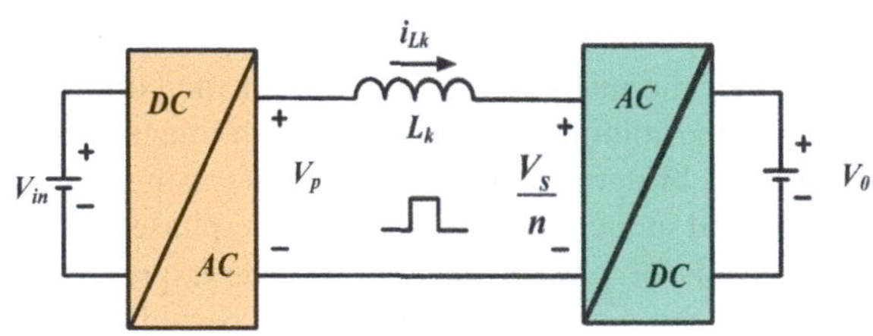

Fig. 13.3 Equivalent circuit of dual active bridge

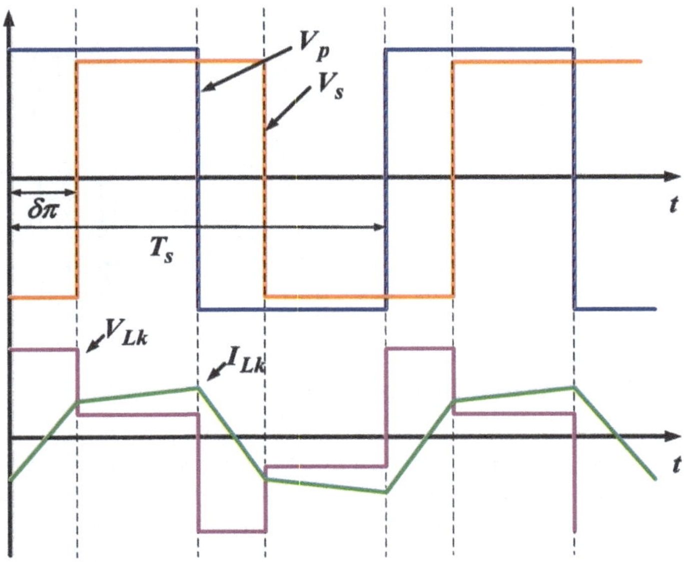

Fig. 13.4 Waveforms for SPS modulation

leading to the lagging bridge. Thus, the power direction can be effortlessly reversed with the phase shift value as the bridges are controlled simultaneously. The duty ratio of all switches is 50%, and the diagonal switches are turned on and off together to obtain a square wave. These voltages, as well as the equivalent inductor current waveforms, operated with single-phase shift (SPS) modulation, is depicted in Fig. 13.4, where f_s is the switching frequency and '$\delta\pi$' denoting the phase lag among the bridge voltages and 'n' denotes the voltage ratio of HFT.

13.2.1 Analysis of DAB Converter Circuit

For the analysis of the DAB converter, a simplified diagram is used as outlined in Fig. 13.3, and the steady-state waveform is depicted in Fig. 13.4 with the phase shift of '$\delta\pi$' in bridge voltage waveforms, inductor voltage 'V_{L_k}' and inductor current 'I_{L_k}'. The inductor voltage and current are symmetrical about the time axis and are dependent on the bridge voltages and the phase shift of two H-bridges. Therefore, the maximum magnitude of the current 'I_{L_k}' in the positive half period matches the minimum magnitude of the negative half period. The power transfer across the bridges is presented in [14] using a Fourier series analysis of bridge voltage and inductor current. The single-phase shift (SPS) modulation is utilized for the analysis of power transfer, and the expression for power flow is derived by averaging the power in one switching period as expressed in (13.1) and (13.2).

$$P_t = \sum_{h=1,3,5\ldots}^{\infty} P_h \tag{13.1}$$

$$P_h = \frac{\left(8.V_{\text{in}}.\frac{V_o}{n}\right)}{\pi^2 \omega_s L_k} \cdot \frac{\sin(h.\delta.\pi)}{h^3} \tag{13.2}$$

Thus, expression (13.2) shows that the power transfer from input to output terminal depends upon the phase shift between the voltage waveform, equivalent inductance and switching frequency. However, for the fixed value of the equivalent inductor and switching frequency, the phase shift value is the sole parameter available to regulate the power. Also, the power flow capacity is inversely proportional to the equivalent inductance 'L_k'. The auxiliary inductance is designed for the DAB operation to handle the specified power after the proper design of the transformer and considering its leakage inductance. The input capacitor has an effect in DAB interface with a renewable energy source to mitigate the voltage ripple and reverse current flow. Therefore, the analysis is focused on the configuration of the equivalent inductor 'L_k' and input capacitor 'C_{in}'. The output capacitor 'C_o' is designed using the standard design process of any dc-link capacitor to handle the voltage ripple. So, the transformer design and output capacitor design are not discussed due to the limited scope of the chapter.

The power extraction from the source depends on the DAB operation and is controlled by the phase shift value '$\delta\pi$'. The current analysis in the input node is essential to design the converter parameters 'L_k' and 'C_{in}'. The current waveforms 'I_{in}', '$I_{C_{\text{in}}}$' and 'I_b' at input node along with the inductor current 'I_{L_k}' are shown in Fig. 13.5 for a single switching period. The Kirchhoff current law at input node is denoted as:

$$i_{\text{in}}(t) = i_{C_{\text{in}}}(t) + i_b(t) \tag{13.3}$$

And the current in the primary bridge (13.4) is formulated with the switching function S_1 in (13.5)

$$i_b(t) = i_{L_k}(t).S_1(t) \tag{13.4}$$

$$S_1(t) = \begin{cases} 1; & 0 < t < \frac{T_s}{2} \\ -1 & \frac{T_s}{2} < t < T_s \end{cases}; \tag{13.5}$$

With the consideration that in the steady-state situation, the average capacitor current is zero, and the average value of input current (13.3) can be obtained as

$$\langle i_{\text{in}} \rangle = I_{\text{in}} = \langle i_{L_k}(t) S_1(t) \rangle \tag{13.6}$$

Therefore, expression (13.6) demonstrates that the input current drawn from the source depends upon the equivalent inductor current. Consequently, the power

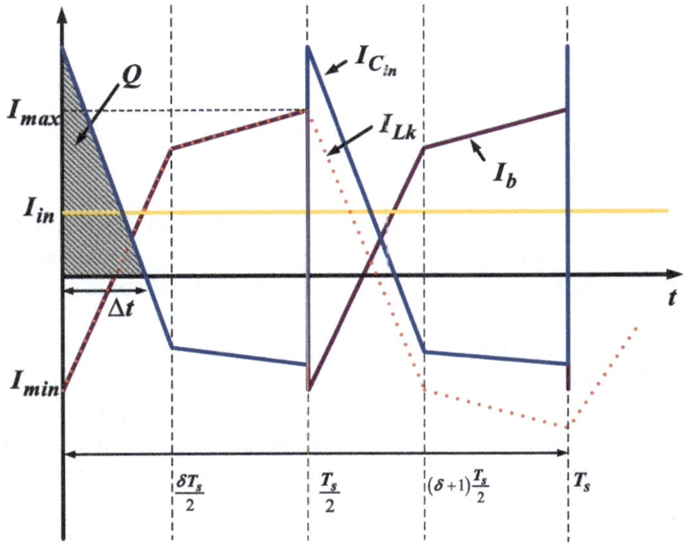

Fig. 13.5 Input terminal currents waveforms for SPS modulation

extraction from the source and its delivery to dc bus thus depends on the value of inductor 'L_k' and phase shift value '$\delta\pi$'. The inductor current analysis is required for the design of an equivalent inductor and input capacitor. The inductor current is shown in Fig. 13.1 with the maximum value denoted as '$I_{\max}$' and minimum value as '$I_{\min}$'. The steady-state inductor current is symmetric around the time axis; thus, the magnitude of the maximum and minimum current can be considered as same in both half cycles, i.e., $|I_{\max}| = |I_{\min}|$. So, in one-half cycle, inductor current can be described by two linear equations as

$$I_{L_k}(t) = I_{\min} + \left(V_{\text{in}} + \frac{V_{\text{o}}}{n}\right) \cdot \frac{t}{L_k}; \quad 0 < t < \frac{\delta T_{\text{s}}}{2} \tag{13.7}$$

$$I_{L_k}(t) = A + \left(V_{\text{in}} - \frac{V_{\text{o}}}{n}\right) \cdot \frac{t}{L_k}; \quad \frac{\delta T_{\text{s}}}{2} < t < \frac{T_{\text{s}}}{2} \tag{13.8}$$

The constant 'A' can be evaluated from (13.8) at $t = \frac{\delta T_{\text{s}}}{2}$ as

$$A = I_{L_{k}\left(\frac{\delta T_{\text{s}}}{2}\right)} - \left(V_{\text{in}} - \frac{V_{\text{o}}}{n}\right) \cdot \frac{\delta T_{\text{s}}}{2L_k} \tag{13.9}$$

Therefore from (13.7) and (13.8), the amount of current at $t = \frac{\delta T_{\text{s}}}{2}$ is

$$I_{L_k\left(\frac{\delta T_s}{2}\right)} = I_{\min} + \left(V_{in} + \frac{V_o}{n}\right).\frac{\delta T_s}{2L_k} \tag{13.10}$$

Replacing the value of $I_{L_k\left(\frac{\delta T_s}{2}\right)}$ from (13.10) in (13.9), the value of constant 'A' is obtained as

$$A = I_{\min} + \frac{\delta T_s V_o}{nL_k} \tag{13.11}$$

Further, the maximum inductor current value can be derived with (13.8) and (13.11) at $t = \frac{T_s}{2}$ and using the condition $|I_{\max}| = |I_{\min}|$

$$I_{\max} = \frac{T_s}{4L_k}\left[V_{in} + (2.\delta - 1)\frac{V_o}{n}\right] \tag{13.12}$$

Substituting (13.12) in (13.10), the value of inductor current at $t = \frac{\delta T_s}{2}$ results in Eq. (13.13)

$$I_{L_k\left(\frac{\delta T_s}{2}\right)} = \frac{T_s}{4L_k}\left[(2.\delta - 1)V_{in} + \frac{V_o}{n}\right] \tag{13.13}$$

Finally, the expression for the inductor current is formed for the half cycle as

$$I_{L_k}(t) = \begin{cases} I_{L_k(1)}; & 0 < t \leq \frac{\delta T_s}{2} \\ I_{L_k(2)}; & \frac{\delta T_s}{2} < t \leq \frac{T_s}{2} \end{cases} \tag{13.14}$$

where

$$\begin{cases} I_{L_k(1)} = \left(V_{in} + \frac{V_o}{n}\right).\frac{t}{L_k} - \left(V_{in} + (2.\delta - 1)\frac{V_o}{n}\right).\frac{T_s}{4L_k} \\ I_{L_k(2)} = \left(V_{in} - \frac{V_o}{n}\right).\frac{t}{L_k} - \left(V_{in} - (2.\delta + 1)\frac{V_o}{n}\right).\frac{T_s}{4L_k} \end{cases} \tag{13.15}$$

13.2.1.1 Equivalent Inductor (L_k) Design

The power transferred by the DAB converter from input to output through the transformer is denoted by expression (13.1). This is further represented in the form as

$$P_t = \frac{8V_{in}\left(\frac{V_o}{n}\right)}{\pi^2.\omega_s.L_k}.\sum_{h=1,3,5\ldots}^{\infty}\frac{\sin(h.\delta.\pi)}{h^3} \tag{13.16}$$

It clearly indicates that the power flow depends upon the input and output voltage, i.e., (V_{in}) and (V_{out}), switching frequency ($\omega_s = 2\pi f_s$) transformer turns ratio (n), phase shift ($\delta\pi$) and equivalent inductance (L_k). The inductor value is denoted from (13.16) as

$$L_k = \frac{8V_{in}\left(\frac{V_o}{n}\right)}{\pi^2.\omega_s.P_t}.\sum_{h=1,3,5\ldots}^{\infty}\frac{\sin(h.\delta.\pi)}{h^3} \tag{13.17}$$

As the power flow equation is inversely related to the equivalent inductor value, the large size of the inductor will make the maximum power extraction from renewable sources impossible. Thus, with a defined maximum power of renewable source denoted as '$P_{t(mpp)}$' a critical inductance value is defined to extract and transfer the maximum possible power. The phase shift condition for the maximum power transfer is also considered from (13.16). As the range of phase shift is defined for one direction power flow as $0<\delta<0.5$, the maximum value of all harmonic components occurs for the phase shift value $\delta = 0.5$. Thus, putting this value of phase shift in (13.17), the maximum or critical inductance value is obtained as

$$L_{k(mpp)} = \frac{V_{in(mpp)}.V_o.\pi}{4.n.\omega_s.P_{t(mpp)}} \tag{13.18}$$

This value of equivalent inductance is, in fact, the total inductance provided by the transformer leakage inductance with an additional inductance in series. The additional inductance, described as an auxiliary inductance, is required for the power transfer. The auxiliary inductor can be avoided if the transformer is properly designed and has the leakage inductance less than or equal to the value mentioned in (13.18).

13.2.1.2 Input Capacitor Design

The input voltage fluctuation is a common problem for the converters interfaced with any renewable energy sources. The input capacitor is an important component as it is directly interfacing the renewable energy source and minimize the voltage ripple at the input terminal. The capacitor also absorbs the reverse current from DAB in each cycle and maintains a constant voltage in the input terminal. The design of the input capacitor is important for DAB converters to provide a constant voltage containing less ripple. The design process initiates with the input terminal current Eq. (13.3) and considering the current waveform illustrated in Fig. 13.4. Thus, the capacitor current is denoted as

$$i_{C_{\text{in}}}(t) = i_{\text{in}}(t) - i_b(t) \tag{13.19}$$

With the input voltage ripple ΔV_{in}, the charge accumulated in the capacitor during the charging process in the interval Δt is

$$Q = 2.\Delta V_{\text{in}}.C_{\text{in}} = \frac{1}{2}\Delta t(I_{\max} + I_{\text{in}}) \tag{13.20}$$

The input current is obtained from (13.6) and (13.15) as

$$I_{\text{in}} = \frac{2}{T_s}\left[\int_0^{\frac{\delta T_s}{2}} I_{L_k(1)} + \int_0^{\frac{T_s}{2}} I_{L_k(2)}\right] \tag{13.21}$$

$$I_{\text{in}} = \frac{T_s V_o \delta(1-\delta)}{2nL_k} \tag{13.22}$$

Now, using (13.4) and (13.22) and considering the maximum power flow situation, the capacitor current is simplified as

$$I_{C_{\text{in}}}(t) = \frac{T_s}{4L_k}\left[V_{\text{in}} - (2.\delta^2 - 4\delta + 1)\frac{V_o}{n} - \left(V_{\text{in}} + \frac{V_o}{n}\right).\frac{4t}{T_s}\right] \tag{13.23}$$

With the initial condition of $i_{C_{\text{in}}} = 0$, the time interval is expressed as

$$\Delta t = \frac{T_s}{4}\left[\frac{V_{\text{in}} - (2.\delta^2 - 4.\delta + 1)\frac{V_o}{n}}{V_{\text{in}} + \frac{V_o}{n}}\right] \tag{13.24}$$

Substituting (13.12), (13.22) and (13.24) in (13.20), the input capacitor value is obtained with the desired input ripple as

$$C_{\text{in}} = \frac{T_s^2}{64.\Delta V_{\text{in}}.L_k}\left[\frac{\left[\frac{V_o}{n}(2.\delta^2 - 4\delta + 1) - V_{\text{in}}\right]^2}{\frac{V_o}{n} + V_{\text{in}}}\right] \tag{13.25}$$

The expression (13.25) shows that the derived capacitor value depends on the phase shift value, and thereby, the capacitor should be designated to mitigate the maximum undulation in the source voltage. The relation of capacitance value with regard to phase shift can be analyzed as

$$\frac{\partial C_{\text{in}}}{\partial \delta} > 0 \tag{13.26}$$

As the maximum ripple occurs at the highest value of ‘δ’, the value of the capacitor is taken as

$$C_{\text{in}} = \frac{T_s^2}{64.\Delta V_{\text{in}}.L_k}\left[\frac{\left(\frac{V_o}{2n} + V_{\text{in}}\right)^2}{\frac{V_o}{n} + V_{\text{in}}}\right] \tag{13.27}$$

13.3 Overview of Control Methods

The dynamic control of DAB utilizes the mathematical modeling of DAB converter, including the small-signal modeling and the average modeling methods [15]. The power transfer equation and output voltage dynamic equation are also used in some control methods, which renders a better dynamic performance with the easy mathematical representation of DAB converter system. The control techniques for improved dynamic behavior are classified broadly as.

Output current feedforward scheme [14, 16],

Inductor current control strategy [17, 18]

Power-based control methods [19, 20].

A brief overview of these control schemes is presented in this chapter, which is suitable for DAB control for the renewable energy system.

13.3.1 Output Current Feedforward Scheme

The feedforward technique is a popular choice for better dynamic performance of power converters [21–23]. In the DAB converter, the dynamic behavior is influenced by the source voltage fluctuation. The performance under sudden load change is also an essential issue for the control scheme and design. The feedforward control is explored in various literature to provide improved dynamic control [21].

An illustration of the current-mode feedforward scheme with the controller and current modulator is showed in Fig. 13.6. The measured voltage value of the input terminal and the measured load current of DAB is utilized to formulate the feedforward control signal, and the voltage error compensation is supplemented for desired output voltage tracking. The current modulator then generates the switching signals to attain the reference voltage. Similar to this strategy, a load current compensation approach, as shown in Fig. 13.7, is discussed to enhance the transient performance of the DAB for sudden load variation situation [14]. The scheme eliminates the inner current modulator as the phase shift ratio (δ) is directly calculated from the measured output current 'i_o' using the expression

$$i_o = \frac{\frac{16}{\pi^2} V_{\text{in}}}{n \sum_{j=0}^{\infty} \left(\frac{\sin[2j+1]\Delta\delta}{[2j+1]^3 \omega_s L_k} \right)} \tag{13.28}$$

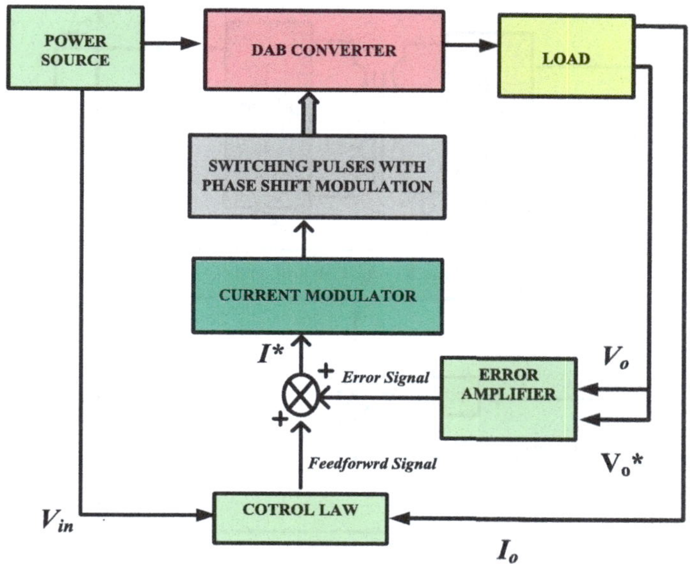

Fig. 13.6 Current-modulated feedforward control of DAB

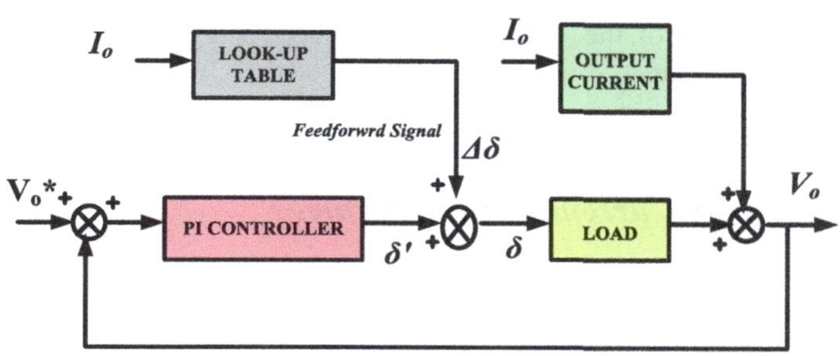

Fig. 13.7 Load current feedforward control scheme with lookup table

The real-time implementation of the method is difficult as the online calculation of $\Delta\delta$ is tough using a DSP or other microprocessors. To simplify the process, a lookup table is created and the phase shift value is precalculated for the required output current 'i_o'. Another feedforward control is proposed with the current-mode modulation utilizing the output and the inductance–current relationship [16].

The block diagram is presented in Fig. 13.8 where PI controller is used to compensate the output current errors between the measured and reference value. The current-mode modulator generates the modulated switching signals with the time shift for the switches in the other bridge. The SPS modulation is considered for

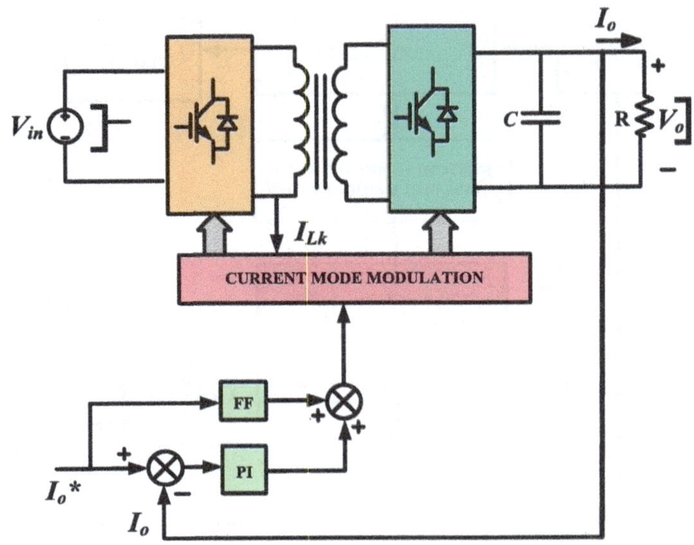

Fig. 13.8 Simplified current feedforward control with current modulation

the proposed method and calculation of reference inductor current. Thus, the dynamic performance of the DAB can be enhanced suggestively with the feed-forward control schemes.

13.3.2 Inductor Current Control Strategy

The inductor current control strategies are a common method of control for DC–DC converters [24, 25]. The inductor current control methods are classified as peak current control, valley current control and average current control in DAB converter [18, 26, 27]. The peak-valley inductor current control is proposed in [17] for a fast transient response. The steady state is attained in a few switching cycles with excellent start-up and voltage tracking performance. However, the method requires five sensors to measure the input and output voltages, transformer current on both the primary and secondary sides, and output current. Similar control of the valley and peak inductor current is proposed in [28] with independent switching signals and a novel current modulation method. The method shows less improvement in dynamic performance as the dependency on the outer voltage loop is more than the current modulation loop. The other novel modulation methods in [18, 29, 30], with desired boundary inductance current in a cycle, are proposed for the power transfer in DAB, which shows enhanced dynamic performance.

To avoid the dc bias current in the transformer during transition, a double-side modulation method is reported. The modulation method uses an optimized coefficient 'x' for defining the phase shift more accurately during the dynamic process, and the switching pulses are modulated for both the bridges simultaneously. The voltage and current waveforms for the modulation process are depicted in Fig. 13.9, with the updated shift ratio 'D'', 'd' is the enhancement in the phase shift, and 'x' is the optimal multiplication factor. The relation between 'D'' and 'x' is formulated as

$$D' = D + xd \tag{13.29}$$

$$x = \frac{nV_{\text{in}}}{nV_{\text{in}} + V_{\text{o}}} \tag{13.30}$$

Similarly, a novel modulation method to remove the dc bias in transformer current is proposed in [30] with a predictive current-mode control. The diagram for the method is shown in Fig. 13.10, which shows the fast-current reference tracking in one switching period. The values 'i_{p1}' and 'i_{p2}' are measured in each switching period; thus, the sampling frequency needs to be doubled. The 'i^*_{p1}' is the reference of the peak value of inductor current and 'i^*_{n1}' is the reference of valley of inductor current.

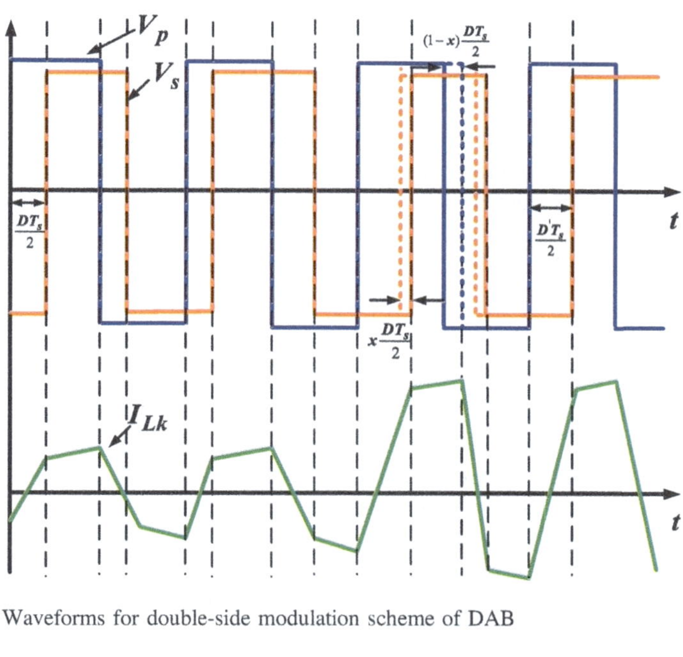

Fig. 13.9 Waveforms for double-side modulation scheme of DAB

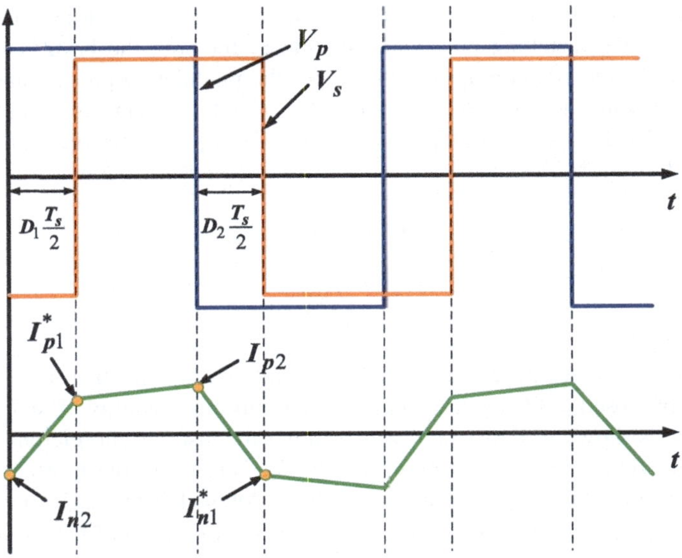

Fig. 13.10 Duty cycle modulation scheme of DAB

13.3.3 Power-Based Control Strategy

In renewable energy systems, direct power control is a well-known strategy applied in active rectifiers and inverters for enhanced transient performance during load change [31–33]. Likewise, the power-based control method is utilized for the DAB converter for improved performance [19, 20, 34]. The power control method is suitable for the DAB converter as the power flow across the transformer can be controlled with specified phase shift control in each switching period [35]. The power transfer in SPS modulation can be expressed as

$$P = \frac{V_{\text{in}} V_{\text{o}} \delta(1 - \delta) T_{\text{s}}}{2nL_k} \tag{13.31}$$

The method can be well adapted for other phase shift modulations, which provides additional advantages like increased soft-switching range, less reactive power flow and zero dc bias transformer current. A virtual direct power control (VDPC) scheme, with a block diagram shown in Fig. 13.11, is proposed [34] to boost the transient performance. A virtual reference voltage is generated using the measured output voltage and a PI-based control. Then the shift ratio denoted as $D*$ is obtained from relation in (13.32) and (13.33) for the bidirectional power flow as

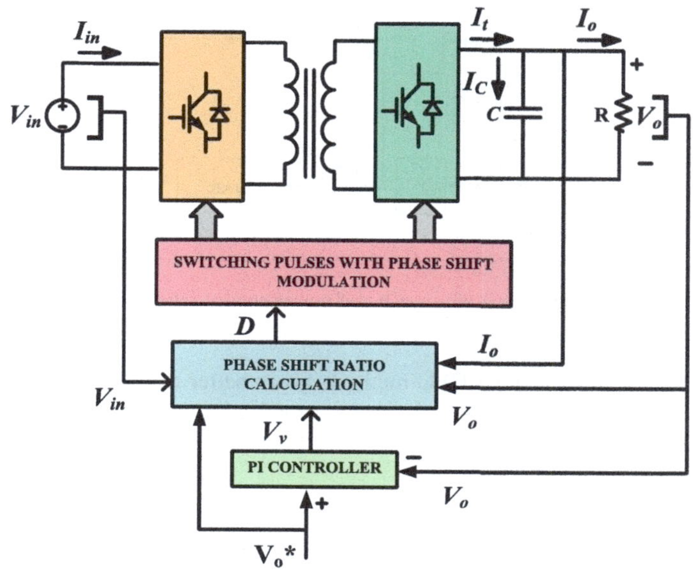

Fig. 13.11 Virtual direct power control (VDPC) of DAB

$$D^* = \frac{1}{2} - \sqrt{\frac{1}{4} - \frac{V_o^* V_v^* i_o}{V_o^2 V_{in}}}, \quad P \geq 0 \tag{13.32}$$

$$D^* = -\frac{1}{2} + \sqrt{\frac{1}{4} + \frac{V_o^* V_v^* i_o}{V_o^2 V_{in}}}, \quad P \leq 0 \tag{13.33}$$

The VDPC scheme provides excellent dynamic response under sudden load change conditions. As the operating frequency is high, accurate current measurement is a difficult task, especially when the DAB is coupled with other power converters having an unknown capacitor in its terminal.

The current sensorless control schemes can be explored to resolve the problem of the current measurement. A nonlinear disturbance observer is used in [36, 37] to estimate the output current and combine the direct power control for the DAB converter. The block diagram for disturbance observer-based power with control method is shown in Fig. 13.12 with '$\hat{i}_o(k)$' as the estimated output current in the kth switching period expressed as

$$\hat{i}_o(k) = z(k) + lV_o(k) \tag{13.34}$$

where 'z' is the control parameter in internal loop of disturbance observer, and 'l' denotes the gain to control the convergence degree. Though the performance with

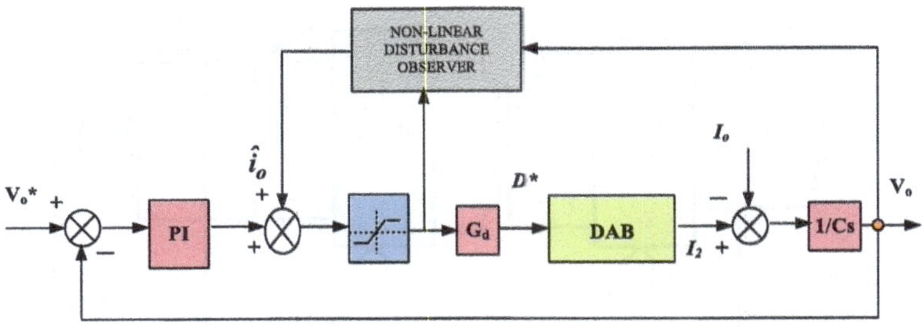

Fig. 13.12 Nonlinear disturbance observer-based power control scheme of DAB

both schemes is similar, the VDPC method shows better results under load change conditions, whereas the disturbance observer method provides a current sensorless option.

13.4 Direct Power Control (DPC) of DAB Converter

The direct power control (DPC) method is developed [38] where the error between output voltage and reference voltage is used to produce a power reference signal using a PI controller. The relation between the output capacitor's energy and power can be stated as

$$\frac{\mathrm{d}E_C}{\mathrm{d}t} = \frac{\mathrm{d}\left(\frac{1}{2}CV_o^2\right)}{\mathrm{d}t} = P_t - V_o i_o \tag{13.35}$$

Accounting for the positive power transfer, the value of phase shift ratio (δ) is considered to be positive. Thereby, using (13.31) in (13.35), the output voltage dynamic equation is obtained as

$$CV_o \frac{\mathrm{d}V_o(t)}{\mathrm{d}t} = \frac{V_{in}V_o\delta(1-\delta)}{2nL_k f_s} - V_o i_o \tag{13.36}$$

Consider a term,

$$P^* = \frac{V_{in}V_o\delta(1-\delta)}{2nL_k f_s} \tag{13.37}$$

The term P^* now becomes the control variable for controlling the power transfer through the transformer by direct control of the phase shift ratio. From (13.36), it can be regarded that under a steady state, the rate of change of output voltage should become zero. A reference output voltage is set for the controller, thus for all fluctuation and load change variation, the controller tends to make the rate of change of output voltage to zero and tracks the reference value with the short response time. The control variable reference $P*$ can be generated using a PI controller as mentioned below (13.38), and the coefficients k_P and k_I are tuned for required dynamic behavior

$$P^* = V_o i_o + k_P(V_o^* - V_o(t)) + k_I \int (V_o^* - V_o(t))dt \quad (13.38)$$

The exact value for these coefficients makes the controller robust and enables it to track the reference voltage with a fast transient response. However, any mismatch in the converter parameter or in the controller coefficients makes the system performance to have unsatisfactory results. The main control schema of DPC is illustrated in Fig. 13.13. To investigate the performance of the suggested DPC scheme for DAB, simulations are conducted using MATLAB/Simulink software. The dynamic and steady-state behavior of the converter relies on the design of DPC under load change conditions. The specifications in Table 13.1 are applied for the simulation study of the converter, operated with the DPC method. The simulation results of DAB under FDPC as described above ensure a stable and fixed output voltage of the DAB stage under load variation.

The result in Fig. 13.14a depicts the load voltage waveform with the load is changed by 1 kW, i.e., from 1 to 2 kW at 0.1 s. The change in load current with the load change is shown in Fig. 13.14b. The change in load disturbs the output voltage, and the value drops with an increase in current at the output terminal.

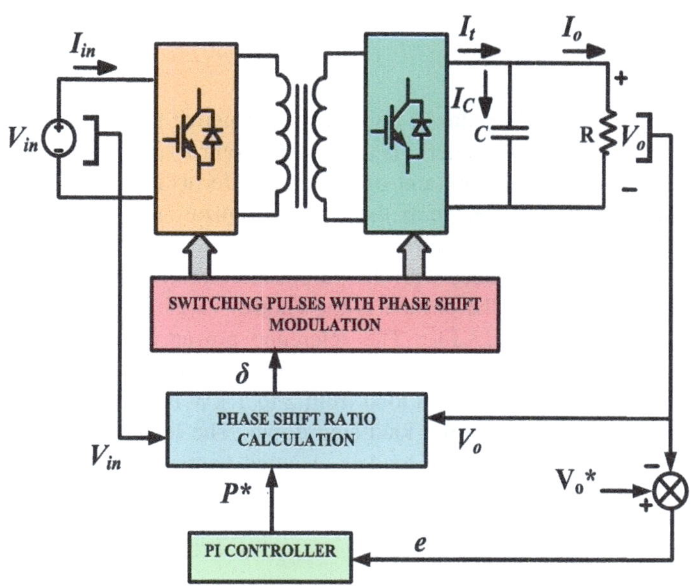

Fig. 13.13 DPC scheme of DAB

Table 13.1 DAB converter parameters

Parameters		Value
Input capacitance	C	11,000 μF
Output capacitance	C	11,000 μF
Source voltage	V_{in}	450 V
Output voltage	V_o	450 V
Equivalent inductance	L_k	950 μH
Voltage ratio	n	1
Operating frequency	f_s	10 kHz
Resistive value		
• 2000 W	R	202.5 Ω
• 1000 W	R	101.25 Ω

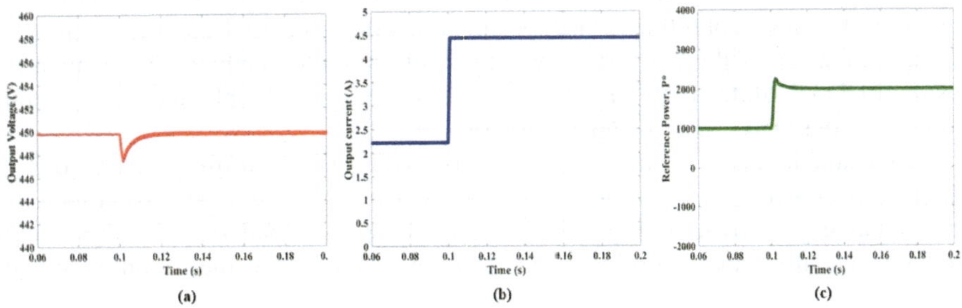

Fig. 13.14 **a** Output voltage, **b** load current and **c** reference power when load is changed from 1000 to 2000 W at 0.1 s

The voltage error is used by the PI controller to generate the new changed power reference. The power reference term is used as the control law to generate the phase shift ratio, thus enabling the amount of power transfer to increase. Thus, the controller tracks to reference voltage as the error minimization is the primary control objective. The fast response, preferably to track the reference voltage and ensure a stable, steady-state performance with the proposed controller. The reference power generated by the PI controller, which is the control variable for phase shift modulation, is shown in Fig. 13.14c. The reference power generation during load change is rapid, which provides a better transient response under large power change conditions. The decrease in load from 2 to 1 kW is introduced to verify the controller performance during the load step down. The load voltage waveform is depicted in Fig. 13.15a when the load is changed from 2 to 1 kW at 0.1 s. The change in current is depicted in Fig. 13.15b, and the reference power change is shown in Fig. 13.15c for the load step down.

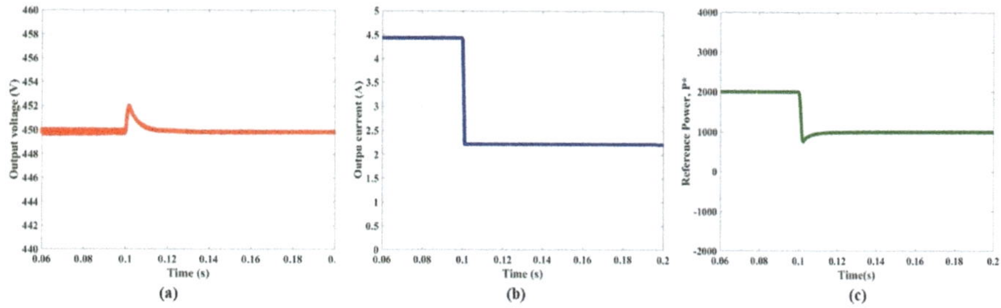

Fig. 13.15 **a** Output voltage, **b** load current and **c** reference power when load is changed from 2000 to 1000 W at 0.1 s

13.5 Conclusion

In this chapter, the DAB converter circuit is analyzed, and the inductor and input capacitor design are presented for the application of the DAB converter in a renewable energy system. The inductor design is crucial for the power handling and performance of the converter, and the input capacitor is responsible for the reduced input voltage ripple. Thus, the proper design of converter parameters is essential for the extraction of maximum power from renewable energy sources and provides efficient control over the dynamic and static performance of the converter. Also, the chapter includes a brief review of various control strategies of DAB converter for the renewable energy system. The power-based control schemes provide better performance under load change and input voltage fluctuation conditions. Moreover, advanced modulation and control strategies like dual-phase shift and triple-phase shift modulations can be implemented in power-based control strategies. The MPPT algorithm can also be adapted with power-based control methods, which provides better suitability in PV systems. Further, the direct power control is suggested in the chapter, and the simulation results are provided, showing an enhanced output voltage regulation during load change conditions.

References

1. Zhao, B., Song, Q., Liu, W., Sun, Y.: Overview of dual-active-bridge isolated bidirectional DC–DC converter for high-frequency-link power-conversion system. IEEE Trans. Power Electron. **29**(8), 4091–4106 (2014)
2. Kheraluwala, M.N., Gascoigne, R.W., Divan, D.M., Baumann, E.D.: Performance characterization of a high-power dual active bridge DC-to-DC converter. IEEE Trans. Ind. Appl. **28** (6), 1294–1301 (1992)
3. Krismer, F., Kolar, J.W.: Efficiency-optimized high-current dual active bridge converter for automotive applications. IEEE Trans. Industr. Electron. **59**(7), 2745–2760 (2012)
4. Shi, Y., Li, R., Xue, Y., Li, H.: Optimized operation of current-fed dual active bridge DC–DC converter for PV applications. IEEE Trans. Industr. Electron. **62**(11), 6986–6995 (2015)

5. Zhou, H., Duong, T., Sing, S.T., Khambadkone, A.M.: Interleaved bi-directional dual active bridge DC-DC converter for interfacing ultracapacitor in micro-grid application. 2010 IEEE International Symposium on Industrial Electronics, Bari, pp. 2229–2234 (2010).
6. Talbi, S., Mabwe, A.M., Hajjaji, A.E.: Control of a bidirectional dual active bridge converter for charge and discharge of a Li-Ion Battery. IECON 2015-41st Annual Conference of the IEEE Industrial Electronics Society, Yokohama, pp. 000849–000856 (2015).
7. Wu, H., Sun, K., Chen, L., Zhu, L., Xing, Y.: High step-up/step-down soft-switching bidirectional DC–DC converter With coupled-inductor and voltage matching control for energy storage systems. In: IEEE Transactions on Industrial Electronics. **63**(5), 2892–2903, May (2016).
8. Joebges, P., Hu, J., De Doncker, R.W.: Design method and efficiency analysis of a DAB converter for PV integration in DC grids. 2016 IEEE 2nd Annual Southern Power Electronics Conference (SPEC), Auckland, pp. 1–6 (2016).
9. Hou, N., Song, W., Li, Y., Zhu, Y., Zhu, Y.: A Comprehensive optimization control of dual-active-bridge DC–DC converters based on unified-phase-shift and power-balancing scheme. In: IEEE Transactions on Power Electronics. **34**(1), 826–839, Jan (2019).
10. Hou, N., Li, Y.W.: Overview and comparison of modulation and control strategies for a nonresonant single-phase dual-active-bridge DC–DC converter. In: IEEE Transactions on Power Electronics. **35**(3), 3148–3172, March (2020).
11. Zhao, B., Song, Q., Liu, W., Liu, G., Zhao, Y.: Universal high-frequency-link characterization and practical fundamental-optimal strategy for dual-active-bridge DC-DC converter under PWM plus phase-shift control. In: IEEE Transactions on Power Electronics. **30**(12), 6488–6494, Dec (2015).
12. Shi, Y., Li, R., Xue, Y., Li, H.: Optimized operation of current-fed dual active bridge DC–DC converter for PV applications. In: IEEE Transactions on Industrial Electronics. **62**(11), 6986–6995, Nov (2015).
13. Hu, J., Joebges, P., Pasupuleti, G.C., Averous, N.R., De Doncker, R.W.: A maximum-output-power-point-tracking-controlled dual-active bridge converter for photovoltaic energy integration into MVDC Grids. In: IEEE Transactions on Energy Conversion. **34**(1), 170–180, Mar (2019).
14. Segaran, D., Holmes, D.G., McGrath, B.P.: Enhanced load step response for a bidirectional DC–DC converter. In: IEEE Transactions on Power Electronics. **28**(1), 371–379, Jan (2013).
15. Mueller, J.A., Kimball, J.W.: Modeling dual active bridge converters in DC distribution systems. In: IEEE Transactions on Power Electronics. **34**(6), 5867–5879, June (2019).
16. Shan, Z., Jatskevich, J., Iu, H.H., Fernando, T.: Simplified load-feedforward control design for dual-active-bridge converters With current-mode modulation. In: IEEE Journal of Emerging and Selected Topics in Power Electronics. **6**(4), 2073–2085, Dec (2018).
17. Oggier, G.G., Ordonez, M., Galvez, J.M., Luchino, F.: Fast transient boundary control and steady-state operation of the dual active bridge converter using the natural switching surface. In: IEEE Transactions on Power Electronics. **29**(2), 946–957, Feb (2014).
18. Huang, J., Wang, Y., Li, Z., Lei, W.: Predictive valley-peak current control of isolated bidirectional dual active bridge DC-DC converter. 2015 IEEE Energy Conversion Congress and Exposition (ECCE), Montreal, QC, pp. 1467–1472 (2015).
19. Ge, J., Zhao, Z., Yuan, L., Lu, T.: Energy feed-forward and direct feed-forward control for solid-state transformer. In: IEEE Transactions on Power Electronics. **30**(8), 4042–4047, Aug (2015).
20. Lin, S.T., Li, X., Sun, C., Tang, Y.: Fast transient control for power adjustment in a dual-active-bridge converter. In: Electronics Letters. **53**(16), 1130–1132, 3 8 (2017).
21. Redl, R., Sokal, N.O.: Near-optimum dynamic regulation of DC-DC converters using feed-forward of output current and input voltage with current-mode control. In: IEEE Transactions on Power Electronics. vol. PE-1, **3**, 181–192, July (1986).
22. Nielsen, J.G., Newman, M., Nielsen, H., Blaabjerg, F.: Control and testing of a dynamic voltage restorer (DVR) at medium voltage level. In: IEEE Transactions on Power Electronics. **19**(3), 806–813, May (2004).

23. He, J., Li, Y.W., Guerrero, J.M., Blaabjerg, F., Vasquez, J.C.: An Islanding microgrid power sharing approach using enhanced virtual impedance control scheme. In: IEEE Transactions on Power Electronics. **28**(11), 5272–5282, Nov (2013).
24. Zhou, G., Xu, J., Jin, Y.: Elimination of subharmonic oscillation of digital-average-current-controlled switching DC–DC converters. In: IEEE Transactions on Industrial Electronics. **57** (8), 2904–2907, Aug (2010). https://doi.org/10.1109/TIE.2009.2035996.
25. Leng, M., Zhou, G., Zhou, S., Zhang, K., Xu, S.: Stability analysis for peak current-mode controlled buck LED driver based on discrete-time modeling. In: IEEE Journal of Emerging and Selected Topics in Power Electronics. **6**(3), 1567–1580, Sept (2018).
26. Nasr, M., Poshtkouhi, S., Radimov, N., Cojocaru, C., Trescases, O.: Fast average current mode control of dual-active-bridge DC-DC converter using cycle-by-cycle sensing and self-calibrated digital feedforward. 2017 IEEE Applied Power Electronics Conference and Exposition (APEC), Tampa, FL, pp. 1129–1133 (2017). https://doi.org/10.1109/APEC.2017.7930837.
27. Shan, Z., Jatskevich, J., Cho, E., Shin, M., Lee, Y.: A feedforward control method of dual-active-bridge dc/dc converter to achieve fast dynamic response. 2014 IEEE 36th International Telecommunications Energy Conference (INTELEC), Vancouver, BC, pp. 1–6 (2014).
28. Takagi, K., Fujita, H.: Dynamic control and performance of a dual-active-bridge DC–DC converter. In: IEEE Transactions on Power Electronics. **33**(9), 7858–7866, Sept (2018).
29. Li, X., Li, Y.: An optimized phase-shift modulation for fast transient response in a dual-active-bridge converter. In: IEEE Transactions on Power Electronics. **29**(6), 2661–2665, June (2014).
30. Dutta, S., Hazra, S., Bhattacharya, S.: A digital predictive current-mode controller for a single-phase high-frequency transformer-isolated dual-active bridge DC-to-DC converter. In: IEEE Transactions on Industrial Electronics. **63**(9), 5943–5952, Sept (2016).
31. CortÉs, P., RodrÍguez, J., Antoniewicz, P., Kazmierkowski, M.: Direct power control of an AFE using predictive control. In: IEEE Transactions on Power Electronics. **23**(5), 2516–2523, Sept (2008).
32. Zhang, Y., Qu, C.: Table-based direct power control for three-phase AC/DC converters under unbalanced grid voltages. In: IEEE Transactions on Power Electronics. **30**(12), 7090–7099, Dec (2015).
33. Malinowski, M., Jasinski, M., Kazmierkowski, M.P.: Simple direct power control of three-phase PWM rectifier using space-vector modulation (DPC-SVM). In: IEEE Transactions on Industrial Electronics. **51**(2), 447–454, April (2004).
34. Nguyen, D., Fujita, G., Bui-Dang, Q., Ta, M.C.: Reduced-order observer-based control system for dual-active-bridge DC/DC converter. In: IEEE Transactions on Industry Applications. **54**(4), 3426–3439, July-Aug (2018).
35. Song, W., Hou, N., Wu, M.: Virtual direct power control scheme of dual active bridge DC–DC converters for fast dynamic response. In: IEEE Transactions on Power Electronics. **33**(2), 1750–1759, Feb (2018).
36. Nguyen, D., Nguyen, D., Funabashi, T., Fujita, G.: Sensorless control of dual-active-bridge converter with reduced-order proportional-integral observer. Energies. **11**, pp. 931–948, Apr (2018).
37. Xiong, F., Wu, J., Liu, Z., Hao, L.: Current sensorless control for dual active bridge DC–DC converter with estimated load-current feedforward. In: IEEE Transactions on Power Electronics. **33**(4), 3552–3566, April (2018).
38. Tiwary, N., Venkataramana, N.N., Panda, A.k., Narendra, A.: Direct power control of dual active bridge bidirectional DC-DC converter. 2019 International Conference on Power Electronics, Control and Automation (ICPECA), New Delhi, India, pp. 1–4 (2019).

Chapter 14
Hardware-in-Loop-Based Reliability Improvement of Power Converter for Critical Electrical Drive Applications

R. Manikandan, R. Raja Singh, G. Edison, and S. Darius Gnanaraj

14.1 Introduction

Safety first-Save Lives/Systems is the main agenda in the design process of any safety critical electrical drive systems. Critical electrical drive systems are those systems, breakdown of which might cause loss of life and significant resource damage. The main application fields are such as automotive, aircraft flight control and railway traction systems. Many modern electric drive systems change into critical electrical drives system to save human life or protect system from failure. In future, critical electrical drive system will become more widespread and dominant [1]. Due to recent advancements in electrical drive and power converters, the industrial process systems and electrical drives received increased attention [2]. The reliable and fault-tolerant operation of these critical electrical drives is main area of interest. In any electrified drive system, the operations like steering control, brake-by-wire system and fuel pumps control are plays vital role in uninterrupted operation of the system. Hence, parallel redundancy has been widely employed in many safety critical electrical drive applications such as space machines, aerospace, defense, electric vehicle, and many other industrial applications. In general, the principal parts of electrical drive systems are source, power electronic converters, AC or DC machines, sensors and controller. For the uninterrupted service of any critical loads, a reliability study (diagnosis, supervision and fault detection) is important. Hence, this chapter focuses on reliable and fault-tolerant power converters in safety electrical drive system. The inverters with good physical condition assure better dependability, efficiency and performance in critical electrical drive applications. These performance metrics can be accomplished by using higher rating components and rendering reconfigurable internal

R. Manikandan · R. Raja Singh (✉)
School of Electrical Engineering, Vellore Institute of Technology, Vellore, India

G. Edison · S. D. Gnanaraj
School of Mechanical Engineering, Vellore Institute of Technology, Vellore, India

N. Priyadarshi et al. (eds.), *DC–DC Converters for Future Renewable Energy Systems*, Energy Systems in Electrical Engineering,
https://doi.org/10.1007/978-981-16-4388-0_14

fault-tolerance topologies [3]. By considering first option, a suitable power switches are investigated for inverter in more than two decades. Since 1990, insulated gate bipolar transistors (IGBTs) are the most popular semiconductor device for the design inverters, which is incorporated in many high-power systems of electrical engineering. Later to simplify the assembly process of inverters, press-fit pins and PC-TIM phase change—thermal interface material) are introduced [4]. At present, integrated gate-commutated thyristors (IGCTs), IGBTs and metal–oxide–semiconductor field-effect transistors (MOSFETs) represent modern switching devices. For the portable, automotive, railway traction and aerospace applications, power-integrated circuits (PICs)-based inverters are developed [5]. The use of silicon carbide (SiC) MOSFET as an alternative to silicon (Si) IGBT brought some considerable changes in the performance of various power converters. Basically, SiC MOSFET based modular multilevel converter has the drawback of high expense and increased dynamic loss during high-voltage and high-current applications. To resolve these drawbacks, a Si IGBT and SiC MOSFET based hybrid MMC are introduced [6]. From the advanced technology in material process, contributed wide band gap SiC and IGBT represents most excellent power devices that can be fabricated by SiC successfully. In SiC IGBT based applications, the experimental findings and current problems are summarized in the literature [7]. The hybrid structure which uses Si and SIC switche combination can take the advantage of fast switching speed and high conduction property of Si and SiC, respectively. However, two separate gate drivers are normally required to operate the hybrid switch, which increases the complexity and expense. Further, to resolve this issue, an efficient gate driver circuit was designed with single control signal for both Si and SiC based hybrid power switch. It makes the Si/SiC switch as a typical three-terminal device [8]. From the literature, SiC IGBT is best suited for high switching frequency applications where it shown low switching losses. Fault-tolerant structures are historically employed in systems where a main function is continuity of operation. The switch faults in the converters are mainly classified into two categories: open-circuit fault (OCF) and short-circuit fault (SCF).This chapter focuses on fault diagnosis and suitable reconfigurable strategy for both faults (OCF, SCF).The essential need for a fault-tolerant control of power converter is the converter structures which has fault ride through ability. Hence, reconfigurable strategies with the redundant structure design that has the benefits of extremely reliable have allured numerous works in last decades. In [9] investigated a power converter structure with redundant triacs and fuses which improves induction motor drive. This paper uses a topology that allows continuous, normal operation even after phase faults in power convert and induction drive. In [10] to diagnose the fault in power transistor, freewheeling diode and bus current are used. In [11], an open-circuit fault identification scheme is proposed for HEV by comparing the duty cycle and inductor current slope. The strategy explored in [12] uses a redundant phase leg, where the neutral point of the motor is connected to the redundant leg to ensure the fault-tolerant operation. In addition, over the past decade, numerous investigation works has been performed to detect fault in the 3-φ permanent magnet synchronous motors (PMSM) inverters. The work in [13] proposes a remedial technique for failures in inverter based on the similar technique in [14] and [15] to attain the torque without disturbance for PMSM drives. It is important to notice that the fault-tolerant strategies discussed in [13–16] will decrease the maximum motor torque and increase the system power loss and current in their post-fault operations and that is inevitable. Another potential fault-tolerant converter design [17, 18] is an additional inverter leg that connects via a bypass TRIAC to the neutral point of the system. In this method, the motor can normally without disturbance but more current will flow through the power switches, and this may create overheat to the power switches. In addition, there are different analytical methods also proposed in based on ANN, FFT, fuzzy logic and genetic algorithms. Due to open-circuit faults (OCFs), the PMSM phase currents will not be in the pure sinusoidal form. Also, DC components lead to an increase in copper losses. It can be shown that as compared to standard working conditions and the other remaining phases, the defective phase has a lesser RMS value. Owing to the non-sinusoidal form of the PMSM phase currents, a significant increase of about 65.8% in total harmonic distortion (THD) values is expressed in their harmonic distortion. Also, the electromagnetic torque value is not constant any more. Due to short-circuit fault, the mechanical torque becomes more pulsating in nature and also constitutes major threat to rotor demagnetization. So, the SCFs are severe than OCFs [19].

Table 14.1 summarizes some important existing works related to fault diagnosis and fault-tolerant mitigation technique for open-circuit fault (OCF) and short-circuit fault (SCF) in power converters. This chapter focuses on fault diagnosis strategy and fault-tolerant scheme for permanent magnet synchronous motor (PMSM) in safety critical electrical drive application. The proposed diagnostic strategy identifies the faults from the high-frequency current signal by inserting a HF voltage signal into the PWM control circuit of the inverter. A fault isolation and system reconfiguration approach is applied with the study of the amplitude and frequency of the HF signal phenomenon after the switch faults. The chapter is organized into several sections. Section 14.2 describes the mathematical modeling of inverter. In Sect. 14.3, PMSM based critical electrical drive system is discussed. Section 14.4 discusses the fault diagnosis and fault ride techniques. Section 14.5 discusses the results of the simulation for different faults in different operating conditions. Finally, important conclusions and future scopes are summarized in Sect. 14.6.

14.2 Mathematical Modeling and Operation of Three-Phase Two-Level VSI

14.2.1 Electrical Model

Inverters are one of main power converter device, which is used to convert dc source into single-phase or three-phase AC source. Three-phase inverters are widely used for applications of medium and high-power. The typical three-phase VSI circuit is depicted in Fig. 14.1. The switches pairs of the three-phase inverter in

Table 14.1 Fault-tolerant power converters summarized

Sl. no	Power converter	Fault type	Mitigation technique	Application	Year
1	IGBT based asymmetrical half-bridge converter	OCF	Using DC bus current and fast FFT algorithm (fault diagnosis only)	Switched reluctance motor system	2016 [17]
2	IGBT based three-phase inverter	OCF	Using phase current and six additional triacs	Two PMSM drives for traction application	2018 [20]
3	IGBT based three-phase inverter	OCF	Using phase angle of stator current and additional leg six additional triacs	PMSM drive system	2018 [21]
4	Four-phase IGBT based asymmetric bridge converter	OCF and SCF phase fault in Inverter/ Machine	Using DC bus current amplitude (fault diagnosis only)	Switched reluctance motors system	2012 [22]
5	Two-level IGBT based voltage source inverter	OCF and SCF	Using redundant leg and isolating thyristors	PMSM drive system	2012 [23]
6	Four-phase IGBT based asymmetric bridge converter	OCF and SCF	Using phase current (fault diagnosis only)	Switched reluctance motors for aircraft and automotive applications	2013 [24]
7	IGBT based three-phase inverter	OCF	Using current residual vector (CRV) (fault diagnosis only)	PMSM drive system	2014 [25]
8	IGBT based DC–DC boost converter	OCF and SCF	Using inductor current slope and one redundant switch	Photovoltaic systems	2015 [26]

OCF Open-circuit fault; *SCF* Short-circuit fault

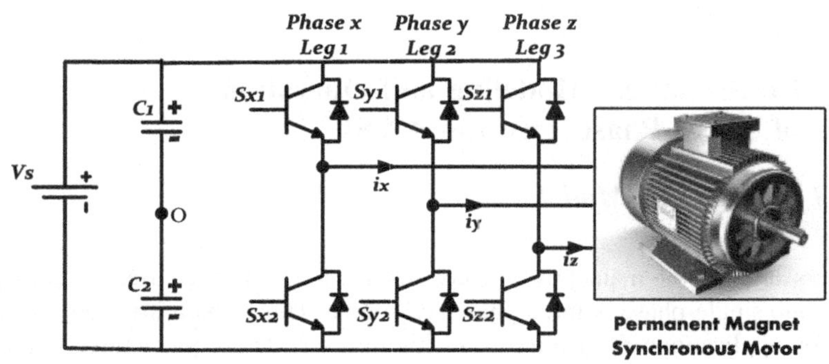

Fig. 14.1 Three-phase two-level VSI topology

Table 14.2 Fault signatures: HF current and phase current during normal/open-circuit switch fault

Fault signatures		Normal operation	Open-circuit fault in any leg	
			Top switch	Bottom switch
HF current	Amplitude (A)	Normal	Zero	Zero
	Frequency (Hz)	f	$<f$	$<f$
Phase current	Amplitude (A)	Normal	$<I_o$	$<I_o$

same leg [(S_{x1}, S_{x2}), (S_{y1}, S_{y2}), (S_{z1}, S_{z2})] should not be turned on concurrently because that will create short-circuit in the leg. To create a proper voltage waveform, switching of states should occur in the inverter. As a result, the AC output L–L voltages are in the form of discrete levels of $+V_s$, 0, and $-V_s$ for the circuit shown below.

Here, V_{xo}, V_{yo}, V_{zo} = inverter output phase voltages, V_{xy} = inverter line-to-line output voltage = $V_{xo} - V_{yo}$, V_{yz} = inverter line -to- line output voltage = $V_{yo} - V_{zo}$ and V_{zx} = inverter line-to-line output voltage = $V_{zo} - V_{xo}$.

The Fourier series form of output phase voltage is:

$$v_{xo} = \sum_{n=1,3,5}^{\infty} V_{xo,n} \sin(n\omega t) \tag{14.3}$$

where V_{xo} is the output phase voltage between phase x and point O in Fig. 14.1 After simplification,

$$v_{xo} = \sum_{n=1,3,5}^{\infty} \frac{2V_{\text{in}}}{n\pi} \sin(n\omega t) \tag{14.4}$$

Similarly, the remaining output phase voltage V_{yo} and V_{zo} can be written as:

$$v_{yo} = \sum_{n=1,3,5}^{\infty} \frac{2V_{\text{in}}}{n\pi} \sin[n(\omega t - 120°)] \tag{14.5}$$

$$v_{zo} = \sum_{n=1,3,5}^{\infty} \frac{2V_{\text{in}}}{n\pi} \sin[n(\omega t - 240°)] \tag{14.6}$$

The output line-to-line voltage V_{ab} will be depicted using Fourier series as follows:

$$v_{xy} = \sum_{n=1,3,5}^{\infty} V_{xy,n} \sin[n(\omega t + 30°)] \tag{14.7}$$

where $V_{xy,n}$ = amplitude of the nth harmonic component of v_{xy}.

After simplification,

$$v_{xy} = \sum_{n=1,5,7,11}^{\infty} \frac{4V_s}{n\pi} \cos\left(\frac{n\pi}{6}\right) \sin n(\omega t + 30°) \quad (14.8)$$

Similarly, the remaining output L–L voltages V_{yz} and V_{zx} can be written as:

$$v_{yz} = \sum_{n=1,5,7,11}^{\infty} \frac{4V_s}{n\pi} \cos\left(\frac{n\pi}{6}\right) \sin n(\omega t + 30° - 120°) \quad (14.9)$$

$$v_{zx} = \sum_{n=1,5,7,11}^{\infty} \frac{4V_s}{n\pi} \cos\left(\frac{n\pi}{6}\right) \sin n(\omega t + 30° - 240°) \quad (14.10)$$

14.2.2 Thermal Model

According to relevant data, almost 60% of the failure of power devices due to higher junction temperatures induced by heat and increased thermal resistance of power devices or excess of power loss [27, 28]. Consequently, the IGBT junction temperature is regarded as one of the main measures of its health and reliability. The popular models used for thermal modeling of power switches are Cauer-type and Foster-type both are based on resistance–capacitance (RC) network (Fig. 14.2). Here, the physical structure of the device can be modeled with the help of Cauer–thermal model [29, 30]. The Foster-type model is calculated by means of the power device temperature dynamics [31]. The thermal module referred to in this chapter consists of a three-phase inverter containing in one package all six IGBTs and diodes. For the thermal network of a 3-φ inverter power module, the Foster-based thermal model is considered.

The Fig. 14.3 shows the cross-sectional view of the IGBT diode module with heat sinks, where each physical layer is expressed as a "cell" consisting of two parallel subcircuits of thermal resistance and capacitance, linked in sequence, as depicted in Fig. 14.4.

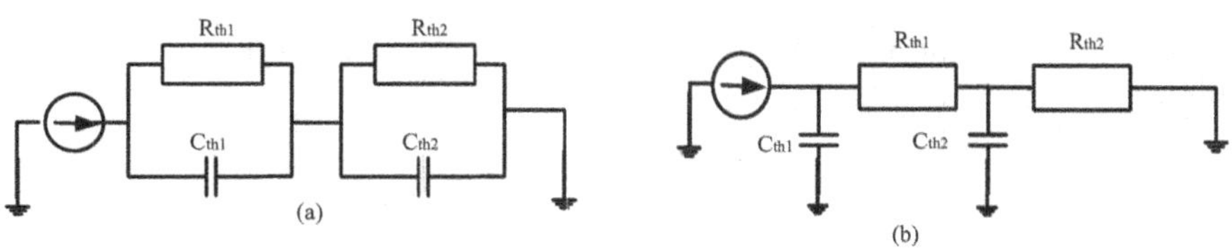

Fig. 14.2 Thermal networks of **a** Foster **b** Cauer

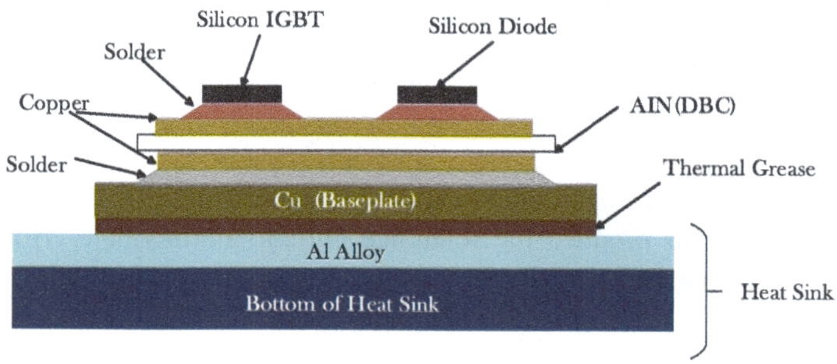

Fig. 14.3 Power module with heat-sink cross section view

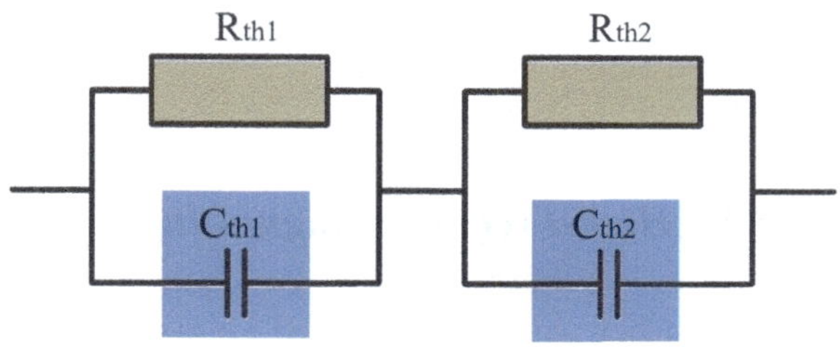

Fig. 14.4 Power module each material layer thermal model

The following equation shows mathematical expression of each cell

$$Z_{\text{th}i}(t) = R_2\left[1 - \exp\left(\frac{t}{R_1C_1}\right)\right] + R_2\left[1 - \exp\left(\frac{t}{R_2C_2}\right)\right] \tag{14.11}$$

The thermal model for an inverter module mounted on a typical heat-sink with multi-IGBT devices is expressed as

$$\begin{bmatrix} T_{j1} \\ T_{j2} \\ \vdots \\ T_{jm} \end{bmatrix} = \begin{bmatrix} Z_{11} & Z_{12} & \cdots & Z_{1m} \\ Z_{21} & Z_{22} & \cdots & Z_{2m} \\ \vdots & \vdots & \ddots & \vdots \\ Z_{m1} & Z_{m2} & \cdots & Z_{mm} \end{bmatrix} \cdot \begin{bmatrix} P_{L1} \\ P_{L2} \\ \vdots \\ P_{Lm} \end{bmatrix} \tag{14.12}$$

where self-impedances of each chip are represented by the diagonal elements ($i = j$), whereas mutual –impedances are represented by off-diagonal elements ($i \neq j$). P_{Li}, $T_{ji}(i = 1, 2, \ldots, m)$ illustrates m heat sources and junction temperatures.

14.3 PMSM Based Critical Electrical Drive System with Fault–Tolerant Topology

In Fig. 14.5, the critical drive system with fault-tolerant topology is shown. The speed estimator is used to sense the speed of the machine, and it is further processed using PI controller to produce the reference current. Further, the error between reference current I_{ref} and the current measured from phase $I_{m,}$ and it is further converted into reference voltage, V_{ref} by processing through a PI controller. The switching signals the inverter switches that are created by regulating the reference voltage V_{ref} with a carrier signal (triangular wave). Filtering and frequency estimation unit, filters out the HF portion from output-phase current and calculates the frequency of HF signal. The fault diagnosis and relay actuating unit generate relay actuating signal based on amplitude of phase current, amplitude and frequency of HF current signal.

14.3.1 High-Frequency Signal Injection Scheme:

High-frequency signal injection (HFSI) is one the widely used approach for sensorless control of low speed machine and also to detect turn fault in electric machines. Here, a sinusoidal HF voltage injection scheme [32] is implemented to diagnose switches faults in three-phase VSI inverter. With the help of the HFSI technique, fault signatures are generated during switches faults (OCF, SCF). The HF voltage signal *VHF* is superimposed to the V_{ref} for the generation of switching signal. The HF voltage signal is specified as follows,

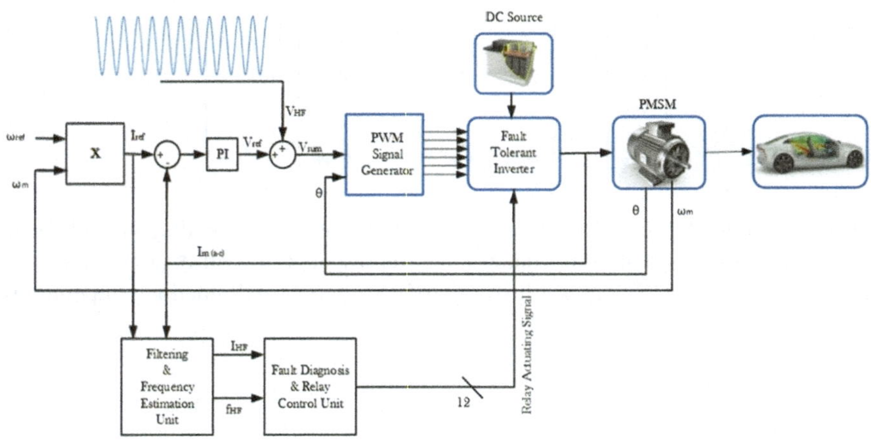

Fig. 14.5 Critical drive system with fault ride through

$$V_{HF} = V_m \sin(\omega t) \tag{14.13}$$

Here, the amplitude of injected HF voltage signal is V_m, and the frequency will be of ω. Now the final reference voltage is obtained as follows,

$$V_{sum} = V_f + V_{HF} \tag{14.14}$$

Now PWM module compares *V*sum with triangular carrier signal to produce switching pulse inverter switches. The PWM duty cycle is controlled by a PWM generator based on the reference voltage, V_{sum}. Since HF signal is introduced to the reference voltage and due to that, the measured phases current also have high-frequency components. These measured phase currents are used to identify the semiconductor switch fault in the power converter.

14.4 Fault Diagnosis and Redundant Strategies

Fault-tolerant inverter is a two-step process, and fault diagnosis operation is performed prior to fault ride through operation. The first challenge in the fault is to reduce the processing time of fault detection and localization. Especially, for a short-circuit (SCF) fault in the inverter, a particular leg has to be detected and isolated less than 10 μsec to save the overall system [33]. Information of fault locations often requires an efficient fault-tolerance technique. Thus, two cascaded mechanisms are fault detection and fault tolerance. In last two decades, several approaches proposed to identify faults in power converters [34, 35]. Based on the output variables of inverter existing fault diagnosis techniques of power converters classified broadly in two categories: (1) current-based approach and (2) voltage-based approach. The current-based approaches are commonly proposed because they do not require extra sensors. In [36], the literature proposed two methods to detect the faults which occurs in the inverter. The first method detects the fault by analyzing of the current waveform values, and the second method identifies the fault type using the frequency of current waveform. Average motor current Park's vector is used to track power devices faults in [37], which detect the OCFs and SCFs of power switches in the inverter. Different fault tolerant techniques have been investigated. The inverter structures shown in Fig. 14.6a, b can bear OCFs and SCFs with exclusive post-fault behaviors [38]. In the below two circuits, triacs are used to connect the additional leg to the main legs. During normal operation, triacs will be in off condition, and the drive system continues to work without any malfunction. When the fault takes place, with the help of fast acting fuses, the faulty phase is removed from the main circuit. The load connected to the particular phase linked to the DC supply through midpoint of the fourth leg by activating the triacs connected to the faulty phase. For a standard two-level 3-φ

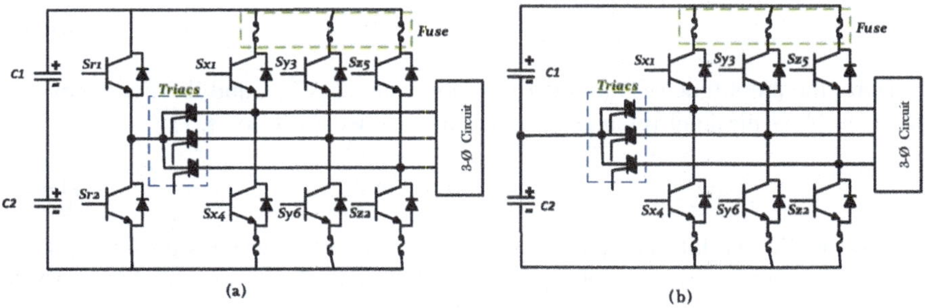

Fig. 14.6 3-φ two-level four-leg fault-tolerant inverters schematic: **a** switch-based; **b** capacitor-based

inverter, the output L–L voltage varies between (0, V_s) or (0, $-V_s$), here V_s is the DC source voltage. But, when a split DC bus capacitor is used as the extra leg as shown in Fig. 14.6b, the output voltage varies between (0, $V_s/2$) or (0, $-V_s/2$).

14.4.1 Fault Diagnosis Scheme

Figure 14.7 presents the block diagram that represents the fault diagnosis scheme this chapter focuses. It depicts the HF current separation from measured phase current and frequency calculation of the HF current. In the focused fault detection scheme, HF voltage signal is added to the reference voltage derived from reference current and measured three-phase currents. Now, the total sum is sent to the PWM block as reference voltage. Since HF components are injected into triggering pulses of power switches, the phase currents will have these high-frequency components. The magnitude and frequency of HF current signal are used as the main fault signatures. In order to find out the frequency of HF current signal, which need to be filtered out from the difference of I_{ref} and I_m using band-pass filter. Next, the filtered signal is sent to the zero-crossing detection unit and further given to counter module. The time period between two consecutive zero crossing is half the time duration of the HF current signal. So, this time period is used to estimate the frequency of HF signal. The fault diagnosis and relay control unit generate relay actuating signal based on the fault signatures (amplitude of phase current and amplitude and frequency of HF signal). Table 14.2 displays the HF signal status and the underlying current during normal service. It can be observed that current sampling introduced delay T_d in the output of BPF as expected. This will affect the frequency calculation based on the zero-crossing detector method and in the fault identification accordingly. So that calculated frequency alone is not taken as the fault diagnosis parameter, therefore changes in HF current amplitude [39] also considered as another fault diagnosis parameter. The OCFs and SCFs will be identified using two threshold values I_o and I_s, respectively. As a consequence, the

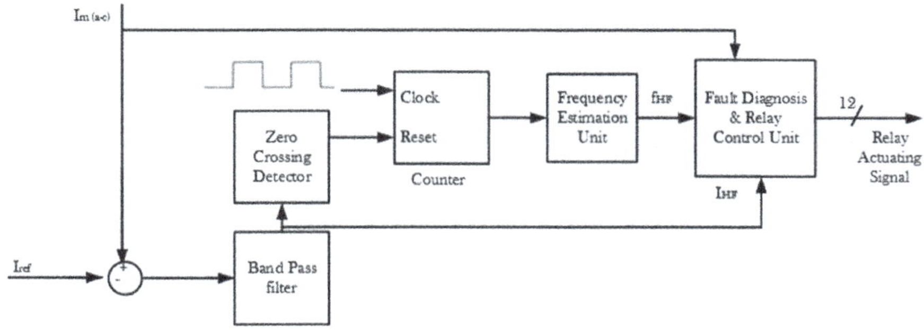

Fig. 14.7 Fault diagnosis scheme

power switches' OCFs and SCFs are observed by measuring the variations in the amplitude and frequency of the HF current after filtering along with the difference in the fundamental phase current. The fault diagnosis methodology of the power switches is covered in detail in the following sections.

14.4.1.1 Fault Diagnosis of Open-Circuit Fault (OCF) in Power Devices:

The moment OCF takes place in any one of the top switch of the three phases; the particular phase leg will be removed from +ve terminal of the power supply as depicted in Fig. 14.8. And the phase current will immediately drop to zero, enabling the OCF to be identified via the threshold value I_o. By observing the amplitude and frequency of the HF current, the particular defective switch is localized. Since the HF signal is inserted into the top switches, after the occurrence of the fault, there will be no HF current portion, and thus, the HF current frequency will drop to zero, indicating an open-circuit fault in the top switch. The occurrence of the OCF in the bottom switch can also be defined in the same way, as depicted in

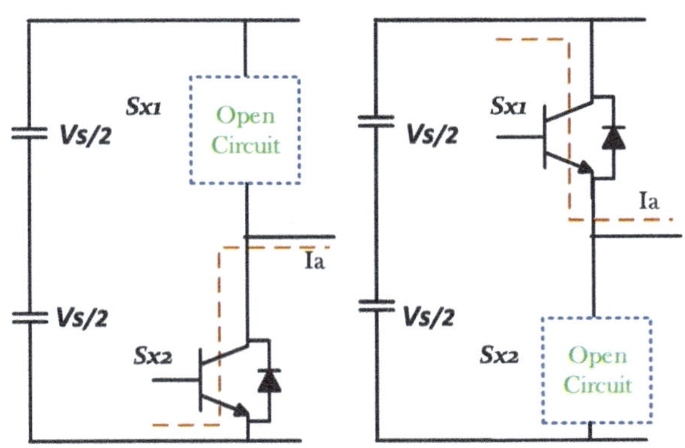

Fig. 14.8 Under OCF condition. **a** OCF in a top switch, **b** OCF in a bottom switch

Fig. 14.8. Table 14.2 summarizes the fault signatures of an OCF in the top and bottom power devices.

14.4.1.2 Fault Diagnosis of Short-Circuit Fault (SCF) in Power Switches

The SCF signatures in the top and bottom semiconductor power devices are given in Table 14.3. The control strategy which is used in PWM does not modulate the defective phase voltage owing to when the top switch of either of the phases is short-circuited, as depicted in Fig. 14.9, and as a result, the phase is continuously excited. As a consequence, the basic current of the defective phase rises exponentially till it exceeds the threshold value of the short-circuit fault current, I_s. At the same time, the HF current should immediately vanish due to the switching signals of PWM which are not available at top power device. The frequency of the HF current would decrease to zero as a result. A short-circuit fault in the top switch is localized by incorporating these fault parameters, as given in Table 14.3. Similarly, when the bottom power device is short-circuited, the same signatures are used to localize the short-circuit switch.

Table 14.3 Fault signatures: HF current signal and phase current during normal/short-circuit switch fault

Fault signatures		Normal operation	Short-circuit fault in any leg	
			Top switch	Bottom switch
HF current	Amplitude (A)	Normal	Zero	Zero
	Frequency (Hz)	f	$<f$	$<f$
Phase current	Amplitude (A)	Normal	$>I_s$	$>I_s$

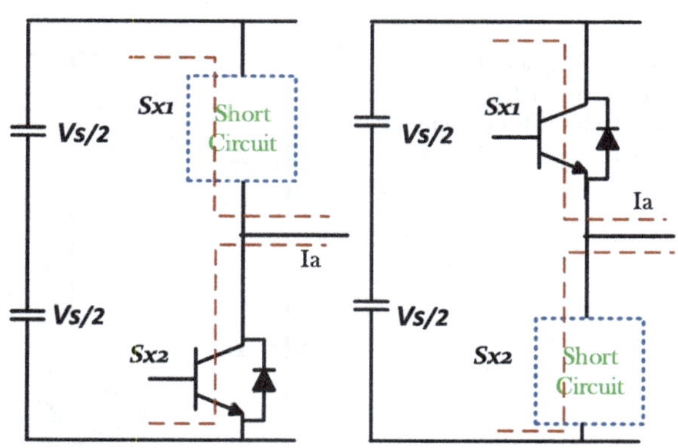

Fig. 14.9 Under SCF conditions **a** SCF in a top switch, **b** SCF in a bottom switch

14.4.2 Fault-Tolerant Topology

A modified 3-φ two-level inverter with fault-tolerant ability is depicted in Fig. 14.10 in order to maintain continuous driving operation after failure. The fault-tolerant system is also made up of four extra power devices (IGBTs) along with a relay circuit. The relay that corresponds to the fault is switched ON when a fault is found in either switch, and the associated fault free switch replace the defective switch to make sure the motor stable performance.

14.4.2.1 Open-Circuit Fault Tolerant

When OCF fault takes place in the top power device as shown in Fig. 14.10, X_1 relay is actuated and to replace the defective power device S_{x1} by S_{r1}. Similarly, S_{r2} is linked via X_4 when the fault occurs in the lower switch. X_1 and X_4 simultaneously switched ON in the event of concurrent fault in S_1 and S_4 switches, and redundant switches S_{r1} and S_{r2} ensure the continuity of driving service.

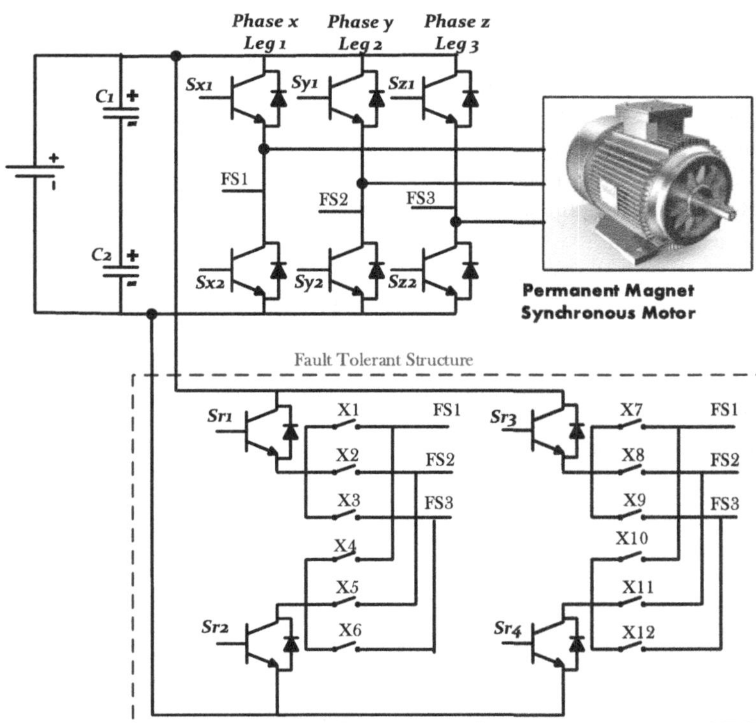

Fig. 14.10 Fault-tolerant inverter structure

14.4.2.2 Short-Circuit Fault Tolerant

The diagnostic algorithm detects the fault when a SCF takes place in the top or bottom switch, the defective switch is isolated immediately, and the redundant switches S_{r1} or S_{r2} continues to drive service via relay X_1 or X_4, respectively. X_1 and X_4 simultaneously switched ON in the event of concurrent fault in S_{x1} and S_{x2} switches, and stable switches S_{r1} and S_{r2} ensure the continuity of driving service.

14.5 Results and Discussion

The fault diagnosis and fault-tolerant methodology focused in this chapter verified using Typhoon HIL simulation environment. Here field-oriented control (FOC)-based PMSM driven by fault-tolerant inverter. Table 14.4 presents the parameters of PMSM. The amplitude of HF voltage signal added with reference modulating signal is 7% of the DC input value; also, the HF signal frequency was 4 kHz. Fault occurs in IGBT switches that are mitigated using redundant IGBTs connected in parallel with main switches through relays. Results are plotted for OCF and SCF which takes place in any one of the three-phase leg. Also, the methodology tested for simultaneous occurrence of both faults in the same leg. This is a two-step process, first the fault leg will be identified based on magnitude of phase current and HF current. Next from the generation of fault flag, the particular relay will be turned on to connect the redundant IGBT to the fault IGBT in parallel.

Figure 14.11 shows the simulation output of phase X current (I_a) and filtered high-frequency current signal during normal operation. Here, two threshold values for phase current are defined to find open-circuit fault and short-circuit fault namely I_o and I_s as follows,

$$I_o = K_O \times I_{ref} \tag{14.14}$$

$$I_s = K_s \times I_{ref} \tag{14.15}$$

Here, the coefficients K_O and K_S were 0.7 and 1.35, respectively, which varies depends on the motor parameters.

Table 14.4 PMSM parameters of PMSM

Parameter	Quantity	Parameter	Quantity
Rated power (P_n)	6.9 kW	d-inductance (L_d)	6.9 mH
Rated voltage (V_L)	400 V	q-inductance (L_q)	6.9 mH
Rated electrical frequency (f)	50 Hz	Moment of inertia (J_m)	7.8 m kgm2
Rated current (I_L)	10.9 A	Rated torque (T_n)	32 Nm
Stator resistance (R_s)	0.42 Ω	Rated speed (N_n)	2000 rpm

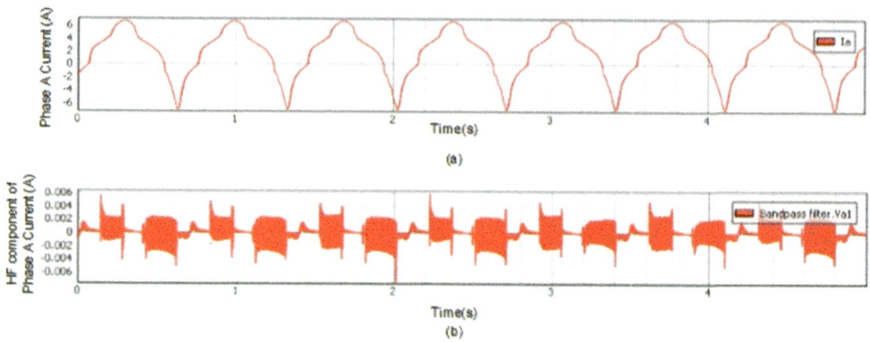

Fig. 14.11 During normal operation of inverter **a** phase *X* line current I_a (A) **b** HF component filtered from phase *X* current

14.5.1 Fault Diagnosis and Tolerant Control for OCF:

Figure 14.12 displays results of the simulation when the motor runs at 955 rpm and rated load for an OCF in the top switch of phase *X*. At the top switch of the power converter, the initially open-circuit fault is created. As anticipated, it can be seen that the phase current (I_a), fell below a certain I_o value, signaling an OCF occurrence, either in the top or bottom switch, in phase *X*. Because of the open switch circuit, the HF signal element is zero, as stated in section III. Flags FTo reflect the fault event and the identification of the OCF. Once the fault found, the fault tag is set and the relay X_1 connected to the redundant IGBT is triggered. Next, the defective switch gate signals passed to the new switch. It will isolate from the circuit the defective IGBT. Similarly, an open-circuit fault was simulated at rated torque below 955 rpm in the phase *X* bottom switch. The results of the simulation show defective phase current decreased immediately after the fault occurred, exceeding the I_o threshold. The defective switches are localized on the basis of these fault signatures mentioned in Table 14.2.

The fault tag is set once the fault is detected and the relay X_2 linked to the redundant IGBT is actuated. Now the defective switch gate signal will be passed to the new switch. Similarly, at rated torque below 955 rpm, a SCF was created in the top switch of phase *X*. The results of the simulation indicate that the defective phase current is increasing exponentially, exceeding the threshold right after the fault occurs. The defective switches are localized on the basis of fault signatures listed in Table 14.3.

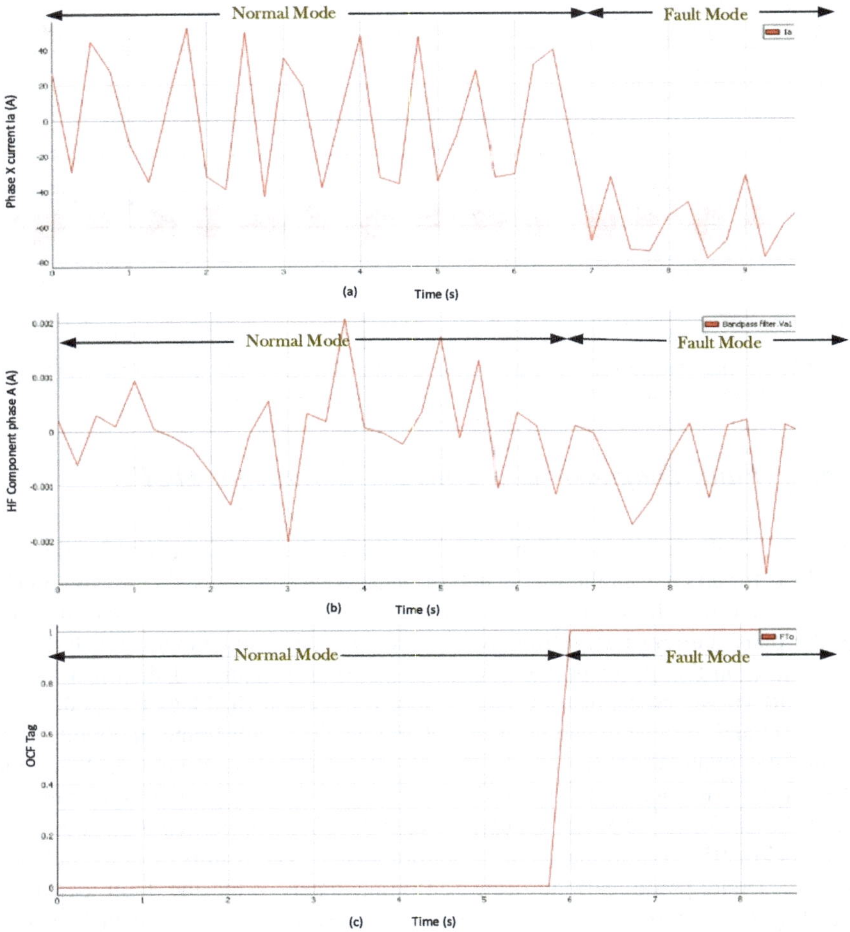

Fig. 14.12 OCF diagnosis and tolerant control of top power devices in phase X leg. **a** Phase X current I_a **b** Filtered high-frequency current **c** OCF flag

14.5.2 Fault Diagnosis and Tolerant Control for SCF

For a SCF in the bottom power device, Fig. 14.13 displays the fault-diagnostic waveforms. The short-circuit fault created at the bottom of the three-phase inverter switch. When the bottom switch is short-circuited, the increase in the defective phase current can be monitored. The amplitude HF current component values are also reduced as compared to normal working condition. As a consequence, the I_s threshold was attained, and a SCF in phase X is found. The FTs flags represent the existence of fault and SCF detection. The fault tag is set once the fault is detected and the relay X_2 linked to the redundant IGBT is actuated. Now the defective switch gate signal will be passed to the new switch. Similarly, at rated torque below 955 rpm, a SCF was created in the top switch of phase X. The results of the simulation indicate that the defective phase current is increasing exponentially, exceeding the threshold right after the fault occurs. The defective switches are localized on the basis of fault signatures as listed in Table 14.3.

When both OCF and SCF takes place simultaneously in upper switch and lower switch, respectively, the fault flags will be generated as show in Fig. 14.14 based on the signatures discussed in Tables 14.2 and 14.3 in Sect. 14.4. The above tests are

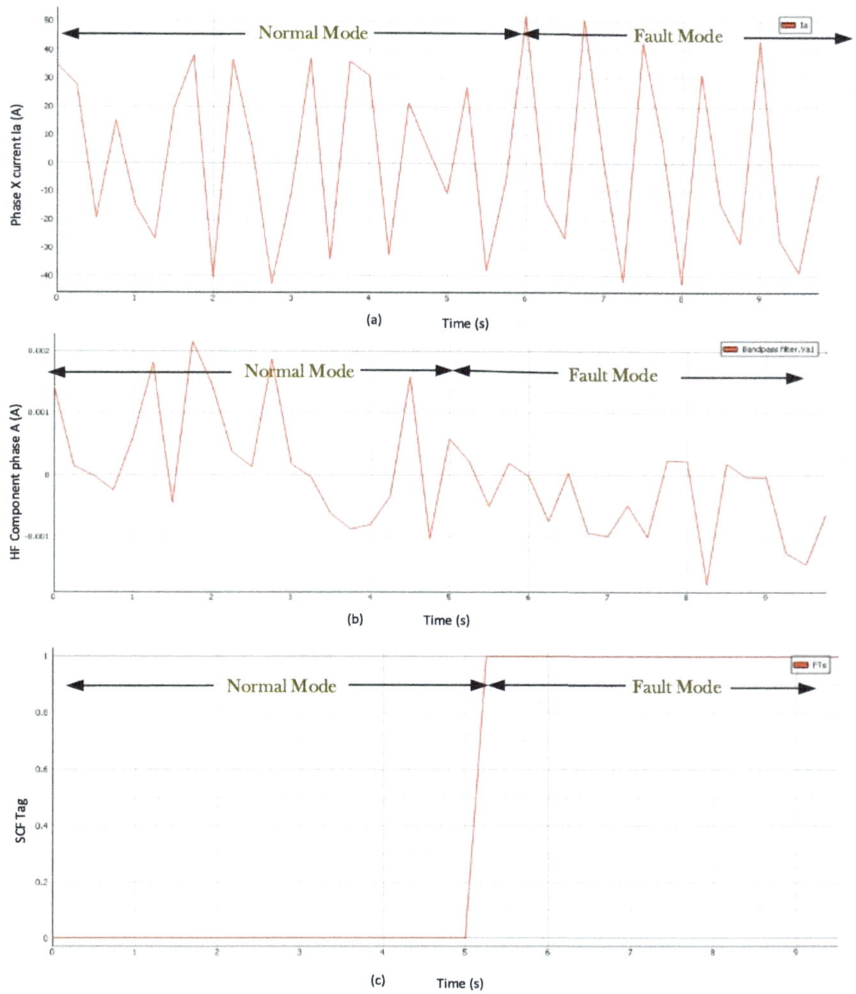

Fig. 14.13 SCF diagnosis and tolerant control of bottom switch in phase X leg. **a** Phase X current I_a **b** Filtered HF current **c** SCF flag

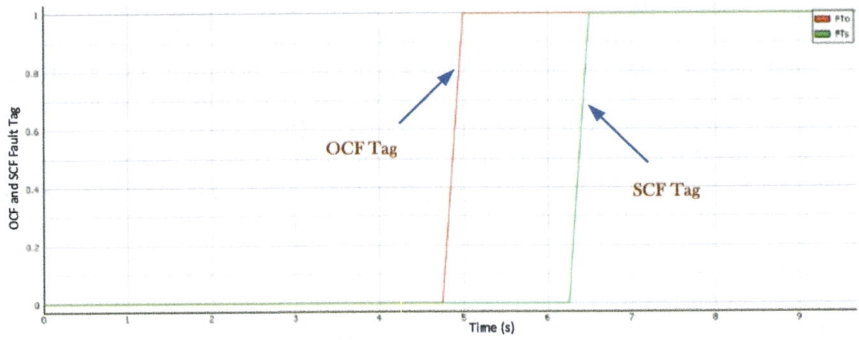

Fig. 14.14 Flag generation during open- and short-circuit faults in top and bottom switch of phase *X* leg

repeated for different working conditions such as running speed, load power and mechanical torque to ensure the robustness of the topology.

14.6 Conclusion and Future Scope

The presented fault-tolerant VSI in this chapter that can handle both SCFs and OCFs in the power switches. The introduced fault-tolerant structure consists of four extra IGBT devices along with a relay circuit connected to the traditional symmetric power converter. In addition, it investigates a fault-diagnostic technique based on HF signal injection that does not need extra sensors, unlike other approaches. This technique only uses input variables which already exist in the primary drive control system. So that increase in cost and complexity due to additional sensors and electronics devices can be avoided. The focused algorithm does not depend on the motor running speed, load power and motor mechanical load conditions. Both OCF and SCF faults are diagnosed based by the changes in amplitude of the inserted HF signal and the output phase current at the instant of the fault. It is simple in construction, robust, and simple in control. Simulation results extracted from Typhoon HIL software shows that the fault diagnosis and tolerant strategy is quick enough such that drive systems will have negligible amount of disturbance. The outputs demonstrate the effectiveness of methodology focused, which successfully isolates the faulted switches for any fault. The machine's post-fault operation is the same as the prefault one. The simulation output results validate that the implemented inverter will be suitable in any safety critical electrical drive applications in which fault tolerance is most essential. Hence, it significantly improves the performance of the PMSM drive system by more efficient and cost-effective system, which is more suitable safety critical electrical drive applications such as aircraft and automotive applications. Future analysis can be extended to implement the fault-tolerant inverter with a smaller number of redundant power switches with suitable control algorithm. Also input and output transient phenomena and non-idealities make the system switch fault diagnosis and mitigation process more complex, so in future investigation under these issues can be considered.

References

1. Knight, J.C.: Safety critical systems: challenges and directions. In: Proceedings of International Conference on Software Engineering, pp. 547–550 (2002)
2. Merabet, A.: Advanced control for electric drives: current challenges and future perspectives. Electronics **9**(1762), 1–4 (2020)
3. Mirafzal, B.: Survey of fault-tolerance techniques for three-phase voltage source inverters. IEEE Trans. Ind. Electron. **61**(10), 5192–5202 (2014)
4. Mitsubishi Electric Semiconductors and Devices: Product Inform, pp. 1–3 (2020). https://www.mitsubishielectric.com/semiconductors/products/powermo
5. Benda, V.: Power semiconductors—state of art and future trends. AIP Conf. Proc. **1337** (June), 16–24 (2011)
6. Yin, T., Xu, C., Lin, L., Jing, K.: A SiC MOSFET and Si IGBT hybrid modular multilevel converter with specialized modulation scheme. IEEE Trans. Power Electron. **35**(12), 12623–12628 (2020)
7. Han, L., Liang, L., Kang, Y., Qiu, Y.: A review of SiC IGBT: models, fabrications, characteristics, and applications. IEEE Trans. Power Electron. **36**(2), 2080–2093 (2020). https://doi.org/10.1109/tpel.2020.3005940
8. Song, X., Zhang, L., Huang, A.Q.: Three-terminal Si/SiC hybrid switch. IEEE Trans. Power Electron. **35**(9), 8867–8871 (2020)
9. Liu, T.H., Fu, J.R., Lipo, T.A.: A strategy for improving reliability of field-oriented controlled induction motor drives. IEEE Trans. Ind. Appl. **29**(5), 910–918 (1993)
10. Chen, H., Lu, S.: Fault diagnosis digital method for power transistors in power converters of switched reluctance motors. IEEE Trans. Ind. Electron. **60**(2), 749–763 (2013).
11. Park, T., Kim, T.: Novel fault tolerant power conversion system for hybrid electric vehicles. 2011 IEEE Veh. Power Propuls. Conf. VPPC 2011. pp. 0–5 (2011).
12. Beltrão, M., Corrêa, D.R., Member, S., Jacobina, C.B., Member, S.: An induction motor drive system. **37**(3), 873–879 (2001).
13. Bolognani, S., Zordan, M., Zigliotto, M.: Experimental fault-tolerant control of a PMSM drive. **47**(5), 1134–1141 (2000).
14. Klima, J., Schreier, L.: Investigation and control of an induction motor drive under inverter fault conditions. IFAC Proc. Vol. **37**(14) 489–494 (2004).
15. Richardeau, F., Mavier, J., Piquet, H., Gateau, G.: Fault-tolerant inverter for on-board aircraft EHA. 2007 European Conference on Power Electronics and Applications. 173–175 (2007).
16. Lin, F., Chau, K.T., Chan, C.C., Liu, C.: Fault diagnosis of power components in electric vehicles. J. Asian Electr. Veh. **11**(2), 1659–1666 (2013).
17. Gan, C., Wu, J., Yang, S., Hu, Y., Cao, W., Si, J.: Fault diagnosis scheme for open-circuit faults in switched reluctance motor drives using fast Fourier transform algorithm with bus current detection. **9**, 20–30 (2016).
18. Arkadan, A.A., Du, P., Sidani, M., Bouji, M.: Performance prediction of SRM drive systems under normal and fault operating conditions using GA-based ANN method. IEEE Trans. Magn. **36**(4 I) 1945–1949 (2000).
19. Estima, J.O., Cardoso, A.J.M.: Performance evaluation of DTC-SVM permanent magnet synchronous motor drives under inverter fault conditions. IECON Proc. (Industrial Electron. Conf. pp. 1228–1233 (2009).

20. Gao, H., Zhang, W. , Wang, Y., Chen, Z.: Fault-tolerant control strategy for 12-phase permanent magnet synchronous motor. Energies. **12**(18), (2019).
21. Abassi, M., Saadaoui, O., Tlili, I.: PMSM DTC drive system fed by fault-tolerant inverter connected to a photovoltaic source. no. Irec. 2–6 (2018).
22. Gameiro, N.S., Cardoso, A.J.M., Member, S.: A new method for power converter fault diagnosis in SRM drives. **48**(2), 653–662 (2012).
23. Errabelli, R.R., Mutschler, P.: Fault-tolerant voltage source inverter for permanent magnet drives. **27**(2), 500–508 (2012).
24. Marques, J.F., Estima, J.O., Gameiro, N.S., Cardoso, A.J.M.: A new diagnostic technique for real-time diagnosis of power converter faults in switched reluctance motor drives. **50**, 1854–1860 (2014).
25. An, Q., Sun, L., Sun, L., Member, S.: Current residual vector based open-switch fault diagnosis of inverters in PMSM drive systems. 8993 (2014).
26. Jamshidpour, E., Poure, P., Saadate, S.: Photovoltaic systems reliability improvement by real-time FPGA-based switch failure diagnosis and fault-tolerant DC-DC converter. IEEE Trans. Ind. Electron. **62**(11), 7247–7255 (2015).
27. Afiq, M.I., Tamchek, N., Hassan, M., Dambul, K., Selvaraj, J., Rahim, N., Sandoghchi, R., Adikan, F.: A fiber Bragg grating-bimetal temperature sensor for solar panel inverters. Sensors. **11**(9), 8665–8673 (2011).
28. Hu, Z., Du, M., Wei, K.: Online calculation of the increase in thermal resistance caused by solder fatigue for IGBT modules. IEEE Trans. Device Mater. Reliab. **17**(4), 785–794 (2017).
29. Ma, K., He, N., Liserre, M., Blaabjerg, F.: Frequency-domain thermal modeling and characterization of power semiconductor devices. IEEE Trans. Power Electron. **31**(10), 7183–7193 (2016).
30. Dutta, S., Parkhideh, B., Bhattacharya, S., Moghaddam, G.K., Gould, R.: Development of a predictive observer thermal model for power semiconductor devices for overload monitoring in high power high frequency converters. Conf. Proc. - IEEE Appl. Power Electron. Conf. Expo. - APEC. 2305–2310 (2012).
31. Chen, M., Hu, A., Yang, X.: Predicting IGBT junction temperature with thermal network component model. Asia-Pacific Power Energy Eng. Conf. APPEEC. 1–4 (2011).
32. Ali, N., Gao, Q., Xu, C., Makys, P., Stulrajter, M.: Fault diagnosis and tolerant control for power converter in SRM drives. J. Eng., vol. 2018. **13**, 546–551 (2018).
33. Richardeau, F., Baudesson, P., Meynard, T.: Failures-tolerance and remedial strategies of a PWM multicell inverter. PESC Rec. - IEEE Annu. Power Electron. Spec. Conf. **2**, 649–654 (2000).
34. Song, Y., Wang, B.: Survey on reliability of power electronic systems. IEEE Trans. Power Electron. **28**(1), 591–604 (2013).
35. Lu, B., Sharma, S.K.: A literature review of IGBT fault diagnostic and protection methods for power inverters. IEEE Trans. Ind. Appl. **45**(5), 1770–1777 (2009).
36. Peuget, R., Courtine, S., Rognon, J.P.: Fault detection and isolation on a pwm inverter by knowledge-based model. IEEE Trans. Ind. Appl. **34**(6), 1318–1326 (1998).
37. Mendes, A.M.S., Marques Cardoso, A.J.: Voltage source inverter fault diagnosis in variable speed AC drives, by the average current Park's vector approach. IEEE Int. Electr. Mach. Drives Conf. IEMDC 1999 - Proc. 704–706 (1999).
38. Lúcio, R., Jacobina, C.B., da Silva, E.R.C., Lima, A.M.N.: Fault-tolerant voltage-fed PWM inverter aC motor drive systems. IEEE Trans. Ind. Electron. **51**(2), 439–446 (2004).
39. Zhang, W., Xu, D., Enjeti, P.N., Li, H., Hawke, J.T., Krishnamoorthy, H.S.: Survey on fault-tolerant techniques for power electronic converters. IEEE Trans. Power Electron. **29**(12), 6319–6331 (2014).

Chapter 15
Modeling and Performance Analysis of Various DC–DC Converters

Subhadip Goswami and Abhik Banerjee

15.1 Introduction

For the most part, the Buck Converter is used in SMPS circuits where the DC yield voltage should be lower than the DC input voltage. The DC supply can be acquired from reviewed AC or from any DC deftly. It is useful where electrical withdrawal is not needed between the trading circuit and the yield, anyway where the data is from a changed AC source, partition between the AC source and the rectifier could be given by a mains disconnecting transformer. The converter between the data and yield of the Buck Converter interminably turns on and off at high recurrence. While the DC contribution to a boost converter can be acquired from numerous sources, for example, batteries, corrected AC from the mains gracefully, or DC from sun based boards, power modules, dynamos and DC generators and so on. The lift converter is not exactly equivalent to the buck converter in that its yield voltage is comparable to, or more critical than its data voltage. At any rate it is basic to remember that, as power (P) = voltage (V) × current (I), if the yield voltage is extended, the available yield current should lessen. Again, a Buck–Boost converter is quite a traded mode power easily which joins the standards of buck converter and lift converter in a singular circuit. It gives an oversaw DC yield voltage from either an AC or a DC input. The Buck converter conveys a DC yield in a range from 0 V to only not exactly the stockpile voltage. The lift converter will make a yield voltage going from a comparable voltage as the commitment, to a level much higher than the information. By solidifying these two regulator plans, it is possible to have a regulator circuit that can adjust to a wide extent of data voltages, both in

S. Goswami · A. Banerjee (✉)
Department of Electrical Engineering, NIT Arunachal Pradesh, Itanagar, Arunachal Pradesh 791112, India
e-mail: abhik@nitap.ac.in

S. Goswami
e-mail: subhadip.phd@nitap.ac.in

N. Priyadarshi et al. (eds.), *DC–DC Converters for Future Renewable Energy Systems*, Energy Systems in Electrical Engineering,
https://doi.org/10.1007/978-981-16-4388-0_15

higher or lower than that required by the circuit. Fundamentally, the CUK converter is such a DC/DC converter that has a yield voltage monstrosity that is either more basic than or not really the information voltage size. It is on a very basic level a lift converter followed by a buck converter with a condenser to couple the energy. It utilizes a condenser as its basic energy-gathering segment, not at all like most different sorts of converters which utilize an inductor. Again, the flyback converter is used in both AC/DC and DC/DC change with galvanic partition between the information and any yields. It is a buck-uphold converter with the inductor split to lay out a transformer, so the voltage degrees are extended with an extra bit of breathing space of detachment. The properties of different converters were confirmed by reenactments, their yield results were appeared for changed obligation cycle and various boundaries, proficiency was determined for every converter too, and a PWM circuit was built for the changing recurrence to drive the switch entryway [1]. Coordinated rectifiers can be utilized to diminish the huge conduction misfortune brought about by diode forward voltage drops in low-voltage applications [2]. This chapter proposed a viable and direct technique to fit the numerical *I*–*V* bend to the three (*V*, *I*) remarkable focuses without the need to figure or to gauge some other boundaries aside from the diode consistent and has proposed a closed answer for the issue of finding the boundaries of the single-diode model condition of a down-to-earth PV cluster [3]. To accomplish nearly swell-free cluster current, we have utilized wave-controlling marvels with the assistance of incorporated inductor in a coupled inductor single-ended primary inductor converter (SEPIC) converter [4] and a *V*2-based maximum power point tracking (MPPT) plot is created utilizing a buck-support change geography where the photovoltaic (PV) generator shows a nonlinear *I*–*V* trademark and its most extreme force MPPT differs with sun powered insolation. Progressed material covers devices for the reproduction of exchanging converters (counting both PSpice and MATLAB recreations) and the essential ideas important to comprehend different real and arising applications for exchanging converters. For example, power factor amendment, LED drivers, low-commotion converters and exchanging converters geographies for sun-based and energy units [5]. The ideal yield voltages can be achieved by choosing legitimate estimations of inductor, capacitor and exchanging recurrence [6]. Investigation of DC buck converter that can be applied for sustainable power application that can be diminished or step down its info voltage to a lower-level contingent on the heap necessity. By fluctuating the obligation pattern of the DC buck converter, the yield voltage can be changed relatively [7]. The reenactment results show that if the variety of inductance esteem was too little in consistent express, the yield voltage worth will have a major diverse with yield voltage of Buck Converter hypothetically. In addition, if the inductance is excessively huge, it will deliver a ton of swaying. The job of regulator is significant while it is utilizing for a specific application. Power electronics is the utilization of strong state hardware to control and change one type of electrical control over to another structure, for example, changing over among AC and DC or changing the greatness and period of voltage and flow or recurrence or blend of these. The guideline of the principle circuit buck-support geography is examined in detail, and the states of delicate exchanging activity shows the circuit outline for the condition of the cycle with related waveforms. The results show that the DC–DC converter accomplishes delicate exchanging, lessens the exchanging misfortune and improves the productivity of the circuit. An arrangement-type buck-help converter is recommended that utilizes a H-connect circuit and a force help circuit with little force rating and trial results affirmed that the proposed circuits could diminish misfortunes by 2/3 at the most extreme effectiveness point. The zero current switching (ZCS) and the zero voltage switching (ZVS) technique is embraced by soft-exchanging converter and research about newton-technique control utilizing buck-support converter that is applied to photovoltaic framework considering delicate exchanging strategy is performed [6]. The specific scientific treatment of the buck-support converter is done to play out an insightful investigation of the force factor of the converter [7]. The nonlinear wonders in the new Buck–Boost converter might be preferably unpredictable over that in the conventional converters, which will decay the exhibition of the converter somewhat, the bifurcation and the disarray in customary DC–DC converters were concentrated widely [8–11]. It was exhibited that there was different sorts of nonlinear wonders in the customary Buck–Boost converter [12–15].

15.2 Buck Converter

This converter is routinely used in networks that implies down the voltage from the data voltage as shown by the essential. The objective for obtaining the voltage is to figure the current and voltage of inductor first when the switch is shut and a short time later for open switch. The qualification in inductor current more than one period should be zero for steady work. It is so named on the grounds that the inductor consistently 'bucks' or acts against the information voltage. The yield voltage of an ideal buck converter is equivalent to the result of the exchanging obligation cycle and the stockpile voltage. At the point when the switch is opened, the inventory current to the inductor is unexpectedly interfered. A simple circuit is given in Fig. 15.1.

This type of converter works in nonstop mode if the inductor current never tumbles to zero during the compensation cycle. This is portrayed in Fig. 15.2.

The development of V and I with respect to time in continuous conduction mode is shown in Fig. 15.3.

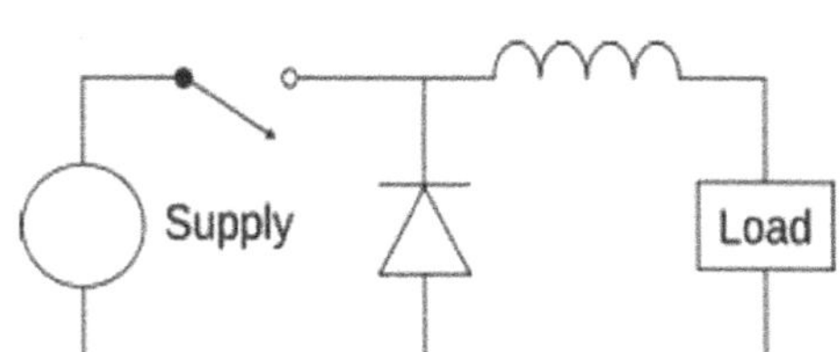

Fig. 15.1 Simple buck converter

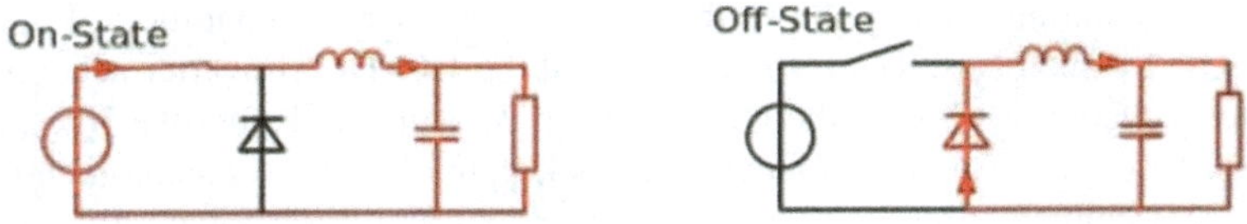

Fig. 15.2 On and off states of buck converter

Fig. 15.3 Advancement of *V* and *I* with respect to time in CCM

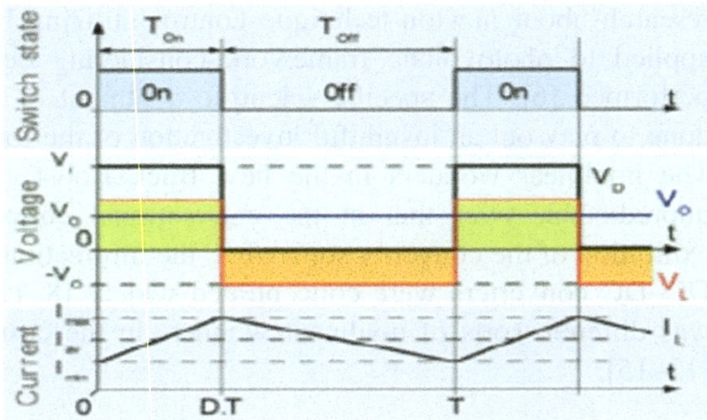

The development of *V* and *I* with respect to time in discontinuous conduction mode is shown in Fig. 15.4.

Less current is taken by load in discontinuous method and higher load current levels in continuous method. Figure 15.5 is shown below.

This converter is modeled and simulated in MATLAB environment. Figure 15.6 depicts the simulation of buck converter.

The parameters are adjusted in such a way that on giving a voltage of 12 V, an output of 4.679 V is obtained. The resulting response of the designed network is given in Fig. 15.7.

Fig. 15.4 Development of *V* and *I* with respect to time in DCM

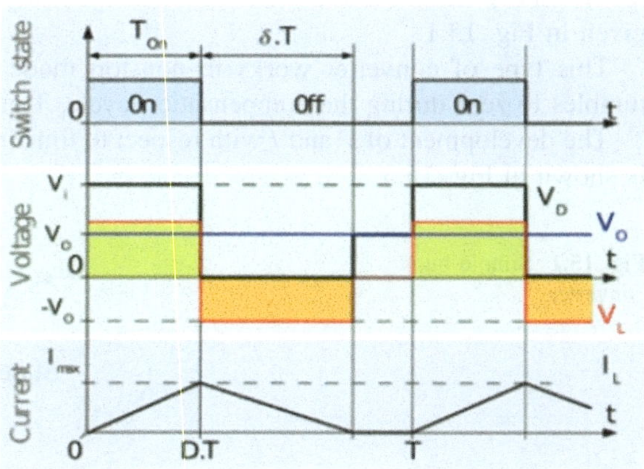

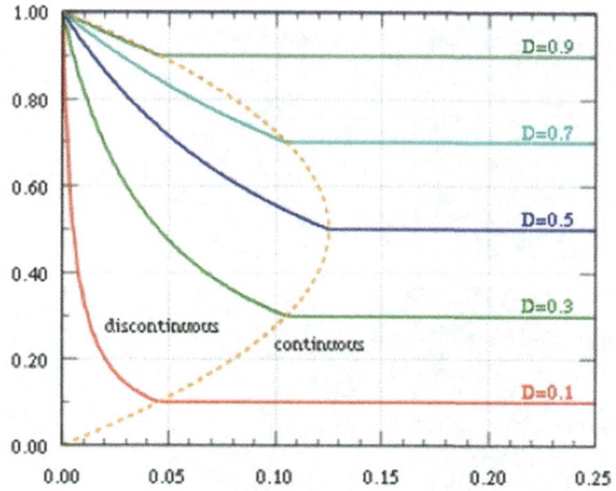

Fig. 15.5 Standardized yield voltage with the standardized yield current

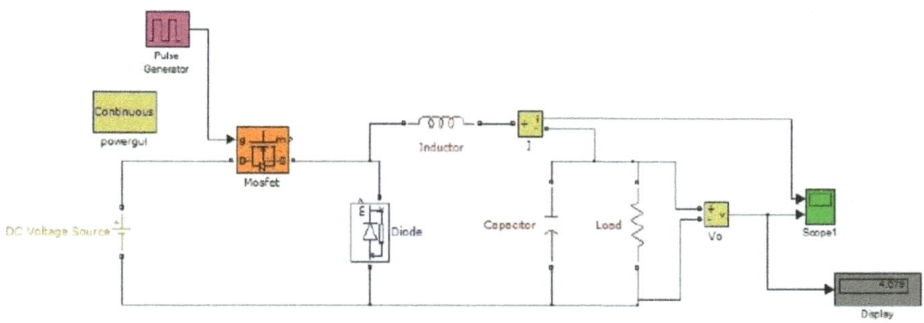

Fig. 15.6 Simulink of buck converter

15.3 Boost Converter

In this sort of converter, voltage is wandered up from its commitment to the yield. Now and again channels made of condensers are added to quite a converter's yield and commitment to lessen swell in voltage. The hypothesis that drives this sort of converter is the affinity of inductor to restrict changes in current by making and crushing an appealing field. In this converter, the yield voltage is reliably considerably more than the data voltage. The two current ways of this converter are shown in Fig. 15.8.

Exactly when such a converter works in predictable mode, the current over the inductor does not tumble to zero. Below diagram shows the graphs of currents and voltages in a converter working in this mode (Fig. 15.9).

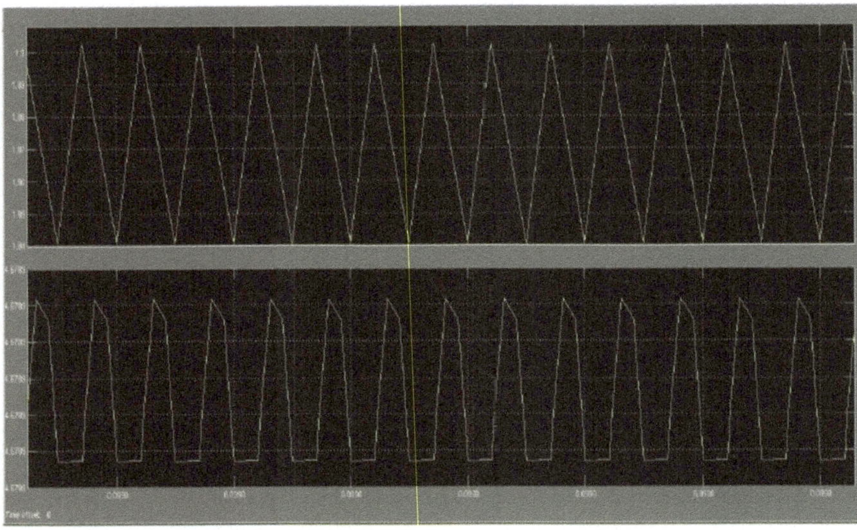

Fig. 15.7 Output response of designed circuit

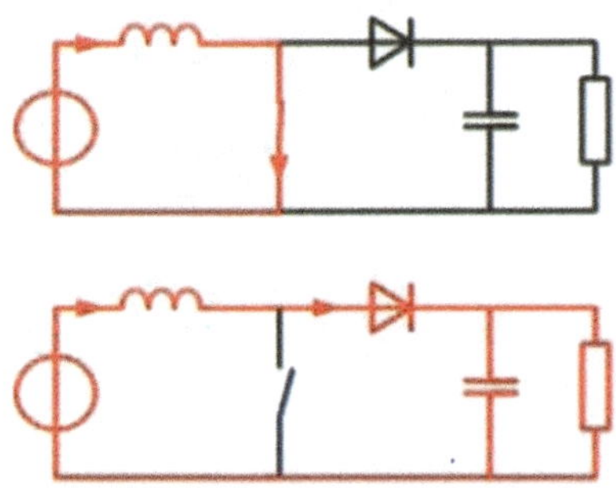

Fig. 15.8 Two current ways of converter

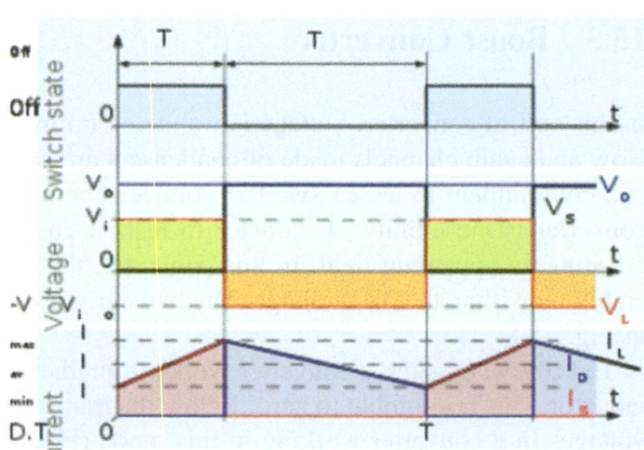

Fig. 15.9 Waveform of current and voltage in continuous conduction mode

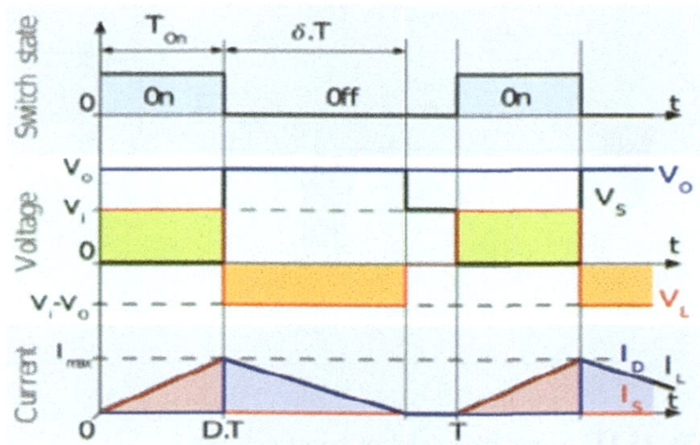

Fig. 15.10 Waveform of current and voltage in discontinuous conduction mode

The inductor might be completely released at high wave current value before a total recompense cycle closes. For this situation, current across inductor tumbles to zero. Despite the fact that there is an almost no distinction, it considerably affects the yield voltage. Figure 15.10 shows the discontinuous mode operation.

The Simulink diagram of above converter is shown in Fig. 15.11. It is simulated in MATLAB environment.

It is found that output voltage is increased to 13.91 V from an input source voltage of 5 V. The output response of the designed circuit is shown in Fig. 15.12.

15.4 Buck–Boost Converter

It is such a DC-to-DC converter that has a yield voltage size that is either more critical than or not actually the information voltage size. It is equivalent to a converter which uses a single inductor instead of a transformer. Both of them can make an extent of yield voltages, going from much greater than the data voltage, down to practically zero. If the current through the inductor never tumbles to zero during a recompense cycle, the converter works in constant mode. Sometimes, the proportion of energy required by the pile is adequately little to be moved in a period more humble than the whole reward time span. For the present circumstance, the current through the inductor tumbles to zero during part of the period. The fundamental differentiation in the standard portrayed above is that the inductor is completely delivered close to the completion of the replacement cycle. Nonetheless, parasitic preventions are available in each organization. In this way, a piece of the force is dispersed by these parasitic checks which is utilized by the converter (Fig. 15.13).

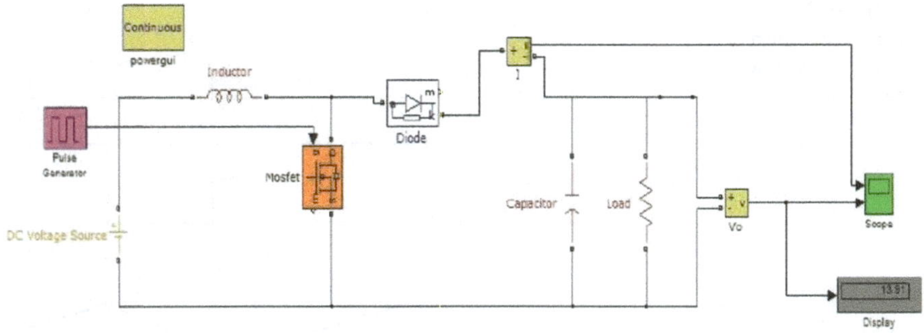

Fig. 15.11 Simulink model of boost converter

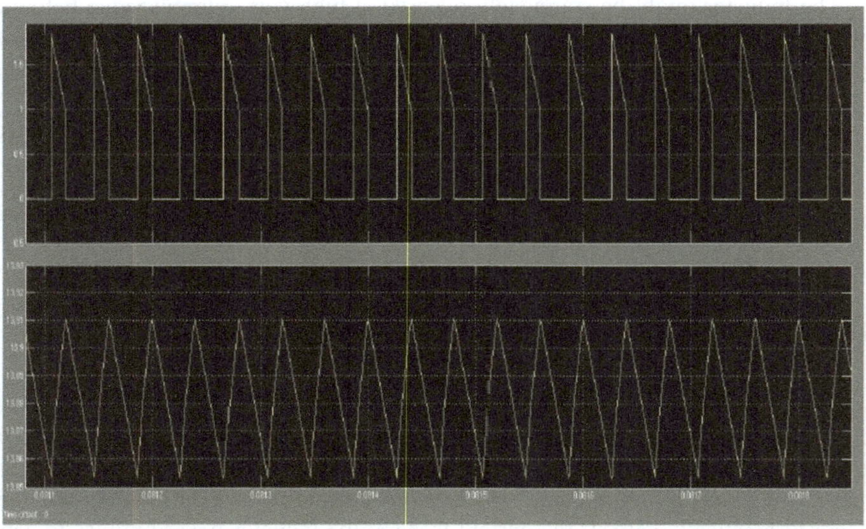

Fig. 15.12 Output response of boost converter

15.5 Flyback Converter

It is utilized in both AC/DC and DC/DC transformation where between the information and yield there is galvanic disconnection. This is a sort of the immediate aforementioned converter having split inductor to deliver a transformer, for which the proportions of voltage are increased (Fig. 15.14).

The two existing control plans are voltage control and current control. Both require a sign relating to the yield voltage. There are numerous cycles to make this voltage. Initially, an optocoupler is used on the optional circuit to confer a sign on controller. Furthermore, the cycle is to wound an alternate contorting on the loop

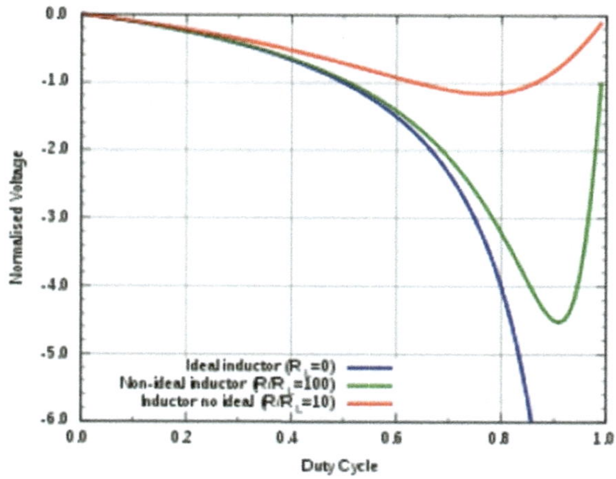

Fig. 15.13 Assessment of yield voltage of a buck-help converter with obligation cycle on increment of the parasitic obstruction

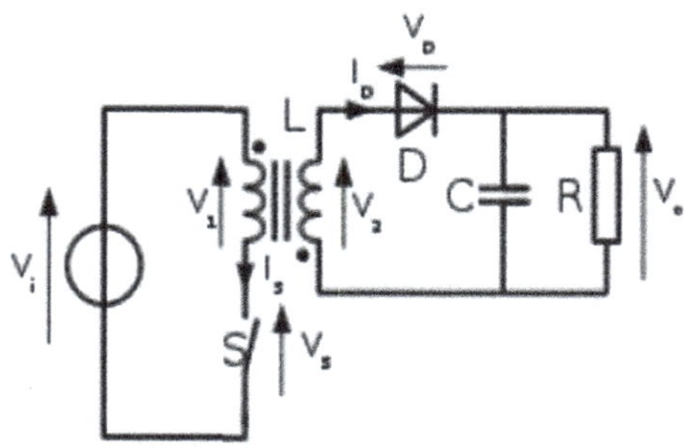

Fig. 15.14 Schematic diagram of a flyback converter

and depends on cross guideline. The third control plan is to investigate the essential side voltage, as shown by the standing essential DC voltage. The essential methodology using an optocoupler is used for getting demanding voltage and current rule, while the ensuing cycle has been created for cost-fragile applications where the yield should not be deliberately controlled, anyway up to 11 portions including the optocoupler could be taken out. The third technique, essential side distinguishing, can be as exact as the underlying one and particularly traditionalist than second, yet needs a base burden with the objective that the delivery event keeps occurring, giving the events to test the 1:N discretionary voltage at the fundamental winding (Fig. 15.15).

Generally, output voltage curve and dissipated power of MOSFET along with load current are shown in Fig. 15.16. With an input voltage of 5 V, the output voltage obtained is 15 V.

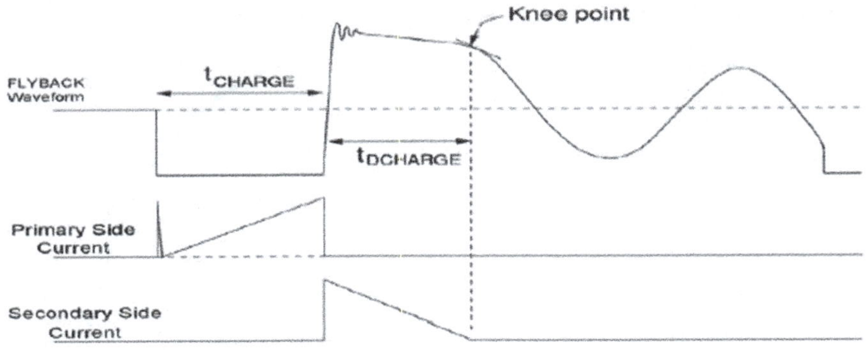

Fig. 15.15 Waveform of 'knee point'

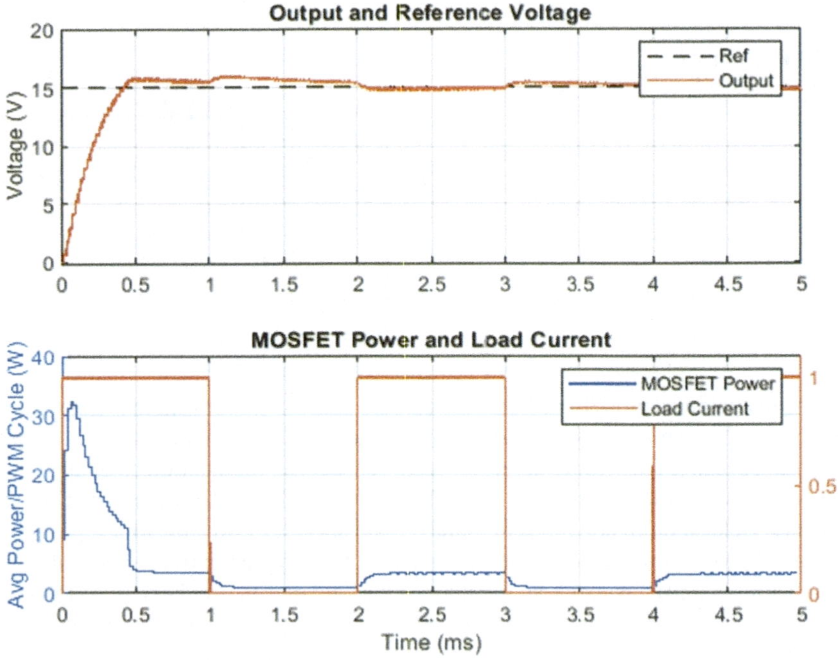

Fig. 15.16 Output voltage response and dissipated power of MOSFET along with I_L

15.6 Push–Pull Converter

It utilizes a transformer to transform the DC voltage. The distinctive component is that the transformer essential is furnished with current from the data line by sets of semiconductors in a fair network. The semiconductors are of course turned on and

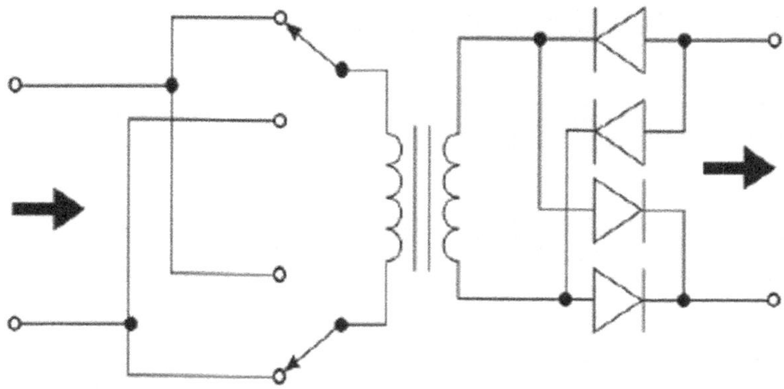

Fig. 15.17 Schematic of a full-bridge converter

off, irregularly exchanging the transformer's current value. This differences with buck-help converters, in which the information current is provided by a solitary semiconductor which is turned here and there, so current is just drawn from the line during a large portion of the exchanging cycle. During the other, a large portion of the yield power is provided by energy put away in inductors or condensers in the power flexibly. They have steadier info current, make less clamor on information line which are more proficient in higher power applications (Fig. 15.17).

The term is once in a while used to for the most part allude with bidirectional excitation of the transformer to any converter. For instance, in a full-bridge converter, the switches substitute the voltage over the flexibly side of the transformer, making it work as for AC force and give a voltage on its yield side. Regardless, the yield is then redressed and moved to the load. Condensers are regularly included at the yield to channel the exchanging clamor. It is imperative to allow a little range between controlling the transformer one way and energizing it the other: The switches are generally similar to semiconductors. Thus, a little stand by is expected to keep away from this issue. This stand by time is classified 'dead time' and is important to stay away from semiconductor shoot-through. The PI regulator subsystem utilizes a basic indispensable control to change over and keep up the ostensible yield voltage. At the hour of beginning, the reference voltage is ventured to yield voltage. In the case of a push–pull buck converter, the full load consists of a constant load and a cyclic load. Here, an input of 400 V is taken, and the yield is 80 V with transformer turns ratio of 2:1. The output curves are simulated and shown in Fig. 15.18.

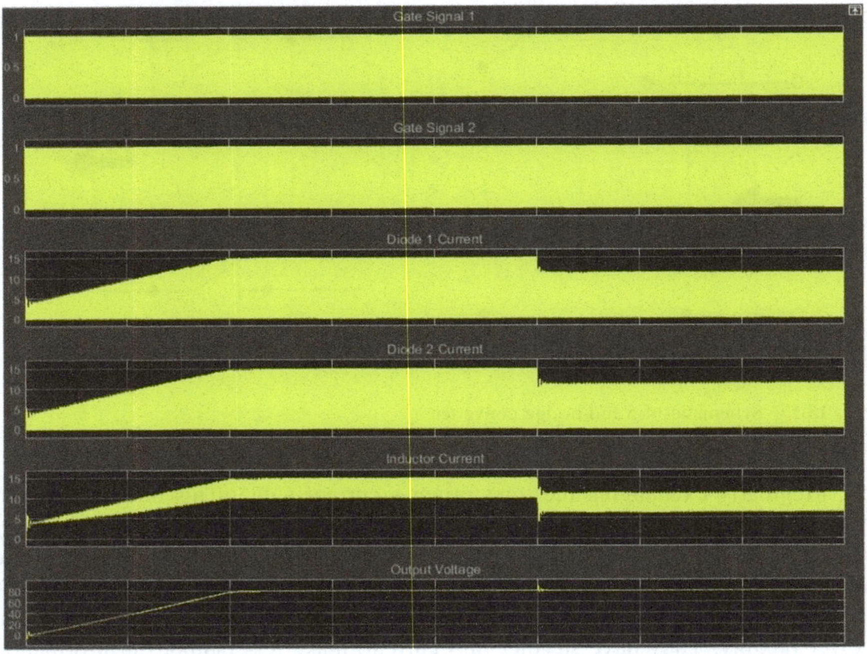

Fig. 15.18 Output curves of a flyback converter

15.7 Cuk Converter

It is a kind of converter that has a yield voltage esteem that is either more important than or not actually the data voltage size. Like the buck–boost converter with inverting geography, the yield voltage of non-disengaged Cuk is commonly likewise reversing, furthermore, can be lower or higher than the information. It utilizes a condenser as its primary energy-accumulating part. The Cuk converter permits energy to stream bi-directionally by utilizing a diode and a switch as shown in (Fig. 15.19).

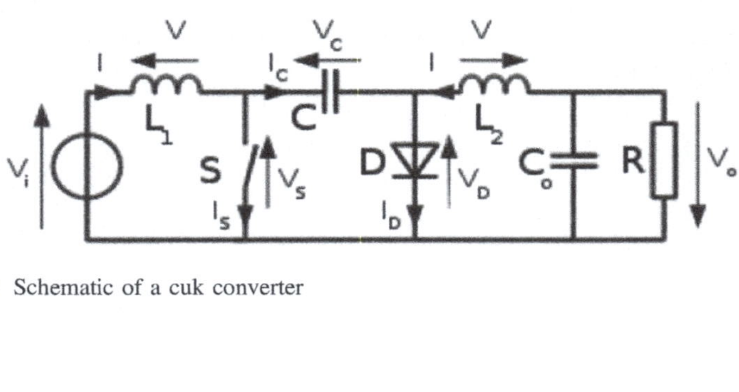

Fig. 15.19 Schematic of a cuk converter

This type of converter is of two sorts: Initial one is non-detached cuk converter, and the subsequent one is segregated cuk converter. The condenser *C* is utilized to move energy and is associated then again to the information and to the yield of the converter through the substitution of the semiconductor and the diode. Two inductors are used to change over separately the data voltage source and the yield voltage source into current sources. At a short period of time scale, an inductor can be considered as a current source as it keeps up a predictable current. This transformation is important since, in such a case that the condenser was associated straightforwardly to the voltage source, the current would be restricted simply by the parasitic resistances, bringing about high energy deprivation. Charging a condenser forestalls resistive current limiting and its related energy misfortune. Essentially likewise with various converters the Cuk converter can either work in ceaseless or unpredictable current mode. Regardless, rather than these converters, it can similarly work in sporadic voltage mode as in (Fig. 15.20).

The Cuk converter can be made in a detached kind. An AC transformer and an extra capacitor should be added. Since the detached converter is segregated, the yield voltage extremity can be picked openly. As the non-detached converter, the secluded version can have a yield voltage extent that is either more prominent than or not exactly the information voltage size, even with a 1:1 AC transformer. Figure 15.21 shows a secluded type of converter.

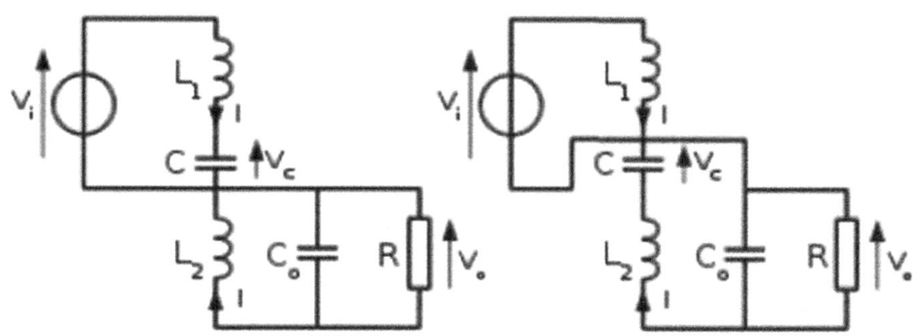

Fig. 15.20 Modes of non-isolated cuk converter

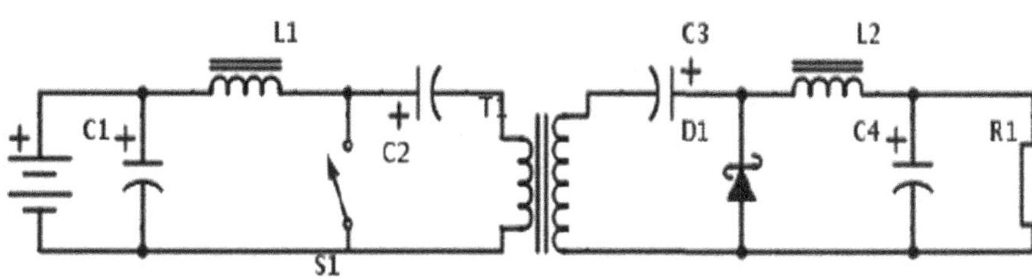

Fig. 15.21 Secluded Cuk converter

15.8 Conclusion

After investigation of each sort of converter, it tends to be seen that for venturing down the voltage, buck converter can be utilized. For venturing up the voltage, boost converter can be utilized. Buck–Boost converter can venture up or venture down the information voltage. Cuk converter is like buck–boost with added advantage is that it decreases the harmonics for a similar contribution as if there should be an occurrence of buck–boost. Flyback is utilized to plan high change proportion because of essence of transformer. Push–Pull converter has consistent info current and makes less commotion on the information line which in turn yields more proficiency in greater power management.

Acknowledgements This work could not be completed without the useful guidance, suggestions and inspirations from my Supervisor at National Institute of Technology, Arunachal Pradesh. I feel inspired to work with them and complete this chapter in a proficient manner.

References

1. Tushar, M.H.: Comparative study on DC–DC converters. Id-09221205
2. Erickson, R.W.: DC–DC power converters. Department of Electrical and Computer Engineering in University of Colorado
3. Villalva, M.G., Gazoli, J.R., Ruppert Filho, E.: Comprehensive approach to modeling and simulation of photovoltaic arrays, May (2009)
4. Priyadarshi, N., Padmanaban, S., Maroti, P.K., Sharma, A.: An extensive practical investigation of FPSO-based MPPT for grid integrated PV system under variable operating conditions with anti-islanding protection. IEEE Syst. J., 1–11 (2018)
5. Priyadarshi, N., Padmanaban, S., Bhaskar, M.S., Blaabjerg, F., Sharma, A.: A fuzzy SVPWM based inverter control realization of grid integrated PV-wind system with FPSO MPPT algorithm for a grid-connected PV/wind power generation system: hardware implementation. IET Electric Power Appl., 1–12 (2018)
6. Padmanaban, S., Priyadarshi, N., Holm-Nielsen, J.B., Bhaskar, M.S., Azam, F., Sharma, A. K.: A novel modified sine-cosine optimized MPPT algorithm for grid integrated PV system under real operating conditions. IEEE Access **7**, 10467–10477 (2019). https://doi.org/10.1109/ACCESS.2018.2890533
7. Padmanaban, S., Priyadarshi, N., Holm-Nielsen, J.B., Bhaskar, M.S., Hossain, E., Azam, F.: A hybrid photovoltaic-fuel cell for grid integration with jaya-based maximum power point tracking: experimental performance evaluation. IEEE Access **7**, 82978–82990 (2019). https://doi.org/10.1109/ACCESS.2019.2924264
8. Veerachary, M.: Power tracking for nonlinear PV sources with coupled inductor SEPTIC converter. IEEE Trans. Aerosp. Electron. Syst. **41**(3), 1019–1029 (2005)
9. Ang, S., Oliva, A.: Power Switching Converters
10. Patra, A., Mukhopadhyay, S., Mondal, P., Kastha, D., Chattopadhyay, S.: Modelling and Control design for DC–DC converter, Power Management group, AVLSI Lab, IIT-Kharagpur, January 2008
11. Durán, E.J., Galán, M., Sidrach-De-Cardona, Andújar, J.M.: Measuring the I–V curve of photovoltaic generators-analyzing different dc–dc converter topologies. September (2009)
12. Solanki, C.S.: Solar Photovoltaics Fundamental, Technologies and Application, 2nd edn. PHI Learning Private Limited (2011)

13. Nak. A., Roy, H., Verma, A.K.: Application of non-conventional and renewable energy sources, Chapter 12, pp. 147–160. Bureau of Energy Efficiency, India
14. Vardia, M., Priyadarshi, N., Ali, I., Azam, F., Bhoi, A.K.: Maximum power point tracking for wind energy conversion system. In: Bhoi, A., Sherpa, K., Kalam, A., Chae, G.S. (eds.) Advances in Greener Energy Technologies. Green Energy and Technology. Springer, Singapore (2020). https://doi.org/10.1007/978-981-15-4246-6_36
15. Rashid, M.H.: Power Electronics Circuits, Devices and Applications, 3rd edn

Chapter 16
Design and Analysis of DC–DC Buck Converter with Drift-Free MPPT Algorithm for a SEIG-Based Wind Energy Generation System

Jyotismita Mishra and Monalisa Pattnaik

16.1 Introduction

Energy is one of the key factors for the development of modern society. The available energy sources can be broadly classified into two types, namely non-renewable and renewable. Non-renewable sources are fossil fuel (oil, natural gas, coal) and nuclear energy. As per the Renewable Energy Policy Network 21 (REN21) 2018 report, the global scenario of energy generation by non-renewable sources is approximately 73.5% and the rest 26.5% generation comes from renewable energy sources [1]. Over the past decades, the energy demand is exponentially growing due to industrialization, globalization, new technologies invention and increased household energy consumption in urban population. This leads to a huge gap between supply and demand in power sector for which non-renewable energy sources become unsustainable and enviable. Nowadays, the demand of renewable energy sources is due to ever-increasing energy consumption, the inadequate nature of fossil fuel and greenhouse effect. Among all renewable energy sources, wind, solar, biomass, geothermal and hydro are popular because of clean and sustainable solution with low environmental impact.

Wind energy is becoming a leading renewable energy sources and is now achieving exponential growth due to the advancement of wind turbine technologies, rapid falling prices and potential environmental benefits. By the end of 2017, according to World Wind Energy Association (WWEA), installed wind power is increased to 539.6 GW, which covers around 5.6% of total power generation. To supply electricity to grid, at least 83 countries are using wind power around the world. Installation of wind power in 2017 is increased by 51,924 MW as compared to 2016 [1]. Moreover, Asia became the largest regional contributor around 48% of

J. Mishra (✉) · M. Pattnaik
Indian Institute of Information Technology Design and Manufacturing, Kancheepuram, Chennai, India

N. Priyadarshi et al. (eds.), *DC–DC Converters for Future Renewable Energy Systems*, Energy Systems in Electrical Engineering,
https://doi.org/10.1007/978-981-16-4388-0_16

added capacity and Europe contributes around 30% followed by 14% of North America and Latin America and 6% of Caribbean. Denmark is achieving the highest penetration of wind power with 43.4% of consumed electricity come from wind by the end of 2017. Gansu Wind Farm in China is the largest farm in world having capacity 6800 MW [2]. However, in India, wind power generation capacity has been increased significantly due to the availability of wind energy. According to the Ministry of New Renewable Energy (MNRE) physical progress achievement data, by June 30, 2018, the total installed wind generation capacity is 34.293 GW, which is the fourth largest installed wind power capacity in the world [3]. Moreover, during auctions for wind projects in December 2017, wind power tariff is reduced to Rs. 2.43 (0.04$) per kWh (without any direct or indirect subsidies) [2].

The primary components of a wind generation system include wind turbine, a gear box, a generator and power electronic converters as shown in Fig. 16.1. Wind turbine converts wind aerodynamic power to mechanical power. Then, low-speed high-torque mechanical power of wind turbine is converted to electrical power by the use of gear box and suitable generator. Electrical power is fed to either load or grid through power electronics converter. Due to the advancement of power electronic converters, generated power can be transferred to end user with high efficiency and improved static and dynamic responses.

Basically, three types of wind turbine (WT) are used, i.e., horizontal axis, vertical axis and ducted WT. Among them, horizontal axis WT is mostly used [4, 5]. Many possible solutions for electrical and mechanical configurations of wind generation system are based on gear box connection, evolution of power electronic converters and the use of wind generators. Mainly induction generator (IG) and permanent magnet synchronous generator (PMSG) are used to convert the wind energy into electrical energy. Induction generator is of two types, i.e., squirrel cage induction generator (SCIG) or self-excited induction generator (SEIG) and wound-rotor induction generator or doubly fed induction generator (DFIG) [8]. Based on gear box connection, wind power generation system is divided into two types, i.e., direct drive wind generation system (without gear box) and indirect drive wind generation system (with gear box). However, in direct drive wind generation system, permanent magnet synchronous generator (PMSG) and synchronous multipole generators [6] are used for energy conversion unit. All other wind generators are used in indirect wind generation system. Further, for high-power wind turbine, DFIG is used as wind

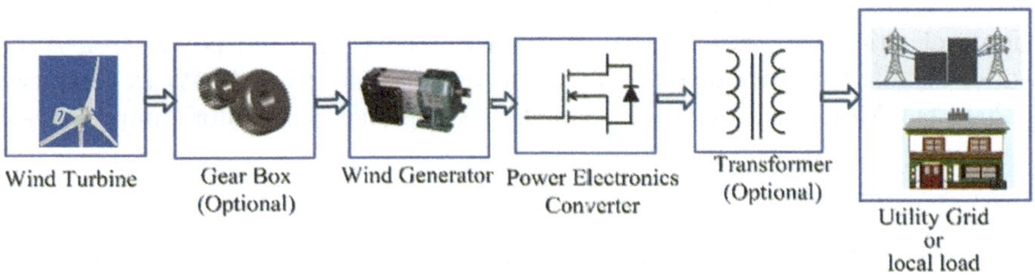

Fig. 16.1 Components of wind generation system

generator. In this, the stator is directly connected to the grid/load and rotor is connected through a AC/DC–DC/AC back-to-back converter [7]. The advantageous of this type of configuration is that only a part of wind power production is fed through converter. So, the nominal power rating of converter is less than the generator rating. For low-power application, SEIG is an appropriate choice due to low cost, easy maintenance, ruggedness and self-protection from short circuit. Usually, AC/DC–DC/AC-controlled voltage source converters are used to control active and reactive power independently. AC–DC-uncontrolled rectifier with DC–DC converter is also used to reduce the converter cost which is shown in Fig. 16.2. Steady-state and transient performance of SEIG has been well established in [8]. Different control techniques like active and reactive power control, sensorless vector control, indirect field-oriented control, grid fault detection control are available in the literature to get the fast dynamic response of the system [9–11].

Nowadays, maximum power extraction from wind turbine is one of the important researches works. This can be achieved by proper selection of maximum power point tracking (MPPT) algorithm [12, 13] using power electronics converter. The design and implementation of DC–DC buck converter-based wind energy generation system (WEGS) will be simple and efficient solution. Among all the available MPPT algorithms [11], P&O is most common type. The conventional P&O algorithm has more tracking time as well as more oscillation which leads power loss [14–16]. Further, the sudden change in wind speed leads to wrong direction which deteriorates its performance [17–19].

This chapter presents a modified P&O MPPT algorithm for DC–DC buck converter-based WEGS. This algorithm is simple with robustness in nature which also does not need any speed sensor, additional current sensor and turbine parameter information.

16.1.1 DC–DC Buck Converter-Based WEGS

The single-line diagram of SEIG-based WGS is shown in Fig. 16.3 with DC–DC converter. This DC–DC buck converter is interfaced between 3-ϕ rectifier and load which makes the WGS to operate at maximum power point. The duty cycle of this converter is generated from the MPPT controller to ensure optimal power point operation.

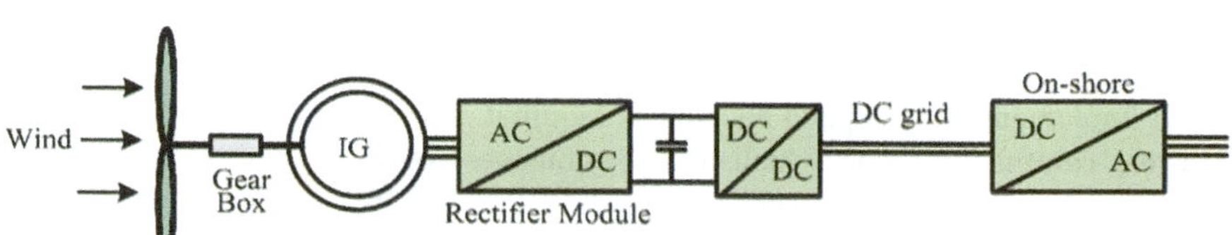

Fig. 16.2 Induction generator with rectifier and DC–DC converter

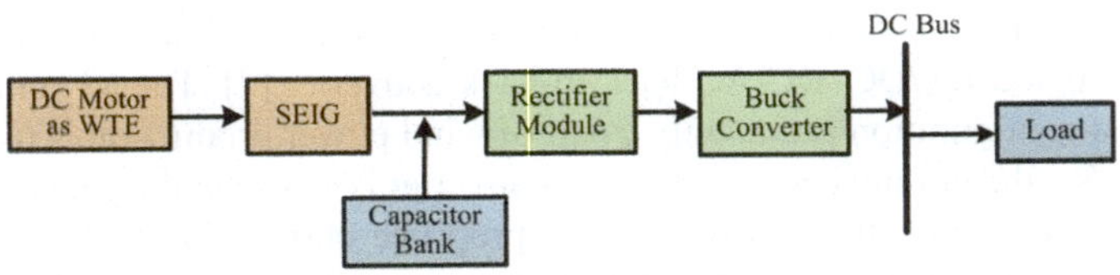

Fig. 16.3 Single-line diagram of WEGS with buck converter

16.1.1.1 Buck Converter Design

The circuit diagram of buck converter with filter inductor (L), capacitor (C) and load is shown in Fig. 16.4.

The output voltage of buck converter is:

$$V_o = DV_{dc} \quad \text{and} \quad I_{dc} = DI_o \tag{16.1}$$

To calculate the value of inductor, the following equations are considered.

$$\begin{aligned} \frac{\Delta i_l}{\Delta t} &= \frac{\Delta i_l}{DT_s} = \frac{V_{dc} - V_o}{L} \\ \Rightarrow L &= \frac{(V_{dc} - V_o)D}{\Delta i_l f_s} \end{aligned} \tag{16.2}$$

And, the filter capacitor value is found from Eq. (16.3).

The capacitor charge change is as follows:

$$\Delta Q = C\mathrm{d}V_o = \frac{1}{2}\frac{T_s}{2}\frac{\Delta i_l}{2} = \frac{\Delta i_l}{8f_s} \tag{16.3}$$

16.1.1.2 Drift-Free MPPT Algorithm

From power–shaft speed characteristics of wind turbine (WT) as shown in Fig. 16.5, it is clear that the maximum power can be extracted from WT by adjusting rotor shaft speed [14] and Eq. (16.4) should be satisfied.

$$\frac{\mathrm{d}P_{WT}}{\mathrm{d}\omega_t} = 0 \tag{16.4}$$

To implement this, a speed encoder is required. This needs extra wiring, space, cost and careful mounting which reduce robustness of the system [15]. To overcome all these drawbacks, speed sensorless P&O MPPT algorithms are used. P_m–ω_r curve is obtained by varying shaft speed, and similarly, P_d–V_d characteristic

Fig. 16.4 Circuit diagram of buck converter

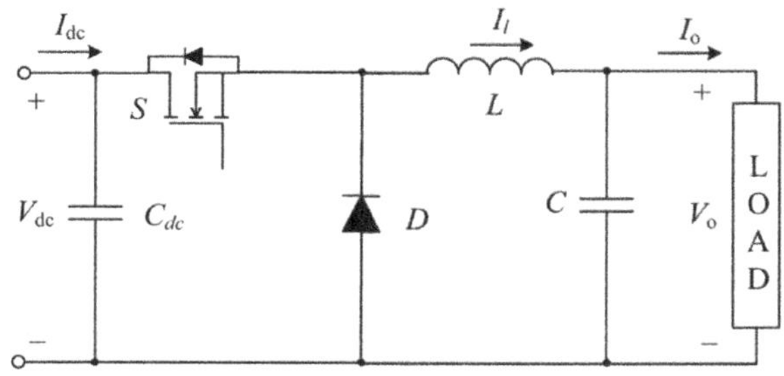

can be drawn by varying duty of the buck converter. Both the theoretical characteristics are shown in Fig. 16.5, and it is observed that the MPP of both the curves coincides. In order to prove it theoretically, the following analysis is done.

Equation (16.4) is rewritten as Eq. (16.5) by using chain rule.

$$\frac{\mathrm{d}P_{\mathrm{WT}}}{\mathrm{d}\omega_{\mathrm{t}}} = \frac{\mathrm{d}P_{\mathrm{WT}}}{\mathrm{d}V_{\mathrm{d}}} \cdot \frac{\mathrm{d}V_{\mathrm{d}}}{\mathrm{d}\omega_{\mathrm{e}}} \cdot \frac{\mathrm{d}\omega_{\mathrm{e}}}{\mathrm{d}\omega_{\mathrm{t}}} = 0 \tag{16.5}$$

After voltage build-up, DC-link voltage is positive $\frac{\mathrm{d}V_{\mathrm{d}}}{\mathrm{d}\omega_{\mathrm{e}}} \neq 0$ and $\frac{\mathrm{d}\omega_{\mathrm{e}}}{\mathrm{d}\omega_{\mathrm{r}}} \neq 0$. Neglecting all machine and converter losses, $P_{\mathrm{m}} = P_{\mathrm{d}}$. Now, Eq. (16.5) can be rewritten as

$$\frac{\mathrm{d}P_{\mathrm{m}}}{\mathrm{d}\omega_{\mathrm{r}}} = 0 \rightarrow \frac{\mathrm{d}P_{\mathrm{d}}}{\mathrm{d}V_{\mathrm{d}}} = 0 \tag{16.6}$$

where V_{d} is the rectifier output voltage, P_{d} is the DC-link power, ω_{e} is the electrical angular speed and p is the number of pole pair of induction machine. From the above analysis, it is clear that optimum DC-link voltage (corresponding to P_{dMPP}) can be obtained by directly regulating the duty cycle which is the output of MPPT controller. As DC-link power–voltage-based P&O algorithm is sensorless and easy to implement, this technique is used to harness maximum power [11].

For algorithm implementation, at first, all variable parameters (V_{d}, I_{d} and P_{d}), MPPT sampling time (Ts_{MPPT}), duty cycle perturbation (ΔD) are initialized. Then, DC-link voltage and current are sampled at the rate of 1 s to get the present iteration value ($V_{\mathrm{d}}(k)$, $I_{\mathrm{d}}(k)$), and the present iteration of DC-link power ($P_{\mathrm{d}}(k)$) is calculated by multiplying these two quantities. After first search, all the present variables are updated as previous iteration ($k - 1$) with the rate of sampling time. Figure 16.6 shows the conventional P&O MPPT algorithm.

$$\frac{\mathrm{d}P_{\mathrm{d}}}{\mathrm{d}V_{\mathrm{d}}} > 0 \Rightarrow \text{Left side of MPP} \Rightarrow D(k) = D(k-1) + \Delta D \tag{16.7}$$

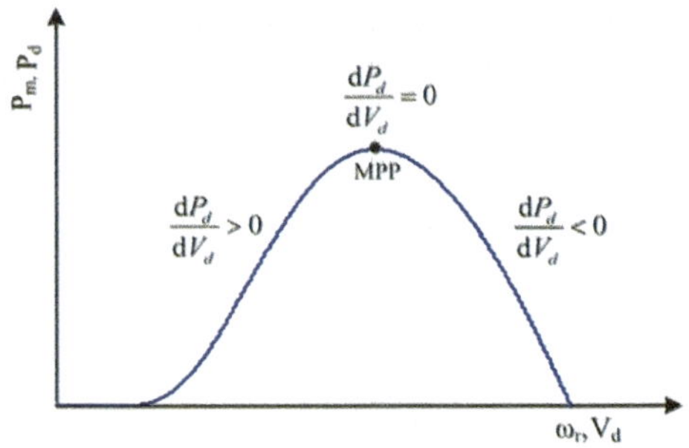

Fig. 16.5 Theoretical P_m–ω_r and P_d–V_d characteristics

$$\frac{dP_d}{dV_d} < 0 \Rightarrow \text{Right side of MPP} \Rightarrow D(k) = D(k-1) + \Delta D \tag{16.8}$$

$$\frac{dP_d}{dV_d} = 0 \Rightarrow \text{at MPP} \tag{16.9}$$

In case of constant average wind speed, the algorithm searches around the MPP which results more oscillation. However, for rapid wind speed change, the algorithm fails to perturb in right direction by only observing P_d–V_d characteristic. In such situation, wrong direction detection leads to drift phenomenon. At this situation, algorithm fails to detect the wind speed variation and confuses in between change in perturbation and wind speed [15–18].

The modified MPPT algorithm solves the above issue by adding the DC-link current information under DC-link voltage loop [14]. The change in DC-link voltage and current effect with respect to wind speed variation, the experiment has been conducted by varying duty of the buck converter. DC-link current and voltage are noted to plot I_d–V_d characteristics. In this algorithm, DC-link current naturally changes with wind speed variation. So, increased wind speed condition can be detected by using I_d information when V_d is positive, and exact MPP track detection is possible with insertion of the DC-link current information.

This drift-free MPPT algorithm does not require any additional current sensor as DC-link current information is already available to calculate power. The proposed MPPT algorithm flow diagram is presented below. As stated above, to overcome the above-mentioned problem of conventional P&O algorithm, a DC-link current loop is added which increases the efficiency of the algorithm at sudden wind speed condition with increased tracking speed. It also reduces oscillation during steady-state operation. Figure 16.6 shows the flowchart of proposed MPPT algorithm.

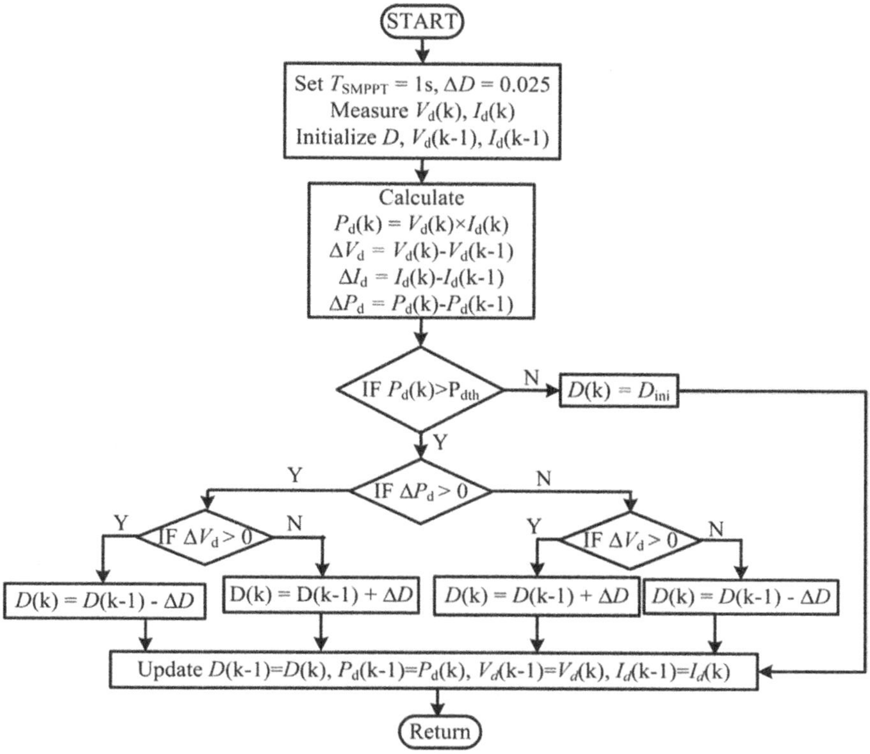

Fig. 16.6 Flowchart of Conventional MPPT algorithm

16.2 Implementation of MPPT Algorithm Using DC–DC Buck Converter

According to the design considerations, IRF460 MOSFET and MUR3060PT fast recovery diode is used for DC–DC buck converter. TLP250 gate driver is connected with the circuit to drive the MOSFET. For the MPPT algorithm implementation, DC-link power is calculated from DC-link voltage and current which is measured by TBV10/25A voltage transducer and LEM55p current transducer. As DC-link current is switched current which is pulsed in nature, a second-order low-pass filter is connected in between current sensor and digital controller input. Low-pass filter circuit is shown in Fig. 16.7 (Figs. 16.8 and 16.9).

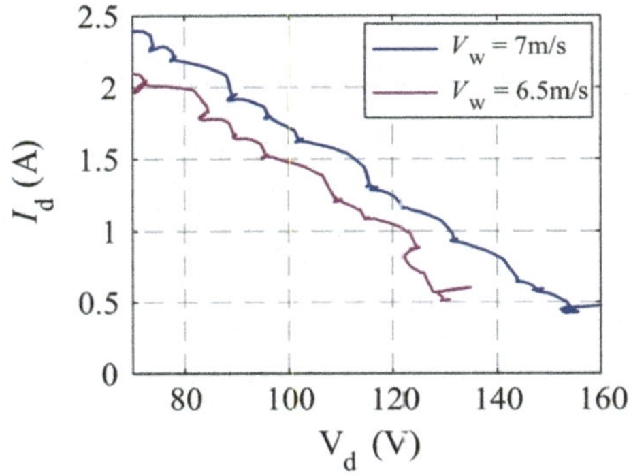

Fig. 16.7 Experimental I_d–V_d curve of WEGS

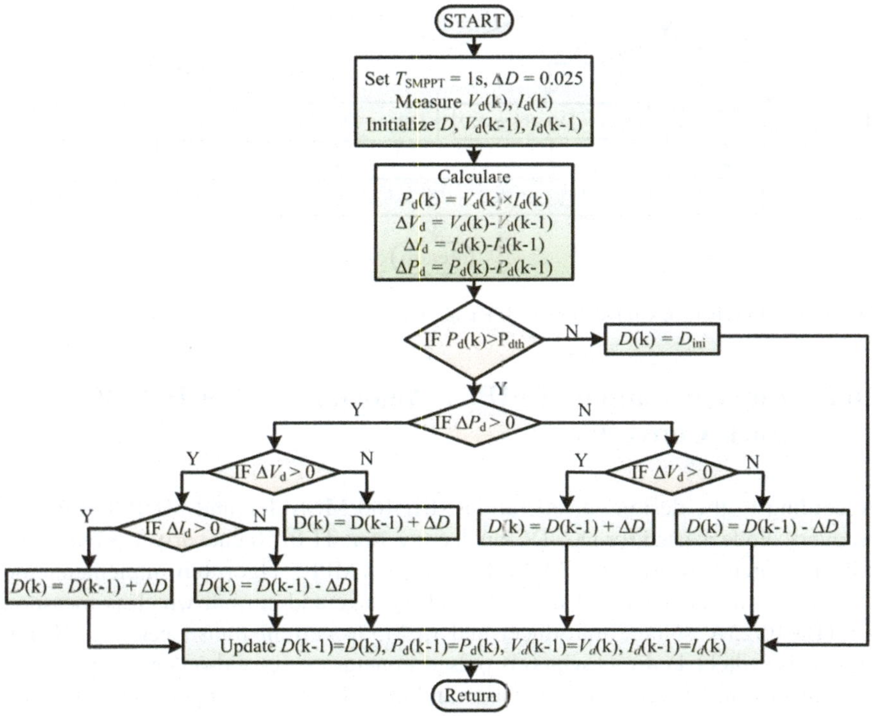

Fig. 16.8 Flowchart of drift-free MPPT algorithm

16.3 Result Analysis

To prove the functionality of the drift-free MPPT algorithm, at first, the experiment is conducted by gradually varying the duty cycle of buck converter to get the operating characteristics of WTE. P_d, V_d, and D are recorded and plotted in Fig. 16.10 for 6.5 m/s wind speed. It is seen that as the duty ratio increases the DC-link voltage decreases and DC-link power increases correspondingly. Then, P_d–V_d characteristics are plotted from right side of the curve. It is observed that the maximum DC-link power is 152 W (Fig. 16.10).

To prove the effectiveness of the algorithm more precisely, the experiment is done for various wind speed conditions, i.e., 6.5, 7, 6, 6.25, 7.5 and 7.25 m/s. The performance of the proposed MPPT algorithm under steady-state condition is shown in Fig. 16.11. By proposing one DC-link current loop under positive slope of DC-link voltage loop, the oscillations around MPP reduces and an average value of 150 W is obtained as maximum power. The efficiency is calculated as 98.68% which is more than the conventional P&O MPPT algorithm. With the addition of DC-link current slope, the algorithm does not confuse and searches to get the power very close to MPP at less tracking time. The change in DC-link voltage can be due to either algorithm perturbation or variation in wind speed. Sometimes, in case of wind speed change, the conventional algorithm cannot differentiate the above change and will deviate from MPP point which is discussed earlier, and it will reduce the efficiency of the system. From I_d–V_d characteristics, as shown in

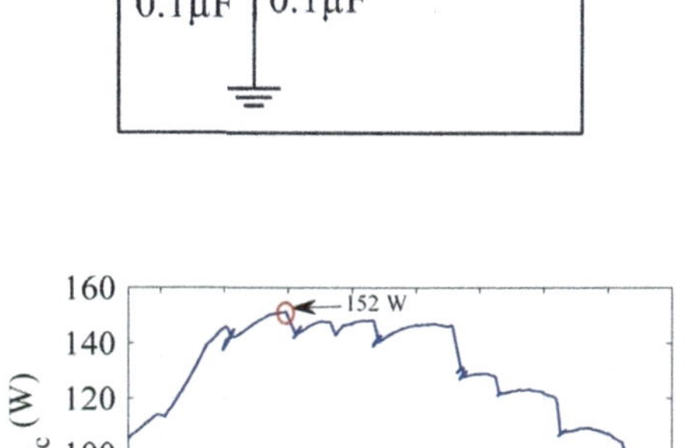

Fig. 16.9 Circuit diagram of low-pass filter

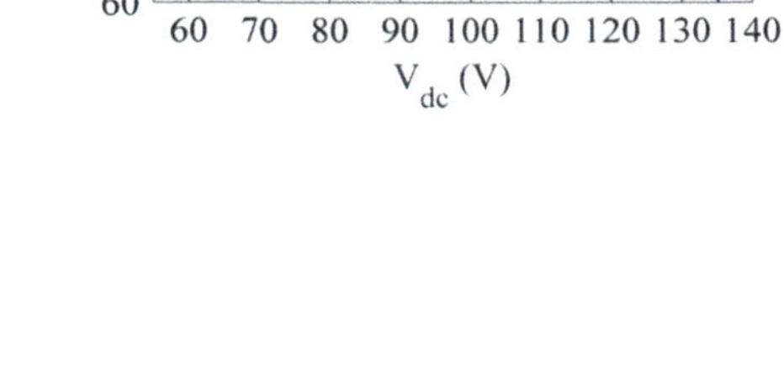

Fig. 16.10 P_d–V_d characteristics of WEGS

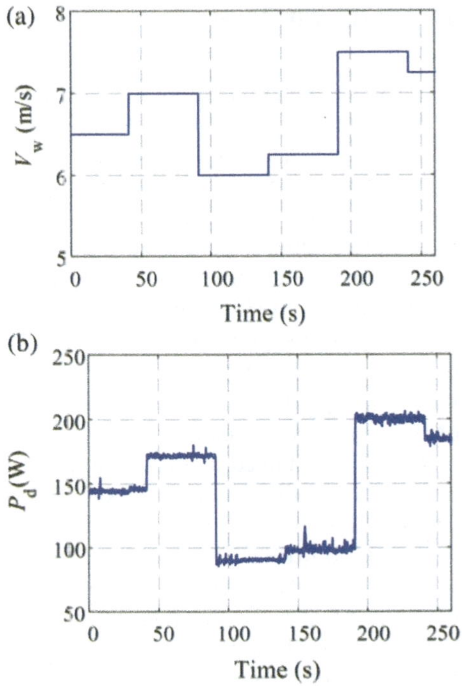

Fig. 16.11 Experimental waveforms of drift-free MPPT algorithm **a** wind speed and **b** DC-link power

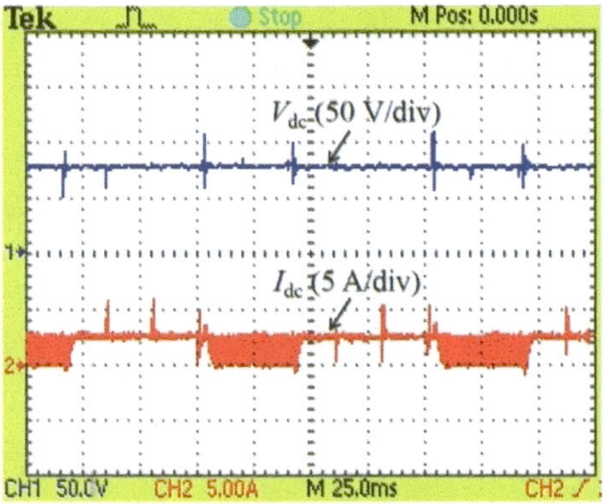

Fig. 16.12 Experimental waveforms of **a** DC-link voltage **a** and **b** DC-link current

Fig. 16.7, it is observed that I_d is only increasing during increase in wind speed. Otherwise, it decreases in nature. So, by including I_d loop, the algorithm will search in appropriate direction with reduced efficiency (Figs. 16.11 and 16.12).

16.4 Summary

This chapter proposes a drift-free P&O algorithm for DC–DC buck converter-based stand-alone WEGS. This algorithm is simple to implement and does not require either speed sensor or extra current sensor to detect the change in wind speed which reduces the system cost. By including ΔI_d information, the dynamic performance is effectively increased in terms of tracking speed. The steady-state performance is also improved by reducing power oscillation around MPP which increases the system efficiency. The effectiveness of the system behavior with random variation of wind speed condition is observed. More DC-link power is harvested by using proposed algorithm as compared to conventional P&O. Thus, the developed system with proposed algorithm could be a reliable candidate for WGS. Significant performance improvement is achieved during steady-state and transient condition with reduction in power loss.

References

1. O'Brien, J.M., Young, T.M., O'Mahoney, D.C., Griffin, P.C.: Horizontal axis wind turbine research: a review of commercial CFD, FE codes and experimental practices. Prog. Aerosp. Sci. **92**, 1–24 (2017)
2. Howell, R., Qin, N., Edwards, J., Durrani, N.: Wind tunnel and numerical study of a small vertical axis wind turbine. Renew. Energy **35**(2), 412–422 (2010)
3. Farret, F.A., Simoes, M.G.: Integration of alternative sources of energy. Wiley (2006)
4. Blaabjerg, F., Liserre, M., Ma, K.: Power electronics converters for wind turbine systems. Ind. Appl. IEEE **48**(2), 708–719 (2012). https://doi.org/10.1109/TIA.2011.2181290
5. Priyadarshi, N., Padmanaban, S., Bhaskar, M.S., Blaabjerg, F., Sharma, A.: A fuzzy SVPWM based inverter control realization of grid integrated PV-wind system with FPSO MPPT algorithm for a grid-connected PV/wind power generation system: hardware implementation. IET Electr. Power Appl. 1–12 (2018)
6. Priyadarshi, N., Sharma, A.K., Azam, F.: A hybrid firefly-asymmetrical fuzzy logic controller based MPPT for PV-Wind-fuel grid integration. Int. J. Renew. Energy Res. **7**(4) (2017)
7. Yaramasu, V., Dekka, A., Durán, M.J., Kouro, S., Wu, B.: PMSG-based wind energy conversion systems: survey on power converters and controls. IET Electr. Power Appl. **11**(6), 956–968 (2017)
8. Liu, J., et al.: Impact of power grid strength and PLL parameters on stability of grid-connected DFIG wind farm. IEEE Trans. Sustain. Energy **11**(1), 545–557 (2019)
9. Shmilovitz, D.: On the control of photovoltaic maximum power point tracker via output parameters. IEE Proc. Electric. Power Appl. **152**(2), 239–248 (2005)
10. Padmanaban, S., Priyadarshi, N., Holm-Nielsen, J.B., Bhaskar, M.S., Azam, F., Sharma, A. K.: A novel modified sine-cosine optimized MPPT algorithm for grid integrated PV system under real operating conditions. IEEE Access **7**, 10467–10477 (2019). https://doi.org/10.1109/ACCESS.2018.2890533
11. Padmanaban, S., Priyadarshi, N., Holm-Nielsen, J.B., Bhaskar, M.S., Hossain, E., Azam, F.: A hybrid photovoltaic-fuel cell for grid integration with Jaya-based maximum power point tracking: experimental performance evaluation. IEEE Access **7**, 82978–82990 (2019). https://doi.org/10.1109/ACCESS.2019.2924264

12. Kasal, G.K., Singh, B.: Voltage and frequency controllers for an asynchronous generator-based isolated wind energy conversion system. IEEE Trans. Energy Convers. **26** (2), 402–416 (2011)
13. Koutroulis, E., Kalaitzakis, K., Voulgaris N.C.: Development of a microcontroller-based, photovoltaic maximum power point tracking control system. IEEE Trans. Power Electron. 46–54 (2001)
14. Vardia, M., Priyadarshi, N., Ali, I., Azam, F., Bhoi, A.K.: Maximum power point tracking for wind energy conversion system. In: Bhoi, A., Sherpa, K., Kalam, A., Chae, G.S. (eds.) Advances in Greener Energy Technologies. Green Energy and Technology. Springer, Singapore (2020). https://doi.org/10.1007/978-981-15-4246-6_36
15. Nayanar, V., Kumaresan, N., Gounden, N.A.: A single-sensor-based MPPT controller for wind-driven induction generators supplying DC microgrid. IEEE Trans. Power Electron. **31** (2), 1161–1172 (2016)
16. Hui, J., Bakhshai, A.: A new adaptive control algorithm for maximum power point tracking for wind energy conversion systems. In: Power Electronics Specialists Conference, PESC, pp. 4003–4007 (2008)
17. Killi, M., Samanta, S.: Modified perturb and observe MPPT algorithm for drift avoidance in photovoltaic systems. IEEE Trans. Ind. Electron. **62**(9), 5549–5559 (2015)
18. Nishida, K., Ahmed, T., Nakaoka, M.: A cost-effective high-efficiency power conditioner with simple MPPT control algorithm for wind-power grid integration. IEEE Trans. Ind. Appl. **47** (2), 893–900 (2011)
19. World Energy Scenarios: Composing energy futures to 2050, howpublished, World Energy, World Energy (2018).

Chapter 17
Grid-Connected RTPV System with Fuzzy-Based Energy Management System

T. Hari Priya, M. Nagajyothi, G. Radhika, and O. Sobhana

17.1 Introduction

With the increase in use of renewable sources, many microgrid and distribution generation interconnections are happening [1–3]. Power quality improvement and battery-based management of energy have become a scope of research. Renewable energy sources can be used both in grid and in isolated mode of working [4, 5]. To integrate RES to AC grid, energy management system plays a vital role, which helps in regulating the flow of power and maintains energy balance [6–8].

In the literature, there are many energy management techniques developed by researchers [9, 10]. One among these strategies are proportional and integral (PI) controller applied for microgrid battery scheduling [11–13]. PI controller is widely used for controller design due to its simplicity and flexibility to operate at various operating points [14]. But the existence of nonlinearities will degrade system performance and is difficult to attain the optimal solution. When compared to PI, 'fuzzy logic controller' offers optimal value for dynamic systems that have wide fluctuations in parameters of the system. It also offers advantages like simple and robust construction, faster response, can deal nonlinearities in model and used in MIMO applications. By using PI and fuzzy controller, the problem of dynamic variation in parameter can be effectively dealt.

The block diagram of RTPV-fed grid with maximum power point tracking (MPPT) and EMS is represented in Fig. 17.1. It includes solar PV array-fed converter (boost), where MPPT is implemented. MPPT helps in operating the system at peak power operating point. The output of it is connected to inverter and to the load. The excess power available is supplied to the grid and receives the power from the grid at low-power generation. EMS that consists of battery and controller will maintain balance between generated power and load demand. In Sect. 17.2, solar

T. Hari Priya (✉) · M. Nagajyothi · G. Radhika · O. Sobhana
EEE Department, VNR VJIET, Hyderabad 500090, India
e-mail: haripriya_t@vnrvjiet.in

N. Priyadarshi et al. (eds.), *DC–DC Converters for Future Renewable Energy Systems*, Energy Systems in Electrical Engineering,
https://doi.org/10.1007/978-981-16-4388-0_17

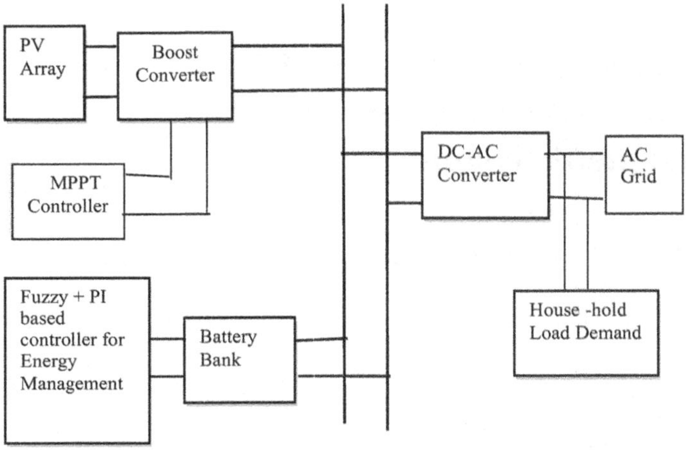

Fig. 17.1 AC grid-connected RES with EMS

cell modeling is explained followed by modeling of battery source in Sect. 17.3. In Sect. 17.4, fuzzy-based PI controller working is explained. Results and analysis is given in Sect. 17.5.

17.2 Modeling of PV Cell

In this work, solar PV on roof top is chosen for power generation. The equivalent circuit model considered for simulation is given in Fig. 17.2. RTPV is constructed by placing PV cells in parallel and series fashion.

The equation relating PV cell parameters is given by [14]

$$I_{\text{phn}} = [I_{\text{se}} + K_{\text{i}}(T_{\text{o}} - 298)] \times I_{\text{r}}/1000 \tag{17.1}$$

The PV cell reverse saturation current is

$$I_{\text{rs}} = I_{\text{se}}/[\exp(qV_{\text{OC}}/N_{\text{S}}knT_{\text{o}}) - 1] \tag{17.2}$$

And, saturation current is

$$I_{\text{o}} = I_{\text{rs}}\left[\frac{T_{\text{o}}}{T_{\text{r}}}\right]^3\left[\exp\frac{qE_{\text{go}}}{nk}\left(\frac{1}{T_{\text{o}}} - \frac{1}{T_{\text{r}}}\right)\right] \tag{17.3}$$

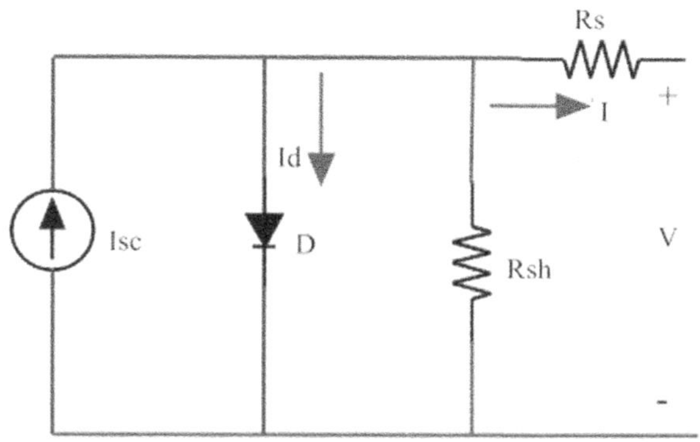

Fig. 17.2 PV cell one-diode equivalent model

The output current is

$$I = p\left(N_p \times I_{phe} - N_p \times I_o\right) \exp\left[\left(\frac{V/N_s + I \times R_s/N_p}{n \times V_t}\right) - 1\right] - I_p \quad (17.4)$$

$$V_t = \frac{kT_o}{q} \quad (17.5)$$

$$I_p = \frac{V \times \left(N_p/N_s + I \times R_s\right)}{R_{sh}} \quad (17.6)$$

where

I_{phn} is photocurrent
I_r is input irradiation
T_o represents operating temperature and
T_r nominal temperatures
I_{se} is short-circuit current
N indicates ideality factor
q is charge acquired by electron
N_s and N_p represent series and parallel connected cells
E_{g0} indicates available band gap energy.

17.3 Battery Model

Battery an energy storage system plays an important part in power balancing of AC system. 'State of charge' (SOC) indicates the status of charge in it, i.e., the power available in it. To predict SOC of battery, the open-circuit model of it is considered.

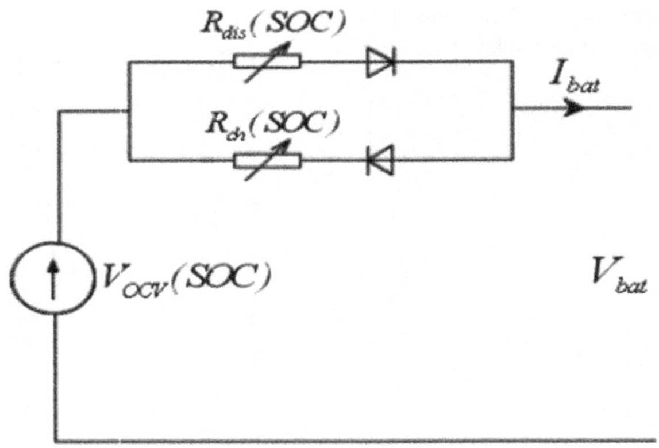

Fig. 17.3 Battery model

In this model, battery is treated as an ideal source of voltage with resistance in series which is internal resistance offered by the circuit, as in Fig. 17.3.

Battery operation during charging mode is influenced by charging currents and internal resistance, SOC. The rate at which battery charges is decided by depth of discharge (DOD) and current available for charging.

Battery characteristics during discharge mode for Li-ion battery is given by:

$$V_b = V_c - \lambda \frac{\mu}{(\mu - it)} + \alpha e^{-\beta it} \tag{17.7}$$

$$V_b = V_{oc} - R_b^i I_b \tag{17.8}$$

where

$$V_{oc} = f_1 \text{ (SOC)} \tag{17.9}$$

$R_b^i = R_{dis} = f_2$ (SOC) during charge

$= R_{dis} = f_3$ (SOC) during discharge

and

$$\text{SOC} = \text{SOC}_i - \int \frac{\eta I_b}{Q} dt \tag{17.10}$$

where

V_{oc} indicates no load voltage in volts
V_c indicates constant voltage
λ indicates polarization potential in volts
μ indicates capacity in Ah
R_b^i is internal resistance
α indicates voltage that changes exponentially in volts
β indicates the battery capacity that changes exponentially in $(Ah)^{-1}$
SOC is state of charge.

17.4 Fuzzy-Based PI Controller for EMS

The controller used in EMS of grid-connected RES is given in Fig. 17.4. It consists of 'fuzzy-based proportional and integral (PI)' controller where the K_p and K_i values of PI controller are found using FLC for various working conditions dynamically. For example, change in error and error will indicate the first found derivatives so as to find K_p value [14]. Here, error indicates the variation between the power supplied and the actual demand. Inference with the rule base provides set of rules using if–then statements which will identify K_p, in terms of two inputs.

For the input variables, the fuzzy subsets are given by Negative Bigger (NG), Negative Lower (NL), Zero (ZE), Positive Lower (PL) and Positive Bigger (PG). The fuzzy subsets for output K_p are Smaller (SM), Moderate (ME) and Larger (LG). Table 17.2 describes the rule base for fuzzy controller. In this controller design, Mamdani's MAX-MIN technique is used for checking the inference. Defuzzification is performed using 'center of gravity' (COG) method. The integral constant K_i is found dynamically by using proportional constant K_p as described in Fig. 17.4.

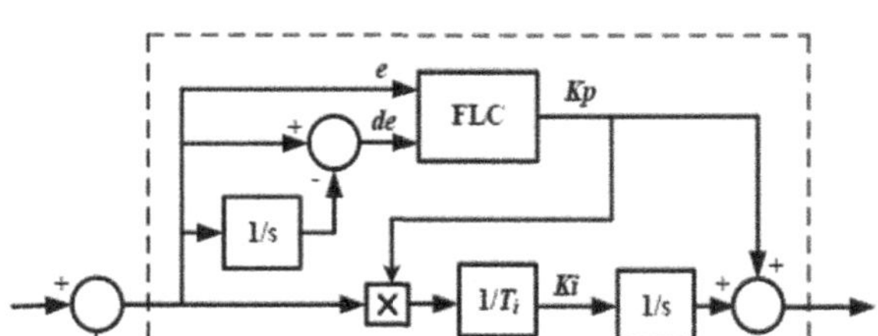

Fig. 17.4 Controller for EMS

17.5 Results and Discussion

The proposed model consists of RTPV and BMS connected to an AC grid through converters setup. The ratings of PV module in RTPV panel are shown in Table 17.1. RTPV panel output is fed to AC grid by using DC–DC converter and inverter setup. BMS used in the proposed system will regulate the power flow and provide the power to the load when RTPV alone does not satisfy the load demand. The PV array output V_o is stepped up to 500 V by DC–DC converter circuit, whose ratings are mentioned in Table 17.2. The *I–V* and *P–V* plots of PV cell used in RTPV panel are given in Figs. 17.5 and 17.6.

The input to RTPV panel, i.e., variable irradiation, is represented in Fig. 17.7. The converter (DC–DC) output voltage using different control techniques is plotted in Fig. 17.8. It is examined from the results that RSM-based MPPT controller offers better tracking efficiency compared to fuzzy, P&O and increment conductance technique.

From the Fig. 17.8, it is noticed that using RSM controller the voltage response is faster with lesser oscillations in output, and optimal MPP is reached. P&O techniques has large number of variations in output response, and noise is high in fuzzy, and INC technique suffers from poor tracking efficiency. For 24 h in a day, the variation of irradiation will be in the form of sinusoid. So, for a sinusoidal change in irradiation for 24 h, the response of RTPV panel, load demand, battery power output, power supplied to the grid are plotted. From Fig. 17.9, it is examined that the PV power output is zero from 19 (7 PM) to 4 (4 AM) hours as there is absence of sun light during that period. Peak power is reached in between 14 and 15 h, i.e., 5 kW, as the irradiation is at its maximum. The output power generated by RTPV is of sinusoidal shape as the irradiation is varying in similar pattern.

Table 17.1 FLC rule base

e/de	NG	NL	ZE	PL	PG
NG	LG	LG	ME	LG	LG
NL	LG	ME	SM	ME	LG
ZO	ME	SM	SM	SM	ME
PL	LG	ME	SM	ME	LG
PG	LG	LG	ME	LG	LG

Table 17.2 PV module rating

S. No.	Parameter	Value
1	V_{occ}, voltage in volts	44.429 V
2	V_{mp}, voltage at P_{max} in volts	36.9 V
3	I_{scc}, current in Amps at short circuit	8.21 A
4	I_{mp}, current at P_{max}	7.61 A
5	Total cells connected	72
6	P_{max}, peak power	280.91 W

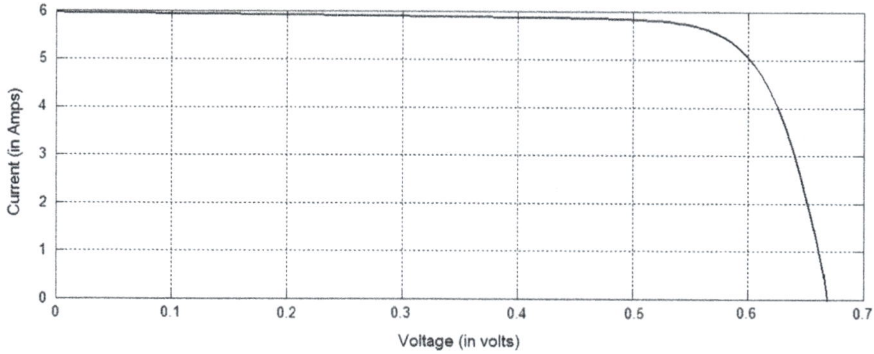

Fig. 17.5 *I–V* curve of PV cell

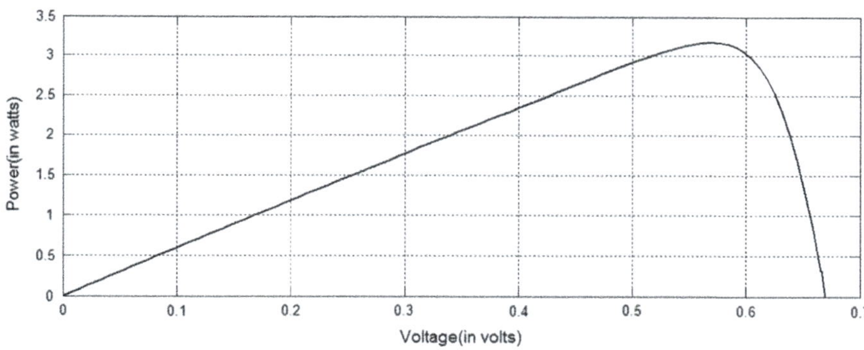

Fig. 17.6 *P–V* curve of PV cell

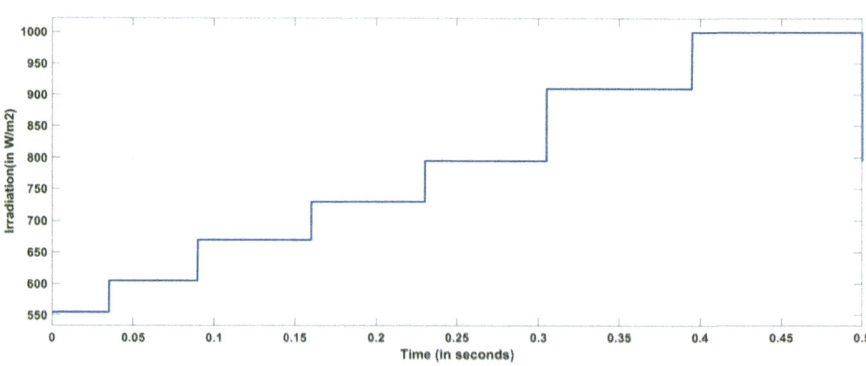

Fig. 17.7 Irradiation to RTPV

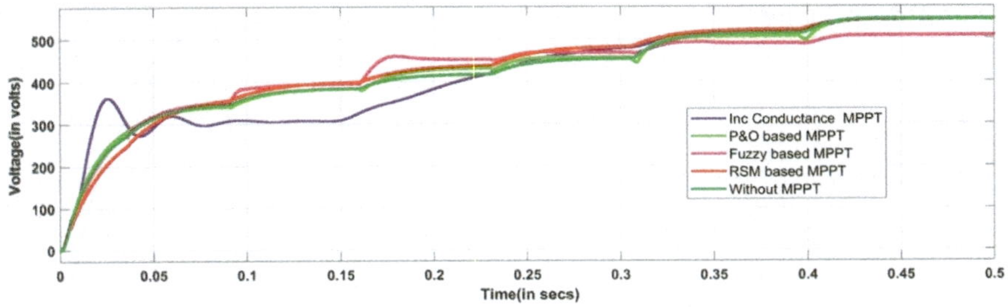

Fig. 17.8 Output voltage of converter (DC–DC) with and without MPPT techniques

Fig. 17.9 Output power of RTPV for variable irradiation in a day

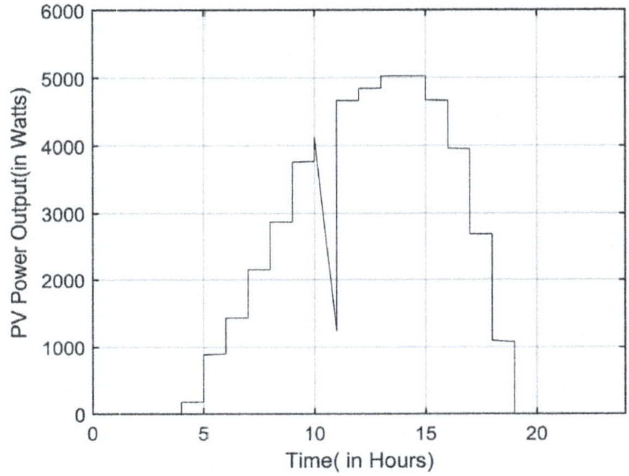

Fig. 17.10 Power delivered to the load

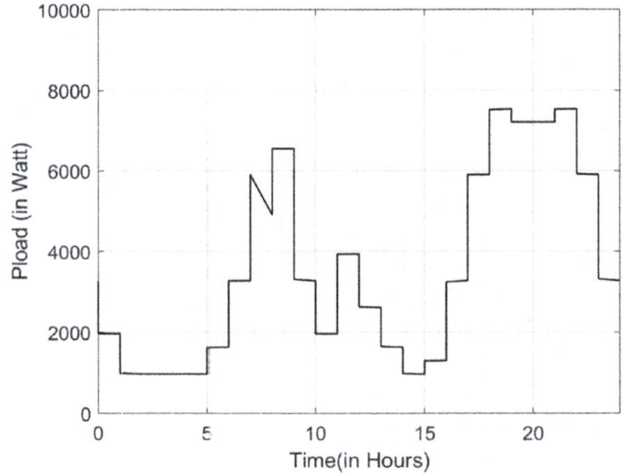

The power drawn by the load connected across RTPV is in Fig. 17.10. It is examined the power drawn is maximum at certain hours in a day and is varying continuously throughout the day. From the characteristics of battery shown in Fig. 17.11, the power charged and power discharged by it are examined. From the plot, it is observed that without the use of EMS at battery from 12 to 18 h, SOC is maintained constant and there is no charge or discharge of power from the battery. If there is reduced power generation, then the balance power is supplied by the grid as in Figs. 17.11 and 17.12. For the rest of the hours, EMS will come into action along with the battery so that there is a continuous power balance between the source and the load.

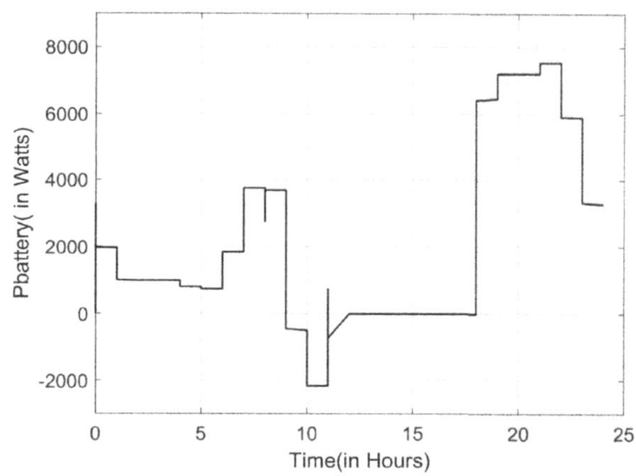

Fig. 17.11 Output power stored/discharged by the battery

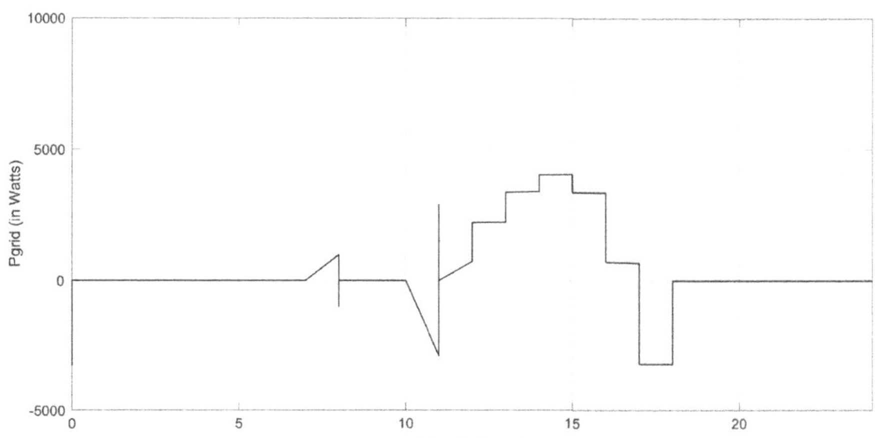

Fig. 17.12 Output power supplied to/from the grid

Table 17.3 Specifications of converter (DC–DC)

S. No.	Parameter	Value
1	L_1, inductance offered	4500 μH
2	C_1, capacitance offered	0.550 mF
3	F_s, switching frequency	2 kHz
4	V_{in}, voltage input	0.250 kV
5	V_{out}, voltage output	0.5 kV

From Fig. 17.10, it is noticed that load power drawn is maximum, i.e., 6.5 kW at 9 AM. But the PV power output is only 2.8 kW. So, the rest of the power is supplied by the battery as in Fig. 17.11. Power delivered to the load in a day for variable irradiation is shown in Fig. 17.12. The power charged and power discharged by battery are shown in Fig. 17.11. From the results, it is examined that the fuzzy-based PI controller along with the battery is able to balance the energy flow from source to the load (Table 17.3).

17.6 Conclusion

In this work, the performance of EMS controller for management of energy in grid-connected RTPV application is validated using simulation. The MPPT controller used at converter (DC–DC) is able to track the maximum power efficiently using RSM controller when compared to other controllers. From the results, it is examined that the fuzzy-based PI controller along with the battery is able to balance the energy flow from source to the load. This EMS helps in maintaining the improved power quality without any discontinuity in power to the load. The EMS when used with battery will regulate the power flow from the battery. The battery used here provides the sufficient power. If the power of the microgrid is not sufficient and absorbs surplus current from the grid, then the battery power surpasses the load power.

References

1. Priyadarshi, N., Padmanaban, N., Holm-Nielsen, J.B., Blaabjerg, F., Bhaskar, M.S.: An experimental estimation of hybrid ANFIS–PSO-based MPPT for PV grid integration under fluctuating sun irradiance. IEEE Syst. J. **14**(1), 1218–1229 (2020). https://doi.org/10.1109/JSYST.2019.2949083
2. Kamalapathi, K., Priyadarshi, N., Padmanaban, S., Holm-Nielsen, J.B., Azam, F., Umayal, C., Ramachandaramurthy, V.K.: A hybrid moth-flame fuzzy logic controller based integrated Cuk converter fed brushless DC motor for power factor correction. Electronics **7**, 288 (2018)

3. Hari Priya, T., Parimi, A.M.: Hybrid controller topology for large solar PV installations in high-voltage DC grid-connected applications. Electr. Eng. **100**, 2537–2552 (2018). https://doi.org/10.1007/s00202-018-0715-7
4. Vardia, M., Priyadarshi, N., Ali, I., Azam, F., Bhoi, A.K.: Design of wind energy conversion system under different fault conditions. In: Bhoi, A., Sherpa, K., Kalam, A., Chae, GS. (eds.) Advances in Greener Energy Technologies. Green Energy and Technology. Springer, Singapore (2020). https://doi.org/10.1007/978-981-15-4246-6_41
5. Choudhary, T., Priyadarshi, N., Kumar, P., Azam, F., Bhoi, A.K.: A fuzzy logic control based vibration control system for renewable application. In: Bhoi, A., Sherpa, K., Kalam, A., Chae, GS. (eds.) Advances in Greener Energy Technologies. Green Energy and Technology. Springer, Singapore (2020). https://doi.org/10.1007/978-981-15-4246-6_38
6. Priya, T.H., Alivelu, M.P., Rao, U.M.: Development of hybrid controller for photovoltaic based DC-DC boost converter in DC grid connected applications. In: 2016 International Conference on Circuit Power and Computing Technologies (ICCPCT) (2016)
7. Sebastián, R.: Application of a battery energy storage for frequency regulation and peak shaving in a wind diesel power system. IET Generation Transmission & Distribution (2016)
8. Haripriya, T., Parimi, A.M.: Comparison between different MPPT controllers for solar PV fed DC grid connected applications. In: 2019 3rd International Conference on Recent Developments in Control, Automation & Power Engineering, pp. 334–339. https://doi.org/10.1109/RDCAPE47089.2019.8979069
9. Mahmood, H., Michaelson, D., Jiang, J.: A power management strategy for pv/battery hybrid systems in islanded microgrids. IEEE J. Emer. Sel. Top. Power Electron. **2**(4), 870–882 (2014)
10. Priya, H.T., Parimi, A.M., Rao, U.M.: Performance evaluation of high voltage gain boost converters for DC grid integration. In: 2016 International Conference on Circuit, Power and Computing Technologies (ICCPCT), pp. 1–6. https://doi.org/10.1109/ICCPCT.2016.7530241
11. Saez, D., Avila, F., Olivares, D., et al.: Fuzzy prediction interval models for forecasting renewable resources and loads in microgrids. Smart Grid IEEE Trans. **6**(2), 548–556 (2015)
12. Hari Priya T., Parimi A.M.: Energy management system for PV—wind and battery-fed DC microgrid using fuzzy-based proportional integral controller. In: Mishra S., Sood Y., Tomar A. (eds.) Applications of Computing, Automation and Wireless Systems in Electrical Engineering. Lecture Notes in Electrical Engineering, vol. 553. Springer, Singapore. https://doi.org/10.1007/978-981-13-6772-4_105
13. Yao, L., Lai, C.C., Lim, W.H.: Home energy management system based on photovoltaic system. In: Proceeding IEEE International Conference on Data Science and Data Intensive Systems, pp. 644–650 (2015). https://doi.org/10.1109/DSDIS.2015.42
14. Padmanaban, S., Priyadarshi, N., Holm-Nielsen, J.B., Bhaskar, M.S., Azam, F., Sharma, A. K.: A novel modified sine-cosine optimized MPPT algorithm for grid integrated PV system under real operating conditions. IEEE Access **7**, 10467–10477 (2019). https://doi.org/10.1109/ACCESS.2018.2890533

Chapter 18
Implementation of the Virtual Synchronous Machine in Grid-Connected and Stand-alone Mode

Gurugubelli Vikash, Diksha Funde, and Arnab Ghosh

18.1 Introduction

18.1.1 Motivation and Incitement

It is time to replace the traditional power grid with MG systems because it has a high maintenance cost, hazardous to the environment, and we lost more than 30% of generated power through transmission and distribution systems. Nowadays, the local power generation systems (i.e., photovoltaics) are available and storage systems (i.e., batteries) are also available to store power. So, one can able to manage the low-cost power generation and storage systems, the burden on the conventional power system may reduce. Therefore, MGs will occupy the whole power system within very less time. The main features of these grids are they are integrated with the local generation and increase the reliability (it can operate in both grid-tied and stand-alone mode of operation). The local generators are photovoltaics, diesel generators, wind turbines, etc. Except for diesel generators, remaining all local loads are environmentally friendly. The difficulty with RESs is lack of inertia so that the power quality, power system stability, as well as the operation, get compromised; this can be solved or minimized by applying the proper control strategies using virtual synchronous machine (VSM). With sufficient control mechanisms, VSM will provide damping properties as well as inertia similar to the regular synchronous generator and help the grid to keep maintain stability [1–3].

G. Vikash (✉) · D. Funde · A. Ghosh
Department of Electrical Engineering, National Institute of Technology, Rourkela, Odisha 769008, India

N. Priyadarshi et al. (eds.), *DC–DC Converters for Future Renewable Energy Systems*, Energy Systems in Electrical Engineering,
https://doi.org/10.1007/978-981-16-4388-0_18

18.1.2 Literature Review

Due to the high penetration of RESs in the global power system, the share of energy from the RESs is dominating the share of conventional power. The RESs like fuel cells, PV are not having rotating mass, unlike SM. Therefore, the grid stability, as well as damping performance, gets affected [4]. In the inverter-based system, the main challenge is to incorporate the inverters with the conventional grid system in grid-tied mode and synchronization of the inverters in islanding mode. In the utility grid, the contribution of the RESs is increasing rapidly such that there are changes in the grid instability and the rate of change of frequency (ROCOF) is also high due to the less inertia in the system. The solution to mitigate these problems is to provide sufficient inertia to the system [5–7]. The inertia is embedded in the microcontroller of the inverter control such that the inverter behaves like a VSM, and one can call the inertia is virtual inertia (VI). The VI can be embedded into the system using a short-term ESS like ultra-capacitors, supercapacitors. We can emulate the behavior of the SMs, by controlling the output of the inverter [8–10].

The design and control of different interfacing converters with innovative control methods for RESs, storage systems and microgrid applications are reported in references. The fifth-order model SM-based VSM is proposed in [11, 12]. In [11], the primary stage control is VSM, the second-stage control is current-to-voltage model, and it requires an energy storage system (ESS). In [12], the second stage is voltage to the current model, and it does not require ESS. The seventh-order model SM-based VSM is proposed in [13–15], and it requires ESS. In [16–19], the authors proposed VSM-based droop control to achieve extra benefits added with droop control, and it does not require ESS. In [20–23], the authors proposed the second-order model SM-based VSM, which is nothing but the swing equation-based VSM model. In [20, 21] the VSM is based on the swing equation, and there is no need of PLL, and it requires ESS. In [22, 23], the VSM is based on the swing equation, and it requires PLL and does not require ESS.

The inverter is controlled in such a manner that if any disturbance occurs, it exhibits performance similar to that of a traditional or conventional SM. This conception is well-known as a 'virtual synchronous machine' [24–26]. The VSM model is anticipated to behave like a conventional SM; with a sufficient VI as well as damping, this can be obtained by controlling the inverter such that one can get the desired frequency and the amplitude of the output terminal voltage [27–30]. Therefore, it can give regulation of the system voltage and frequency. Also, the phase-locked loop (PLL) can be removed which results in VI can give support for maintaining the system without losing stability with DGs/RESs [31–33].

18.1.3 Contribution and Paper Organization

This paper validates the aforesaid control strategy in the framework of an MG system as shown in Fig. 18.1 in which the RESs like solar farms and wind farms are the DGs in the system. Sometimes, one can use battery energy storage systems also. In this work, the battery provides DC supply to the inverter, so that no need for a storage system in the system. Here, the inverter converts the DC power into AC power using pulse width modulation (PWM). The PWM signals are generated from the control strategy such that the inverter acts like conventional SM. The steady-state operation of the system with the VSM control technique is almost similar to the well-known droop control, but there is a remarkable difference observed in the ROCOF. The simulation results corroborate the arguments that VSM gives a better performance compared to droop control.

The significant contributions of this work are (a) in this authors clearly explained how the proposed VSM inspired by the conventional SM; (b) the detailed analysis and implementation of VSM in both grid-tied mode and stand-alone mode have presented; (c) in previous papers, authors showed only the rate of change of frequency, but in this paper, authors have presented all the parameters in the MG, and real-time experimental studies are accomplished here.

This work is structured into seven sections. The introduction is the first section, and the second section gives the background of the VSM. VSM control and its implementation are presented in the third section. The fourth section presents the VSM in grid-connected mode and its case studies. The fifth section includes the VSM in islanding mode and its case studies. The sixth section includes real-time experimental results. In section seven, concluding remarks are made on the VSM control strategy.

18.2 Background

18.2.1 Traditional Power System

In an overall conventional power system, the electrical power is produced by the centralized electric power station which is interconnected with the transmission and distribution system, which carries the electric power from generation unit to the load, e.g., industries, home, etc. Most of the power generation plants use fossil fuel like coal, gas, oil, etc., as the primary source of energy, which drives the turbine or engine called a prime mover. An automatic load frequency control loop (ALFC) controls the frequency by controlling the power output of the generator via control valves and speed governor. The secondary control loop of (ALFC) maintains proper adjustment of the frequency and also the active power based on the power–frequency droop characteristics. An AVR is used to regulate the terminal voltage and also the reactive power by controlling the generator field excitation [34].

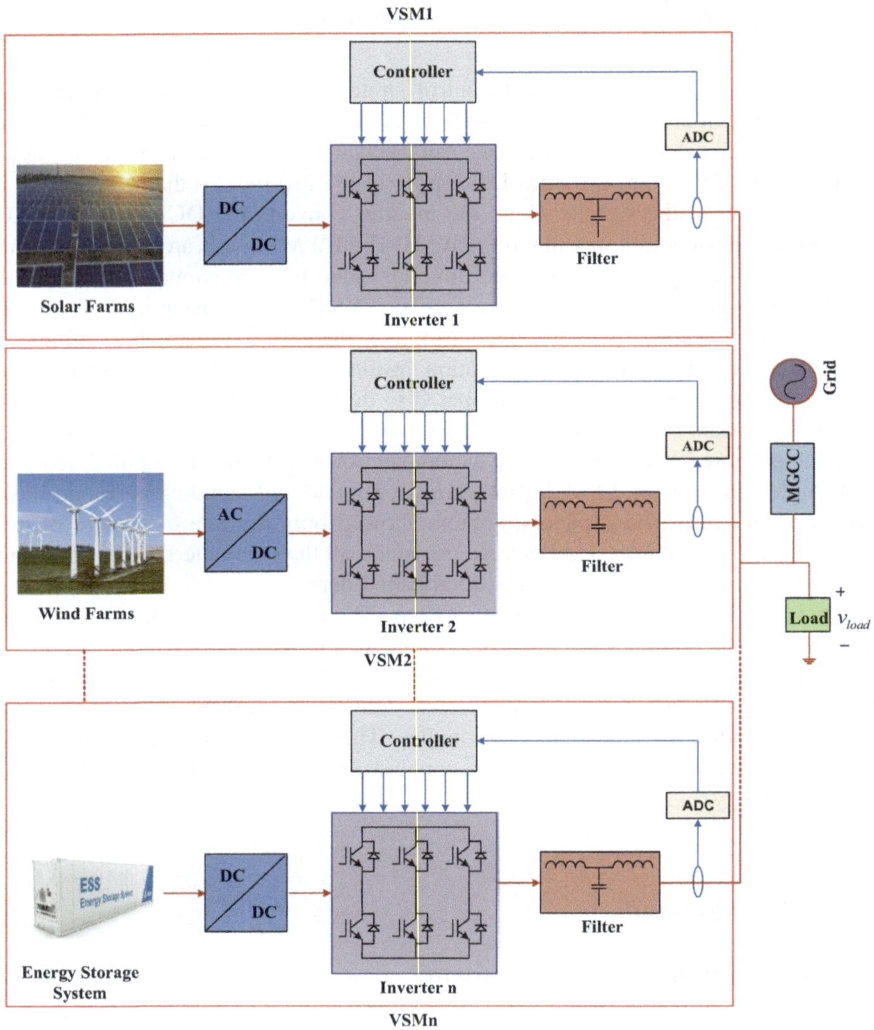

Fig. 18.1 Architecture of MG with several virtual synchronous machines

In the global power system, the bulk synchronous plays a very active role in supporting or maintaining the system operation as well as system stability of the power generation unit. All because of the damping property and the inertia, the unit contributes to the voltage control or the active power control through the rotor excitation control as well as primary frequency control/regulation. These features are not there in the conventional control of renewable energy sources power electronic interfaces, due to lack of inertia.

The power system is having various types of disturbances, some are like a line to line fault, a short circuit between line and ground, the sudden change in demand of load, three-phase faults, etc.; during sudden and small change or disturbance, it is expected that power system will adopt the situation and start to continue operating. However, in case of a major disturbance, like sudden loss or failure of a generator or short circuit in the transmission system could isolate the damaged or faulted part from the remaining or rest of the grid, which causes the structural change in the power system. Various types of stability are categorized in power system study to find out the causes of stability and work on the methods to enhance the stability.

18.2.2 Swing Equation

This is the most important component of the stability of the power system that makes it possible to assess transient power system stability. It explains the relative movement of the rotor to the stator field in the following time function

$$\tau_{\mathrm{m}}(t) - \tau_{\mathrm{em}}(t) = J\frac{\mathrm{d}^2}{\mathrm{d}t^2}\delta_1(t) + D\frac{\mathrm{d}}{\mathrm{d}t}\delta_1(t) \tag{18.1}$$

Here, τ_{m} is the mechanical torque of the SM rotor, τ_{em} is the electromagnetic torque of an SM, δ_1 is the difference between rotor angular position and the reference position, t is the time in second, D is damping constant, J is inertia constant. The inertia of the generator shows a main role in the system stability, as it reacts to the disturbance, which is the most important property of the synchronous generator.

18.3 Virtual Synchronous Machine

18.3.1 Concept of VSM

The VSM's main objective is to embed the dynamic and the inertia properties of a traditional SM in the microcontroller of a power inverter to achieve great system stability. In RESs/DGs units do not have inertia like conventional SM. The transient behavior of the SM gives the possibility of managing the sudden changes in active power and ROCOF of the system due to its inherent capability. With several VSMs, one can achieve a good parallel inverter system that can operate as a conventional parallel SMs system.

The VSM includes an inverter, the ESS and the control topology as given in Fig. 18.2. In the VSM-based DG arrangement, VSM plays a major role to transfer the power from the input DC side to the output AC side. The VI is incorporated to manage the momentary active power mismatch due to sudden changes in the load. There can be no difference between the conventional SM and VSM, excluding the

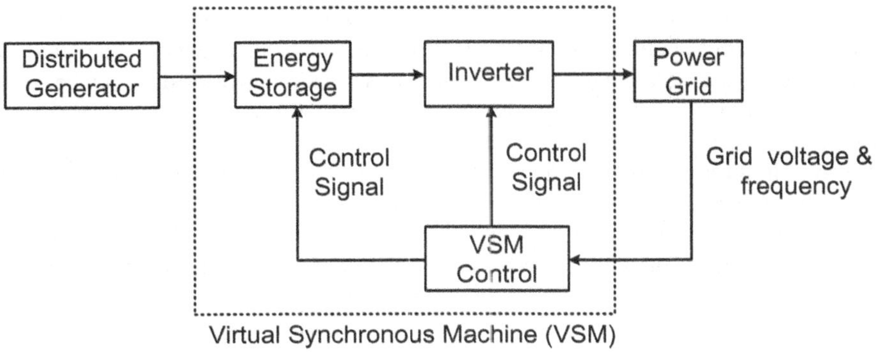

Fig. 18.2 VSM structure

high switching frequency noise due to the PWM techniques used in the inverter control. The dynamic balance is achieved in the VSM due to the swing equation, and the damping properties of traditional SM are embedded in the microcontroller of the power inverter, which defines the output power based on the ROCOF of the system.

18.3.2 VSM Inertia Emulation and Droop Characteristics

VSM idea is taken from the characteristics of the traditional SM to adopt the main characteristics of the conventional SM like the moment of inertia of the rotor, damping property of the rotor winding, and conventional droop characteristics. The investigated VSM implementation is mainly based on traditional swing equation and damping as represented in Eq. (18.2) as follows

$$\frac{\mathrm{d}\omega_{\mathrm{VSM}}}{\mathrm{d}t} = \frac{P^{r^*}}{\tau_{\mathrm{a}}} - \frac{P}{\tau_{\mathrm{a}}} - \frac{P_{\mathrm{d}}}{\tau_{\mathrm{a}}} \tag{18.2}$$

Here, the input mechanical power represented virtually as P^{r^*}, the P is measured electrical power flowing from VSM to the load, P_{D} is the damping power, τ_{a} is the virtual mechanical time constant (equivalent to the 2H in an SM), and ω_{vsm} is the angular velocity in rad/s, and phase angle θ_{vsm} is calculated by the integration of the speed.

The damping power of the VSM represents the damping effect of the traditional synchronous generator, which is well-defined in Eq. (18.3) and the governor model of the SG including with the all above equation becomes Eq. (18.4) as shown below

$$P_{\mathrm{D}} = k_{\mathrm{D}}\left(\omega_{\mathrm{vsm}} - \omega^{*}_{\mathrm{vsm}}\right) \tag{18.3}$$

By including the droop controller, the swing Eq. (18.1) changes as follows:

$$\frac{\mathrm{d}\omega_{\mathrm{vsm}}}{\mathrm{d}t} = \frac{1}{\tau_{\mathrm{a}}}\left[P^{r^{*}} - P \pm K_{\mathrm{D}}\left(\omega_{\mathrm{vsm}} - \omega^{*}_{\mathrm{vsm}}\right) - K_{\mathrm{W}}\left(\omega_{\mathrm{vsm}} - \omega^{*}_{\mathrm{vsm}}\right)\right] \tag{18.4}$$

Here, K_{D} is the damping constant and K_{W} is the droop constant of the VSM. The above equation is incorporated in the controller of the inverter. From Eq. (18.4), we can calculate the ω_{vsm}. The obtained θ_{vsm} is used for controlling the real power by inverter. Finally, the real power is controlled by proper control of load angle of voltage at the output of the inverter. Load angle is controlled by the PWM technique with the help of the θ_{vsm}.

18.3.3 Current Control in the VSM

In the *d-q* setting, the active power and the imaginary power of the system as shown in Fig. 18.3 are explained as follows:

$$P_s(t) = \frac{3}{2}\left[V_{sd}(t)i_d(t) + V_{sq}(t)i_q(t)\right] \tag{18.5}$$

$$Q_s(t) = \frac{3}{2}\left[-V_{sd}(t)i_q(t) + V_{sq}(t)i_d(t)\right] \tag{18.6}$$

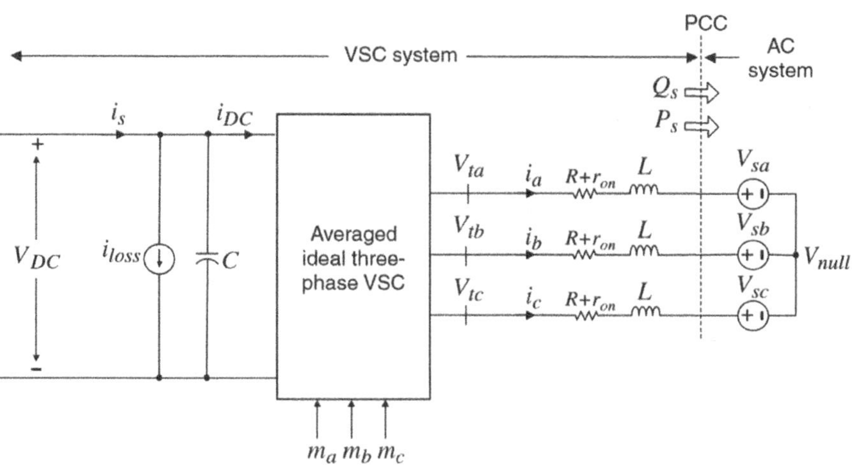

Fig. 18.3 Block diagram of grid-tied VSC

$P_s(t)$ and $Q_s(t)$ are the real and the reactive power, respectively, which are varying to the time. V_{sd} and V_{sq} are the AC voltage component of the *d*-*q* frame. To synchronize the voltage source converter (VSC) with the grid, phase-locked loop can be used, and for such cases, V_{sq} component should be zero. Equations 18.5 and 18.6 can be modified as follows:

$$P_s(t) = \frac{3}{2} V_{sd}(t) i_d(t) \tag{18.7}$$

$$Q_s(t) = -\frac{3}{2} V_{sd}(t) i_q(t) \tag{18.8}$$

From the above equation, it is clear that with the control of the current component of the *d*-*q* frame, one can control real and reactive power.

After application of the KVL in Fig. 18.3, we obtained the following equation:

$$L_f \frac{\mathrm{d}I_d(t)}{\mathrm{d}t} + (R_f + r_{\mathrm{on}}) I_d(t) = V_{td}(t) - V_{sd}(t) + \omega L_f I_q(t) \tag{18.9}$$

$$L_f \frac{\mathrm{d}I_q(t)}{\mathrm{d}t} + (R_f + r_{\mathrm{on}}) I_q(t) = V_{tq}(t) - V_{sq}(t) + \omega L_f I_d(t) \tag{18.10}$$

where, L_f—filter inductance, R_f—internal resistance of filter inductance, r_{on}—ON-state resistance of the switch. V_{td}, V_{sd} and V_{tq}, V_{sq} are the inverter output and the grid voltages in the *d*-*q* reference frame, respectively. The relationship between inverter output voltages in the *d*-*q* frame is represented as follows:

$$V_{td}(t) = \frac{V_{\mathrm{DC}}}{2} m_{1d}(t) \tag{18.11}$$

$$V_{tq}(t) = \frac{V_{\mathrm{DC}}}{2} m_{1q}(t) \tag{18.12}$$

Here, m_{1d} and m_{1q} are the modulation index in the *d*-*q* frame. Dependence between the states, one can say that there is a coupling between the dynamics of the I_d and I_q for the decoupling of the dynamics the modulation indices are introduced, i.e., m_{1d} and m_{1q} as follows

$$m_{1d} = \frac{2}{V_{\mathrm{DC}}} (U_{1d} - \omega L_f I_q + V_{sd}) \tag{18.13}$$

$$m_{1q} = \frac{2}{V_{\mathrm{DC}}} (U_{1q} - \omega L_f I_d + V_{sq}) \tag{18.14}$$

Here, U_{1d} and U_{1q} are the two new control inputs. The modulation indices are m_{1d} and m_{1q} obtained to decouple Eqs. (18.9), (18.10), and replacing them in Eqs. (18.11) and (18.12), for the V_{td} and V_{tq}, we obtained the following equations:

$$L_f \frac{\mathrm{d}I_d}{\mathrm{d}t} = -(R + r_{\mathrm{on}})I_d + U_{1d} \tag{18.15}$$

$$L_f \frac{\mathrm{d}I_q}{\mathrm{d}t} = -(R + r_{\mathrm{on}})I_q + U_{1q} \tag{18.16}$$

Equations (18.15) and (18.16) state that I_d and I_q can be controlled by using U_{1d} and U_{1q}, correspondingly. Figure 18.4 demonstrates the illustration of the d-axis and the q-axis current controllers of VSC systems where U_{1d} and U_{1q} are the output of the two corresponding compensators [33]. The d-axis compensator proceeds like $e_d = i_{d\mathrm{ref}} - i_d$ and gives U_{1d}. Then, based on Eq. (18.13), U_{1d} has been contributed to the m_{1d}. Similarly, the q-axis compensator proceeds like $e_q = i_{q\mathrm{ref}} - i_q$ and gives U_{1q}, which is based on Eq. (18.14) which is contributed to m_{1q}. Further, the VSC amplifies the m_{1d} and m_{1q} by the factor $V_{\mathrm{DC}}/2$ which further generates the V_{td} and V_{tq}, that controls the I_d and I_q based on Eqs. (18.9) and (18.10). Based on the above-discussed control process, one can draw the simplified control block diagram which is shown in Fig. 18.4, which is equivalent to the control process of Figs. 18.3 and 18.5 shows the control plants in both d, as well as q-axis current control loops, are similar. Therefore, the corresponding components can be similar. Now, consider the d-axis control loop; K_d can be a PI compensator for enabling the tracking of the DC reference command.

The compensators are shown in Eq. (18.17), where $K_d(s), K_q(s)$ are P-I compensators

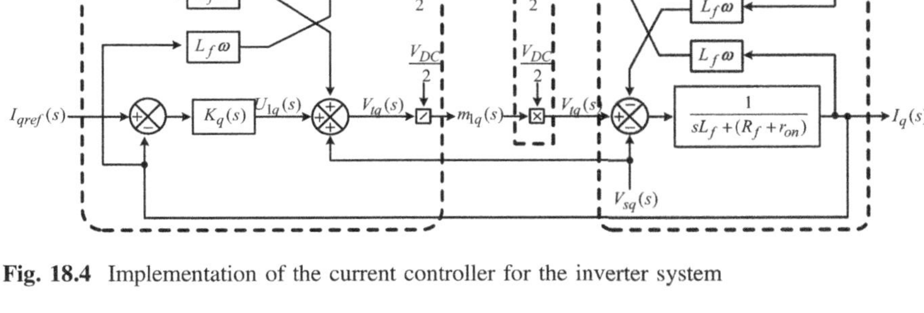

Fig. 18.4 Implementation of the current controller for the inverter system

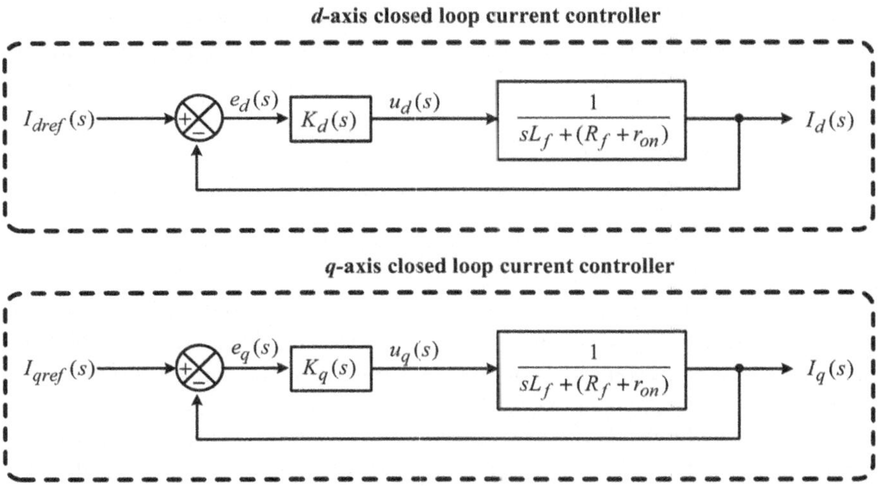

Fig. 18.5 Individual current controller loops in the d-q-axis

$$K_d(s) = K_q(s) = K_{P_c}\left(\frac{s+z_c}{s}\right) = K_{P_c} + \frac{K_{I_c}}{s} \tag{18.17}$$

The open-loop gain of the closed-loop system is shown in Eq. (18.18) and the closed-loop transfer functions are as shown in Eq. (18.19).

$$l_c(s) = \left(K_{P_c} + \frac{K_{I_c}}{s}\right)\left(\frac{1}{sL_f + (R_f + r_{\text{on}})}\right) = \left(\frac{K_{P_c}}{sL_f}\right)\left(\frac{s + \frac{K_{I_c}}{K_{P_c}}}{s + \left(\frac{R_f + r_{\text{on}}}{L_f}\right)}\right) \tag{18.18}$$

$$\left(\frac{I_d(s)}{I_{d_{\text{ref}}}(s)}\right) = \left(\frac{I_q(s)}{I_{q_{\text{ref}}}(s)}\right) = \left(\frac{l_c(s)}{1 + l_c(s)}\right) = \frac{1}{1 + \left(\frac{L_f}{K_{P_c}}\right)s} = \frac{1}{1 + \tau_i s} \tag{18.19}$$

where τ_i is the desired time constant for the closed-loop system, and it can be represented in Eq. (18.20). The choice of the time constant τ_i is the part which is important. The time constant is kept small for the quick current control response. It is selected in the range (0.5–5 ms).

$$\tau_i == \frac{L_f}{K_{P_c}} \tag{18.20}$$

The following are the final expressions for the parameters of the P-I compensator:

$$K_{P_c} = \frac{L_f}{\tau_i}; \quad K_{I_c} = \frac{R_f + r_{\text{on}}}{\tau_i} \tag{18.21}$$

18.3.4 VSM Control Implementation

VSG control realized in a single inverter is as shown in Fig. 18.6. The power measurement block can measure the average real power and imaginary power from the bus voltage and the output current. From Eq. (18.4), one can generate the reference angular displacement. The reference voltage signal is generated from the voltage-reactive power droop law. Finally, using the reference angular displacement signal and the output from the voltage control droop law, the pulses to the inverter are generated. The output of the inverter is joined to the load via an LCL filter, which eliminates/removes harmonics and regulates the shape of the output waveforms.

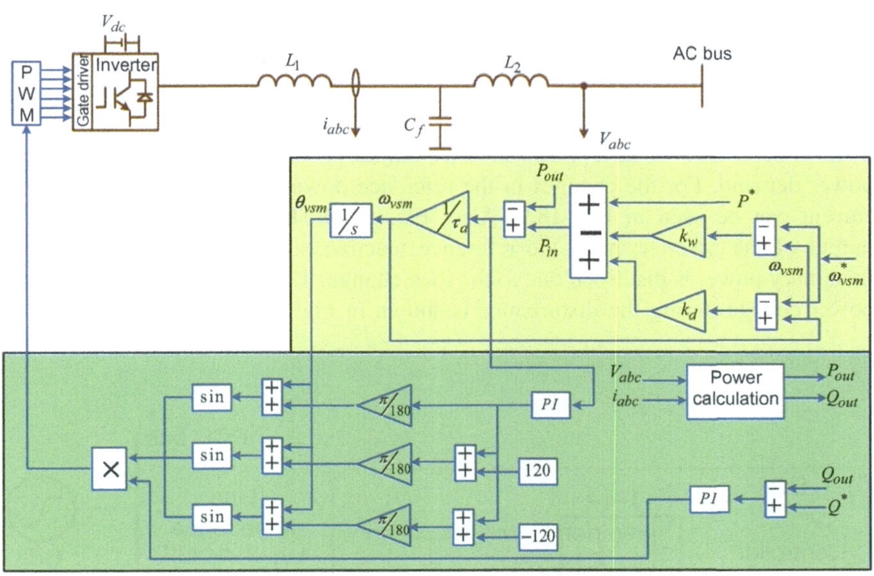

Fig. 18.6 Virtual synchronous machine control implementation

18.4 VSM in Grid-Tied Mode

The main challenges in the grid-connected system are to integrate the inverters to the grid and control the real and imaginary power of the inverters. The investigated VSM control arrangement for the VSC differs from the traditional control arrangement for the VSC. In the VSM control arrangement, the inverter will act as a conventional SM. This control structure is mainly founded on the swing equation as well as the damping of a traditional synchronous machine.

18.4.1 Case Studies

Figure 18.7 shows the architecture of the simulation of grid-tied VSM. This consists of a DC source as the input to the inverter, a filter connecting the output of the inverter to the grid. The filter used here is the LCL filter which is used to eliminate the harmonics. The controlling of VSM is done here to track the real and imaginary power changes due to unpredictable load variations. The controller implementation is showed in Fig. 18.6.

The grid voltage (phase to phase) is set as the desired voltage, Fig. 18.8 shows the inverter output voltage in the grid-tied mode of operation when the inverter is controlled by the VSM control strategy. Figure 18.9 shows that the current fed to the grid from the inverter with various reference powers with the VSM control strategy. From Fig. 18.10, one can observe that the reference real power is changing from 0 to 500 W at the instant of 2 s and from 500 to −300 W at 4 s and from −300 to 500 W at 6 s. The inverter output power is trying to meet the load power demand. For the changes in the reference power, corresponding changes in current can be seen in Fig. 18.9. Also, Fig. 18.11 shows that imaginary power output of the inverter; here, the reference reactive power is zero, but the actual imaginary power is disturbed due to the load change. The variation in the imaginary power change during the disturbance is shown in Fig. 18.11.

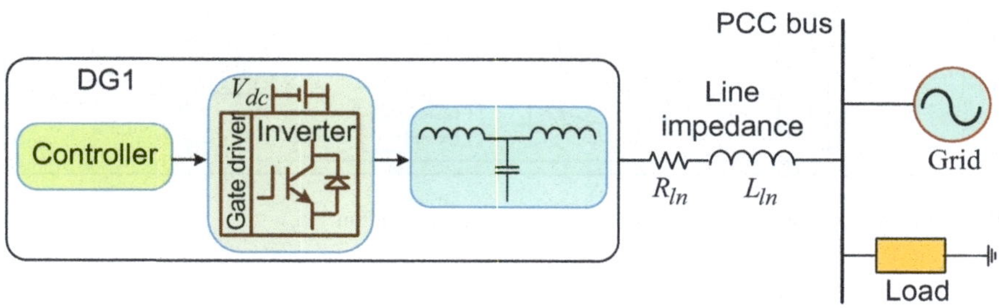

Fig. 18.7 Architecture of the model of the grid-connected VSM

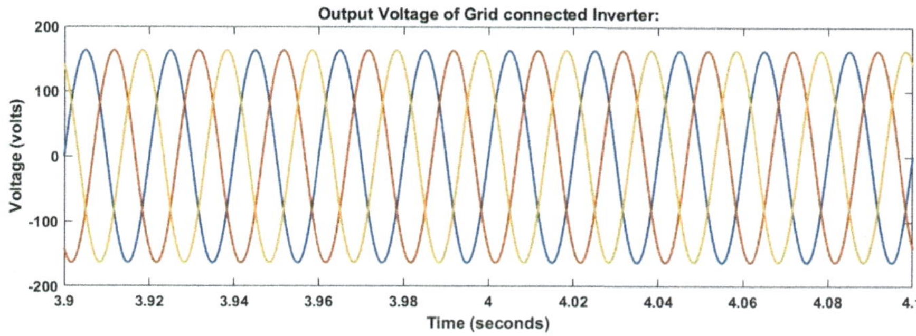

Fig. 18.8 Output voltage of the grid-tied inverter with VSM control strategy

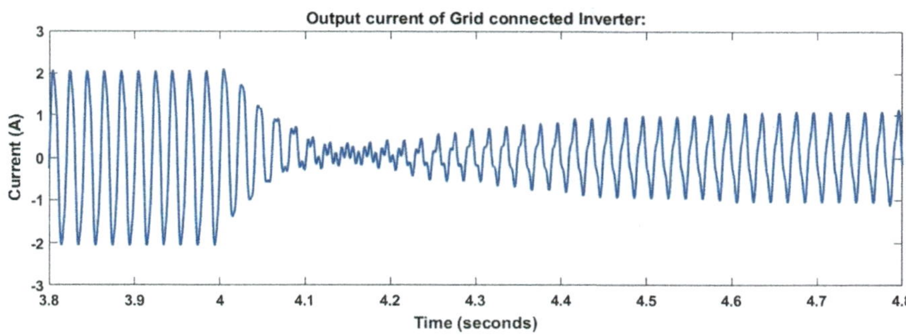

Fig. 18.9 Inverter output current with VSM control strategy for various reference powers in the grid-tied mode

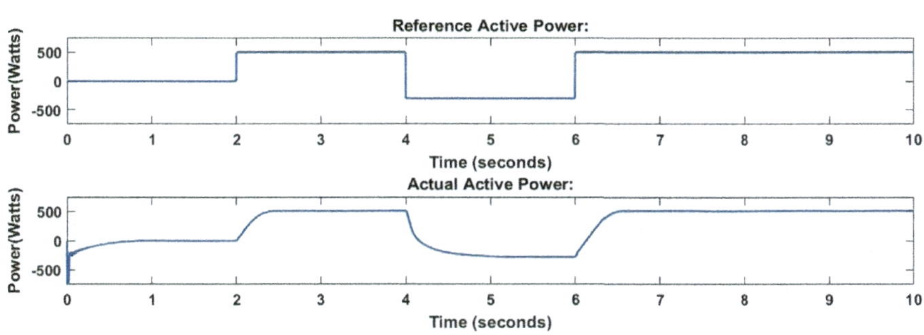

Fig. 18.10 Active power tracking with VSM

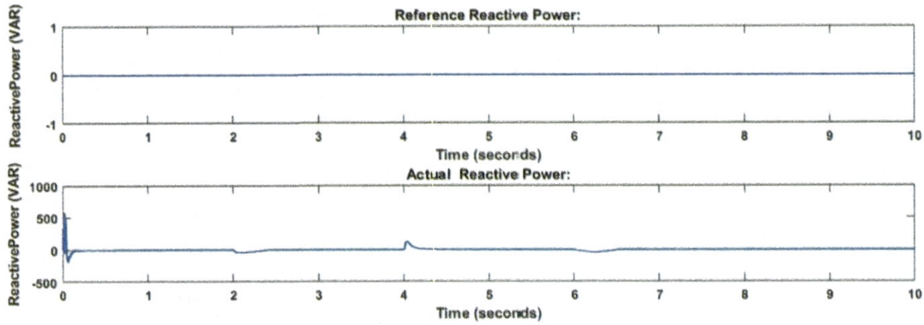

Fig. 18.11 Imaginary power tracking of the inverter with the VSM control strategy

18.5 VSM in Stand-alone Mode

In this section, one can understand the performance of the VSG control structure in islanding mode. In the islanding mode, the main challenge is to synchronize the inverters which are connected in parallel. For proper synchronization of the inverters, we need to maintain the output voltage magnitude of each inverter should be the same and the frequency of each inverter kept constant.

18.5.1 Case Studies

Figure 18.12 shows the simulation model of the VSM in islanding mode. This consists of a DC source as the input to the inverter and an LCL filter connecting the output of the inverter which is used to eliminate the harmonics. VSM control strategy is implemented here to control the output voltage and frequency of the inverter with the sudden change in load change. Here two inverters are connected in parallel to share the load according to their power ratings. The controller implementation is showed in Fig. 18.6.

When the extra load is added for 0.4–0.6 s, the changes occur in power as well as current with unchanged voltage. The output voltage of the inverters is synchronized, and the output voltage of the inverter is shown in Fig. 18.13a. The load current is the sum of the individual inverter currents which are participating in parallel operation in which the current sharing of the inverts depends upon load. The load current of the inverters operated in parallel is shown in Fig. 18.13b.

The current sharing depends on the droop coefficients in VSM controllers. The dynamic current sharing is very good in VSM compared to the well-known droop control. The main advantage of the VSM is a good dynamic performance compared to droop control. The current sharing in the system using VSM is shown in Fig. 18.14. In stand-alone mode, the real power is changing from time to time; accordingly, the actual power also tries to meet the load power. The output power

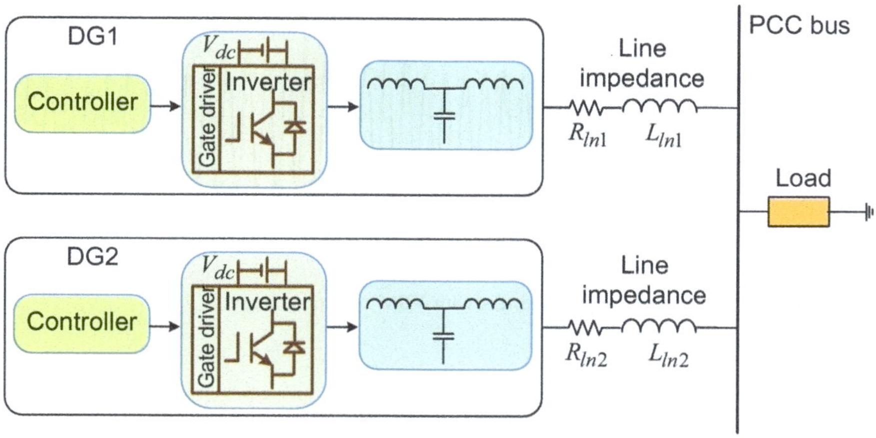

Fig. 18.12 Block diagram of the model of VSM in stand-alone mode

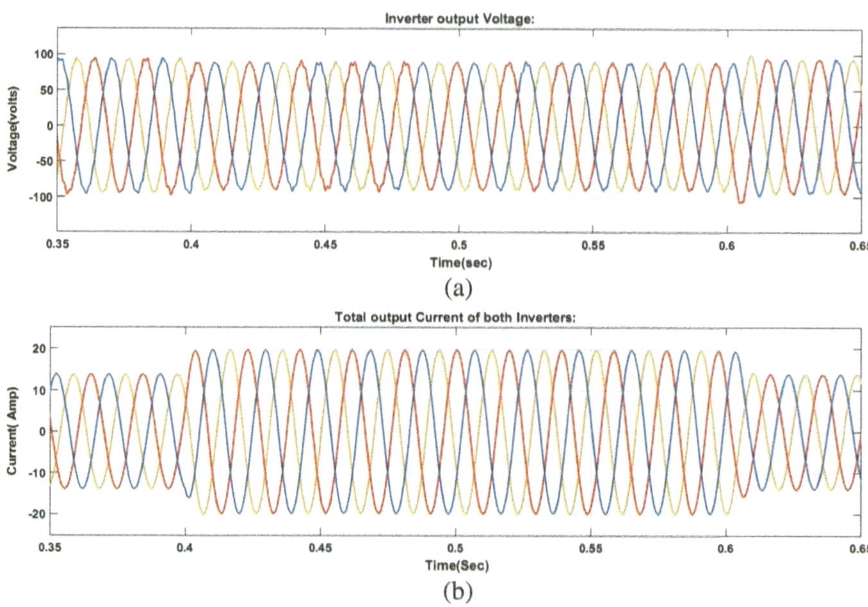

Fig. 18.13 **a** Load voltage and **b** load current of the inverter

of the inverters follows the load changes from time to time. If one can change the inertia of the system, the performance of the inverter is also changed. The active power and imaginary power of the load are shown in Figs. 18.15 and 18.16

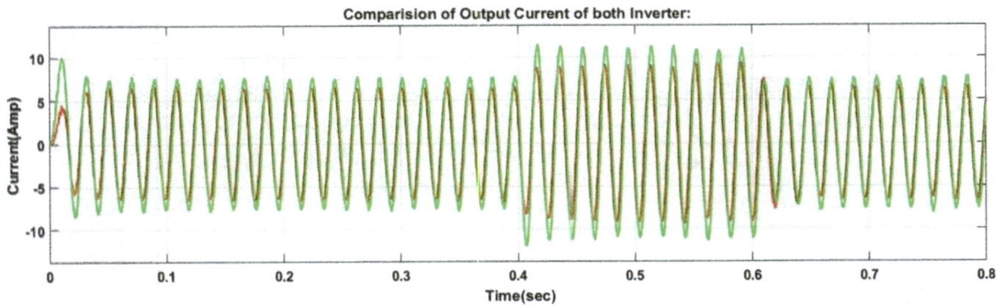

Fig. 18.14 Current sharing in the inverters with VSM control strategy

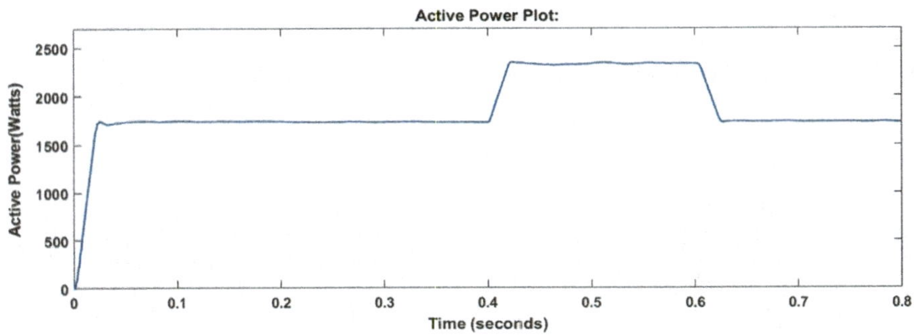

Fig. 18.15 Output real power of load

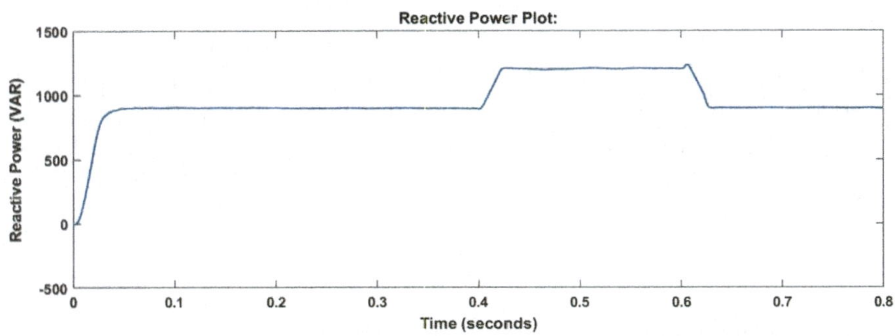

Fig. 18.16 Output reactive power of load

exclusively. The main advantage of the VSM control scheme is ROCOF in the system is decreased due to the virtual inertia of the inverters. The frequency of the inverters is shown in Figs. 18.17 and 18.18, respectively.

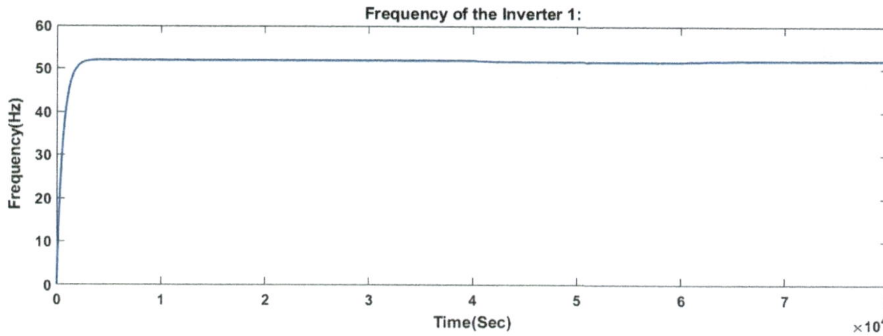

Fig. 18.17 Inverter 1 output frequency

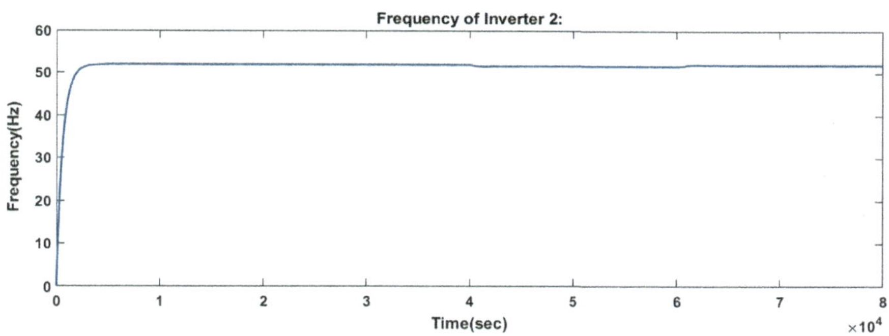

Fig. 18.18 Inverter 2 output frequency

18.6 Conclusion

A simulation of the control of the power inverter with VSM control strategy, operating in the grid-tied as well as islanding mode, has been developed. Without rotating inertia, these inverters based systems cause instability issues. The VSM control structure is applied to the inverter of the DG unit which is operating in both grid-connected as well as islanding mode for supporting the system stability by giving the similar behavior of the SM. The VSM combines the swing equation of the SM to express the term VI. By using this VSM controller, the system stability increases because of the inertia of the system increased. Since the generated power from RESs/DGs is not suitable for the direct grid connection due to the absence of VI, if we incorporate the VI into the power inverters, then one can easily integrate the DGs into MG. There is no such confirmation of the proper satisfactory operation of a grid for supporting inverter operates as both grid forming and the grid feeding inverter when both real and reactive loads are considered. The VSM idea connected with inverter-based DGs is the solution to such problems. In the VSM, the inverter is controlled in such a way so that it can easily emulate the behavior similar to the SM efficiently to obtain stability and controllability. The VSM control strategy is efficiently working for active and reactive power tracking in grid-tied mode and also capable to work in stand-alone mode. The simulation results corroborate that VSM gives a superior performance.

References

1. Lamell, J.O., Trumbo, T., Nestli, T.F.: Offshore platform powered by the new electrical motor drive system. Petroleum and Chemical Industry Conference, Industry Applications Society 52nd Annual, pp. 259–266 (2005).
2. Rocabert, J., Luna, A., Blaabjerg, F., Rodriguez, P.: Control of power converters in AC microgrids. IEEE Trans. Power Electron. **27**, 4734–4749 (2012).
3. Bevrani, H., Ise, T., Miura, Y.: Virtual synchronous generators: A survey and new perspectives. Int. J. Electr. Power Energy Syst. **54**, 244–254 (2014).
4. Blaabjerg, F., Teodorescu, R., Liserre, M., Timbus, A.: Overview of control and grid synchronization for distributed power generation systems. IEEE Trans. Ind. Electron. **53**, 1398–1409 (2006)
5. Beck, H.P., Hesse, R.: Virtual synchronous machine. In: Proceedings IEEE 9th International Conference Electrical Power Quality Utilisation, Barcelona, Spain, pp. 1–6 (2007)
6. D'Arco, S., Suul, J.A., Fosso, O.B.: Small-signal modeling and parametric sensitivity of a virtual synchronous machine in islanded operation. Int. J. Electr. Power Energy Syst. **72**, 3–15 (2015)
7. Yong, C., Hesse, R., Turschner, D., Beck, H.P.: Improving the grid power quality using virtual synchronous machines. In: 2011 IEEE International Conference on Power Engineering, Energy and Electrical Drives, pp. 1–6 (2011)
8. Alipoor, J., Miura, Y., Ise, T.: Power system stabilization using virtual synchronous generator with alternating moment of inertia. IEEE J. Emer. Selec. Topic. Power Electron. **3**(2), 451–458 (2014)
9. Van Wesenbeeck, M.P.N., de Haan, S.W.H., Varela, P., Visscher, K.: Grid tied converter with virtual kinetic storage. In: PowerTech, 2009 IEEE Bucharest, pp. 1–7 (2009)
10. Zhong, Q.C., Nguyen, P.L., Ma, Z., Sheng, W.: Self-synchronized synchronverters: inverters without a dedicated synchronization unit. IEEE Trans. Power Electron. **29**, 617–630 (2014)
11. Chen, Y., Hesse, R., Turschner, D., Beck, H.P.: Comparison of methods for implementing virtual synchronous machine on inverters. In: International Conference on Renewable Energies and Power Quality. pp. 414–424 (2012).
12. Chen, Y., Hesse, R., Turschner, D., Beck, H.P. : Dynamic properties of the virtual synchronous machine (VISMA). Proc. ICREPQ, **11**, (2011).
13. Hesse, R., Turschner, D., Beck, H.P.: Micro grid stabilization using the virtual synchronous machine (VISMA). In: Proceedings of the International Conference on Renewable Energies and Power Quality (ICREPQ'09). Valencia, Spain, pp. 15–17 (2009).
14. Pulendran, S., Tate, J.E.: Hysteresis control of voltage source converters for synchronous machine emulation. In: 2012 15th International Power Electronics and Motion Control Conference (EPE/PEMC). pp. LS3b-2, IEEE (2012).
15. Pogaku, N., Prodanovic, M., Green, T.C.: Modeling, analysis and testing of autonomous operation of an inverter-based microgrid. IEEE Trans. Power Electron. **22**, 613–625 (2007).
16. Guerrero, J. M., Matas, J., De Vicuna, L.G.D.V., Castilla, M., Miret, J.: Wireless-control strategy for parallel operation of distributed-generation inverters. IEEE Trans. Ind. Electron. **53**, 1461–1470 (2006).

17. Yao, W., Chen, M., Matas, J., Guerrero, J.M., Qian, Z.M.: Design and analysis of the droop control method for parallel inverters considering the impact of the complex impedance on the power sharing. IEEE Trans. Ind. Electron. **58**, 576–588 (2010).
18. Vasquez, J.C., Guerrero, J.M., Luna, A., Rodríguez, P., Teodorescu, R.: Adaptive droop control applied to voltage-source inverters operating in grid-connected and islanded modes. IEEE Trans. Ind. Electron. **56**, 4088–4096 (2009).
19. D'Arco, S., Suul, J.A.: Equivalence of virtual synchronous machines and frequency-droops for converter-based microgrids. IEEE Trans. Smart Grid. **5**, 394–395 (2013).
20. D'Arco, S., Suul, J.A., Fosso, O.B.: Automatic tuning of cascaded controllers for power converters using eigenvalue parametric sensitivities. IEEE Trans. Ind. Appl. **51**, 1743–1753 (2014).
21. Delille, G., Francois, B., Malarange, G.: Dynamic frequency control support by energy storage to reduce the impact of wind and solar generation on isolated power system's inertia. IEEE Trans. Sustain. Energy. **3**, 931–939 (2012).
22. Soni, N., Doolla, S., Chandorkar, M.C. (2016). Inertia design methods for islanded microgrids having static and rotating energy sources. IEEE Trans. Ind. Appl. **52**, 5165–5174 (2016).
23. Vikash, G., Ghosh, A., Rudra, S.: Integration of distributed generation to microgrid with virtual inertia. In: IEEE 17th India Council International Conference (INDICON). IEEE. (2020).
24. Zhang, L., Harnefors, L., Nee, H.P.: Power-synchronization control of grid-connected voltage-source converters. IEEE Trans. Power Syst. **25**, 809–820 (2010).
25. Panda, S.K., Ghosh, A.: A computational analysis of interfacing converters with advanced control methodologies for microgrid application. Technol. Econ. Smart Grids Sustain. Energy. **5**(1), 1–18 (2020).
26. Jiang, K., Su, H., Lin, H., He, K., Zeng, H., Che, Y.: A practical secondary frequency control strategy for virtual synchronous generator. IEEE Trans. Smart Grid. **11**, 2734–2736 (2020).
27. Chen, J., O'Donnell, T.: Parameter constraints for virtual synchronous generator considering stability. IEEE Trans. Power Syst. **34**, 2479–2481 (2019).
28. Karimi, A., Khayat, Y., Naderi, M., Dragičević, T., Mirzaei, R., Blaabjerg, F., Bevrani, H.: Inertia response improvement in AC microgrids: A fuzzy-based virtual synchronous generator control. IEEE Trans. Power Electron. **35**, 4321–4331 (2019).
29. Shi, K., Song, W., Ge, H., Xu, P., Yang, Y., Blaabjerg, F.: Transient analysis of microgrids with parallel synchronous generators and virtual synchronous generators. IEEE Trans. Energy Convers. **35**, 95–105 (2019).
30. Cheema, K.M.: A comprehensive review of virtual synchronous generator. Int. J. Electr. Power Energy Syst. 120:106006 (2020).
31. Asrari, A., Mustafa, M., Ansari, M., Khazaei, J.: Impedance analysis of virtual synchronous generator-based vector controlled converters for weak AC grid integration. IEEE Trans. Sustain. Energy. **10**, 1481–1490 (2019).
32. Liu, J., Miura, Y., Bevrani, H., Ise, T.: A unified modeling method of virtual synchronous generator for multi-operation-mode analyses. IEEE J. Emer. Selec. Topic. Power Electron. (2020).
33. Muftau, B., Fazeli, M., Egwebe, A.: Stability analysis of a PMSG based virtual synchronous machine. Electr. Power Syst. Res. **180**, 160–170 (2020).
34. Yazdani, A., Iravani, R.: Voltage-sourced converters in power systems: modeling, control, and applications. John Wiley & Sons. (2010).

Chapter 19
A DC–DC Converter for Green Energy Applications

M. Nandhini Gayathri

19.1 Introduction

Electrical loads are energized from the renewable energy sources instead of the utility line, known as a stand-alone system [1–3]. As the stochastic nature of energy sources, batteries are used to assure the reliability of the power supply as shown in Fig. 19.1.

The renewable sources can be solar, wind or the combination of multiple renewable sources [4, 5]. The combination of two or more sources is known as hybrid sources [6–8]. Hybrid sources are preferred in stand-alone applications to increase the system efficiency and balance the supply–demand. In a stand-alone system, the electrical load is fixed or bounded. So, there is a need for maintaining supply–demand balance [9, 10]. It is achieved with the help of an energy storage system made up of battery [11–13]. The maximum power delivery and maximum power consumption are fixed for renewable sources and electrical loads, respectively [14, 15]. But, the change in weather conditions causes a change in maximum power delivery from the renewables sources [16–18]. The characteristic waveform represents the variation in the power delivery in terms of voltage for different weather conditions as shown in Fig. 19.2 for a typical PV module 1Soltech STH-215-P [19]. Similarly, the random use of electrical loads also results in variation in electrical power consumption.

In this condition, to transfer maximum power from the source and to supply the loads with constant voltage, battery and DC–DC converters are employed with maximum power point tracking (MPPT) technique.

The battery charging condition is represented by the term 'state of charge' (SOC). It denotes the present capacity as a function of the rated capacity. The SOC value ranging from 0 to 100% represents no charge to full charge,

M. Nandhini Gayathri (✉)
Department of Electrical and Electronics Engineering, Schoolof Electrical and Electronics Engineering, SASTRA Deemed University, Tirumalaisamudrum, Thanjavur 613401, India

N. Priyadarshi et al. (eds.), *DC–DC Converters for Future Renewable Energy Systems*, Energy Systems in Electrical Engineering,
https://doi.org/10.1007/978-981-16-4388-0_19

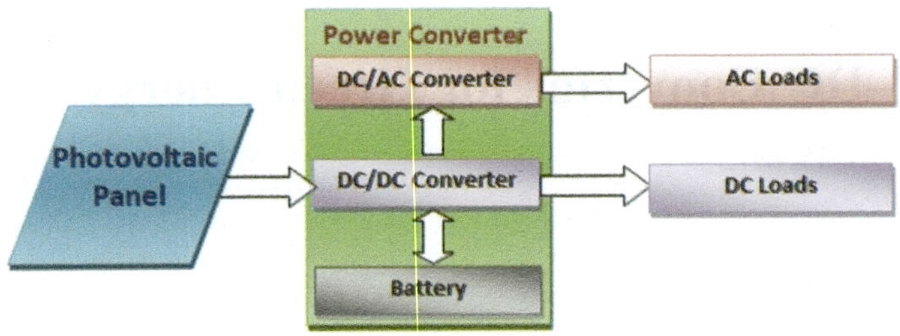

Fig. 19.1 Generalized block diagram of a PV stand-alone system

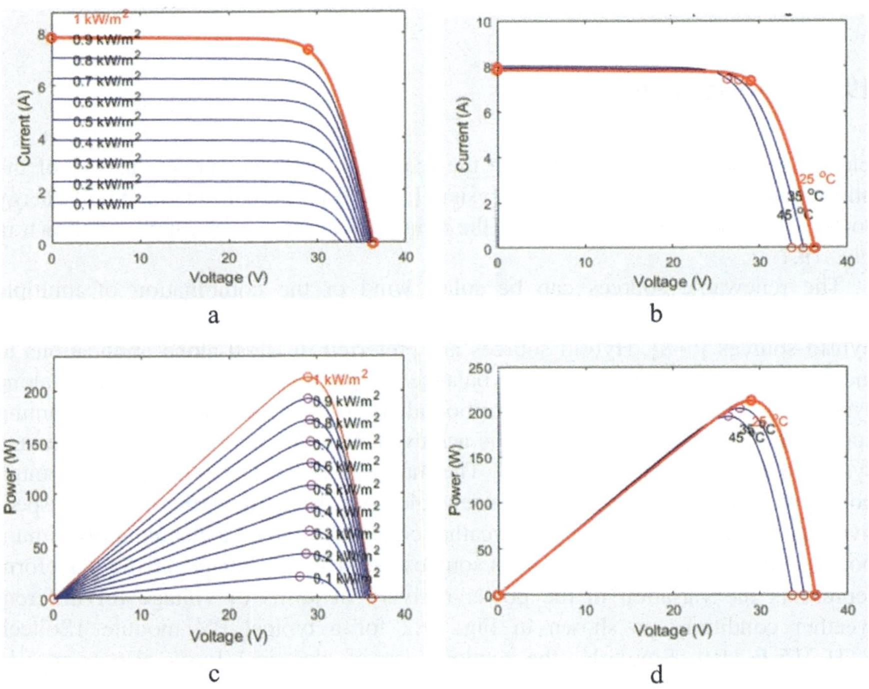

Fig. 19.2 IV and PV characteristics of 1Soltech 1STH-215-P module, **a** IV characteristics for different solar irradiance, **b** IV characteristics for different temperature, **c** PV characteristics for different solar irradiance and **d** IV characteristics for different

respectively. In practical applications, SOC would be equivalent to a 75–80% SOC of a new one. The battery charging cycle involves constant current (CC) and voltage source (CV). The constant current, constant voltage and SOC of a typical battery [16] are shown in Fig. 19.3. In the constant current stage, the battery is connected to a current-limited power supply. The current is set to 50–70% of the nominal battery capacity, and it is enabled until the battery charge is about 80%.

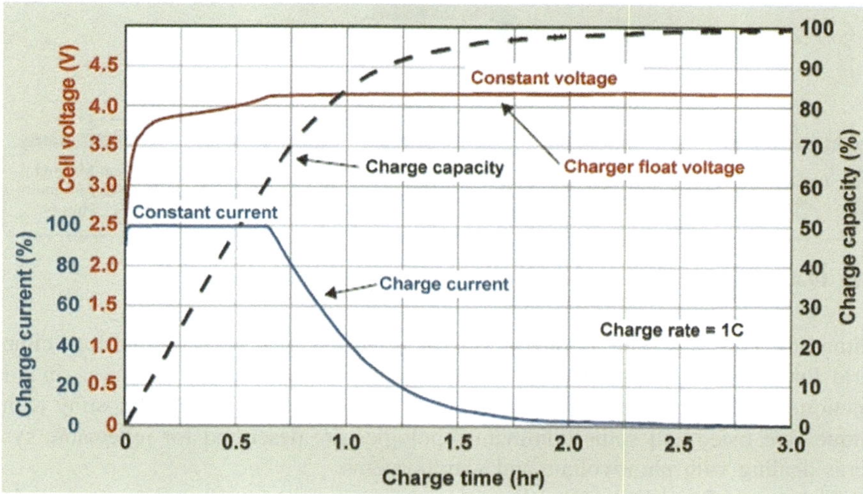

Fig. 19.3 Battery charging curve

Then, the CV mode is enabled and in this mode, the charger acts as a voltage-controlled power supply. The charging current lies in between 3 and 10% of the rated capacity. Now, the battery is treated as fully charged.

Based on the battery SOC and power supply–demand the charging, discharging takes place. If the renewable sources power output is equal to load demand, then there is no charging or discharging takes place on the battery. Mostly, the renewable sources power output is always different from the loads. Hence, there is a power difference between these two always. If excess power output available from the source, then that can be used to charge the battery.

In case of a fully charged battery, the excess power available from the renewable sources cannot be utilized completely. Similarly, when there is a need for additional power to drive the load apart from renewable sources, then battery power can be used to supply. If there is no battery power, then the loads are operated with reduced power rating, i.e., the reduced number of electrical appliances.

19.2 Power Conversion for Renewable Systems: State of the Art

A comprehensive review of step-up/boost DC–DC converter is reported in the literature survey (1) with voltage boosting technique. The DC–DC converters are categorized into four types based on the different aspect as shown in Fig. 19.4. Some of them are high-voltage transfer ratio DC–DC converters, but some others are medium and high-power transfer ratio DC–DC converters. Based on the power

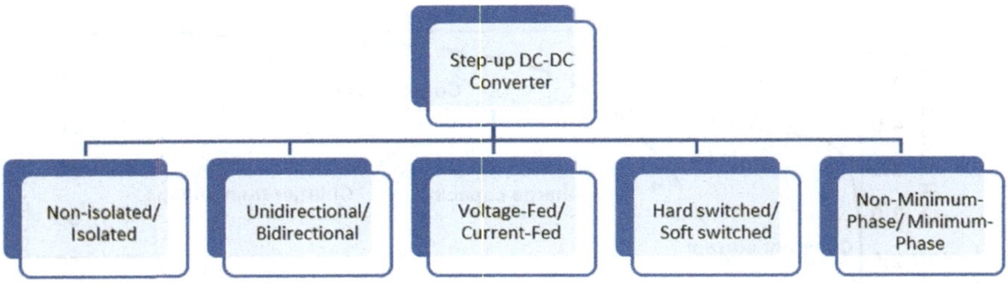

Fig. 19.4 Classification of step-up DC–DC voltage converter topologies

rating, the corresponding topologies are preferred. Irrespective of output load either grid line or stand-alone systems, all the topologies are used in common. In this section, a different type of step-up DC–DC converters and voltage boosting techniques are listed and some dominant topologies are described for renewable systems dealing with photovoltaic and wind systems.

The non-isolated DC–DC voltage converter is made up of energy storage element like an inductor, capacitor with power semiconductor switch. The major advantage of this type is its size and weight. Single-stage energy conversion results in good power conversion efficiency. But the absence of galvanic isolation between input and output limits its voltage and power transfer ratio. In isolated DC–DC converter circuit, an isolation transformer used to provide the galvanic isolation. It serves the step-up operation also. To use the transformer, AC-link is used in between DC input and DC output. So, the number of stages involved in the conversion becomes two. High-frequency AC-link is preferred to keep the transformer size minimum. The two-stage conversion provides good voltage gain and power gain when compared with non-isolated converters. The voltage transfer ratio can be increased by adding voltage boosting circuits like voltage multiplier.

Unidirectional power converters allow power flow in one direction, i.e., from load to source. In this converter type, the unidirectional controllable switch is used. It will not allow the power flow in the reverse direction. It is simple in construction and control and suitable for the power sources that cannot receive the power back like grid and battery, whereas the bidirectional DC–DC power converter allows bidirectional power flow control. It allows the power flow in both directions. So, the excess power from the load can be returned to grid; similarly power from the renewable sources can be stored in the battery, and it can be utilized back using the same converter. It supports regenerative braking in electric drives. It increases the utilization of available sources more effectively. Various control techniques are used to perform the bidirectional power flow. The only drawback of the bidirectional power converter is an increase in cost, size and control complexity.

The bulky capacitor and inductor are connected to the sources, and the sources are known as a voltage source and current source, respectively. So, based on the input circuitry, the converters are classified into voltage-fed and current-fed converters. Voltage-fed converters can be used to supply constant output voltage, and current-fed converters are used to supply constant current output, respectively.

For low-power PV applications, current-fed DC–DC converters are recommended to supply ripple-free current at the input side. Current-fed DC–DC converters overcome the shoot-through problem of VSI. The major features of different types of DC–DC converter are summarized in Table 19.1.

To inject the power to the grid, the voltage level of the renewable sources should be increased. It can be achieved by either a step-up transformer or voltage boost/step-up converters. The energy storage elements like inductor and capacitor are the key elements used to increase the voltage/current magnitude. Inductors and or capacitors used with active devices like diode and power transistor. The active power semiconductor switches are operated in such a way to boost the input voltage. These methods eliminate the need for transformers and high-frequency AC-link. The different switching topologies used in practice are shown in Fig. 19.5.

Table 19.1 Features of various DC–DC voltage converter circuits

DC-DC converter type	Features
Non-isolated	• Uncomplicated structure and low price • Good enough for low and medium power levels • No galvanic isolation
Isolated	• Low noise and EMI problems • Good enough for high-power and grid-connected applications • Galvanic isolation • Complex design
Unidirectional	• Unidirectional power flow • Uncomplicated construction and control • Low price compared to bidirectional
Bidirectional	• Bidirectional power flow • Good enough for regenerative and energy storage application • Complex driver and control circuitry
Voltage fed	• Ripple current input (often discontinues) • Built-in buck characteristics • Good dynamic response
Current fed	• Low ripple • Built-in boost characteristics • Poor dynamic response
Hard switched	• Large switching less and high EMI • Finite switching frequency • Poor power density and efficiency
Soft-switched	• Negligible switching loss (ZVS and ZCS) • Partly complex • High operating frequency • Enhanced power density and efficiency
Non-minimum phase	• Poor dynamic response • Small stability margins • Complex control
Minimum phase	• Good dynamic response • Large stability margins • Uncomplicated design

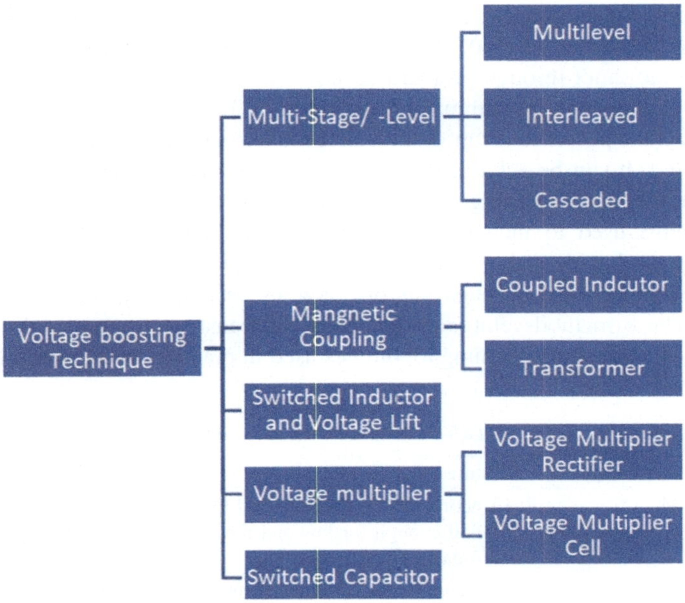

Fig. 19.5 Voltage boosting techniques

Table 19.2 Features of various DC–DC voltage converter topologies

Voltage boosting techniques	Advantages	Limitations
Switched capacitors (Charge pump)	• Small size, lightweight and low cost • Higher power density and good dynamic response	• Inrush start-up current • Sensitive to the ESR
Voltage multiplier	• Higher voltage transfer ratio • Modular structure • Easy to integrate with various structure	• High voltage (dv/dt) stress on components • More components and bulky structure
Switched inductor and voltage lift	• High voltage transfer ratio • Easy to use in many converters	• Increase in passive components count • Low power handling capability
Magnetic coupling	• High design freedom • Versatile to boost the ability • Reduced conduction loss • High efficiency	• Large voltage spike due to leakage inductance • Complex design • Bulky structure
Multi-stage/-level	• Modular structure • High voltage and power capability	• Bulky structure and expensive • The trade-off between efficiency and the number of stages/levels

The voltage and power transfer ratio varies for different techniques available, and it determines the suitable application. The advantages, limitations and the suitable renewable energy applications are listed in Table 19.2.

19.2.1 Grid-Connected PV Systems

Gird-connected systems provide the required power to the AC mains obtained from the photovoltaic (PV) and/or wind system (2). Different control techniques are employed to transfer maximum power from the renewables sources. Some of them are improved perturbation and observation method (3), sliding mode observer technique (4), others (5). The system efficiency is increased by operating the solar cell array (6) and also some wind systems on its maximum power delivery point. Figure 19.6a, b shows the topology of the grid-tied inverter with the two-stage conversion. The front-end is DC–DC step-up converter, and the back-end is the DC–AC inverter. The first stage step-up converter adjusts the input in such a way to extract the maximum power by running the MPPT algorithm.

The second stage is the grid-tied DC–AC inverter and is operated by grid parameters to control the active and reactive power flow. In Fig. 19.6a, basic inverter is presented, and in Fig. 19.6b, multilevel inverter is presented with an isolation transformer. The multilevel inverter produces output voltage with reduced harmonic content when compared with basic inverter topology.

Multi-input converters (8) are the most recommended topology for combining multiple renewable sources and or battery energy storage system. It combines the available power from the multiple sources and provides uninterruptable/sustainable power output. Figure 19.7a, b shows multiple input DC/DC converters of non-isolated (9) and isolated (10) type, respectively. Both types have step-up/step-down capability and allow the two input sources at different voltage levels.

19.2.2 Stand-alone Systems

For stand-alone systems, there is a need for battery energy storage to provide uninterruptable power supply from renewable energy sources. The power available from the source, power supply demand and battery SOC determine the charging/discharging condition of the battery.

Figure 19.8a–c shows the multi-input DC–DC converter (11) for stand-alone systems with bidirectional power flow. It supports bidirectional power flow control for the battery so that the battery can be charged through the converter from the renewables sources and at the same time the charged battery power can be transferred to the load using the same converter in the absence of renewable energy source. In Fig. 19.8a, b, the battery is continuously charged and results in deterioration of its useful life. But in Fig. 19.8c, topology supports the power transfer

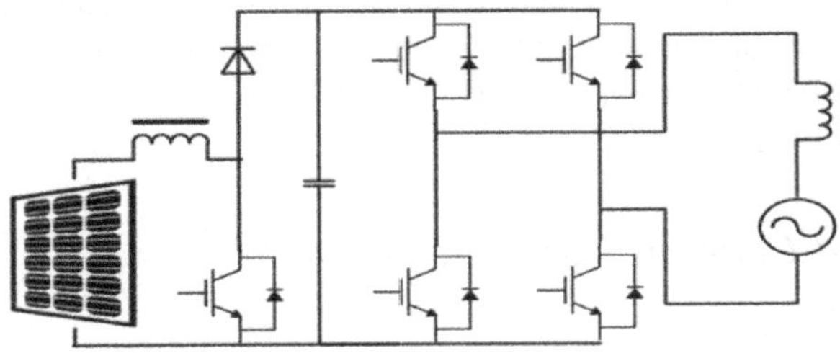

a

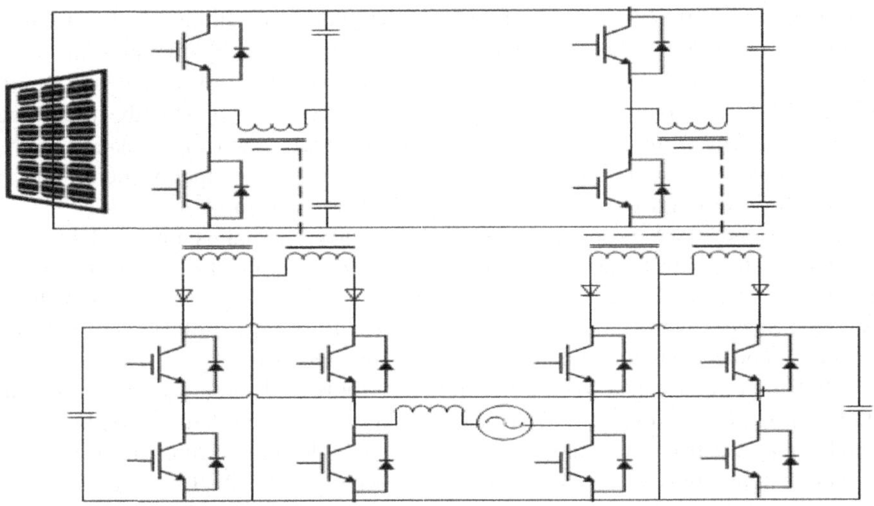

b

Fig. 19.6 Grid-connected DC–DC converter topologies, **a** DC–DC converter and inverter and **b** multiple DC–DC converters and multilevel inverter

from the sources independently. The switching combination determines the selection of power source, whether renewable source or battery. The PWM signal is used to regulate the output voltage and power transfer.

19.3 MATLAB Simulation of a DC–DC Converter for Renewable Energy System

In MATLAB demo, the functionality of the DC–DC step-up converter and grid-connected 100 kW PV array is simulated mathematically. Both the simulation circuits are discussed below.

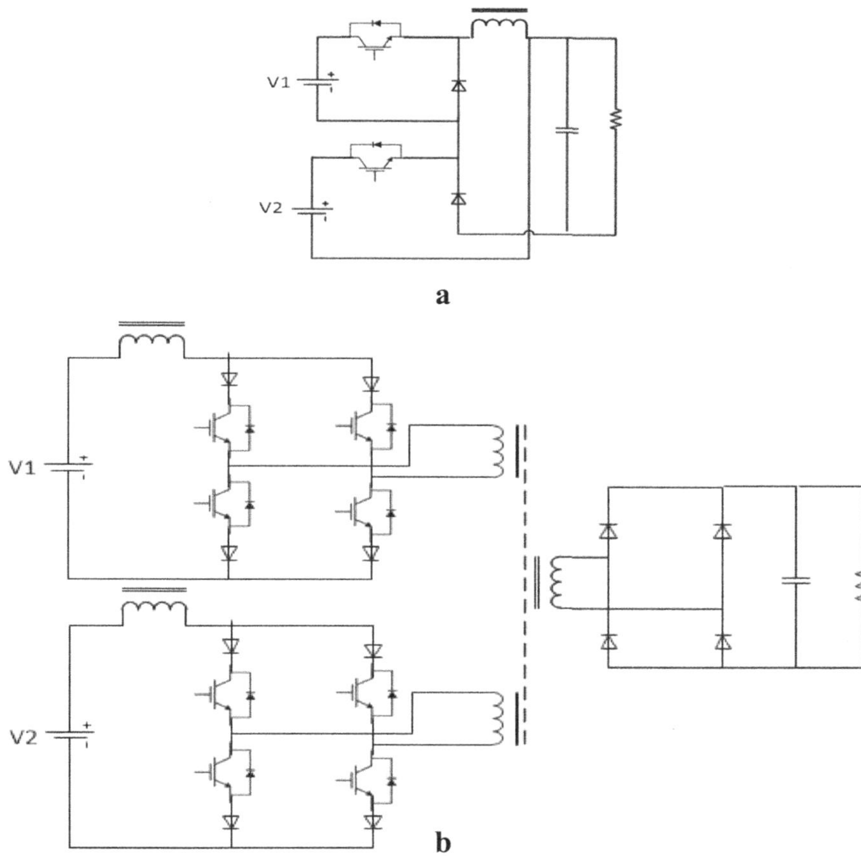

Fig. 19.7 Multi-input converters and **a** buck converter and buck-boost converter and **b** isolated converter

19.3.1 Solar PV System with MPPT Using a Boost Converter [4]

Figure 19.9 a represents the MATLAB/Simulink model of the solar PV system with MPPT using a boost converter. The basic step-up converter topology is used in PV energy conversion, and it is operated with MPPT control. As the PV belongs to a semiconductor device, it is designed with the help of Simscape components. The solar panel and boost converter of MATLAB model are shown in Fig. 19.9b, c, respectively. The solar PV plant parameters are given in Table 19.3.

Figure 19.10 shows the input and output waveforms of the proposed converter simulation circuit. Results are obtained for two different solar irradiation profiles. The four results are output voltage, PV current, PV voltage and PV power output.

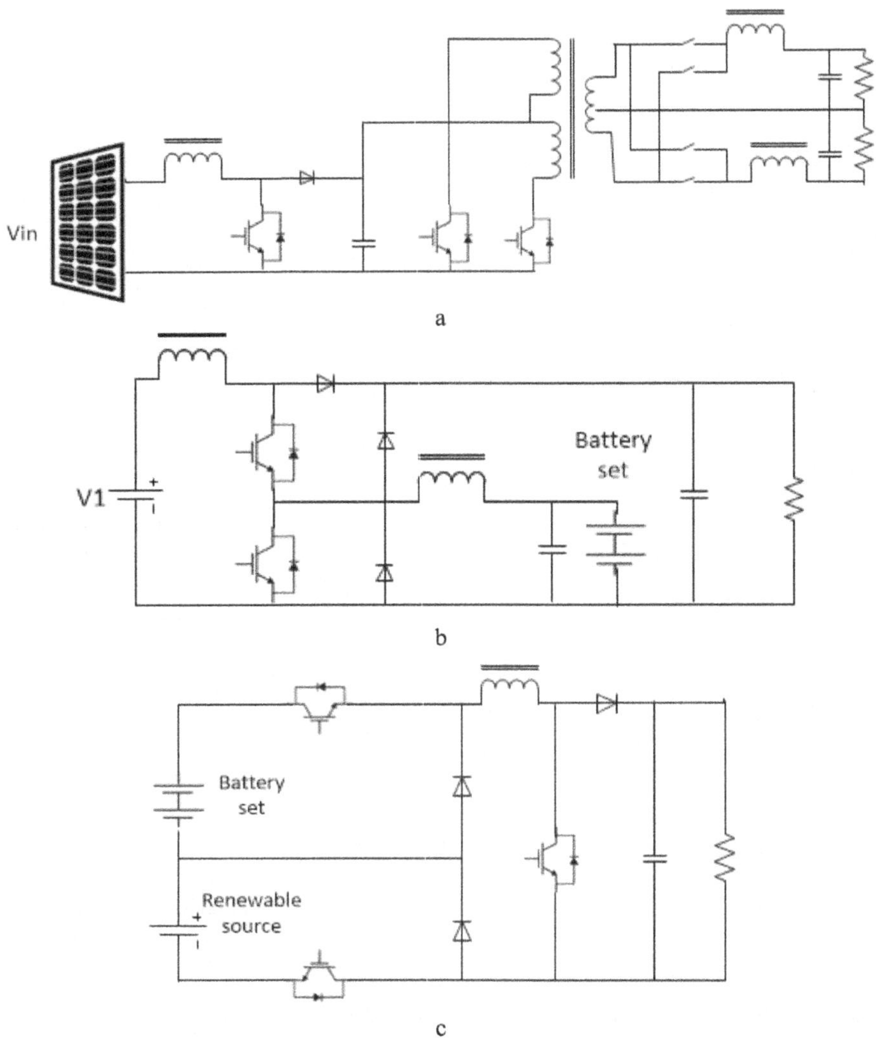

Fig. 19.8 Converter for stand-alone systems, **a** isolated converter, **b** integrated converter and **c** multi-input converter

Solar irradiance varies from the minimum value to the maximum. The DC bus voltage is kept constant at 400 V. Solar current and voltage profile vary and result in a variation of solar power output. The maximum power 2 kW is extracted when maximum solar irradiation available.

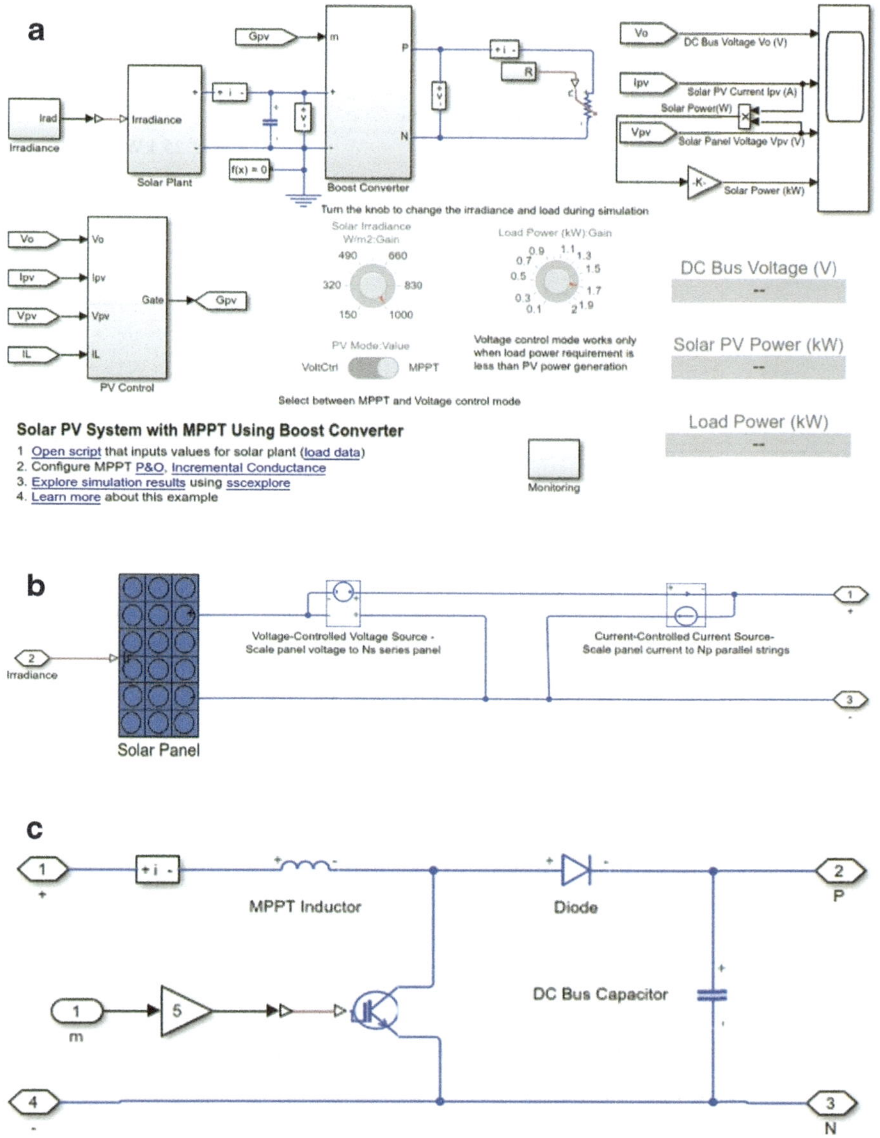

Fig. 19.9 **a** MATLAB/Simulink model [19] of the solar PV system with MPPT using a boost converter, **b** solar panel construction using simscape, **c** DC-DC boost converter

Table 19.3 PV plant parameters

Parameters	Rating
Input power rating	2.00 kW
Minimum and maximum number of panel per string	8/10
Solar PV plant	1.80/2.25 kW

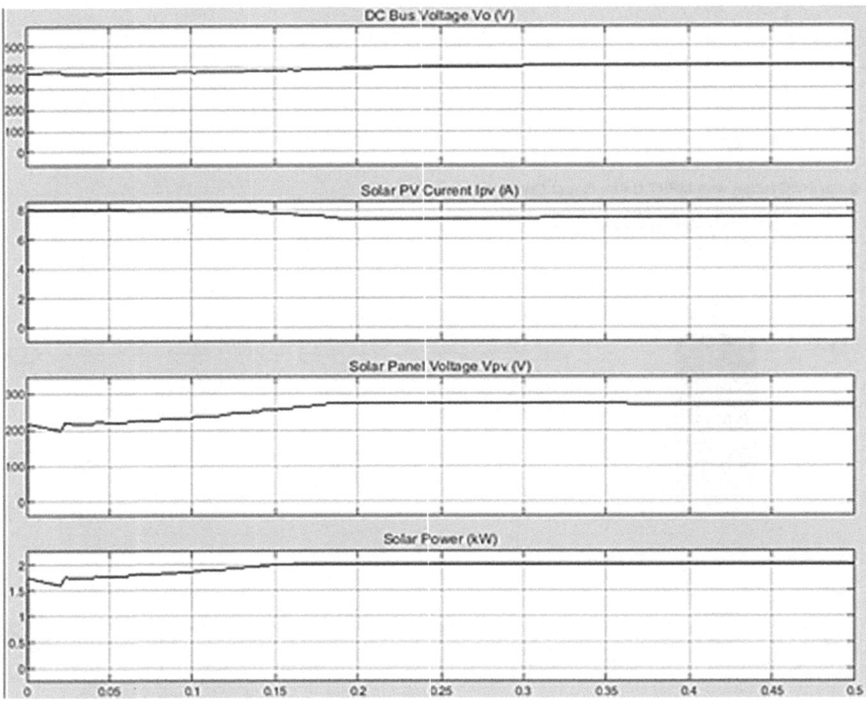

Fig. 19.10 Simulation output: DC bus voltage, PV current, PV voltage and PV power output

19.3.2 MATLAB Model of a 100-KW Grid-Connected PV Array [5]

Figure 19.11 shows a detailed model of a 100-kW array connected to a 25-kV grid via a DC-DC boost converter and a three-phase three-level VSC (Table 19.4).

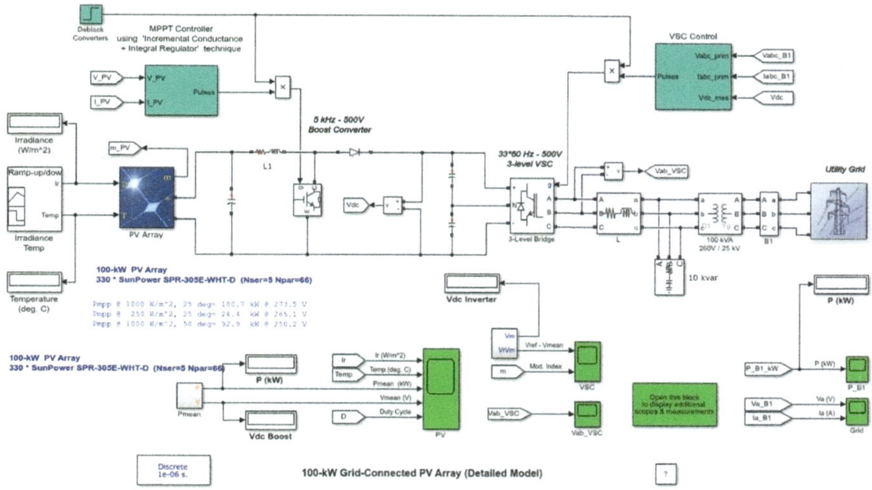

Fig. 19.11 100-kW grid-connected PV array

Table 19.4 Simulation parameters

Parameters	Rating
Solar PV plant power	100 kW
Grid voltage	25 kV
Switching frequency of the DC–DC converter	5 kHz
Inverter switching frequency	1980 Hz
Capacitor bank	10-kvar
Coupling transformer	100-kVA (260 V/25 kV)
Distribution feeder	120 kV
PV String	66
Series connected PV modules	5
Number of cells connected in series	96
Open-circuit voltage (Voc)	64.2 V
Short-circuit current (Isc)	5.96A

Figure 19.12 shows the PV power delivery to change in solar irradiation profile. From top to bottom, solar irradiation profile, temperature variation, average power output from solar, boost converter output voltage and duty ratio variation are displayed in the waveform.

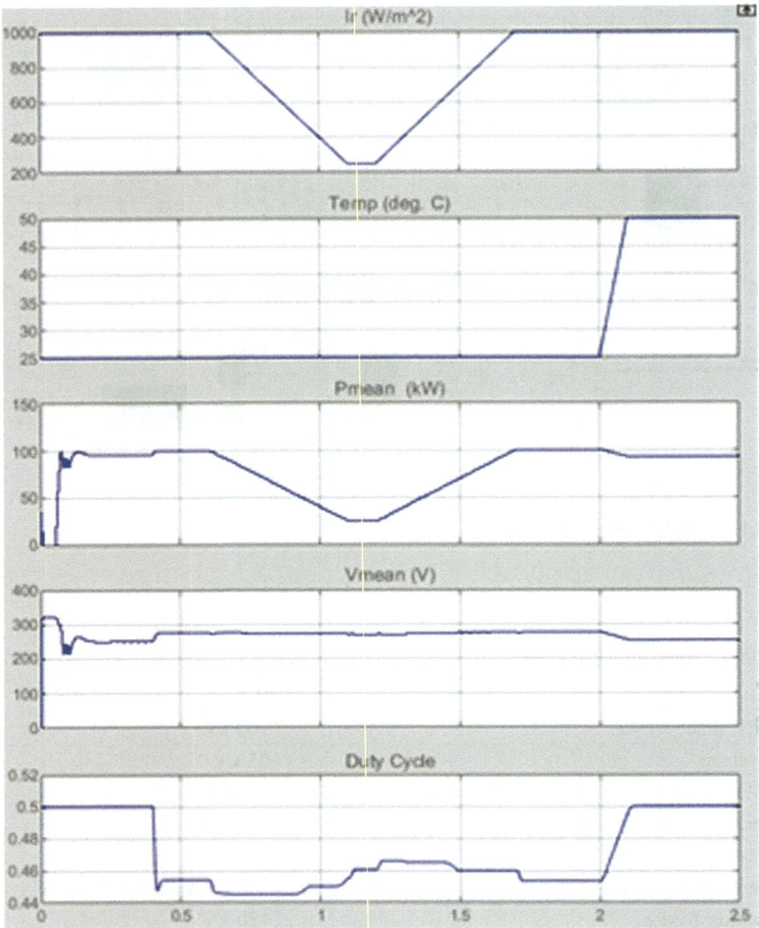

Fig. 19.12 Solar irradiation profile, temperature variation, average power output from solar, boost converter output voltage and duty ratio variation

19.4 Conclusion

The current research in the high-voltage step-up DC–DC converter for green energy conversion has the following targets: high efficiency, high power density, low cost, low complexity and high reliability. There is always a trade-off among the above-mentioned points, and there is a need for compromise also. A comparative summary of dominant topologies used in renewable energy conversion with grid and stand-alone system is provided. Each topology has its unique features and serves specific applications, and there is no common solution for all. The advent growth of power semiconductor devices, artificial control method and advanced design and manufacturing enables more powerful power converter solutions for renewable energy conversion.

References

1. Forouzesh, M., Siwakoti, Y.P., Gorji, S.A., Blaabjerg, F., Lehman, B.: Step-Up DC–DC converters: a comprehensive review of voltage-boosting techniques, topologies, and applications. IEEE Trans. Power Electron. **32**(12), 9143–9178 (2017). https://doi.org/10.1109/TPEL.2017.2652318
2. Carrasco, J.M., Garcia, L., Bialasiewicz, J.T., Galván, E., Portillo, R.C., Martín, Ma. A., León, J.I., Moreno-Alfonso, N.: Power-electronic systems for the grid integration of renewable energy sources: a survey. IEEE Trans. Ind. Electron. **53**(4), 1002–1016, ISSN 0278-0046 (2006)
3. Femia, N., Petrone, G., Spagnuolo, G., Vitelli, M.: A technique for improving P&O MPPT performances of double-stage grid-connected photovoltaic systems. IEEE Trans. Ind. Electron. **56**(11), 4473–4482, ISSN 0278-0046 (2009)
4. Padmanaban, S., Priyadarshi, N., Holm-Nielsen, J.B., Bhaskar, M.S., Azam, F., Sharma, A. K.: A novel modified sine-cosine optimized MPPT algorithm for grid integrated PV system under real operating conditions. IEEE Access **7**, 10467–10477 (2019). https://doi.org/10.1109/ACCESS.2018.2890533
5. Kim, I.S., Kim, M.B., Youn, M.J.: New maximum power point tracker using sliding-mode observer for estimation of solar array current in the grid- connected photovoltaic system. IEEE Trans. Ind. Electron. **53**(4), 1027–1035, ISSN 0278-0046
6. Park, J.H., Ahn, J.Y., Cho, B.H., Yu, G.J.: Dual-module-based maximum power point tracking control of photovoltaic systems. IEEE Trans. Ind. Electron. **53**(4), 1036–1047, ISSN 0278-0046 (2006)
7. Ertl, H., Kolar, J.W., Zach, F.C.: A novel multicell DC–AC converter for applications in renewable energy systems. IEEE Trans. Ind. Electron. **49**(5), 1048–1057, ISSN 0278-0046 (2002)
8. Kwon, J.M., Nam, K.H., Kwon, B.H.: Photovoltaic power conditioning system with line connection. IEEE Trans. Ind. Electron. **53**(4), 1048–1054, ISSN 0278-0046 (2006)
9. Padmanaban, S., Priyadarshi, N., Holm-Nielsen, J.B., Bhaskar, M.S., Hossain, E., Azam, F.: A hybrid photovoltaic-fuel cell for grid integration with Jaya-based maximum power point tracking: experimental performance evaluation. IEEE Access **7**, 82978–82990 (2019). https://doi.org/10.1109/ACCESS.2019.2924264
10. Priyadarshi, N., Padmanaban, N., Holm-Nielsen, J.B., Blaabjerg, F., Bhaskar, M.S.: An experimental estimation of hybrid ANFIS–PSO-based MPPT for PV grid integration under fluctuating sun irradiance. IEEE Syst. J. **14**(1), 1218–1229 (2020). https://doi.org/10.1109/JSYST.2019.2949083
11. Priyadarshi, N., Padmanaban, N., Bhaskar, M.S., Blaabjerg, F., Holm-Nielsen, J. B., Azam, F., Sharma, A.K.: A hybrid photovoltaic-fuel cell-based single-stage grid integration with Lyapunov control scheme. IEEE Syst. J. https://doi.org/10.1109/JSYST.2019.2948899
12. Kamalapathi, K., Priyadarshi, N., Padmanaban, S., Holm-Nielsen, J.B., Azam, F., Umayal, C., Ramachandaramurthy, V.K.: A hybrid moth-flame fuzzy logic controller based integrated Cuk converter fed brushless DC motor for power factor correction. Electronics **7**, 288 (2018)
13. Kobayashi, K., Matsuo, H., Sekine, Y.: Novel solar-cell power supply system using a multiple-input DC-DC converter. IEEE Trans. Ind. Electron. **53**(1), 281–286, ISSN 0278-0046 (2006)

14. Chen, Y.M., Liu, Y.C., Lin, S.H.: Double-Input PWM DC/DC converter for high-/low-voltage sources. IEEE Trans. Ind. Electron. **53**(5), 1538–1545, ISSN 0278-0046 (2006)
15. Chen, Y.M., Liu, Y.C., YuWu, F.: Multi-Input DC/DC converter based on the multiwinding transformer for renewable energy applications. IEEE Trans. Ind. Appl. **38**(4), 1096–1104, ISSN 0093-9994 (2002)
16. Song, Y.J., Enjeti, P.N.: A high frequency link direct DC-AC converter for residential fuel cell power systems. In: Proceedings of IEEE Power Electronics Specialists Conference, pp. 4755–4761, Germany, June 2004, IEEE, Aachen, ISBN 0-7803-8399-0 (2004)
17. Pacheco, V.M., Freitas, L.C., Vieira Jr., J.B., Coelho, E.A.A., Farias, V.J.: A DC-DC converter adequate for alternative supply system applications. In: Proceedings of IEEE Applied Power Electronics Conference and Exposition, pp. 1074–1080, USA, March 2002, IEEE, Dallas, ISBN 0-7803-7404-5 (2002)
18. https://in.mathworks.com/help/physmod/sps/ug/solar-pv-system-maximum-power-point-tracking-using-boost-converter.html
19. https://in.mathworks.com/help/physmod/sps/ug/average-model-of-a-100-kw-grid-connected-pv-array.html

Chapter 20
An Internet of Things-Inspired Dual-Level Boost Converter for BLDC-Driven Photovoltaic Water Pumping Applications

Neeraj Priyadarshi, Farooque Azam, P. Sanjeevikumar, and Jens Bo Holm-Nielsen

20.1 Internet of Things-Based PV Water Pumping System

The photovoltaic (PV) system is the most promising renewable energy technology from last decades because of significant economical operation [1, 2]. It is noted that the monitored platform is required for fault detection, monitoring and control of photovoltaic power system [3, 4]. Internet of things allows the communication between devices through Internet without person's interface which provides connections of entire equipment for data sharing [5, 6]. The Internet of things provides data sensing and remote control using wireless network which interface real-time world to computerize system and has high accuracy, enhanced efficiency and economical operation with minimized person's interaction [7, 8]. Figure 20.1 depicts the IoT inspired PV system emulator, IoT based MPPT operation, IoT gateway and wireless-based integration for energy management system [9, 10].

The IoT based technology employed to observe status of plant. Safety, optimization and energy efficiency are the major issues of wireless architecture [11, 12]. Moreover, for water pumping application in remote places, the wireless sensors are employed where sensor nodes work for less battery requirement and large duration [13, 14]. Figure 20.2 depicts the structure of Internet of things-based PV water pumping control [15, 16].

The proposed architecture comprises soil moisture, color, light and temperature sensors [17, 18]. The moisture and temperature sensor processed data is passed to

N. Priyadarshi (✉) · P. Sanjeevikumar · J. B. Holm-Nielsen
Department of Energy Technology, Aalborg University, 6700 Esbjerg, Denmark
e-mail: san@et.aau.dk

J. B. Holm-Nielsen
e-mail: jhn@et.aau.dk

F. Azam
School of Computer Science and Engineering, REVA University, Bangalore, Karnataka 560077, India

N. Priyadarshi et al. (eds.), *DC–DC Converters for Future Renewable Energy Systems*, Energy Systems in Electrical Engineering,
https://doi.org/10.1007/978-981-16-4388-0_20

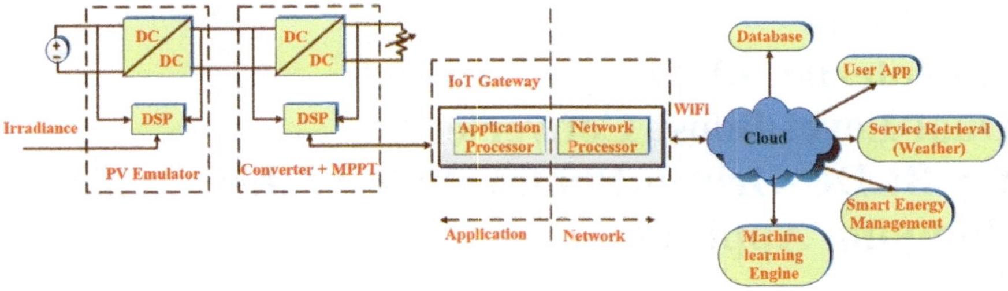

Fig. 20.1 IoT inspired PV system emulator

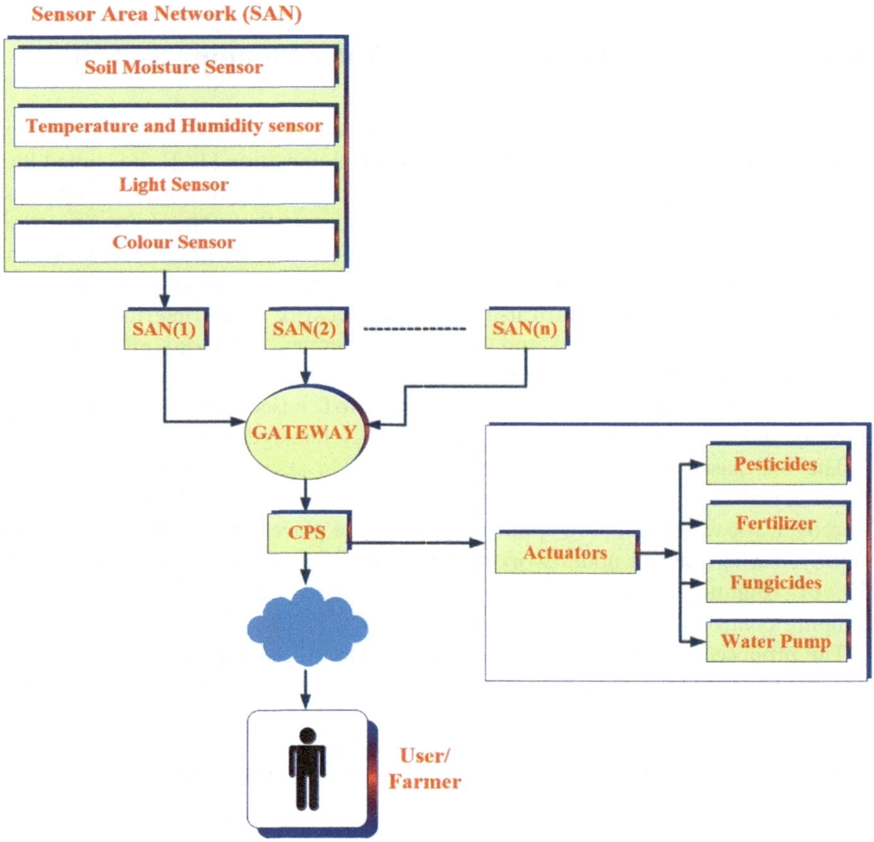

Fig. 20.2 Structure of Internet of things-based PV water pumping control

central processing server which is equated to threshold parameters which will decide starting and ending operation of water pumping system [19, 20]. Moreover, color and light sensors also measured the data which can be processed using central processing server and equated through threshold values for required decision [21, 22].

The IoT based system provides interface of applications and devices using single port and acts as communicating platform among cloud and energy management module [23, 24]. IoT based gateway acts as handling interface of application and devices [25, 26]. The Internet of things-based system comprises ARM design-based processors which are used for collecting and computing data through application as well as communication through cloud [27–29].

For practical implementation of Internet of things-based water pumping applications, Arduino based ATmega 2560, voltage|current sensor, and ESP8266 Wi-Fi model are the major used components [30, 31]. Figure 20.3 presents the block diagram of IoT based PV monitoring system [32, 33].

20.2 Mathematical Modeling of Photovoltaic Generator

Using photoelectric effect phenomena, the solar energy is converted to electrical power [29, 30]. Figure 20.4 describes a basic PV cell equivalent circuit and *V*/*I* relations which are expressed mathematically as:

$$I_{\mathrm{PVG}} = I_{\mathrm{PH}} - I_{\mathrm{OD}}\left\{\exp\left[\frac{Q_{\mathrm{E}}}{N_{\mathrm{D}}K_{\mathrm{B}}T_{\mathrm{A}}}(V_{\mathrm{PVG}} + R_{\mathrm{SL}}I_{\mathrm{PVG}})\right] - 1\right\} - \frac{V_{\mathrm{PVG}} + R_{\mathrm{SL}}I_{\mathrm{PVG}}}{R_{\mathrm{PL}}} \tag{20.1}$$

where

I_{PVG}, V_{PVG} = Output current and output voltage of PVG, respectively
I_{PH} = Photon current
I_{OD} = Diode reverse saturation current
N_{D} = Number of PV cells in series

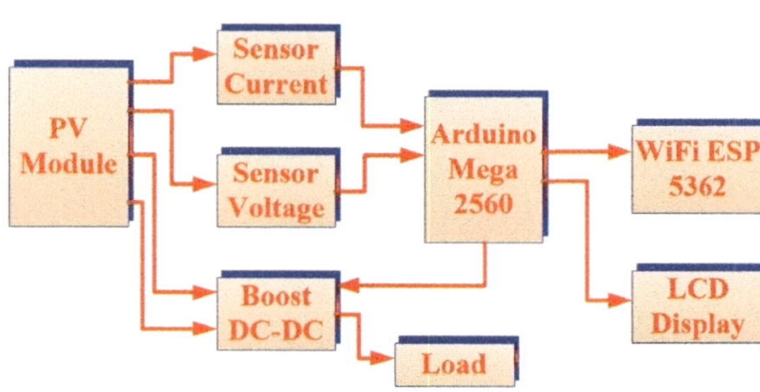

Fig. 20.3 Block diagram of IoT based PV monitoring system

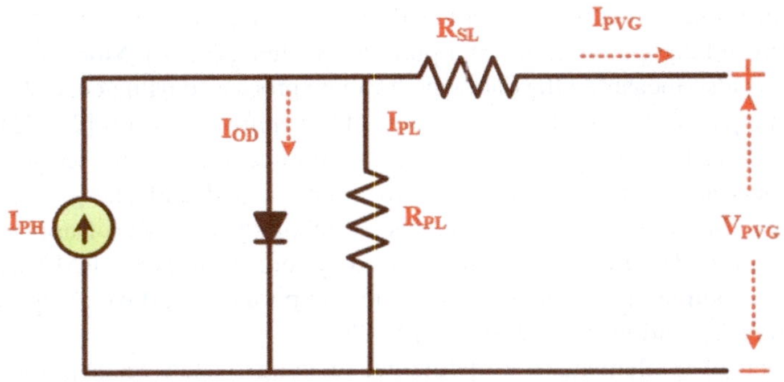

Fig. 20.4 Basic PV cell equivalent circuit

K_B = Boltzman constant
T_A = Ambient temperature
R_{SL} = Series resistance
R_{PL} = Parallel resistance.

It is noted that Eq. (20.1) represents transcendental equation in which entire coefficients are not measurable and has direct correlation with *V*/*I* characteristics. Hence, following points are needed to consider as:

a. Since, $\frac{V_{PVG}+R_{SL}I_{PVG}}{R_{PL}} \ll I_{PH}$ which is very large and may be neglected
b. $R_{SL} \ll R_{PL}$ and I_{PH} is treated as short-circuited current.

Here, assuming $G_B = 1000$ W/m^2 and $T_B = 25$ °C as standard test condition, the PV generator modeling can be expressed through environmental correction technique as:

$$I_{PVG} = I_{short}\left\{1 - K_1\left[\exp\left(\frac{V_{PVG}}{K_2 V_{open}}\right)\right]\right\} \tag{20.2}$$

where

$$K_1 = \left(1 - \frac{I_{max}}{I_{short}}\right)\exp\left(\frac{-V_{max}}{K_2 V_{open}}\right)$$

$$K_2 = \frac{{}^{V_{max}}/_{V_{open}} - 1}{\ln\left(1 - \frac{I_{max}}{I_{short}}\right)}$$

$$\Delta G = \frac{G}{G_B} - 1 \tag{20.3}$$

$$\Delta T_t = T_t - T_B \tag{20.4}$$

$$V'_{\text{open}} = V_{\text{open}}(1 - r\Delta T_t)(1 + q\Delta G) \tag{20.5}$$

$$I'_{\text{short}} = I_{\text{short}} * \frac{G}{G_{\text{B}}}(1 + p\Delta T_t) \tag{20.6}$$

$$I'_{\max} = I_{\max} * \frac{G}{G_{\text{B}}}(1 + p\Delta T_t) \tag{20.7}$$

$$V'_{\max} = V_{\max}(1 - r\Delta T_t)(1 + q\Delta G) \tag{20.8}$$

where

$I_{\text{short}}, V_{\text{open}} \equiv$ Short-circuit current and open-circuit voltage, respectively
$I_{\max}, V_{\max} \equiv$ Current and voltage at MPP, respectively
$V'_{\text{open}}, I'_{\text{short}}, I'_{\max} V'_{\max} \equiv$ Instantaneous correlative variable
P = 25×10^{-4} °C
q = 5×10^{-1}
r = 288×10^{-5} °C^{-1}.

20.3 Mathematical Modeling of BLDC Motor

Dynamic BLDC motor model can be expressed as a function of current, speed and positions derivative with respect to time. Also, phase voltages are derived mathematically as:

$$\begin{bmatrix} V_{\text{A}} \\ V_{\text{B}} \\ V_{\text{C}} \end{bmatrix} = \begin{bmatrix} R_{\text{Ph}} & 0 & 0 \\ 0 & R_{\text{Ph}} & 0 \\ 0 & 0 & R_{\text{Ph}} \end{bmatrix} \begin{bmatrix} I_{\text{A}} \\ I_{\text{B}} \\ I_{\text{C}} \end{bmatrix} + \begin{bmatrix} L_{\text{s}} & M_{\text{u}} & M_{\text{u}} \\ M_{\text{u}} & L_{\text{s}} & M_{\text{u}} \\ M_{\text{u}} & M_{\text{u}} & L_{\text{s}} \end{bmatrix} \frac{\text{d}}{\text{d}t} \begin{bmatrix} I_{\text{A}} \\ I_{\text{B}} \\ I_{\text{C}} \end{bmatrix} + \begin{bmatrix} E_{\text{A}} \\ E_{\text{B}} \\ E_{\text{C}} \end{bmatrix} \tag{20.9}$$

where

$V_{\text{A}}, V_{\text{B}}, V_{\text{C}} \equiv$ Phase voltage
$I_{\text{A}}, I_{\text{B}}, I_{\text{C}} \equiv$ Phase currents
$E_{\text{A}}, E_{\text{B}}, E_{\text{C}} \equiv$ Back EMFs
$R_{\text{Ph}} \equiv$ Phase resistance
$L_{\text{S}} \equiv$ Self-inductance of BLDC stator winding
$M_u \equiv$ Mutual inductance of BLDC stator winding.

Also, it is noted that for star connection of BLDC motor

$$I_A + I_B + I_C = 0 \tag{20.10}$$

And, inductance matrix is simplified as:

$$MI_B + MI_C = -MI_A \tag{20.11}$$

From Eqs. (20.9) and (20.11)

$$\begin{bmatrix} V_A \\ V_B \\ V_C \end{bmatrix} = \begin{bmatrix} R_{Ph} & 0 & 0 \\ 0 & R_{Ph} & 0 \\ 0 & 0 & R_{Ph} \end{bmatrix} \begin{bmatrix} I_A \\ I_B \\ I_C \end{bmatrix} + \left\{ \begin{bmatrix} L_s - M_u & 0 & 0 \\ 0 & L_s - M_u & 0 \\ 0 & 0 & L_s - M_u \end{bmatrix} \frac{d}{dt} \begin{bmatrix} I_A \\ I_B \\ I_C \end{bmatrix} + \begin{bmatrix} E_A \\ E_B \\ E_C \end{bmatrix} \right\} \tag{20.12}$$

Then,

$$\frac{d}{dt} \begin{bmatrix} I_A \\ I_B \\ I_C \end{bmatrix} = \begin{bmatrix} {}^1/L_s - M_u & 0 & 0 \\ 0 & {}^1/L_s - M_u & 0 \\ 0 & 0 & {}^1/L_s - M_u \end{bmatrix} \left\{ \begin{bmatrix} \begin{bmatrix} V_A \\ V_B \\ V_C \end{bmatrix} - \end{bmatrix} \begin{bmatrix} R_{Ph} & 0 & 0 \\ 0 & R_{Ph} & 0 \\ 0 & 0 & R_{Ph} \end{bmatrix} \begin{bmatrix} I_A \\ I_B \\ I_C \end{bmatrix} - \begin{bmatrix} E_A \\ E_B \\ E_C \end{bmatrix} \right\} \tag{20.13}$$

Additionally, electromagnetic torque is expressed mathematically as:

$$T_{EM} = \frac{E_A I_A + E_B I_B + E_C I_C}{\omega_{Rt}} \tag{20.14}$$

Also, back EMF is expressed mathematically as:

$$E_X = F_X(Q_s)\lambda_X \omega_{Rt} \tag{20.15}$$

where

$\lambda_X \equiv$ Flux
$F_X(Q_S) \equiv$ Shape of back EMF
$X \equiv$ Phases A, B, and C.

From Eqs. (20.14) and (20.15)

$$T_{\mathrm{EM}} = \lambda_{\mathrm{A}} F_{\mathrm{A}}(Q_{\mathrm{S}}) I_{\mathrm{A}} + \lambda_{\mathrm{B}} F_{\mathrm{B}}(Q_{\mathrm{S}}) I_{\mathrm{B}} + \lambda_{\mathrm{C}} F_{\mathrm{C}}(Q_{\mathrm{S}}) I_{\mathrm{C}} \tag{20.16}$$

Moreover, mathematically, the motion equation is expressed as:

$$\frac{\mathrm{d}\omega_{\mathrm{Rt}}}{\mathrm{d}t} = \frac{T_{\mathrm{EM}} - T_{\mathrm{M}} - B_{\mathrm{f}}\omega_{\mathrm{Rt}}}{J_I} \tag{20.17}$$

where

T_{M} = Load torque
J_I = Moment of inertia of motor
B_{f} = Frictional constant.

$$\frac{\mathrm{d}Q_{\mathrm{Rt}}}{\mathrm{d}t} = \omega_{\mathrm{Rt}}.$$

20.4 Mathematical Modeling of Centrifugal Water Pumping

The PV-driven BLDC motor-based water pumping provides rated capacity under changing weather situation, so that entire water can be processed. Mathematically, the Torque/Speed behavior of centrifugal pump is expressed as:

$$T_{\mathrm{M}} = K_1\omega_s^2 + \mathrm{sign}(\omega_{\mathrm{s}}).\left(K_2.e^{-K_3|\omega_{\mathrm{s}}|} + K_4\right) \tag{20.18}$$

Also,

$$P_{\mathrm{M}} = C_1\omega_{\mathrm{s}}^3 - C_2\omega_{\mathrm{s}}^2 + C_3\omega_{\mathrm{s}} \approx C_1\omega_{\mathrm{s}}^3$$

where

C_1, C_2, C_3 are dependent of discharge valve setting. And under high speed, the lower order factors can be omitted.

20.5 Mathematical Modeling of Dual-Level Boost Converter

Figure 20.5 illustrates the power circuit of dual-level boost converter employed to achieve high output voltage. Compared to traditional boost converter, the proposed dual-level boost converter has higher efficiency transformation which has no requisite of isolation transformer with economical operation. It is noted that the voltage level is enhanced using addition of capacitors and diodes equated with classical boost converter.

The DC-link capacitors C_A and C_B produces voltage because of varying duty ratio of power switch.

$$V_{C_A} = V_{C_B} = \frac{V_{PVG}}{(1 - D_{duty})} \quad (20.19)$$

Therefore,

$$V_{D_C} = V_{C_A} = \frac{2V_{PVG}}{(1 - D_{duty})} \quad (20.20)$$

Equation 20.20 represents the dual-level DC-link output voltage, where D_{duty} is duty ratio of switch.

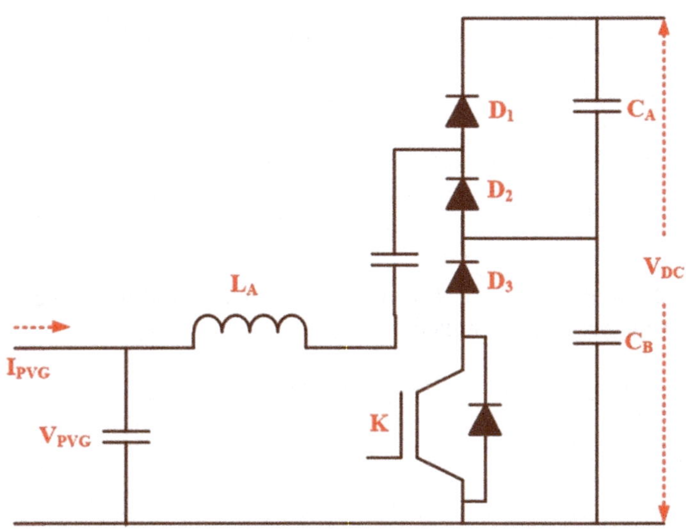

Fig. 20.5 Power circuit of dual-level boost converter

20.6 Conclusion

This research paper explained the Internet of things inspired BLDC motor-driven PV water pumping applications. The detailed mathematical modeling of photovoltaic generator and brushless DC motor has been presented in this research work. The Internet of things based maximum power point tracker has been proposed which provides optimal PV power tracking with high accuracy and efficiency. Equated with conventional DC–DC boost converter, the employed dual-level boost converter gives better efficiency conversion, high output voltage, and there is no requirement of isolation transformer. The proposed IoT based MPPT delivers data sensing and remote control using wireless network system.

References

1. Priyadarshi, N., Padmanaban, S., Maroti, P.K., Sharma. A.: An extensive practical investigation of FPSO-based MPPT for grid integrated PV system under variable operating conditions with anti-islanding protection. IEEE Syst. J. 1–11 (2018)
2. Priyadarshi, N., Padmanaban, S., Bhaskar, M.S., Blaabjerg, F., Sharma, A.: A fuzzy SVPWM based inverter control realization of grid integrated PV-wind system with FPSO MPPT algorithm for a grid-connected PV/Wind power generation system: hardware implementation. IET Electric Power Appl. 1–12 (2018)
3. Priyadarshi, N., Kumar, V., Yadav, K., Vardia, M.: An experimental study on zeta buck-boost converter for application in PV system. In: Handbook of Distributed Generation. Springer https://doi.org/10.1007/978-3-319-51343-0_13
4. Priyadarshi, N., Sharma, A.K., Priyam, S.: An experimental realization of grid-connected PV system with MPPT using dSPACE DS 1104 control board. In: Advances in Smart Grid and Renewable Energy. Lecture Notes in Electrical Engineering, vol. 435, Springer, Singapore (2018)
5. Priyadarshi, N., Sharma, A.K., Priyam, S.: Practical realization of an improved photovoltaic grid integration with MPPT. Int. J. Renew. Energy Res. **7**(4) (2017)
6. Priyadarshi, N., Sharma, A.K., Azam, F.: A hybrid firefly-asymmetrical fuzzy logic controller based MPPT for PV-wind-fuel grid integration. Int. J. Renew. Energy Res. **7**(4) (2017)
7. Priyadarshi, N., Anand, A., Sharma, A.K., Azam, F., Singh, V.K., Sinha, R.K.: An Experimental implementation and testing of GA based maximum power point tracking for PV system under varying ambient conditions using dSPACE DS 1104 controller. Int. J. Renew. Energy Res. **7**(1), 255–265 (2017)
8. Priyadarshi, N., Padmanaban, S., Mihet-Popa, L., Blaabjerg, F., Azam, F.: Maximum power point tracking for brushless DC motor-driven photovoltaic pumping systems using a hybrid ANFIS-FLOWER pollination optimization algorithm. MDPI Energies **11**(1), 1–16 (2018)
9. Priyadarshi, N., Azam, F., Bhoi, A.K., Alam, S.: An artificial fuzzy logic intelligent controller based MPPT for PV grid utility. In: Lecture Notes in Networks and Systems 46. https://doi.org/10.1007/978-981-13-1217-5_88
10. Padmanaban, S., Priyadarshi, N., Holm-Nielsen, J.B., Bhaskar, M.S., Azam, F., Sharma, A. K.: A novel modified sine-cosine optimized MPPT algorithm for grid integrated PV system under real operating conditions. IEEE Access **7**, 10467–10477 (2019). https://doi.org/10.1109/ACCESS.2018.2890533

11. Padmanaban, S., Priyadarshi, N., Holm-Nielsen, J.B., Bhaskar, M.S., Hossain, E., Azam, F.: A hybrid photovoltaic-fuel cell for grid integration with JAYA-based maximum power point tracking: experimental performance evaluation. IEEE Access **7**, 82978–82990 (2019). https://doi.org/10.1109/ACCESS.2019.2924264
12. Priyadarshi, N., Padmanaban, N., Holm-Nielsen, J.B., Blaabjerg, F., Bhaskar, M.S.: An experimental estimation of hybrid ANFIS–PSO-based MPPT for PV grid integration under fluctuating sun irradiance. IEEE Syst. J. **14**(1), 1218–1229 (2020). https://doi.org/10.1109/JSYST.2019.2949083
13. Priyadarshi, N., Padmanaban, N., Bhaskar, M.S., Blaabjerg, F., Holm-Nielsen, J.B., Azam, F., Sharma, A.K.: A hybrid photovoltaic-fuel cell-based single-stage grid integration with Lyapunov control scheme. IEEE Syst. J. https://doi.org/10.1109/JSYST.2019.2948899
14. Priyadarshi, N., Bhaskar, M.S., Padmanaban, N., Blaabjerg, F., Azam, F.: New CUK–SEPIC converter based photovoltaic power system with hybrid GSA–PSO algorithm employing MPPT for water pumping applications. IET Power Electron. 1–0 (2020). https://doi.org/10.1049/iet-pel.2019.1154
15. Priyadarshi, N., Padmanaban, N., Holm-Nielsen, J.B., Bhaskar, M.S., Azam, F.: Internet of things augmented a novel PSO-employed modified zeta converter-based photovoltaic maximum power tracking system: hardware realisation. IET Power Electron. 1–0 (2020). https://doi.org/10.1049/iet-pel.2019.1121
16. Kamalapathi, K., Priyadarshi, N., Padmanaban, S., Holm-Nielsen, J.B., Azam, F., Umayal, C., Ramachandaramurthy, V.K.: A hybrid moth-flame fuzzy logic controller based integrated Cuk converter fed brushless DC motor for power factor correction. Electronics **7**, 288 (2018)
17. Priyadarshi, N., Padmanaban, S., Ionel, D., Mihet-Popa, L., Azam, F.: Hybrid PV-wind, micro-grid development using quasi-Z-source inverter modeling and control—experimental investigation. Energies **11**(9), 2277 (2018). https://doi.org/10.3390/en11092277
18. Priyadarshi, N., Ramachandaramurthy, V.K., Padmanaban, S., Azam, F.: An ant colony optimized MPPT for standalone hybrid PV-wind power system with single Cuk converter. Energies **12**(1), 167 (2019). https://doi.org/10.3390/en12010167
19. Priyadarshi, N., Sharma, A.K., Bhoi, A.K., Ahmad, S.N., Azam, F., Priyam, S.: A practical performance verification of AFLC based MPPT for standalone PV power system under varying weather condition. Int. J. Eng. Technol. **7**(2), 338–343 (2018)
20. Priyadarshi, N., Azam, F., Bhoi, A.K., Sharma, A.K.: A multilevel inverter-controlled photovoltaic generation. In: Bhoi, A., Sherpa, K., Kalam, A., Chae, G.S. (eds) Advances in Greener Energy Technologies. Green Energy and Technology. Springer, Singapore (2020). https://doi.org/10.1007/978-981-15-4246-6_8
21. Priyadarshi, N., Azam, F., Bhoi, A.K., Sharma, A.K.: Dynamic operation of grid-connected photovoltaic power system. In: Bhoi, A., Sherpa, K., Kalam, A., Chae, G.S. (eds) Advances in Greener Energy Technologies. Green Energy and Technology. Springer, Singapore (2020). https://doi.org/10.1007/978-981-15-4246-6_13
22. Priyadarshi, N., Azam, F., Bhoi, A.K., Sharma, A.K.: A proton exchange membrane-based fuel cell integrated power system. In: Bhoi, A., Sherpa, K., Kalam, A., Chae, G.S. (eds) Advances in Greener Energy Technologies. Green Energy and Technology. Springer, Singapore (2020). https://doi.org/10.1007/978-981-15-4246-6_18
23. Priyadarshi, N., Azam, F., Bhoi, A.K., Sharma A.K.: A closed-loop control of fixed pattern rectifier for renewable energy applications. In: Bhoi, A., Sherpa, K., Kalam, A., Chae, G.S. (eds) Advances in Greener Energy Technologies. Green Energy and Technology. Springer, Singapore (2020). https://doi.org/10.1007/978-981-15-4246-6_25
24. Priyadarshi, N., Azam, F., Bhoi, A.K., Sharma, A.K.: A four-switch-type converter fed improved photovoltaic power system. In: Bhoi, A., Sherpa, K., Kalam, A., Chae, G.S. (eds) Advances in Greener Energy Technologies. Green Energy and Technology. Springer, Singapore (2020). https://doi.org/10.1007/978-981-15-4246-6_29

25. Vardia, M., Priyadarshi, N., Ali, I., Azam, F., Bhoi, A.K.: Maximum power point tracking for wind energy conversion system. In: Bhoi, A., Sherpa, K., Kalam, A., Chae, G.S. (eds) Advances in Greener Energy Technologies. Green Energy and Technology. Springer, Singapore (2020). https://doi.org/10.1007/978-981-15-4246-6_36
26. Azam, F., Priyadarshi, N., Nagar, H., Kumar, S., Bhoi, A.K.: An overview of solar-powered electric vehicle charging in vehicular Adhoc network. In: Patel, N., Bhoi, A.K., Padmanaban, S., Holm-Nielsen, J.B. (eds) Electric Vehicles. Green Energy and Technology. Springer, Singapore (2021). https://doi.org/10.1007/978-981-15-9251-5_5
27. Vardia, M., Priyadarshi, N., Ali, I., Azam, F., Bhoi, A.K.: Design of wind energy conversion system under different fault conditions. In: Bhoi, A., Sherpa, K., Kalam, A., Chae, G.S. (eds) Advances in Greener Energy Technologies. Green Energy and Technology. Springer, Singapore (2020). https://doi.org/10.1007/978-981-15-4246-6_41
28. Choudhary, T., Priyadarshi. N., Kumar, P., Azam, F., Bhoi, A.K.: A fuzzy logic control based vibration control system for renewable application. In: Bhoi, A., Sherpa, K., Kalam, A., Chae, G.S. (eds) Advances in Greener Energy Technologies. Green Energy and Technology. Springer, Singapore (2020). https://doi.org/10.1007/978-981-15-4246-6_38
29. Yusof, Z,M., Billah, M.M., Kadir, K., Ali, A.M.M.: A temperature-based omnidirectional solar tracking system for IoT application. In: Proceedings of IEEE International Conference on Smart Instrumentation, Measurement and Application (ICSIMA), Kuala Lumpur, Malaysia, pp. 1–4 (2019). https://doi.org/10.1109/ICSIMA47653.2019.9057301
30. Meng, X., An, Y., Wang, H., Yao, Q., Liang, C.: Tracking the maximum power point of photovoltaic power generation based on self-coding neural network. In: Proceedings of Chinese Control And Decision Conference (CCDC), Nanchang, China, pp. 592–597 (2019). https://doi.org/10.1109/CCDC.2019.8832919
31. Altun, S.N., Dörterler, M., Dogru, I.A.: Fuzzy logic based lighting system supported with IoT for renewable energy resources. In: Proceedings of Innovations in Intelligent Systems and Applications Conference (ASYU), Adana, pp. 1–4 (2018). https://doi.org/10.1109/ASYU.2018.8554026
32. İNCİ, M.: Design and analysis of dual level boost converter based transformerless grid connected PV system for residential applications. In: Proceedings of 4th International Conference on Power Electronics and their Applications (ICPEA), Elazig, Turkey, pp. 1–6 (2019). https://doi.org/10.1109/ICPEA1.2019.8911177
33. Kumar, R., Singh, B.: Grid Interactive solar PV-based water pumping using BLDC motor drive. IEEE Trans. Ind. Appl. **55**(5), 5153–5165 (2019). https://doi.org/10.1109/TIA.2019.2928286

Chapter 21
Switched Mode Fourth-Order Buck–Boost Converter Using Type II and Type III Controllers in DC Grid Applications

S. Saurav and Arnab Ghosh

21.1 Introduction

The power electronic components play a wide variety of role in conversion and control of electric power [1, 2]. These components also play a vital role in renewable energy power generation as well especially solar power [3–5]. According to the data as population is increasing so is the need for electrical energy and its quality is also improving day by day [6, 7]. Nowadays, the maximum utilization of solar power to create sufficient electricity is being targeted by majority of the countries to achieve a renewable green energy as a replacement for non-renewable energy source [8–10]. The main reason to target the solar energy is because of the immense supply of light from the sun which can be adequately harnessed through photovoltaic systems [11, 12]. These photovoltaics provide a renewable source of electricity which can be recharged during the day and replenished during the night in case of household applications. Photovoltaics are also called as solar panels [13–15]. It is one of the most widely used green energy system [16, 17]. Therefore, high-performing electrical and electronic components are needed for the maximum utilization of the power from these sources [18–20]. The current/voltage output of the solar energy generation is DC based [21, 22]. In order to convert this DC output to be able to use it in a variety of application we require the use of power converters to make the output voltage follow desired value [23–25]. Thus, the DC–DC power converters are vital in improving the quality of the generated electric power [26]. Such devices are not always functioning according to the desired value we need, and we require an algorithm to control these devices.

To match the voltage requirement and the load to which the power is used, voltage converters are used. The most commonly used is buck–boost converter which can be used to increase or decrease the amount of voltage extracted from the

S. Saurav (✉) · A. Ghosh
Department of Electrical Engineering, National Institute of Technology Rourkela, Rourkela, Odisha, India

N. Priyadarshi et al. (eds.), *DC–DC Converters for Future Renewable Energy Systems*, Energy Systems in Electrical Engineering,
https://doi.org/10.1007/978-981-16-4388-0_21

photovoltaic cells. The photovoltaic cells are arranged in series like manner and the voltages extracted are stored via a capacitor, and it acts as an input to the converter. This is similar to connecting a battery as input to the converter. The circuit arrangement is described in Fig. 21.1. Here the buck–boost converter topology used is Fourth order. The output voltage from the two converters is sent to a dc bus. From this DC bus, the voltage is extracted and given to DC grids through a converter depending on the load criteria. Again, the converter used here is the same as above. It is used for either bucking or boosting operation according to the convenience and requirement of the load used.

There are several literatures in existence which deal with the renewable source of energy especially solar power. We can manage to find several topological converters used in the industry as well as current research topics. There are several power converter topologies used in the industry, and we can understand the power electronic applications in a variety of application. Discussions related to the different technological advancements have been made with regards to the power usage and the newly developed topologies related to boost and buck–boost converters [10]. A control algorithm to achieve the maximum power that is generated from the existing systems can be done with the use of real and reactive power reference selection process [11]. This area helps in the better understanding of how the solar power is utilized. Load regulation and line regulation are two most important criteria to understand these different power converters can be utilized in an effective manner in different applications [12]. Novel ideas related to maximum power utilizations in renewable power generation are cascading of buck and boost converters [12]. The proposed control algorithm of cascaded systems has a synchronization loop (voltage tracking). This voltage tracking is done with the help of two nested current and voltage loops along with virtual control loop. These cascaded components provide us with robustness in the converter system with parametric uncertainties. Modeling of a fourth-order system is huge task and one such

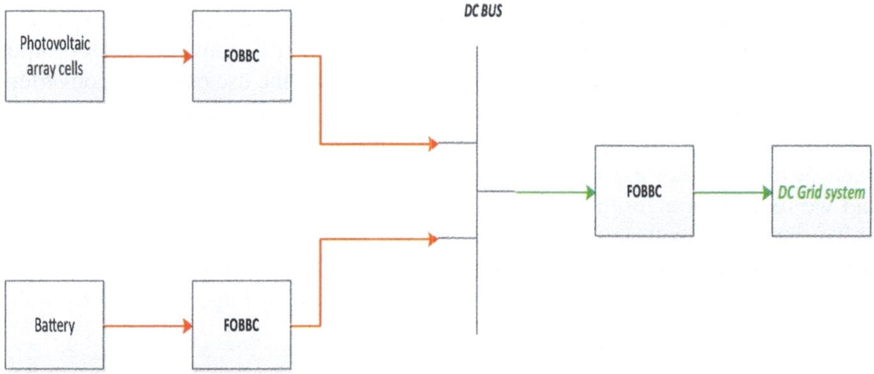

Fig. 21.1 Overall representation of a dual supply bus with solar and battery. (FOBBC-fourth-order buck–boost converter)

approach which helps us to get a dynamic behavior close to second-order converter which helps us to simplify the voltage compensation design is the principal component analysis [12].

Many power converter topologies are discussed, and one such topology is the tri-state boost converter [7]. Here the introduction of an extra degree of freedom is done using a freewheeling interval which has proven to remove the right half plane zero. In a similar way, several discussions have been made on the better controller implementation for the second-order boost converter [7]. The controller which yields better performance according to the load variation as well as source ripple is Type II and Type III converters rather than classical PID controllers [7, 8]. From these literatures, we can conclude that the Type II and Type III are the better classical controllers regarding this application. A more profound is due to the fact that the source ripple/inductor ripple current intervenes and causes disturbances in maintaining a constant output voltage in different converters. A solution can be achieved for this problem by increasing the order of the circuit elements, and this topology can be controlled using the classical converters (Type II and Type III) which yield a better result when compared with the famous PID controller. There are several techniques used to determine the closed loop performance. Transient response using time domain analysis can be implemented by connecting the buck–boost converter in a closed-loop system with the controller.

In this chapter, we have considered the Fourth-Order Buck–Boost converter. It is usually known for its non-minimum phase. So similar to the second-order converter, the fourth-order converter also has a pole in the right half of s-plane. The characteristics of this non minimum phase behavior are shown in [1–3]. Here in this paper we have modeled the fourth-order buck–boost converter using the state space averaging techniques which are more convenient, and it is derived in Sect. 21.2. After modeling is done, various controllers suited for the fourth-order buck–boost converters are discussed in Sect. 21.3. The steady-state analysis and the transient response are also analyzed in the result Sect. 21.4, and the result obtained from the experimentation is compared for the two types of controllers designed. Section 21.5 is conclusion derived from this experiment we have conducted.

21.2 State Space Modeling of the Converter

21.2.1 Topology

The topology which is to be used for the converter is shown in Fig. 21.2. This buck–Boost converter comprises of two controllable switches S_1 and S_2 and two uncontrollable switches D_1 and D_2. It consists of an inductor L_1 and a capacitor C_1. Its function is to switch the voltage, and it is placed in the center of a switch bridge. This bridge also consists of diodes D_1 and D_2. The capacitor connects either to the source voltage V_g or to the inductor L_2 depending on the mode of operation. So, during on time it connects to the source voltage, and during off time it connects to

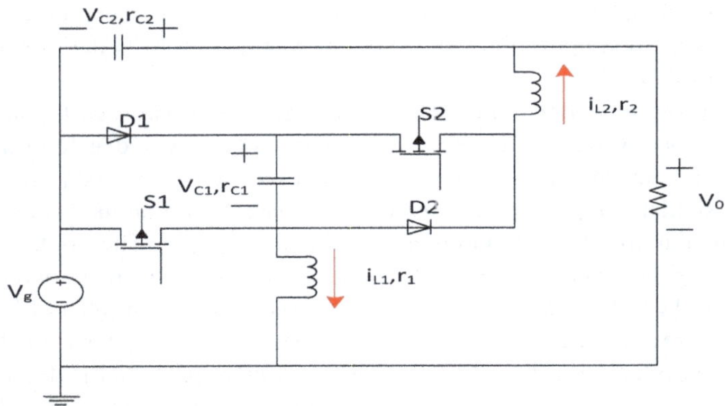

Fig. 21.2 Fourth-order buck–boost converter (general configuration)

the inductor L_2 supplying power. So, the energy is transferred from the source to the load due to the capacitor connection changeover. This topology also reduces the ripple current due to the interconnection of C_1 and L_1 which maintains the source current continuity. This source current directly affects the output voltage in the circuit, and hence, maintaining the ripple current to a minimum is important for the design of this converter. The lower ripple current is achieved with the help of the controllable (S_1 and S_2) and uncontrollable switches (D_1 and D_2). This has been extensively discussed in the literature by Veerachary [1–3].

21.2.2 *Modeling of Fourth-Order Buck–Boost Converter*

The first step in designing any converter is modeling. Here state space averaging technique is used. In this technique, two modes of operation are involved depending on the switch states. In mode I, switch S_1 and switch S_2 are both On and the diode D_1 and diode D_2 are turned off. In mode II, the reverse operation is carried out (D_1 and D_2 are on and S_1 and S_2 are off).

21.2.2.1 Mode I Operation of Fourth-Order Buck–Boost Converter

By applying Kvl and Kcl in Fig. 21.3, we get

$$V_{L_1} = V_g - i_{L_1} r_1 \tag{21.1}$$

$$V_{L_2} = V_{C_1} - V_{C_2} - i_{L_2} r_2 - i_{L_2} r_{C_1} - i_{C_2} \tag{21.2}$$

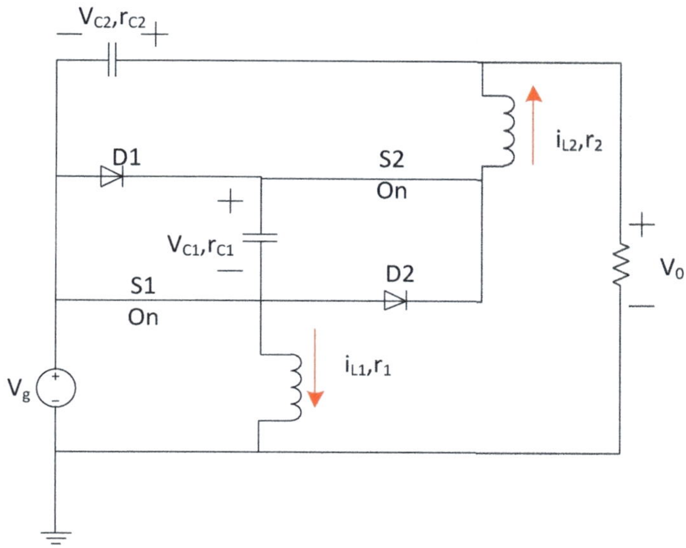

Fig. 21.3 Mode I operation of fourth-order buck–boost converter

$$i_{C_1} = -i_{L_2} \tag{21.3}$$

$$i_{C_2} = i_{L_2} - \frac{V_0}{R} \tag{21.4}$$

The state space equations in mode I are given from the above equations as

$$\dot{i_{L_1}} = \frac{V_g}{L_1} - \frac{i_{L_1} r_1}{L_1} \tag{21.5}$$

$$\dot{i_{L_2}} = -\frac{i_{L_2}(r_{C_2} r_2 + r_{C_2} r_{C_1} + r_2 R + r_{C_1} R + r_{C_2} R)}{L_2} + \frac{V_{C_1}}{L_2} - \frac{V_{C_2} R}{L_2(R + r_{C_2})} \tag{21.6}$$

$$V_{\dot{C}_1} = -\frac{i_{L_2}}{C_1} \tag{21.7}$$

$$V_{\dot{C}_2} = \frac{R i_{L_2}}{(R + r_{C_2}) C_2} - \frac{V_{C_2}}{(R + r_{C_2}) C_2} \tag{21.8}$$

The state space modeling during ON time is given below:

$$\dot{x} = A_1 x + B_1 V_g$$
$$y = C_1 x + D_1 V_g$$

where

$$x = \begin{bmatrix} i_{L_1} & i_{L_2} & v_{C_1} & v_{C_2} \end{bmatrix} \text{ and } y = v_0$$

$$A_1 = \begin{bmatrix} \frac{-r_1}{L_1} & 0 & 0 & 0 \\ 0 & \frac{-\left(r_{C_2}r_2 + r_{C_2}r_{C_1} + r_2R + r_{C_1}R + r_{C_2}R\right)}{L_2\left(R + r_{C_2}\right)} & \frac{1}{L_2} & \frac{-R}{L_2\left(R + r_{C_2}\right)} \\ 0 & \frac{-1}{C_1} & 0 & 0 \\ 0 & \frac{R}{\left(R + r_{C_2}\right)C_2} & 0 & \frac{-1}{\left(R + r_{C_2}\right)C_2} \end{bmatrix}$$

$$B_1 = \begin{bmatrix} \frac{1}{L_1} & \frac{r_{C_2}}{\left(R + r_{C_2}\right)L_2} & 0 & \frac{-1}{\left(R + r_{C_2}\right)C_2} \end{bmatrix}$$

$$C_1 = \begin{bmatrix} 0 & \frac{Rr_{C_2}}{\left(R + r_{C_2}\right)L_2} & 0 & \frac{R}{\left(R + r_{C_2}\right)} \end{bmatrix}$$

$$D_1 = \left[\frac{R}{(R + r_{C_2})}\right]$$

21.2.2.2 Mode II Operation of Fourth-Order Buck–Boost Converter

By applying kvl and kcl to the loops in Fig. 21.4, we get the following

$$V_{L_1} = V_g - V_{C_1} - i_{L_1}r_1 - (i_{L_1} + i_{L_2})r_{C_1} \tag{21.9}$$

$$V_{L_2} = -i_{L_2}r_2 - V_{C_1} - (i_{L_1} + i_{L_2})r_{C_1} - V_{C_2} - i_{C_2}r_{C_2} \tag{21.10}$$

$$i_{C_1} = i_{C_1} + i_{C_2} \tag{21.11}$$

$$i_{C_2} = i_{L_2} - \frac{V_0}{R} \tag{21.12}$$

The state space equations during the off mode (mode II) are given by

$$i_{\dot{L}_1} = \frac{V_g}{L_1} - \frac{i_{L_1}r_1}{L_1} \tag{21.13}$$

$$i_{\dot{L}_2} = -\frac{i_{L_2}(r_{C_2}r_2 + r_{C_2}r_{C_1} + r_2R + r_{C_1}R + r_{C_2}R)}{L_2} - \frac{V_{C_1}}{L_2} - \frac{V_{C_2}R}{L_2(R + r_{C_2})} \tag{21.14}$$

$$V_{\dot{C}_1} = \frac{i_{L_1}}{C_1} + \frac{i_{L_2}}{C_1} \tag{21.15}$$

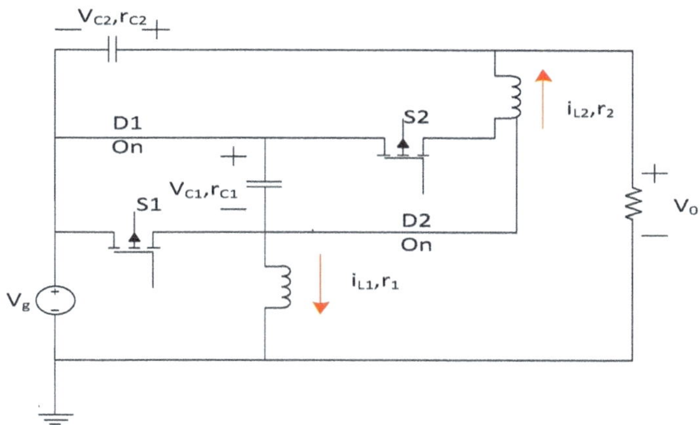

Fig. 21.4 Mode II operation of fourth-order buck–boost converter

$$V_{\dot{C}_2} = \frac{Ri_{L_2}}{(R+r_{C_2})C_2} - \frac{V_{C_2}}{(R+r_{C_2})C_2} \tag{21.16}$$

The steady-state model during Off Time is given by

$$\dot{x} = A_2x + B_2V_g$$

$$y = C_2x + D_2V_g$$

where

$$A_2 = \begin{bmatrix} \frac{-(r_1+r_{C_1})}{L_1} & \frac{-r_{C_1}}{L_1} & \frac{-1}{L_1} & 0 \\ \frac{-r_{C_1}}{L_2} & \frac{-(r_{C_2}r_2 + r_{C_2}r_{C_1} + r_2R + r_{C_1}R + r_{C_2}R)}{L_2(R+r_{C_2})} & \frac{-1}{L_2} & \frac{-R}{L_2(R+r_{C_2})} \\ \frac{1}{C_1} & \frac{1}{C_1} & 0 & 0 \\ 0 & \frac{R}{(R+r_{C_2})C_2} & 0 & \frac{-1}{(R+r_{C_2})C_2} \end{bmatrix}$$

$$B_2 = \begin{bmatrix} \frac{1}{L_1} & \frac{r_{C_2}}{(R+r_{C_2})L_2} & 0 & \frac{-1}{(R+r_{C_2})C_2} \end{bmatrix}$$

$$C_2 = \begin{bmatrix} 0 & \frac{Rr_{C_2}}{(R+r_{C_2})L_2} & 0 & \frac{R}{(R+r_{C_2})} \end{bmatrix}$$

$$D_2 = \begin{bmatrix} \frac{R}{(R+r_{C2})} \end{bmatrix}$$

The state space averaging for this plant is represented as below

$$\dot{x} = Ax + BV_g$$

$$y = Cx + DV_g$$

where the matrices are given by

$A = A_1d + A_2d'$, $B = B_1d + B_2d'$, $C = C_1d + C_2d'$, $D = D_1d + D_2d'$ and $d' = 1 - d$

$$A = \begin{bmatrix} \frac{-(r_1 + r_{C_1}d')}{L_1} & \frac{-r_{C_1}d'}{L_1} & \frac{-d'}{L_1} & 0 \\ \frac{-r_{C_1}d'}{L_2} & \frac{-(r_{C_2}r_2 + r_{C_2}r_{C_1} + r_2R + r_{C_1}R + r_{C_2}R)}{L_2(R + r_{C_2})} & \frac{d-d'}{L_2} & \frac{-R}{L_2(R + r_{C_2})} \\ \frac{d'}{C_1} & \frac{-d+d'}{C_1} & 0 & 0 \\ 0 & \frac{R}{(R + r_{C_2})C_2} & 0 & \frac{-1}{(R + r_{C_2})C_2} \end{bmatrix}$$

$$B_1 = B_2 = B = \begin{bmatrix} \frac{1}{L_1} & \frac{r_{C_2}}{(R + r_{C_2})L_2} & 0 & \frac{-1}{(R + r_{C_2})C_2} \end{bmatrix}$$

$$C_1 = C_2 = C = \begin{bmatrix} 0 & \frac{Rr_{C_2}}{(R + r_{C_2})L_2} & 0 & \frac{R}{(R + r_{C_2})} \end{bmatrix}$$

$$D_1 = D_2 = D = \begin{bmatrix} \frac{R}{(R + r_{C_2})} \end{bmatrix}$$

21.2.3 Small Signal Analysis

This representation of the averaged model is done by splitting the instantaneous value into steady-state value and perturbed value. The final form of the small signal analysis after considering the perturbation and simplifying is given by

$$\dot{\hat{x}} = A\hat{x} + B\hat{u} + [(A_1 - A_2)x + (B_1 - B_2)u]\hat{d} \tag{21.17}$$

$$\hat{y} = C\hat{x} + D\hat{u} + [(C_1 - C_2)x - (D_1 - D_2)u]\hat{d} \tag{21.18}$$

The final expression of output represented in Laplace domain is given by

$$\hat{y}(s) = \left[C(sI - A)^{-1}B + D\right]\hat{u}(s) + \left[C(sI - A)^{-1}P + Q\right]\hat{d}(s)$$

where

$$P = (A_1 - A_2)x + (B_1 - B_2)u \text{ and } Q = (C_1 - C_2)x + (B_1 - B_2)u \quad (21.19)$$

The transfer function in control to output format is obtained as follows

$$\frac{\hat{y}(S)}{\hat{d}(S)} = C(sI - A)^{-1}P + Q \text{ when } \hat{u}(s) = 0 \quad (21.20)$$

The audio susceptibility is given by

$$\frac{\hat{y}(S)}{\hat{d}(S)} = C(sI - A)^{-1}B + D \text{ when } \hat{d}(s) = 0 \quad (21.21)$$

A 24 V–15/36 V converter is designed, and the parameters for deriving the transfer functions are listed in Table 21.1 to demonstrate the controller design. The transfer functions G_{vd} and G_{vg} both in continuous and discrete models are given in the next section.

Transfer functions of the converter discussed:

Transfer function (Control to output) (Figs. 21.5, 21.6 and 21.7):

$$G_{Vd}(s) = \frac{0.99s^4 - 867.7s^3 - 5.09 * 10^8 s^2 - 4.75 * 10^{11} s - 4.86 * 10^{15}}{s^4 + 2928s^3 + 4.611 * 10^7 s^2 + 4.87 * 10^{10} s + 3.604 * 10^{14}}$$

Audio susceptibility (Input to output):

$$G_{Vg}(s) = \frac{0.99s^4 + 2079s^3 + 4.425 * 10^7 s^2 + 3.275 * 10^{10} s + 2.32 * 10^{14}}{s^4 + 2928s^3 + 4.611 * 10^7 s^2 + 4.87 * 10^{10} s + 3.604 * 10^{14}}$$

Table 21.1 Converter parameter for transfer function design

Parameter	Value
V_g (source voltage)	24 V
V_0 (output voltage)	15 V/36 V
L_1, L_2 (inductors)	800 μH
C_1, C_2 (capacitors)	40 μF
R (load resistor)	30 Ω
f_s (switching frequency)	50 kHz

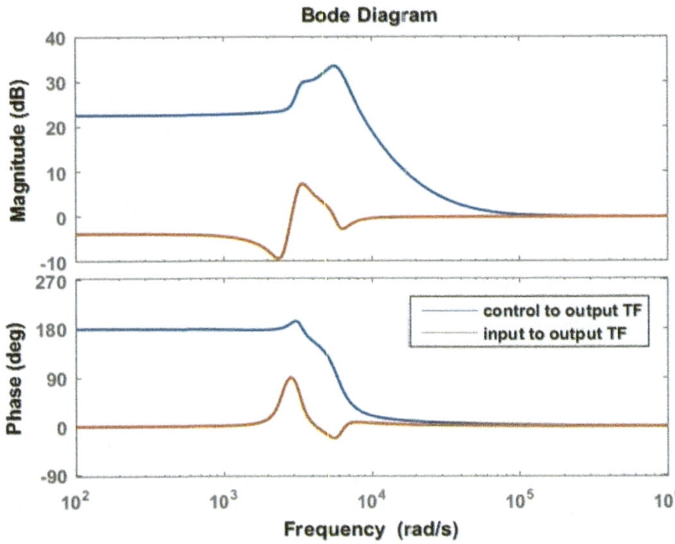

Fig. 21.5 Bode plot of the transfer function

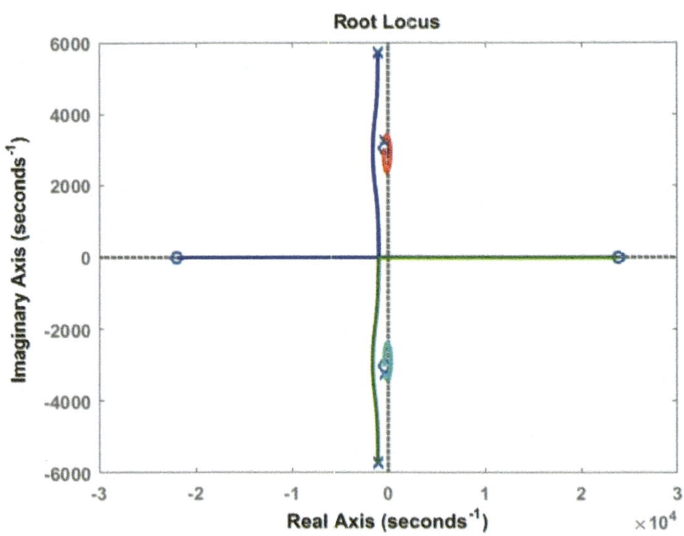

Fig. 21.6 Root locus of G_{vd} function

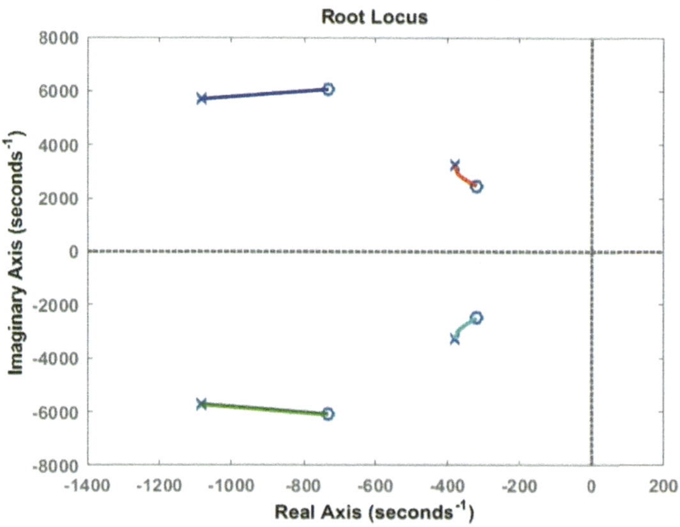

Fig. 21.7 Root locus of G_{vg} function

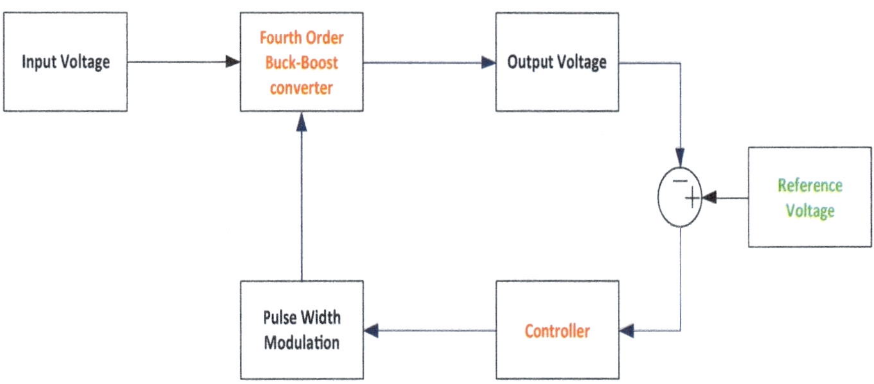

Fig. 21.8 Block diagram of controller implementation

21.3 Controller Design

The main objective of the controller deign is to maintain the output to the desired value depending whether the operation is bucking or boosting. Several challenges are faced while maintaining the output voltage constant irrespective of changes in load as well as external disturbance. It should also be capable of isolating the input from the output. The block diagram (Fig. 21.8.) represents the function of the controller in closed loop. It continuously calculates the error between desired and measured values and applies the correction.

21.3.1 Type II Controller

It is a combination of a phase lead circuit and an inductor. It provides a maximum phase boost of 90°. Even though the discussed converter exhibits non minimum phase, this type of controller exhibits better performance compared to PID controller and reduces the steady-state error.

Let us optimize it into a Type II controller by using the SISOTOOL in MATLAB. Its transfer function is represented as:

$$C(s) = \frac{(1 + s/w_z)}{s/w_{p0}(1 + s/w_p)}$$

where w_z and w_p are the respective zero and pole of Type II controller.

With proper tuning and optimization, we have arrived at the Type II controller's transfer function which can be used for this plant.

21.3.1.1 Design Procedure for Type II Controller

The first step in the design is the derivation of 'k'. Here 'k' in Type II controller is defined as the pole frequency (w_p) to zero frequency (w_z) ratio. The pole zero combination provides an adjustable phase boost ($\emptyset$) from 0 to 90° at crossover frequency (f_c). The relation between k and phase boost provided by the controller is given by

$$k = \tan\left(\frac{\emptyset}{2} + 45^\circ\right)$$

as $k = \frac{f_p}{f_z}$, the location of the pole frequency will be given by

$$f_p = k * f_c = \tan\left(\frac{\emptyset}{2} + 45^\circ\right) * f_c$$

Now the phase boost is calculated from the frequency-phase plot of the Type II controller. The corresponding frequency is the cross-over frequency. The frequency at which maximum boost occurs is calculated from the geometric mean of pole location frequency and zero location frequency.

$$f_{\emptyset\,\max} = \sqrt{f_p * f_z}$$

Now the zero frequency is given by

$$f_z = \frac{f_p}{\tan\left(\frac{\emptyset}{2} + 45^\circ\right)}$$

After substituting the required design value, we can get the following transfer function:

$$G_C(s) = \frac{-58.42(1 + 0.0043s)}{s(1 + 0.033s)}$$

21.3.2 *Type III Controller*

It is capable of giving a maximum phase boost of 180°. When compared with the transfer function in 3.1, this has an extra pair of a pole and a zero [4]. Even though the plant exhibits non minimum phase behavior, this type of controller exhibits better performance compared to Type III controller and reduces the steady state error.

Let us try to optimize the Type II controller into a Type III controller by using the SISOTOOL in MATLAB. Its transfer function is represented as:

$$C(s) = \frac{(1 + s/w_z)}{s/w_{p0}(1 + s/w_p)}$$

where w_{z_1}, w_{z_2} and $w_{p_0}, w_{p_1}, w_{p_2}$ are the respective zero and pole of Type II controller.

With proper tuning and optimization, we have arrived at the Type II controller's transfer function which can be used for this plant.

21.3.2.1 Design Procedure for the Controller

The design procedure is similar to the procedure followed in 3.1.1. Like the Type II controller, the first step is to design the value of 'k'. Here 'k' is the double pole frequency (w_{p_1}, w_{p_2}) to double zero frequency (w_{z_1}, w_{z_2}) ratio. The maximum phase boost $(\emptyset)$ obtained from this pole zero combination is 180° at cross-over frequency (f_c).

$$\sqrt{k} = \tan\left(\frac{\emptyset}{2} + 45^{\circ}\right)$$

Now the pole location is given by

$$f_{p_1}(f_{p_2}) = \sqrt{k} * f_c = \tan\left(\frac{\emptyset}{2} + 45^{\circ}\right) * f_c$$

Now the phase boost is calculated from the bode graph of the controller. The corresponding frequency is the cross-over frequency. The frequency at which maximum boost occurs is calculated from the geometric mean of both the pole location frequency and the zero location frequency.

$$f_{\emptyset\,\max} = \sqrt{f_{p_1}(f_{p_2}) * f_{z_1}(f_{z_2})}$$

Now the zero frequency is given by

$$f_{z_1}(f_{z_2}) = \frac{f_{p_1}(f_{p_z})}{\tan\left(\frac{\emptyset}{2} + 45^{\circ}\right)}$$

After substituting the required design value, we can get the following transfer function:

$$G_C(s) = \frac{-101.18(1 + 6.9 * 10^{-5}\mathrm{s})(1 + 0.0023\ \mathrm{s})}{s(1 + 0.00047\ \mathrm{s})(1 + 0.0033\ \mathrm{s})}$$

21.4 Result and Discussions

The closed-loop system performance of this converter is realized with two different controllers, and their results are shown below. The results are based on simulation performed in MATLAB, and we have achieved satisfied performance. The Type II controller performed very well in terms of the bode response and step response of the uncompensated system. The Type III controller was optimized based on the Type II controller as a base for construction. Although we have achieved very good system response when compared with Type II controller, the order of the controller increases (Figs. 21.9, 21.10 and 21.11).

The step response of the fourth-order buck–boost converter involving Type II controller is given in Fig. 21.11. The transient response gives us the conclusion that this controller performs well in terms of the overshoot as well as the rise time (0.002 s) and settling time (0.01 s). The steepness in the magnitude plot which disturbed the system is also improved (Figs. 21.12, 21.13 and 21.14).

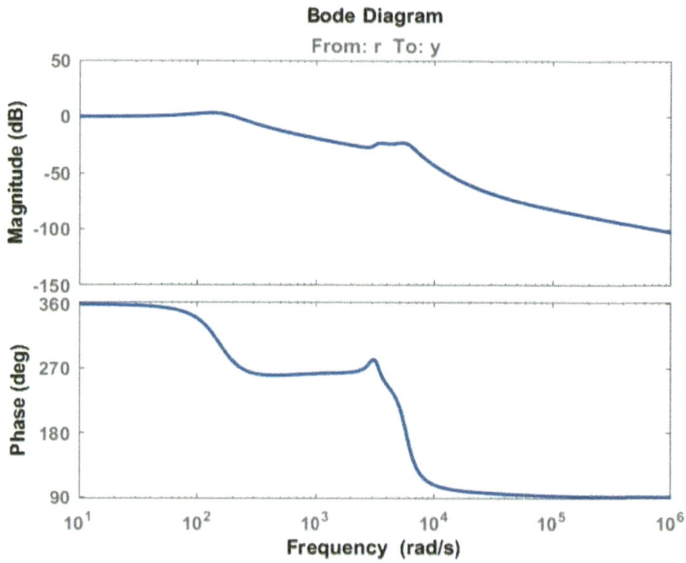

Fig. 21.9 Bode diagram of system response with Type II controller (closed-loop system)

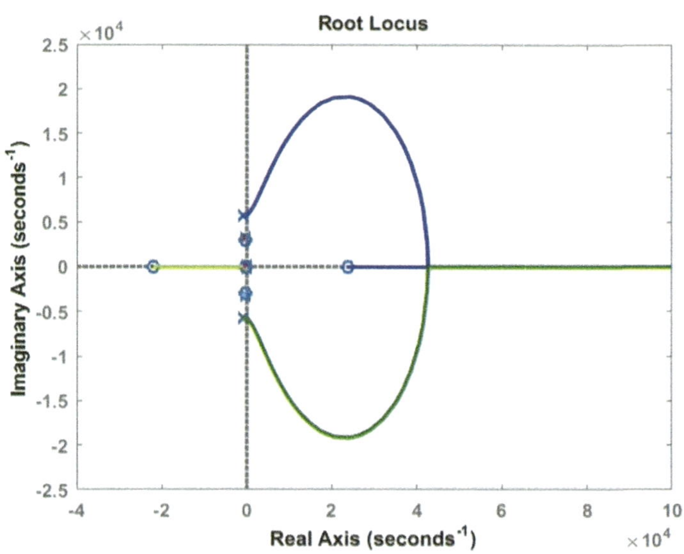

Fig. 21.10 Root locus of system response with Type II controller (closed loop)

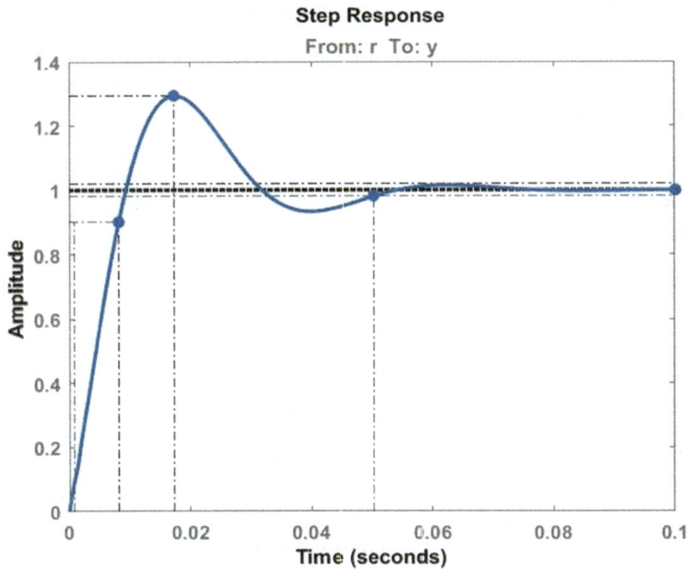

Fig. 21.11 Step response with Type II controller

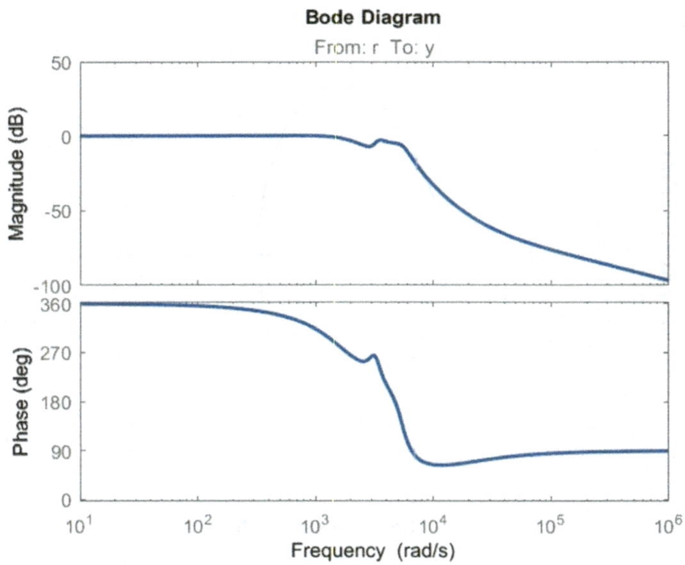

Fig. 21.12 Bode diagram of system response with Type III controller (closed-loop system)

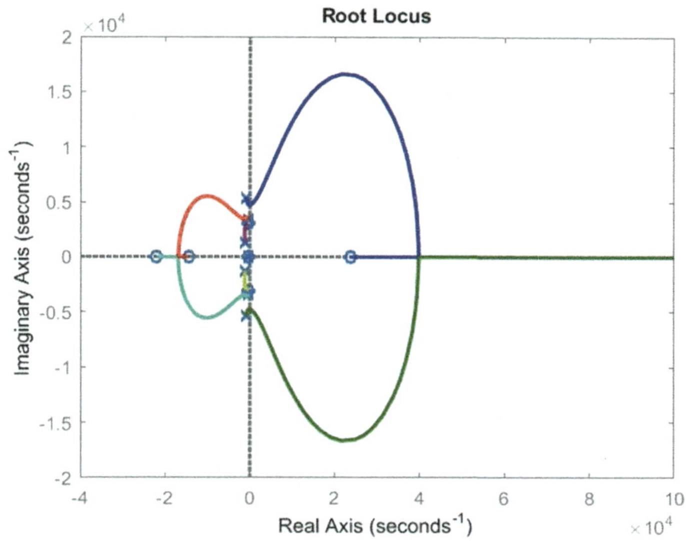

Fig. 21.13 Root locus of the system response with Type III controller (closed-loop system)

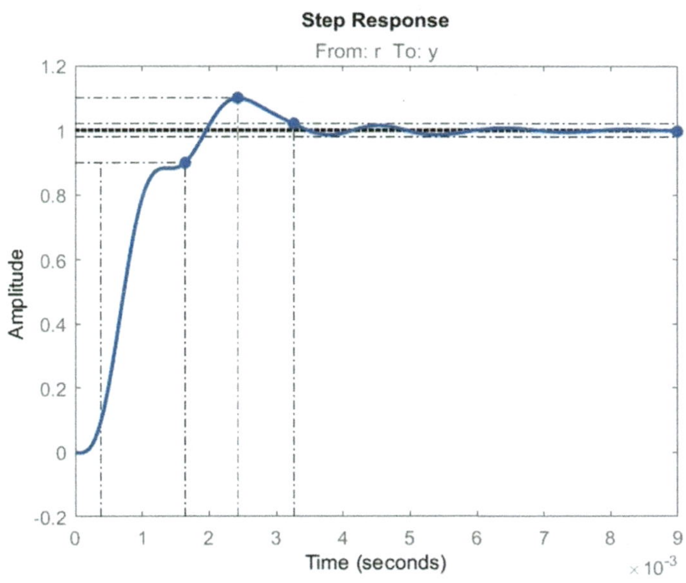

Fig. 21.14 Step response with Type III controller

Table 21.2 Comparison of the two converters

Parameter	Type II controller	Type III controller
Peak time	0.0173 s	0.00242 s
Rise time	0.0073 s	0.00126 s
Settling time	0.0502 s	0.00326 s
Maximum peak overshoot	29%	10%
Gain margin	24.2 dB	8.86 dB
Phase margin	46°	62.5°
Steady-state error	0.0	0.0

As we can see from the above plots, the settling time (less than 0.005 s) and the rise time (<0.002 s) have very much been reduced and hence we can achieve a faster response when compared with the Type III controller. So, we can conclude from our observation that Type III controllers yield a far better result when compared with other controller (Type II) discussed in this paper. We can also observe from Figs. 21.11 and 21.14 that certain small amount of fluctuations present in Fig. 21.11 has been considerably reduced. Thus, Type II and Type III provide a more optimized design when compared with PID controllers.

From Table. 21.2, we can infer that the Type III controller performs well when compared with the Type II controller. The overshoot decreases considerably well in the second case. There is a large possible margin of phase correction that can be done in the case of implementation with Type III controller but in case of the other controller, the gain margin is comparatively higher. The steady-state error is almost same in both the cases of implementation. The difference lies in the settling time which is slower in the case of Type II controller. Hence, the rise time is also quickly achieved according to the desired value in Type III controller minimizing the possible case of fluctuations in output voltage.

21.5 Conclusion

It is observed that the discussed converter is stable and agrees with the reference voltage. It is clear from the graphs in Fig. 21.15 that converter is underdamped initially and after approximately 6.9 ms it settles down in the case of Type II controller and in approximately 4 ms for a Type III controller. The tolerance of duty ratio for a load of 30 Ω is from 0.4 to 0.7. The output voltage in this fourth-order buck-boost converter changes like any other buck–boost converter. When the value of duty ratio is less than 0.5, the output voltage is observed to be 15 V and 36 V when the duty ratio is greater than 0.5. The former is bucking operation while the latter is boosting operation. The maximum peak overshoot in the converter in the transient analysis is also greatly reduced using the Type II controller which is also considerably reduced when operated with Type III controller. The step response of

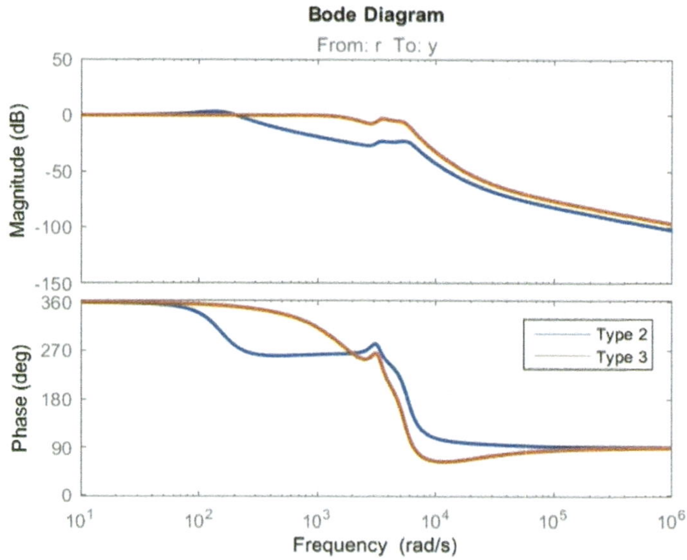

Fig. 21.15 Bode plot comparison with the two controllers

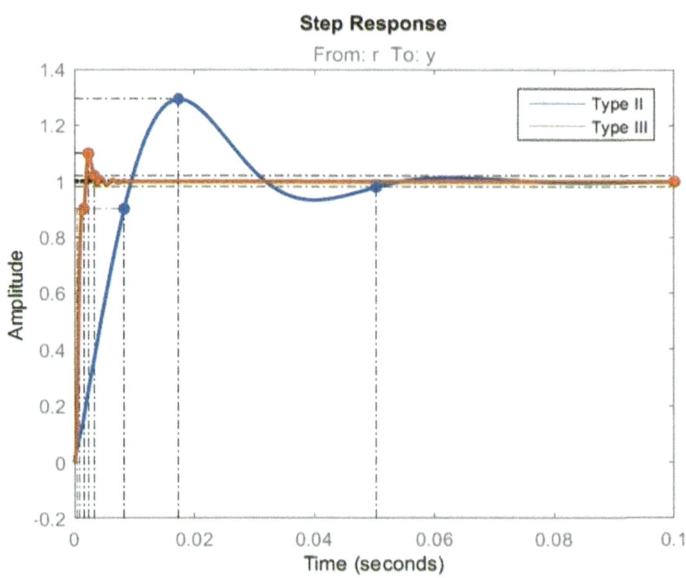

Fig. 21.16 Comparison of step response with the two controllers

the closed-loop converter involving the two controllers is shown in Fig. 21.16. The simulation results are according to the desired value and is improved in terms of quicker response and source fluctuations.

References

1. Veerachary, M., Khubchandani, V.: Control of fourth-order buck-boost converter. In: 2019 IEEE Transportation Electrification Conference (ITEC-India), Bengaluru, India, pp. 1–6 (2019). https://doi.org/10.1109/ITEC-India48457.2019.ITECINDIA2019-139
2. Veerachary, M., Saxena, A.R.: Design of robust digital stabilizing controller for fourth-order boost DC-DC converter: a quantitative feedback theory approach. IEEE Trans. Industr. Electron. **59**(2), 952–963 (2012). https://doi.org/10.1109/TIE.2011.2158040
3. Clerk Maxwell, J.: A Treatise on Electricity and Magnetism, 3rd ed., vol. 2. Oxford, Clarendon, pp. 68–73 (1892)
4. Sudhakar Babu, C., Veerachary, M.: Predictive valley current control for two inductor boost converter. In: Proceedings of the IEEE International Symposium on Industrial Electronics, 2005. ISIE 2005, Dubrovnik, Croatia, vol. 2, pp. 727–731 (2005). https://doi.org/10.1109/ISIE.2005.1529005
5. Priyadarshi, N., Padmanaban, S., Maroti, P.K., Sharma. A.: An extensive practical investigation of FPSO-based MPPT for grid integrated PV system under variable operating conditions with anti-islanding protection. IEEE Syst. J. 1–11 (2018)
6. Priyadarshi, N., Padmanaban, S., Bhaskar, M.S., Blaabjerg, F., Sharma, A.: A fuzzy SVPWM based inverter control realization of grid integrated PV-wind system with FPSO MPPT algorithm for a grid-connected PV/wind power generation system: hardware implementation. IET Electric Power Appl. pp. 1–12 (2018)
7. Ghosh, A., Banerjee, S., Sarkar, M.K., Dutta, P.: Design and implementation of type-II and type-III controller for DC-DC switched-mode boost converter by using K-factor approach and optimisation techniques. In: IET Power Electronics, vol. 9, no. 5, pp. 938–950 (2016). https://doi.org/10.1049/iet-pel.2015.0144
8. Saxena, R., Veerachary, M.: Non-linear sampled-data model for fourth order boost DC-DC converter. In: INTELEC 2009—31st International Telecommunications Energy Conference, Incheon, pp. 1–6 (2009). https://doi.org/10.1109/INTLEC.2009.5351832
9. Veerachary, M., Saxena, A.R.: Optimized power stage design of low source current ripple fourth-order boost DC–DC converter: a PSO approach. IEEE Trans. Industr. Electron. **62**(3), 1491–1502 (2015). https://doi.org/10.1109/TIE.2014.2361316
10. Padmanaban, S., Priyadarshi, N., Holm-Nielsen, J.B., Bhaskar, M.S., Azam, F., Sharma, A. K.: A novel modified sine-cosine optimized MPPT algorithm for grid integrated PV system under real operating conditions. IEEE Access **7**, 10467–10477 (2019). https://doi.org/10.1109/ACCESS.2018.2890533
11. Padmanaban, S., Priyadarshi, N., Holm-Nielsen, J.B., Bhaskar, M.S., Hossain, E., Azam, F.: A hybrid photovoltaic-fuel cell for grid integration with Jaya-based maximum power point tracking: experimental performance evaluation. IEEE Access **7**, 82978–82990 (2019). https://doi.org/10.1109/ACCESS.2019.2924264
12. Ghosh, A., Banerjee, S.: Design of type-III controller for DC-DC switch-mode boost converter. In: 2014 6th IEEE Power India International Conference (PIICON), Delhi, pp. 1–6 (2014) https://doi.org/10.1109/POWERI.2014.7117679
13. Ghosh, A., Banerjee, S.: Control of switched-mode boost converter by using classical and optimized type controllers. J. Control Eng. Appl. Inf. **17**(4), 114–125, Romanian Society of Control Engineering (2015)

14. Ghosh, A., Banerjee, S.: Design and implementation of Type-II compensator in DC-DC switch-mode step-up power supply. In: Proceedings of the 2015 Third International Conference on Computer, Communication, Control and Information Technology (C3IT), Hooghly, pp. 1–5 (2015). https://doi.org/10.1109/C3IT.2015.7060164
15. Sarkar, S., Ghosh, A. Banerjee, S.: Design and implementation of type-III controller in tri state boost converter. In: 2015 Annual IEEE India Conference (INDICON), New Delhi, pp. 1–6 (2015). https://doi.org/10.1109/INDICON.2015.7443152
16. Chatterjee, K., Ghosh, A., Saha, P.K., Das, A.: An analysis of power-factor-correction boost converter's nonlinear dynamics through bifurcation diagrams. In: 2016 International Conference on Intelligent Control Power and Instrumentation (ICICPI), Kolkata, pp. 142–147 (2016). https://doi.org/10.1109/ICICPI.2016.7859691
17. Banerjee, S., Ghosh, A., Rana, N.: Design and fabrication of closed loop two-phase interleaved boost converter with Type-III controller. IECON 2016—42nd Annual Conference of the IEEE Industrial Electronics Society, Florence, pp. 3331–3336 (2016). https://doi.org/10.1109/IECON.2016.7793377
18. Ghosh, A., Banerjee, S.: A comparison between classical and advanced controllers for a boost converter. In: 2018 IEEE International Conference on Power Electronics, Drives and Energy Systems (PEDES), Chennai, India, pp. 1–6 (2018). https://doi.org/10.1109/PEDES.2018.8707911
19. Rana, N., Ghosh, A., Banerjee, S.: Development of an improved Tristate buck-boost converter with optimized type-3 controller. IEEE J. Emerg. Sel. Topics Power Electron. **6**(1), 400–415 (2018). https://doi.org/10.1109/JESTPE.2017.2724847
20. Erickson, R.W., Maksimovic, D.: Fundamentals of power electronics. Springer (2007)
21. Ogata, K.: Modern control engineering. Pearson Education, India (2010)
22. Vardia, M., Priyadarshi, N., Ali, I., Azam, F., Bhoi, A.K.: Maximum power point tracking for wind energy conversion system. In: Bhoi, A., Sherpa, K., Kalam, A., Chae, G.S. (eds) Advances in Greener Energy Technologies. Green Energy and Technology. Springer, Singapore (2020). https://doi.org/10.1007/978-981-15-4246-6_36
23. Choudhary, T., Priyadarshi, N., Kumar, P., Azam, F., Bhoi, A.K.: A fuzzy logic control based vibration control system for renewable application. In: Bhoi, A., Sherpa, K., Kalam, A., Chae, G.S. (eds) Advances in Greener Energy Technologies. Green Energy and Technology. Springer, Singapore (2020). https://doi.org/10.1007/978-981-15-4246-6_38
24. Davoudi, A., Kong, N., Behjati, H., et al.: Automated system identification and controller tuning for digitally controlled dc–dc converters. IET Power Electron. **5**(6), 765–772 (2012). https://doi.org/10.1049/iet-pel.2011.0085
25. Sarkar, S., Ghosh, S.S.: Comparison of advanced analog controllers for a DC-DC boost converter. In: 2020 IEEE 9th Power India International Conference (PIICON), SONEPAT, India, pp. 1–6 (2020). https://doi.org/10.1109/PIICON49524.2020.9113013
26. Veerachary, M., Suresh, M.: Digital voltage-mode control of higher order boost converter. In: 2008 Joint International Conference on Power System Technology and IEEE Power India Conference, New Delhi, pp. 1–6 (2008). https://doi.org/10.1109/ICPST.2008.4745157

Chapter 22
Fractional-Order PI-Lead Controller Design of DC–DC Power Converter for Renewable Energy Applications

Saurav Prajapati and Man Mohan Garg

22.1 Introduction

In the present world of modernization and industrialization, the increase in emission, continuous depletion of old natural resources and rapid hike in demand are the major concern for energy consumption. Research toward solving these issues has given an idea to incorporate renewable sources such as wind energy using wind turbine, photovoltaic cell (PV), hydropower, etc., as alternate resources [1–15]. Integration of these renewable sources in power system can be done using microgrid concept that may vary depending upon our requirements and constraints. The microgrid concept has advantages like reduction of transmission losses, power quality improvement, reliability, etc. However, undesired fluctuations in voltages generated from renewable sources give an idea to incorporate power electronics converters for steady flow of power. The DC–DC converter used in DC microgrid [1, 2] is mainly responsible for interfacing of various renewable resources to the DC grid. Therefore, the field of various DC–DC power converters topologies and its precise control have become a very interesting research area.

Figure 22.1 depicts a typical DC microgrid concept. For sake of simplicity, only specific resources connection is shown. In this chapter, we will mainly be discussing about control of power converter for interfacing of solar PV energy sources to DC microgrid. Though, there are different power converter topologies existing in the literature, but depending upon PV array voltage and DC grid voltage, a specific topology may be chosen. In this chapter, we have assumed that PV output voltage is greater than DC grid voltage. Therefore, buck topology has been used, and its controller design will be studied.

S. Prajapati (✉)
Department of Electrical Engineering, NIT, Rourkela 769008, India

M. M. Garg
Department of Electrical Engineering, MNIT, Jaipur 302017, India

N. Priyadarshi et al. (eds.), *DC–DC Converters for Future Renewable Energy Systems*, Energy Systems in Electrical Engineering,
https://doi.org/10.1007/978-981-16-4388-0_22

In the present world of modernization, the field of control system has become a very promising research area. Recently, a new control strategy called fractional-order control system [3] has been reported by the researchers. With the advent of fractional calculus [4] and its application in control system, the mathematical modeling of the real-time physical system became quite simpler and popular. This control strategy has been found to be more robust [5], accurate and precise as compared to existing control strategies. Though fractional-order calculus has received popularity, it also creates a need for new developments in the stability analysis [6] and controller design for the system.

Though many fractional-order controller techniques are present in the literature, they are either applied to a specific type of system or mathematically very complex and tedious [7–10]. The basic controllers conventionally used in control systems are proportional, derivative, integral, lead, lag or combination of these. These conventional controllers have certain limitations. For example, to speed up the response, proportional controller is generally used, but it affects the steady-state accuracy as well as relative stability. Similarly, integral control action is used to reduce the closed-loop steady-state error (SSE), but it decreases the relative stability. Derivative is used to improve the transient response, but it is sensitive to noise at high frequency. Lag compensator is used to improve SSE like integral controller, but it adds negative phase at desired gain cross-over frequency (GCF). Lead controller adds positive phase at desired GCF as well as improves phase margin (PM) but does not help in improving steady-state error. Therefore, a new generalized technique based on frequency-domain specification has been implemented which is mathematically simpler and gives improved results in comparison with the existing techniques.

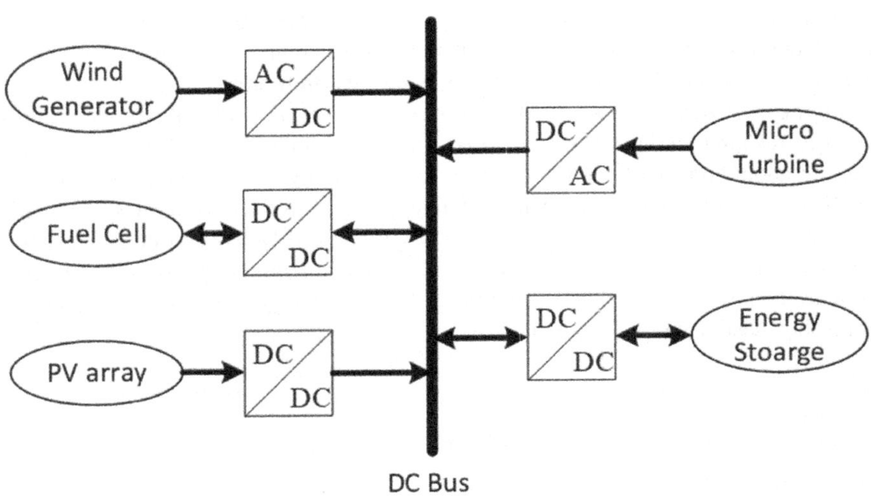

Fig. 22.1 DC microgrid

The main control objective in our case is to remove the closed-loop SSE and to have the specified gain and phase at a specific cross-over frequency ω_c (kHz) with the help of compensator. For this purpose, we used fractional-order PI-Lead compensator (FOPIL), which is nothing but cascade form of fractional-order PI and Lead controllers.

FOPIL controller can be divided into two parts, i.e., PI controller part and FO lead controller part. To remove the SSE of closed-loop system, PI controller is used whose parameters calculation is discussed in this paper, and to improve the transient performance as well as achieve desired phase margin at given GCF, FO lead controller is used. The great flexibility in fractional-order PI-Lead controller makes it possible to attain the required frequency-domain performance indices.

Section 28.1 gives the basic introduction about importance of power electronics converters in DC microgrid application and different types of controllers which exist in the literature, its limitation and further importance of FOPIL controller over conventional one. The transfer function models for non-ideal DC–DC buck converter are presented in Sect. 28.2. In this section, an effort has been made to include all types of non-idealises present in electrical components, i.e., ESRs of capacitors, inductor, etc. Sections 28.3 and 28.4 describe the closed-loop control strategy for tuning fractional-order PI-Lead controller parameters. The. FOPI-Lead controller parameters calculation for non-ideal buck converter is discussed in Sect. 28.5 followed by its simulation results and conclusion in Sects. 28.6 and 28.7.

22.2 Mathematical Modeling of Non-ideal Buck Converter

Before designing any closed-loop control, the mathematical model is necessary. It does not only help in designing a closed-loop controller but also helps in stability analysis. In the literature, there are several techniques present for mathematical modeling of DC–DC converters like circuit averaging, current-injected approach, state space averaging (SSA), etc. [11, 12]. The simplest and most frequently used technique is SSA technique. In this chapter, SSA technique is used for determining the duty cycle to output voltage transfer function of a practical buck converter.

Circuit configuration of a buck converter having non-idealises is presented in Fig. 22.2. The converter's input voltage is taken as PV array *voltage, i.e.,* $V_{PV} = V_g$. The circuit consists of MOSFETS, diode D_d, charge storing element capacitor C, inductor L and load resistor R. Apart from that to study the real case scenario of converter's behavior, we have included non-idealises, i.e., ESRs of inductor and capacitor be r_L and r_C, diode resistance r_D and switch resistance r_{SW}. The operating frequency of converter is taken as f and duty cycle be D. The converter's two operating modes in continuous conduction mode (CCM) are discussed below.

Mode-1: On Duration

In this case, the diode is reversed biased, so it acts as open circuited. The inductor (L) stores energy from input voltage V_g, and the current through it (i_L) increases linearly till it reaches maximum value.

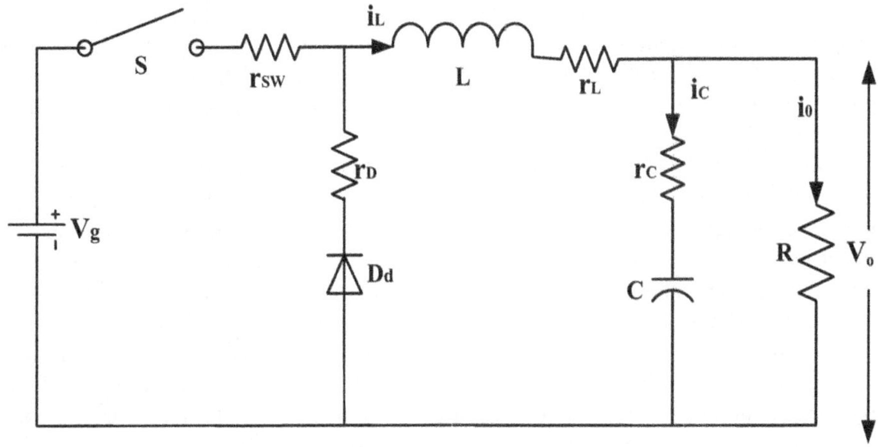

Fig. 22.2 Non-ideal DC–DC buck converter

The state space equations for this period are

$$L\frac{\mathrm{d}i_L(t)}{\mathrm{d}t} = -\left(r_{SW} + r_L + \frac{r_C R}{R + r_C}\right) i_L(t) - \frac{R}{R + r_C} v_c(t) + v_g(t) \tag{22.1}$$

$$C\frac{\mathrm{d}v_C(t)}{\mathrm{d}t} = \frac{R}{R + r_C} i_L(t) - \frac{1}{R + r_C} v_C(t) \tag{22.2}$$

$$v_0(t) = \frac{r_C R}{R + r_C} i_L(t) + \frac{R}{R + r_C} v_C(t) \tag{22.3}$$

In matrix form,

$$A_1 = \begin{pmatrix} \frac{-1}{L}\left(r_{SW} + r_L + \frac{r_C R}{R + r_C}\right) & -\frac{R}{L(R + r_C)} \\ \frac{R}{C(R + r_C)} & \frac{-1}{C(R + r_C)} \end{pmatrix}, \quad B_1 = \begin{pmatrix} \frac{1}{L} \\ 0 \end{pmatrix}, \\ C_1 = \begin{pmatrix} \frac{r_C R}{R + r_C} & \frac{R}{R + r_C} \end{pmatrix} \tag{22.4}$$

Mode-2: OFF Duration

In this case, the switch is off, and diode is on having forward voltage drop. The stored charge of inductor is now discharged via path D_d, C and R.

$$L\frac{\mathrm{d}i_L(t)}{\mathrm{d}t} = -\left(r_d + r_L + \frac{r_C R}{R + r_C}\right) i_L(t) - \frac{R}{R + r_C} v_c(t) \tag{22.5}$$

$$C\frac{\mathrm{d}v_C(t)}{\mathrm{d}t} = \frac{R}{R + r_C} i_L(t) - \frac{1}{R + r_C} v_C(t) \tag{22.6}$$

In matrix form

$$A_2 = \begin{pmatrix} \frac{-1}{L}\left(r_{SW} + r_L + \frac{r_C R}{R + r_C}\right) & -\frac{R}{L(R + r_C)} \\ \frac{R}{C(R + r_C)} & \frac{-1}{C(R + r_C)} \end{pmatrix}, \quad B_2 = \begin{pmatrix} 0 \\ 0 \end{pmatrix}, \quad C_2 = \begin{pmatrix} \frac{r_C R}{R + r_C} & \frac{R}{R + r_C} \end{pmatrix} \tag{22.7}$$

The duty cycle to output voltage transfer function is given as [13]:

$$\frac{v_O(s)}{d(s)} = C(sI - A)^{-1} B_d \tag{22.8}$$

where

$$A = A_1 D + A_2(1 - D), \quad C = C_1 D + C_2(1 - D), \quad B_d = (A_1 - A_2)\begin{pmatrix} I_L \\ V_o \end{pmatrix} - (B_1 - B_2)V_g \tag{22.9}$$

Here, D, V_o and I_L are the steady-state values.
After solving

$$A = \begin{pmatrix} \frac{-1}{L}\left(r_a + r_L + \frac{r_C R}{R + r_C}\right) & -\frac{R}{L(R + r_C)} \\ \frac{R}{C(R + r_C)} & \frac{-1}{C(R + r_C)} \end{pmatrix}, \quad B_d = \begin{pmatrix} \frac{1}{L}\left((r_d - r_{SW})I_L + v_g\right) \\ 0 \end{pmatrix}, \quad C = \begin{pmatrix} \frac{r_C R}{R + r_C} & \frac{R}{R + r_C} \end{pmatrix}, \quad r_a = D r_{SW} + (1 - D) r_d \tag{22.10}$$

Placing the matrices A, B_d, C in (22.8), the required transfer function is derived as

$$G(s) = \frac{R\left((r_d - r_{SW})I_L + v_g\right)(r_C C s + 1)}{(R + r_C)LCs^2 + [L + C(r_C R + (r_a + r_L)(R + r_C))]s + (R + r_a + r_L)} \tag{22.11}$$

22.3 Closed-Loop Control Analysis for PV System

This section gives an overview of closed-loop control analysis for PV array-fed buck converter using FOPI-lead controller. Figure 22.3 shows that the output voltage V_o is compared with desired reference DC bus voltage and calculated error signal is given to the controller. The controller gives controlled output voltage which in turn compared with high-frequency repeating signal (sawtooth waveform) to generate PWM signal that will control the switching operation of switch S. For

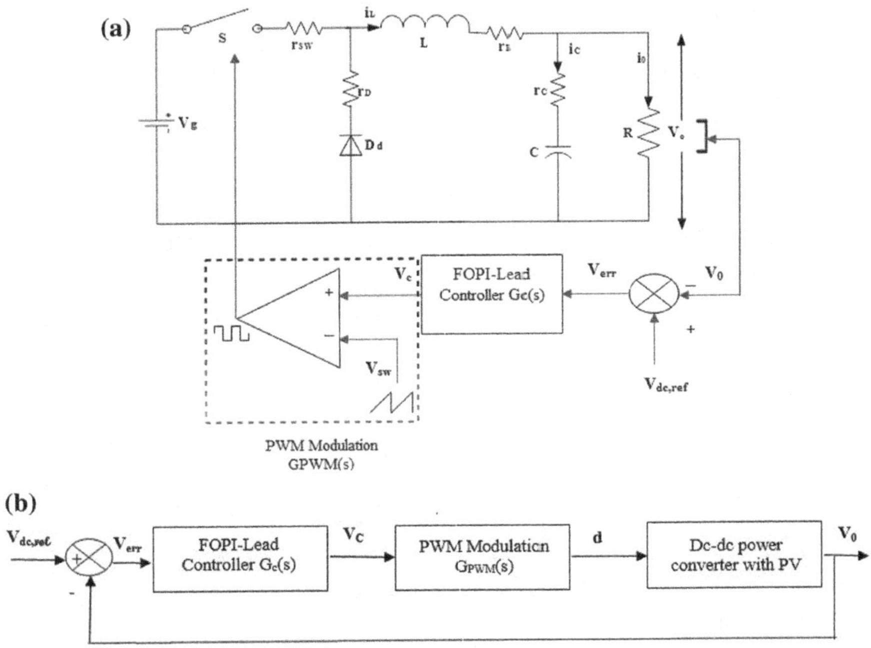

Fig. 22.3 **a** Closed-loop control of FOPI-Lead controlled PV system. **b** Block diagram

simplicity, the magnitude of repeating reference signal, i.e., sawtooth signal is taken as 1 V, and frequency is taken as 20 kHz. In order to check the output voltage regulation of the converter in the presence of any disturbance, i.e., load variation, input voltage variation, etc., the parameter of FOPI-Lead compensator should be tuned properly. In the next section of this chapter, we will discuss the designing of controller parameters in detail.

22.4 FOPI-Lead Controller Design

This section describes a control technique for calculating the fractional-order PI-Lead controller parameters to achieve desired frequency performance-based specification [13]. In the next section, this control technique is applied for a given non-ideal buck converter.

The FOPIL controller is cascaded form of PI and FO lead compensator. The transfer function for FOPI-Lead controller is given as

$$G_{\text{lead}}(s) = \frac{K_{\text{lead}}\left(s^{\lambda} + \alpha\right)}{\left(s^{\lambda} + \beta\right)} \cdot \frac{K_p s + K_i}{s}, \quad \alpha < \beta \tag{22.12}$$

$$G_{\text{lead}}(s) = \frac{K_{\text{lead}} \cdot K_i \left(\frac{s}{K_i/K_p} + 1\right)\left(s^{\lambda} + \alpha\right)}{\left(s^{\lambda} + \beta\right)} \tag{22.13}$$

$$G_{\text{lead}}(s) = \frac{K\left(\frac{s}{\omega_z} + 1\right)\left(s^{\lambda} + \alpha\right)}{\left(s^{\lambda} + \beta\right)} \tag{22.14}$$

where $\omega_z = \frac{K_i}{K_p}$ and $K = K_{\text{lead}}.K_i$.

Therefore, K, α, β, λ are unknown quantities which have to be determined. The procedure for finding unknown quantities is as follows.

22.4.1 Fractional-Order PI (FOPI) Controller Design

Let the generalized transfer function for non-ideal uncompensated buck converter be

$$G(s) = \frac{a_1 s + a_0}{b_2 s^2 + b_1 s + b_0} \tag{22.15}$$

The main control objective in our case is to eliminate the closed-loop steady-state error and to provide the required gain and phase at a desired cross-over frequency ω_c (kHz) with the help of compensator. For this purpose, we used fractional-order PI-Lead compensator, which can be divided into two sections (a) PI section and (b) Fractional lead section. Also, this transfer function has type zero with poor gain at lower frequencies.

The role of PI compensator in our case is to improve the gain at lower frequencies. Let the PI controller as

$$G_{\text{PI}}(s) = \frac{\frac{s}{\omega_z} + 1}{s} \tag{22.16}$$

where ω_z is the cutoff frequency of PI controller. The value of ω_z is chosen sufficiently low (1/10th of gain cross-over frequency) to increase low-frequency gain.

The transfer function of compensated system, i.e., with PI controller is

$$G_1(s) = G_{\text{PI}}(s) \times G(s) = \frac{\frac{a_1}{\omega_z} s^2 + \left(a_1 + \frac{a_0}{\omega_z}\right) s + a_0}{b_2 s^3 + b_1 s^2 + b_0 s} \tag{22.17}$$

PI controller eliminates the closed-loop SSE. Now, FO-Lead controller is designed to improve transient response.

Let the desired PM and GCF of overall compensated system be ϕ_{margin} (deg) and ω_{gc} (rad/s). Also, the magnitude and phase of compensated system $G_1(s)$ (i.e., with PI controller) at gain cross-over frequency ω_{gc} are K_1 and ϕ_1, respectively, i.e.,

$$\phi_1 = \angle G_1(j\omega_{\text{gc}}), \quad K_1 = |G_1(j\omega_{\text{gc}})| \tag{22.18}$$

To achieve the desired PM and GCF exactly, the FO-Lead compensator is to, be designed such that it must contribute magnitude K_{req} and phase angle ϕ_{req} at frequency ω_{gc}.

$$\begin{aligned} K_{\text{req}} &= 1/K_1 \\ \phi_{\text{req}} &= -180 - \phi_1 + \phi_{\text{desired}} \end{aligned} \tag{22.19}$$

22.4.2 Fractional-Order PI-Lead Controller Design

Let the fractional-order lead compensator [13–15] be

$$G_{\text{lead}}(s) = K\frac{(s^{\lambda} + \alpha)}{(s^{\lambda} + \beta)} \tag{22.20}$$

The magnitude and phase of above transfer function are given below:

$$G_{\text{lead}}(j\omega) = K\frac{(j\omega)^{\lambda} + \alpha}{(j\omega)^{\lambda} + \beta} = K\frac{\omega^{\lambda}(\cos\frac{\lambda\pi}{2} + j\sin\frac{\lambda\pi}{2}) + \alpha}{\omega^{\lambda}(\cos\frac{\lambda\pi}{2} + j\sin\frac{\lambda\pi}{2}) + \beta} \tag{22.21}$$

The magnitude part:

$$|G(j\omega)| = K\frac{\sqrt{(\omega^{\lambda}\cos\frac{\lambda\pi}{2} + \alpha)^2 + (\omega^{\lambda}\sin\frac{\lambda\pi}{2})^2}}{\sqrt{(\omega^{\lambda}\cos\frac{\lambda\pi}{2} + \alpha)^2 + (\omega^{\lambda}\sin\frac{\lambda\pi}{2})^2}} \tag{22.22}$$

and the real part:

$$\phi_{\text{lead}}(s) = \tan^{-1}\frac{(\omega^{\lambda}\sin(\frac{\lambda\pi}{2}))}{(\alpha + \omega^{\lambda}\cos(\frac{\lambda\pi}{2}))} - \tan^{-1}\frac{(\omega^{\lambda}\sin(\frac{\lambda\pi}{2}))}{(\beta + \omega^{\lambda}\cos(\frac{\lambda\pi}{2}))} \tag{22.23}$$

Solving above equation

$$\phi_{\text{lead}}(s) = \tan^{-1}\frac{(\beta - \alpha)\omega^{\lambda}\sin\frac{\lambda\pi}{2}}{\left(1 + \frac{\omega^{2\lambda}\sin^2\frac{\lambda\pi}{2}}{(\alpha + \omega^{\lambda}\cos\frac{\lambda\pi}{2})(\beta + \omega^{\lambda}\sin\frac{\lambda\pi}{2})}\right)} \tag{22.24}$$

$$\phi_{\text{lead}}(s) = \tan^{-1}\frac{(\beta-\alpha)\omega^{\lambda}\sin\frac{\lambda\pi}{2}}{\left(\alpha\beta+\omega^{2\lambda}+(\alpha+\beta)\omega^{\lambda}\cos\frac{\lambda\pi}{2}\right)} \tag{22.25}$$

Let

$$c = \left(\alpha\beta+\omega^{2\lambda}+(\alpha+\beta)\omega^{\lambda}\cos\frac{\lambda\pi}{2}\right) \tag{22.26}$$

For finding optimum (ω), i.e., the frequency at which fractional-order lead controller gives maximum phase angle, take derivate of (22.25) with respect to 'ω',

$$c\lambda\omega^{\lambda-1}\sin\frac{\lambda\pi}{2}(\beta-\alpha) = \omega^{\lambda}\sin\frac{\lambda\pi}{2}(\beta-\alpha)\left(2\lambda\omega^{2\lambda-1}+\lambda\omega^{\lambda-1}(\alpha+\beta)\cos\frac{\lambda\pi}{2}\right) \tag{22.27}$$

Substituting the value of 'c' in (22.27), we get

$$\begin{aligned}\alpha\beta &= \omega^{2\lambda}\\ \omega^{\lambda} &= \sqrt{\alpha\beta}\end{aligned} \tag{22.28}$$

The magnitude and phase value at maximum phase angle frequency are

$$\phi_{\text{lead}}(s) = \tan^{-1}\frac{\left(\sqrt{\alpha\beta}\sin\frac{\lambda\pi}{2}(\beta-\alpha)\right)}{\left(2\alpha\beta+(\alpha+\beta)\sqrt{\alpha\beta}\cos\frac{\lambda\pi}{2}\right)} \tag{22.29}$$

After solving,

$$\phi_{\text{lead}}(s) = \tan^{-1}\frac{\left(\sin\frac{\lambda\pi}{2}(\beta-\alpha)\right)}{\left(2\sqrt{\alpha\beta}+(\alpha+\beta)\cos\frac{\lambda\pi}{2}\right)} \tag{22.30}$$

In terms of $\sin\varphi_{\text{lead}}$, above equation can be expressed as

$$\sin\phi_{\text{lead}}(s) = \frac{\left(\sin\frac{\lambda\pi}{2}(\beta-\alpha)\right)}{\left(2\sqrt{\alpha\beta}\cos\frac{\lambda\pi}{2}+(\alpha+\beta)\right)} \tag{22.31}$$

Similarly, above equation can be expressed in terms of $\tan\varphi_{\text{lead}}$, which will give

$$\tan\phi_{\text{lead}}(s) = \frac{\left(\sin\frac{\lambda\pi}{2}(\beta-\alpha)\right)}{\left(2\sqrt{\alpha\beta}+(\alpha+\beta)\cos\frac{\lambda\pi}{2}\right)} \tag{22.32}$$

22.4.3 Calculation of β

For finding the value of β, substitute the value of α from Eq. (22.28) into (22.32), and we get

$$\tan\phi_{\text{lead}}(s) = \frac{\left(\sin\frac{\lambda\pi}{2}\left(\beta - \frac{\omega^{2\lambda}}{\beta}\right)\right)}{\left(2\omega^{\lambda} + \left(\frac{\omega^{2\lambda}}{\beta} + \beta\right)\cos\frac{\lambda\pi}{2}\right)} \tag{22.33}$$

Solving the above equation for the β, we will get quadratic equation in terms of β which can be solved as follows:

$$2\omega^{\lambda}\beta\tan\phi_{\text{lead}} + \beta^2\cos\frac{\lambda\pi}{2}\tan\phi_{\text{lead}} + \omega^{2\lambda}\cos\frac{\lambda\pi}{2}\tan\phi_{\text{lead}} = \beta^2\sin\frac{\lambda\pi}{2} - \omega^{2\lambda}\sin\frac{\lambda\pi}{2} \tag{22.34}$$

$$\beta^2\left(\cos\frac{\lambda\pi}{2}\tan\phi_{\text{lead}} - \sin\frac{\lambda\pi}{2}\right) + \beta\left(2\omega^{\lambda}\tan\phi_{\text{lead}}\right) + \omega^{2\lambda}\left(\cos\frac{\lambda\pi}{2}\tan\phi_{\text{lead}} + \sin\frac{\lambda\pi}{2}\right) = 0. \tag{22.35}$$

On simplification of the quadratic equation,

$$a = \left(\cos\frac{\lambda\pi}{2}\tan\phi_{\text{lead}} - \sin\frac{\lambda\pi}{2}\right), \quad b = \left(2\omega^{\lambda}\tan\phi_{lead}\right), \\ c = \omega^{2\lambda}\left(\cos\frac{\lambda\pi}{2}\tan\phi_{\text{lead}} + \sin\frac{\lambda\pi}{2}\right) \tag{22.36}$$

The value of β can be expressed as

$$\beta = \frac{-b \pm \sqrt{b^2 - 4ac}}{2a}$$

22.4.4 Calculation of α

Similarly, for finding the value of α, substitute the value of β from Eq. (22.28) into (22.32), and we get

$$\tan\phi_{\text{lead}}(s) = \frac{\left(\sin\frac{\lambda\pi}{2}\left(\frac{\omega^{2\lambda}}{\alpha} - \alpha\right)\right)}{\left(2\omega^{\lambda} + \left(\frac{\omega^{2\lambda}}{\alpha} + \alpha\right)\cos\frac{\lambda\pi}{2}\right)} \tag{22.37}$$

Solving the above equation, we get

$$2\omega^{\lambda}\alpha\tan\phi_{\text{lead}} + \alpha^2\cos\frac{\lambda\pi}{2}\tan\phi_{\text{lead}} + \omega^{2\lambda}\cos\frac{\lambda\pi}{2}\tan\phi_{\text{lead}} = \omega^{2\lambda}\sin\frac{\lambda\pi}{2} - \alpha^2\sin\frac{\lambda\pi}{2} \tag{22.38}$$

On simplification of the quadratic equation,

$$\alpha^2\left(\cos\frac{\lambda\pi}{2}\tan\phi_{\text{lead}} + \sin\frac{\lambda\pi}{2}\right) + \alpha\left(2\omega^{\lambda}\tan\phi_{\text{lead}}\right) + \omega^{2\lambda}\left(\cos\frac{\lambda\pi}{2}\tan\phi_{\text{lead}} - \sin\frac{\lambda\pi}{2}\right) = 0 \tag{22.39}$$

Substituting below values:

$$\begin{aligned} a_1 &= \left(\cos\frac{\lambda\pi}{2}\tan\phi_{\text{lead}} + \sin\frac{\lambda\pi}{2}\right), \\ b_1 &= \left(2\omega^{\lambda}\tan\phi_{\text{lead}}\right), \\ c_1 &= \omega^{2\lambda}\left(\cos\frac{\lambda\pi}{2}\tan\phi_{\text{lead}} - \sin\frac{\lambda\pi}{2}\right) \end{aligned} \tag{22.40}$$

The value of α can be expressed as

$$\alpha = \frac{-b_1 \pm \sqrt{b_1^2 - 4a_1c_1}}{2a_1}$$

22.4.5 Calculation of K_m

The magnitude of the FO lead compensator transfer function is

$$|G(j\omega)| = K\frac{\sqrt{\left(\omega^{\lambda}\cos\frac{\lambda\pi}{2} + \alpha\right)^2 + \left(\omega^{\lambda}\sin\frac{\lambda\pi}{2}\right)^2}}{\sqrt{\left(\omega^{\lambda}\cos\frac{\lambda\pi}{2} + \alpha\right)^2 + \left(\omega^{\lambda}\sin\frac{\lambda\pi}{2}\right)^2}} \tag{22.41}$$

Substituting the value of ω from (22.28), we get

$$K_m = K\sqrt{\frac{\alpha}{\beta}}\sqrt{\frac{\left(\sqrt{\beta}\cos\frac{\pi\lambda}{2} + \sqrt{\alpha}\right)^2 + \left(\sqrt{\beta}\sin\frac{\lambda\pi}{2}\right)^2}{\left(\sqrt{\alpha}\sin\frac{\pi\lambda}{2} + \sqrt{\beta}\right)^2 + \left(\sqrt{\alpha}\sin\frac{\lambda\pi}{2}\right)^2}} \tag{22.42}$$

$$K_{\text{req.}} = K_m = K\sqrt{\frac{\alpha}{\beta}}\sqrt{\frac{\left(\sqrt{\beta}\cos\frac{\pi\lambda}{2} + \sqrt{\alpha}\right)^2 + \left(\sqrt{\beta}\sin\frac{\lambda\pi}{2}\right)^2}{\left(\sqrt{\alpha}\sin\frac{\pi\lambda}{2} + \sqrt{\beta}\right)^2 + \left(\sqrt{\alpha}\sin\frac{\lambda\pi}{2}\right)^2}} \tag{22.43}$$

Summary

FOPI-lead controller has mainly two sections, i.e., PI controller and fractional-order lead controller, whose parameters can be decided as follows:

1. Determine the transfer function of uncompensated open-loop converter and specify the corresponding desired PM (ϕ_{margin}) and GCF (ω_{gc} in rad/s).
2. For designing FOPI controller parameters, first thing that we require is placement of open-loop zero (ω_z) for uncompensated transfer function. In general, we keep its value as low as possible (1/10th of gain cross-over frequency).
3. For a mentioned desired PM (ϕ_{margin}) and GCF (ω_{gc} in rad/s), determine the K_{req} and ϕ_{req} for obtaining lead parameters.
4. Calculate the value of α, β, Kusing (22.28) and (22.32), and finally, substitute all the calculated parameters in (22.20) to get lead section of the controller.

22.5 Controller Parameters Design for DC–DC Buck Converter

The component values of buck converter under test are given in Table 22.1 [16]. The calculation of PI-Lead compensator parameters is given below, and corresponding performance analysis of closed-loop compensated system for f_z = 40 Hz, f_{gc} = 2 kHz with PM = 30°, 45°and 75° is as follows. By varying the value of λ, we can achieve any desired PM (ϕ_{margin}) and GCF (ω_{gc} in rad/s). The performance analysis of compensated buck converter system by varying controller parameters is mentioned below:

Substituting the $\omega_z = 2 \times \pi \times 40$ into Eq. (22.16), the transfer function of PI section is

$$G_{\text{PI}}(s)|_{f_z=40Hz} = \frac{0.003979s + 1}{s} \tag{22.44}$$

and corresponding lead parameters for $\lambda = 1.0$, PM $= 30°$ are

$$K = 235.36, \quad \alpha = 11278, \quad \beta = 1.4 \times 10^4 \tag{22.45}$$

Table 22.1 Parameters of buck converter

Parameters	Value
Input voltage (V_g)	16–24 V
Inductance (L), r_L	1.1 mH, 0.18 Ω
Capacitance (C), r_C	84 μF, 0.3 Ω
Switching frequency (f), r_{SW}	10 kHz, 0.044 Ω
Load resistance (R)	11–22 Ω
Output voltage (V_o)	12 V
Duty cycle (D), r_D	0.5, 0.024 Ω

Table 22.2 Performance analysis of closed-loop buck converter system f_z = 40 Hz, f_{gc} = 2 kHz and PM = 30°

Lambda (λ)	PM (deg.)	GM (dB)	Rise time t_r (ms)	Settling time t_s (ms)	Overshoot (%M_p)	S.S error (%)
0.9	28.3	Inf	98.1	4.6	40.58	0.8418
1.0	30.0	Inf	97.41	4.05	39.0	0.7981
1.1	32.28	Inf	97.08	4.44	36.9	0.7543
1.2	35.41	Inf	96.8	4.41	34.40	0.7354

Table 22.3 Performance analysis of closed-loop buck converter system f_z = 40 Hz, f_{gc} = 2 kHz and PM = 45°

Lambda (λ)	PM (deg.)	GM (dB)	Rise time t_r (ms)	Settling time t_s (ms)	Overshoot (%M_p)	S.S error (%)
0.9	39.34	Inf	109.38	5.87	29.54	1.3676
1.0	45.0	Inf	104.5	5.48	24.84	1.1498
1.1	52.56	Inf	102.1	5.26	19.07	1.0305
1.2	62.53	Inf	100.6	5.13	11.81	0.9732

Table 22.4 Performance analysis of closed-loop buck converter system f_z = 40 Hz, f_{gc} = 2 kHz and PM = 75°

Lambda (λ)	PM (deg.)	GM (dB)	Rise time t_r (ms)	Settling time t_s (ms)	Overshoot (%M_p)	S.S error (%)
0.9	62.28	Inf	$4*10^3$	18.4	10.9	6.2980
1.0	75.0	Inf	167.97	17.57	3.84	3.2230
1.1	90.33	Inf	129.07	17.0	0	2.2905
1.2	107.25	Inf	115.19	16.7	0	1.9746

The detailed analysis of closed-loop system for different phase margin is given in Tables 22.2, 22.3 and 22.4.

22.6 Simulation Results

The bode plot and step response of compensated system by varying the value of λ from 0.9 to 1.2 for PM = 30°, PM = 45°, PM = 75° are portrayed in Figs. 22.4 and 22.5, respectively. From the bode plot, it is clear that as λ value is increasing from 0.9 to 1.2, PM is also changing at desired cutoff frequency. For desired PM of 30°, just by varying the value of λ from 0.9 to 1.2, PM of 35.4° can be obtained at 2 kHz desired cutoff frequency. Similarly, for desired PM of 45° and 75° cases, by varying the value λ from 0.9 to 1.2, PM of 62.5° and 107° is obtained, respectively. So, stability margins also increase.

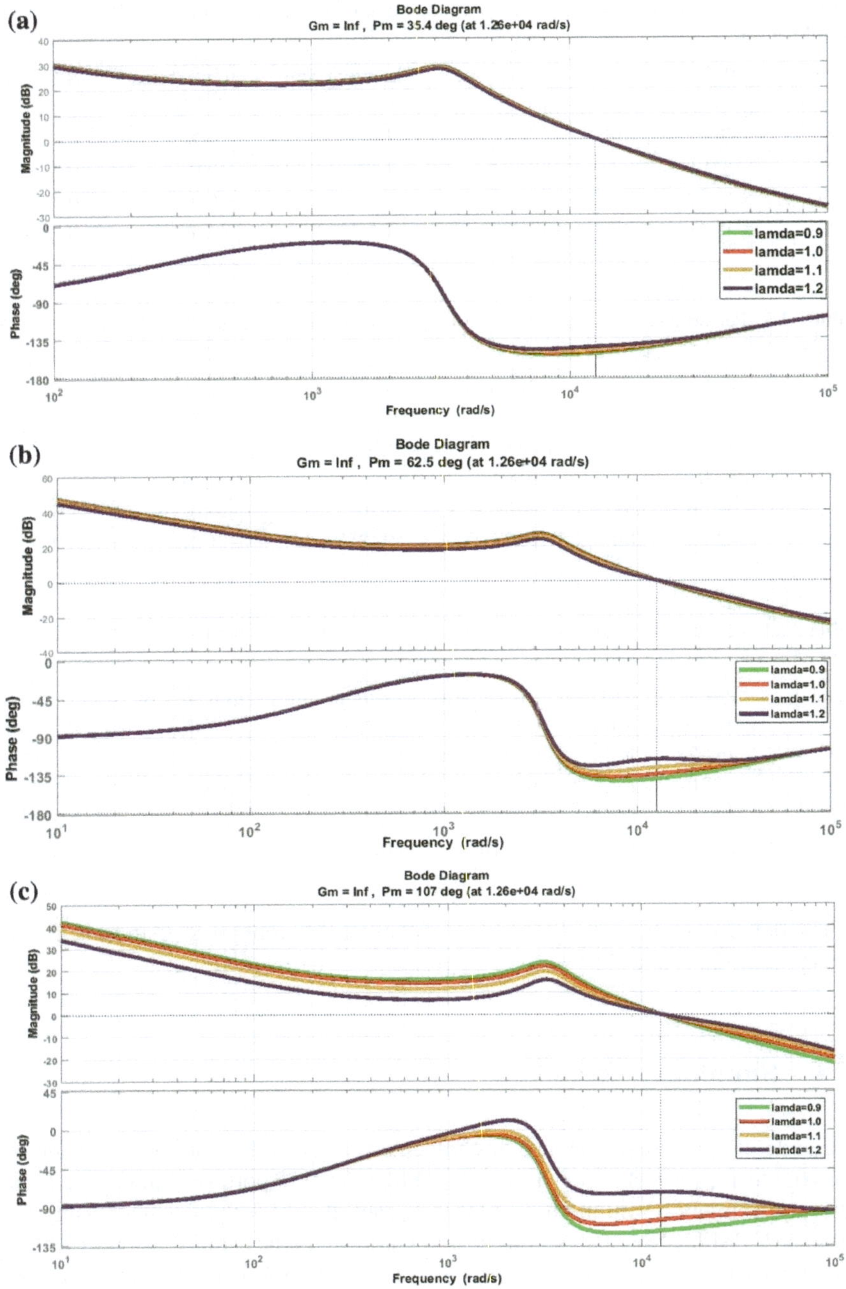

Fig. 22.4 Bode plot of compensated system by varying the value of λ from 0.9 to 1.2 for **a** PM = 30°, **b** PM = 45°, **c** PM = 75°

(a)

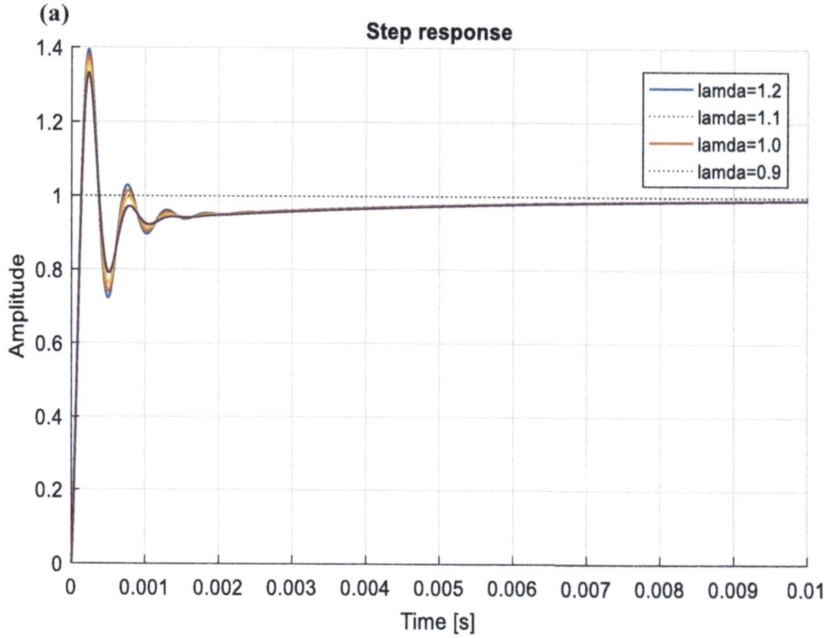

(b)

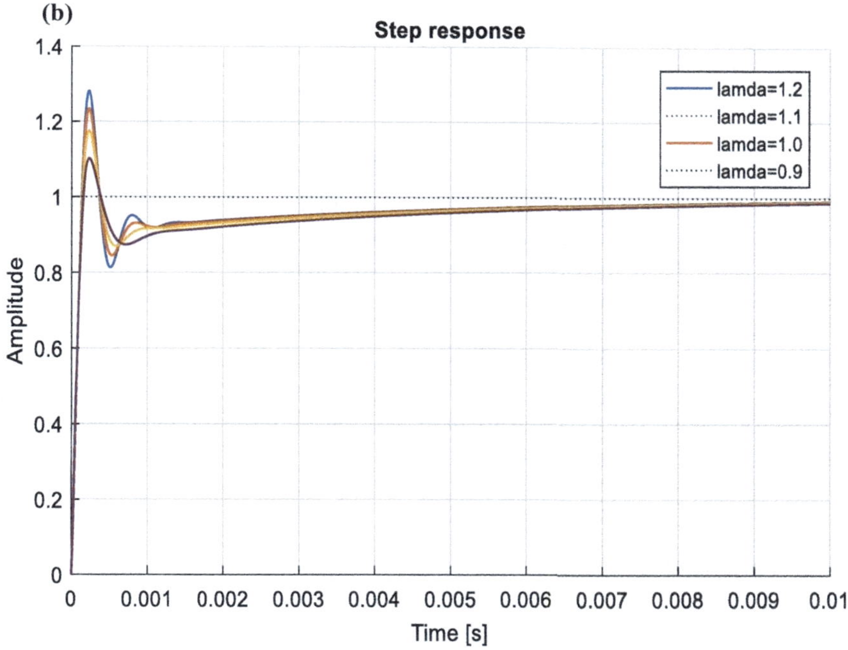

Fig. 22.5 Step response of compensated system by varying the value of λ from 0.9 to 1.2 for **a** PM = 30°, **b** PM = 45°, **c** PM = 75°

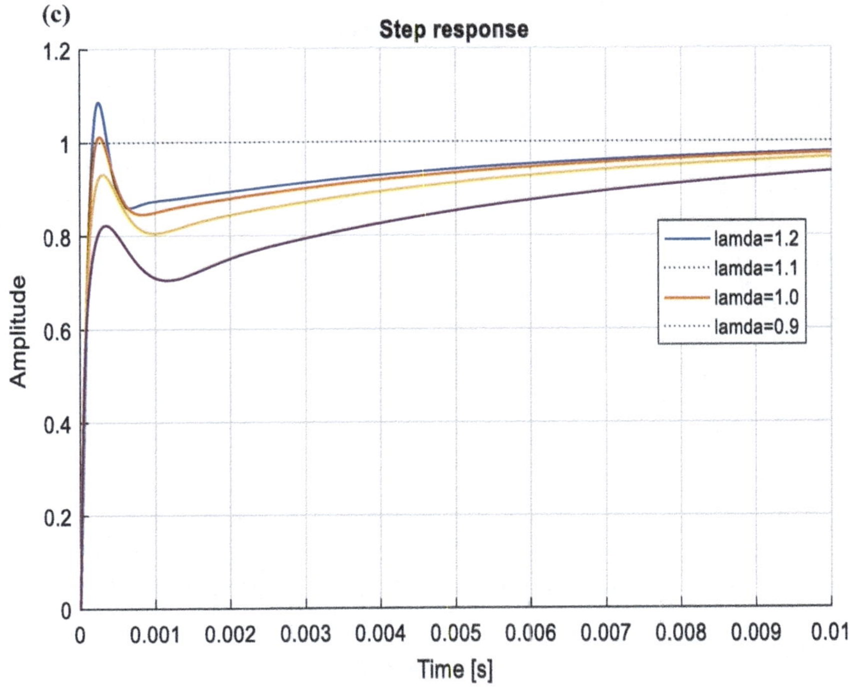

Fig. 22.5 (continued)

The step response of compensated system for different PM by varying the value of λ is shown below. From the step plot, it is clear that as the value of λ is increasing from 0.9 to 1.2, corresponding steady-state error and overshoot values are decreasing. For desired P.M of 30°, just by varying the value of λ from 0.9 to 1.2, steady-state error is changing from 0.8418 to 0.7354%, and overshoot is changing from 40.58 to 34.4% which is a significant change. Similarly, for desired PM of 45° and 75° cases, steady-state error and overshoot are decreased by varying the value of λ from 0.9 to 1.2. This indicates that by the varying the value of λ, stability region can be increased.

22.7 Conclusion

In this chapter, an effort has been made to introduce easy and efficient technique of controller design for DC–DC power converter interfacing with PV array and microgrid application. The importance of FOPIL controller has been illustrated from modeling perspective. The design of FOPIL parameters for non-ideal power converters has been discussed. Using the simulation results, it is clear that the desired frequency-domain performance parameters can be achieved using FOPI-Lead controller-based technique. Though this chapter gives an idea about controller design for PV array-fed DC–DC buck converter in microgrid for low power application, same control design technique can be used for other application where various DC–DC converter topologies are key ingredients.

References

1. Shahbazi, M., Khorsandi, A.: Power Electronic Converters in Microgrid Applications. Elsevier Inc. (2016)
2. Lenz, E., Pagano, D.J., Stramosk, V.: Nonlinear control applied to a dc-dc power converter and the load sharing problem ina dc microgrid. In: IFAC Proceeding, vol. 19, no. V, pp. 534–539 (2014). https://doi.org/10.3182/20140824-6-za-1003.02300
3. El-Khazali, R.: Fractional-order PIλDμ controller design. Comput. Math. with Appl., **66**(5), 639–646, (2013). https://doi.org/10.1016/J.CAMWA.2013.02.015
4. Petráš, I.: Fractional calculus, pp. 7–42 (2011)
5. Shabani, H., Vahidi, B., Ebrahimpour, M.: A robust PID controller based on imperialist competitive algorithm for loadfrequencycontrol of power systems. ISA Trans., **52**(1), 88–95 (2013). https://doi.org/10.1016/J.ISATRA.2012.09.008
6. Petras, I., Chen, Y.Q., Vinagre, B.M., Podlubny, I.: Stability of linear time invariant systems with interval fractionalorders and interval coefficients. In: Second IEEE International Conference on Computational Cybernetics. ICCC 2004., pp.341–346 (2004). https://doi.org/10.1109/ICCCYB.2004.1437745
7. Maâmar, B., Rachid, M.: IMC-PID-fractional-order-filter controllers design for integer order systems. ISA Trans., **53**(5), 1620–1628 (2014). https://doi.org/10.1016/J.ISATRA.2014.05.007
8. Calderón, A.J., Vinagre, B.M., Feliu, V. (2006) Fractional order control strategies for power electronic buck converters. SigProc **86**(10), 2803–2819 (2006). https://doi.org/10.1016/J.SIGPRO.2006.02.022
9. Luo, R., Chen, L.W., Ren, H.B., Liu, C.Y.: Research on self-tuning fuzzy PID control strategy of metallurgy furnacetemperature. Adv. Mater. Res., 472–475, pp. 3063–3066 (2012). https://doi.org/10.4028/www.scientific.net/AMR.472-475.3063
10. Prajapati, S., Garg, M.M., Prithvi, B.: Design of fractional-order PI controller for DC-DC power converters. In: 2018 8thIEEE India International Conference on Power Electronics (IICPE), pp. 1–6 (2018)
11. Garg, M.M., Hote, Y.V., Pathak, M.K. (2012) Leverriers algorithm based modeling of higher-order Dc-Dc converters. IEEE 5th IndiaInternational Conference on Power Electronics (IICPE 2012) by IEEE at DTU Delhi, India / 1–6 / 2012
12. Kanimozhi, G., Meenakshi, J., Sreedevi, V.T.: Small signal modeling of a DC-DC type double boost converterintegrated with SEPIC converter using state space averaging approach. Energy Proc. **117**, 835–846 (2017). https://doi.org/10.1016/j.egypro.2017.05.201
13. Garg, M.M., Hote, Y.V., Pathak, M.K.: Design and performance analysis of a PWM dc–dc buck converter using PI–lead compensator. Arab. J. Sci. Eng. **40**(12) (2015). https://doi.org/10.1007/s13369-015-1838-z.
14. Tavazoei, M.S., Tavakoli-Kakhki, M.: Compensation by fractional-order phase-lead/lag compensators. IET Cont.Theo. Appl. **8**(5), 319–329 (2014). https://doi.org/10.1049/iet-cta.2013.0138

15. Wang, F.-Y.: The exact and unique solution for phase-lead and phase-lag compensation. IEEE Trans. Educ., **46**(2), 258–262 (2003). https://doi.org/10.1109/TE.2002.808279
16. Garg, M.M., Hote, Y.V., Pathak, M.K., Behera, L.: An approach for buck converter PI controller design usingstability boundary locus. Proc. IEEE Power Eng. Soc. Transm. Distrib. Conf., 2018-April, 1–5 (2018). https://doi.org/10.1109/TDC.2018.8440291

Chapter 23
Power Management of Battery Integrated PV System with SMC-Controlled Bidirectional Converter

Shruti Pandey, Neeraj Priyadarshi, and Sanjeevikumar Padmanaban

23.1 Introduction

Non-renewable resources such as fossil fuels are depleting fast, and maximum electricity consumed in India is generated by thermal power plants [1, 2]. Though the power production is cheap, it comes at the cost of lower efficiency and major negative environmental impacts [3, 4]. Renewable energy sources like solar, wind, etc., have emerged as an effective solution to handle the energy crisis of the future electrical power generation [5, 6]. The main features of these resources include low maintenance, natural availability and environmental friendliness [7, 8]. The problem with their integration is due to their variable nature that poses challenges in continuous power supply [9, 10].

The intermittency of renewable resources and fast load changes are circumvented to a great extent by incorporating storage devices along with them [11, 12]. Energy storage devices, therefore, are necessarily needed with the renewable resources to accommodate the variations of available power generation and demand, thus improving the reliability of the system [13, 14]. In the low-voltage low-power DC distribution systems (LVDDS), the energy storage devices are implemented with the help of a bidirectional converter which charges and discharges the battery to maintain the DC bus voltage, to smoothen the transients and to provide backup power to critical loads [15, 16]. The bidirectional converter operates in buck mode and helps the battery to store the energy whenever there is a surplus supply of energy from the renewable energy sources [17, 18]. It helps to deliver the energy whenever renewable energy resources deliver insufficient power and are incapable of satisfying the load [19, 20].

S. Pandey
Department of Electrical and Electronics Engineering, KIET, Ghaziabad, India

N. Priyadarshi (✉) · S. Padmanaban
Department of Energy Technology, Aalborg University, Aalborg, Denmark

N. Priyadarshi et al. (eds.), *DC–DC Converters for Future Renewable Energy Systems*, Energy Systems in Electrical Engineering,
https://doi.org/10.1007/978-981-16-4388-0_23

The control capability is the most important and the decisive features of a distributed system [21, 22]. The main objectives of the control are voltage regulation that is required to guarantee the proper operation of the loads connected [23, 24]. The other one is the battery management that can take care of intermittency of renewable resources and can help to prevent overburdening of any source within a system [25, 26]. The presence of renewable energy source and tightly regulated DC–DC converters also termed as CPLs makes the system highly nonlinear [27, 28]. Therefore, the selection of an appropriate nonlinear controller is desired for the proper functioning of the LVDDS. Sliding mode control (SMC) which is a part of variable structure control scheme is found to be a suitable controller for DC–DC converters [29, 30]. This nonlinear technique of control is supercilious when it comes to stability, simplicity, regulation and robustness under wide range of operating conditions. SMC is developed for the control of variable structure systems, and it is comparatively easier to implement [31, 32].

This paper consists of an MPPT interfaced photovoltaic (PV) source, in combination with a battery whose charging and discharging are regulated by sliding mode-controlled bidirectional converter. This parallel combination is connected to a common DC bus [33, 34]. The DC bus delivers power to the load connected to it. Whenever there is variation in solar irradiation or load, this parallel combination effectively manages this variation and regulates the DC bus voltage [35, 36]. The controlled bidirectional converter steps down and steps up the bus and battery voltage, respectively, and the flow of power is realized in a bidirectional way [37]. Here a modified battery algorithm is presented considering the difference of solar and load power along with SOC to decide the control modes of charging and discharging of the battery. A comparative analysis of SMC and proportional–integral (PI)-controlled bidirectional converter is presented to demonstrate that robust charging/discharging of the battery along with grid voltage regulation can be achieved by SMC [38].

23.2 SMC Control for Two Switch Bidirectional Converter in DC Microgrid

The structure of a DC microgrids is shown in Fig. 23.1. The SMC control has been formulated taking the non-ideal model of bidirectional converter which gives more accurate values of sliding gains.

23.2.1 Modeling of Two Switch Bidirectional Converter

The state space average equations of bidirectional converter shown in Fig. 23.1 in charging or buck mode are

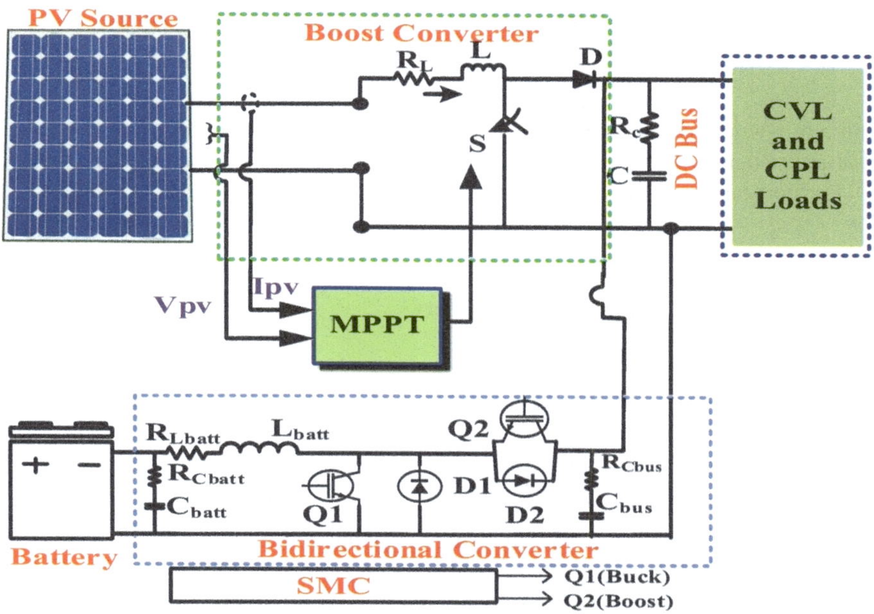

Fig. 23.1 SMC bidirectional converter in DC microgrid

$$
\begin{aligned}
\dot{x}_1 &= \frac{\mu_1}{L}[E_{\text{bus}} + 2V_{\text{d}} - i_L R_{\text{d}}] - \frac{1}{L}[V_{\text{d}} + V_0 + i_L R_{\text{batt}}] \\
\dot{x}_2 &= \frac{1}{C}\left[i_L - \frac{V_0}{R_{\text{load}}}\right]
\end{aligned} \tag{23.1}
$$

Similarly, the state space average equations of bidirectional converter in discharging or boost mode are

$$
\begin{aligned}
\dot{x}_1 &= \frac{\mu_2}{L}[-i_L R_{\text{on}} + V_0] + \frac{1}{L}[E_{\text{batt}} - i_L R_{\text{Lbatt}} - V_{\text{d}} - V_0] \\
\dot{x}_2 &= \frac{1}{C}\left[i_L - \frac{V_0}{R_{\text{load}}} - i_L \mu_2\right]
\end{aligned} \tag{23.2}
$$

Here i_L and V_c are state variables. R_{Lbatt} is internal resistance of inductor. R_{on} is resistance of diode of the switch. V_{d} is forward voltage drop, and R_{d} is resistance of diode. L and C are values of inductor and capacitor. E_{batt} is the battery voltage, and E_{bus} is the bus voltage.

23.2.2 Design *of Sliding Mode Control for Bidirectional Converter*

Initially, switching functions σ_{cc}, σ_{cv}, σ_{boost} for each mode of control are selected that are obtained through state space equations of bidirectional converter.

For constant current charging of battery, it is given by

$$\sigma_{cc} = \gamma(x_1 - x_{1r}) \tag{23.3}$$

For constant voltage charging of battery

$$\sigma_{cv} = x_1 x_{2batt} - x_{1r} x_{2rbatt} + \alpha_{buck}(x_{2batt} - x_{2rbatt}) \tag{23.4}$$

For discharging of battery

$$\sigma_{boost} = x_1 x_{2bus} - x_{1r} x_{2rbus} + \alpha_{boost}(x_{2bus} - x_{2rbus}) \tag{23.5}$$

Here γ, α_{buck} and α_{boost} are the sliding mode control parameters which are obtained from Eqs. (23.1) to (23.5). $x_{1r,}$ x_{2rbus} and x_{2rbatt} are the reference values of inductor current, bus voltage and battery voltage, respectively.

Next, a control law is formulated which helps the trajectory to reach the sliding surface and retain it on the surface. The time taken by the trajectory to reach from any initial state to equilibrium of sliding surface can be easily obtained in reaching coordinates. Therefore, Eq. (23.6) represents the reaching dynamics in reaching coordinated rather than original coordinates.

$$\dot{\sigma} = -\lambda s - K\operatorname{sgn}(s) \tag{23.6}$$

Here λ and K are reaching phase acceleration parameters.

The control laws ($u_{cc,}$ u_{cv}, u_{boost}) where cc and cv control laws are for charging of battery and boost control law is for discharging of battery which are given in Eqs. (23.7)–(23.9):

$$u_{cc} = \frac{(-\lambda\sigma_{scc} - k\operatorname{sgn}\sigma_{scc})L + (v_d + E_{batt} + i_L R_{Lbatt}) - 2v_d\gamma + i_L r_d \gamma}{\gamma E_{bus}} \tag{23.7}$$

$$u_{cv} = \frac{(-\lambda\sigma_{scv} - k\operatorname{sgn}\sigma_{scv}) - \frac{\alpha_{buck}}{C}\left(i_L - \frac{E_{batt}}{R_{ldbu}}\right) - \frac{i_L}{C}\left(i_L - \frac{E_{batt}}{R_{ldbu}}\right) + \frac{V_c}{L}(v_d + E_{batt} + i_L R_{Lbatt}) - \frac{1}{L}(2v_d V_c \alpha_{buck} + i_L r_d \alpha_{buck})}{\alpha_{buck} E_{bus}\left(\frac{V_c}{L}\right)} \tag{23.8}$$

$$u_{boost} = \frac{(-\lambda\sigma_{boost} - k\operatorname{sgn}\sigma_{boost}) - \frac{\alpha_{boost}}{C}\left(i_L - \frac{E_{bus}}{R_{ldbst}}\right) - \frac{V_c}{L}(E_{batt} - i_L R_{batt} - v_d - E_{bus}) - \frac{i_L}{C}\left(i_L - \frac{E_{bus}}{R_{ldbst}}\right))}{\left(\frac{i_L^2}{C} + \frac{V_c E_{bus}}{L} - \frac{V_c i_L R_{ON}}{L} - \frac{\alpha_{boost} i_L}{C}\right)} \tag{23.9}$$

Here R_{ldbu} and R_{ldbst} are the respective load resistances in buck and boost modes.

23.3 Implementation of Power Management

Figure 23.2 represents the algorithm to the control modes of operation. It takes difference of power (PV power and load power) and SOC as input. It switches ON the respective modes according to the given conditions. For example, boost discharging will take place when power difference will be negative, and SOC of battery will be greater than 60%.

23.4 Results and Discussion

MATLAB simulation is done for DC microgrid having a PV panel giving a maximum power (Pmp) of 180 W, Vmp of 24 V and Imp of 7.6 Amp at 1000 W/m^2. The irradiation is decreased to 800 W/m^2 between t = 0.44 and t = 0.84 s, due to which the power is reduced to 145 W. The load connected to the system is of 153 W. The bidirectional converter which is used to control charging and discharging of battery is controlled through SMC and PI controller, respectively. The results are obtained to validate the robustness of SMC over PI in Figs. 23.3, 23.4 and 23.5. Figure 23.3a represents the difference in source and load power which becomes negative with decrease in irradiation. Figure 23.3b shows the DC bus voltage where a small dip of 1.5 V approx. can be seen between t = 0.44 and t = 0.6 s, and the voltage attains constant value of 48 V thereafter. Figure 23.3c shows that the battery current is positive, i.e., battery is discharging with decrease in source power. With the reduction in irradiation, the duty also gets reduced to 0.7

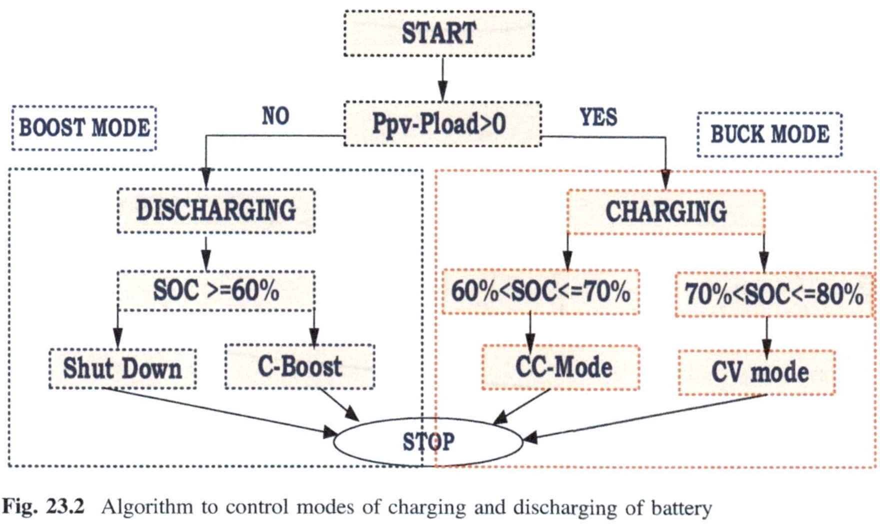

Fig. 23.2 Algorithm to control modes of charging and discharging of battery

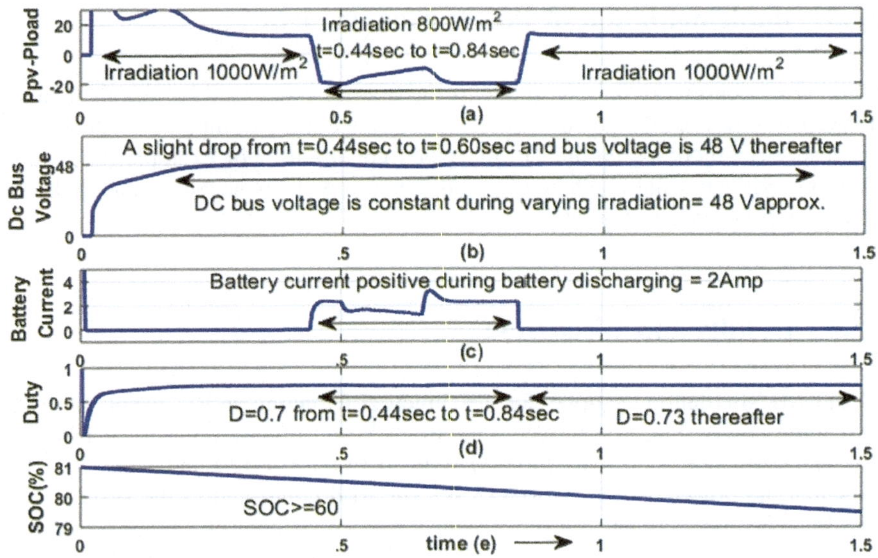

Fig. 23.3 Output waveforms in discharge mode

from 0.73 which can be seen in Fig. 23.3d. Figure 23.3e shows the SOC during discharging mode which according to the algorithm should be greater than 60%.

Figure 23.4a shows the ability of sliding mode control during charging of battery. It helps to charge the battery with desired constant current of 2.2 Amp, and the

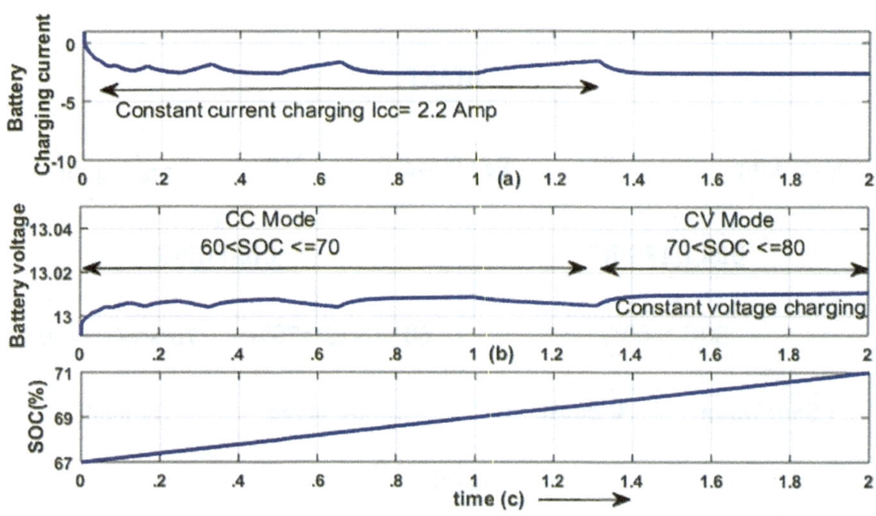

Fig. 23.4 Output waveform in charging mode

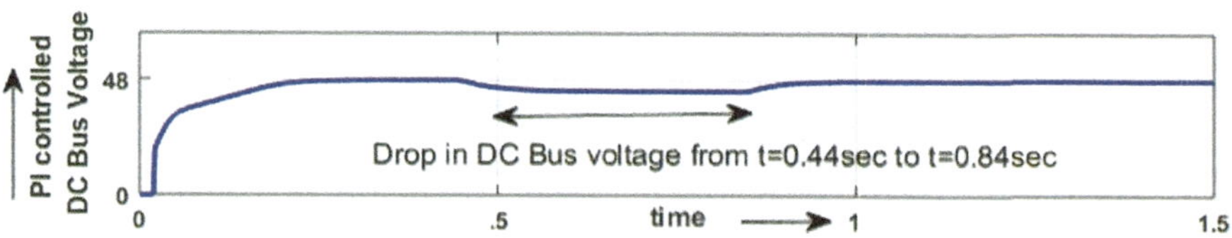

Fig. 23.5 DC bus voltage waveform of PI-controlled bidirectional converter

SOC during this duration is increasing from 67 to 70%. Figure 23.4b represents the battery voltage which gets constant when the SOC is raised over 70% at t = 1.3 s. Figure 23.4c shows the SOC during charging mode.

Figure 23.5 shows the DC bus voltage waveform of PI-controlled bidirectional converter. Here the voltage is significantly reduced (around 5 V) throughout the duration when there is decrease in source irradiation. Therefore, the superiority of SMC-controlled bidirectional converter over PI in tracking the DC bus voltage can be seen from Figs. 23.3b and 23.5.

23.5 Conclusion

An approach which comprises of the incorporation of LVVDS deploying MPPT-controlled PV source, PI and SMC-controlled bidirectional converter with CPLs and CVL has been carried out in this paper. The performance analysis shows that the SMC control is efficiently able to charge/discharge the battery and regulate DC bus voltage maintaining the robustness under wide operating range with the help of a control algorithm. The steady-state error in DC bus voltage is approximately 70% more with PI-controlled converter in comparison with that of SMC-controlled converter during varying irradiation. Therefore, it is interesting to note that SMC controller can satisfactorily charge and discharge the battery to balance the intermittency of photovoltaic source and can be a better choice for fitting it to the microgrid systems.

References

1. Emadi, A., Khaligh, A., Rivetta, C., Williamson, G.: Constant power loads and negative impedance instability in automotive systems: definition, modeling, stability, and control of power electronic converters and motor drives. Veh. Technol. IEEE Trans. **55**(4), 1112–1125 (2006).
2. Seok, B.J., Kim, J.S.: The design of the pi compensator for the bidirectional dc-dc converter in the dc distributed power system. In: International Conference on Electrical Machines and Systems (ICEMS), pp. 372–376, Oct 2010
3. Utkin, V.: Sliding Modes in Control Optimization. Springer, Berlin (1992)

4. Perruquetti, W., Barbot, J.P.: Sliding Mode Control in Engineering. Marcel Dekker, New York (2002)
5. Tan, S.-C., Lai, Y.-M., Tse, C.K.: Sliding Mode Control of Switching Power Converters. Taylor & Francis Group, LLC. (2012)
6. Sharkh, S.M., et al.: Power Electronic Converters for Microgrids. Wiley (2014)
7. Priyadarshi, N., Padmanaban, S., Maroti, P.K., Sharma, A.: An extensive practical investigation of FPSO-based MPPT for grid integrated PV system under variable operating conditions with anti-islanding protection. IEEE Syst. J. 1–11 (2018)
8. Priyadarshi, N., Padmanaban, S., Bhaskar, M.S., Blaabjerg, F., Sharma, A.: A fuzzy SVPWM based inverter control realization of grid integrated PV-wind system with FPSO MPPT algorithm for a grid-connected PV/wind power generation system: hardware implementation. IET Electr. Power Appl. 1–12 (2018)
9. Priyadarshi, N., Kumar, V., Yadav, K., Vardia, M.: An experimental study on zeta buck-boost converter for application in PV system. In: Handbook of Distributed Generation. Springer, Berlin. https://doi.org/10.1007/978-3-319-51343-0_13
10. Priyadarshi, N., Sharma, A.K., Priyam, S.: An experimental realization of grid-connected PV system with MPPT using dSPACE DS 1104 control board. In: Advances in Smart Grid and Renewable Energy. Lecture Notes in Electrical Engineering, vol. 435, Springer, Singapore (2018)
11. Priyadarshi, N., Sharma, A.K., Priyam, S.: Practical realization of an improved photovoltaic grid integration with MPPT. Int. J. Renew. Energy Res. **7**(4) (2017)
12. Priyadarshi, N., Sharma, A.K., Azam, F.: A hybrid firefly-asymmetrical fuzzy logic controller based MPPT for PV-wind-fuel grid integration. Int. J. Renew. Energy Res. **7**(4) (2017)
13. Priyadarshi, N., Anand, A., Sharma, A.K., Azam, F., Singh, V.K., Sinha, R.K.: An experimental implementation and testing of GA based maximum power point tracking for PV system under varying ambient conditions using dSPACE DS 1104 controller. Int. J. Renew. Energy Res. **7**(1), 255–265 (2017)
14. Priyadarshi, N., Padmanaban, S., Mihet-Popa, L., Blaabjerg, F., Azam, F.: Maximum power point tracking for brushless DC motor-driven photovoltaic pumping systems using a hybrid ANFIS-FLOWER pollination optimization algorithm. MDPI Energies **11**(1), 1–16 (2018)
15. Priyadarshi, N., Azam, F., Bhoi, A.K., Alam, S.: An Artificial Fuzzy Logic Intelligent Controller Based MPPT for PV Grid Utility, Lecture Notes in Networks and Systems 46, https://doi.org/10.1007/978-981-13-1217-5_88
16. Padmanaban, S., Priyadarshi, N., Holm-Nielsen, J.B., Bhaskar, M.S., Azam, F., Sharma, A. K.: A novel modified sine-cosine optimized MPPT algorithm for grid integrated PV system under real operating conditions. IEEE Access **7**, 10467–10477 (2019). https://doi.org/10.1109/ACCESS.2018.2890533
17. Padmanaban, S., Priyadarshi, N., Holm-Nielsen, J.B., Bhaskar, M.S., Hossain, E., Azam, F.: A hybrid photovoltaic-fuel cell for grid integration with jaya-based maximum power point tracking: experimental performance evaluation. IEEE Access **7**, 82978–82990 (2019). https://doi.org/10.1109/ACCESS.2019.2924264
18. Priyadarshi, N., Padmanaban, N., Holm-Nielsen, J.B., Blaabjerg, F., Bhaskar, M.S.: An experimental estimation of hybrid ANFIS–PSO-based MPPT for PV grid integration under fluctuating sun irradiance. IEEE Syst. J. **14**(1), 1218–1229 (2020). https://doi.org/10.1109/JSYST.2019.2949083
19. Priyadarshi, N., Padmanaban, N., Bhaskar, M.S., Blaabjerg, F., Holm-Nielsen, J.B., Azam, F., Sharma, A.K.: A hybrid photovoltaic-fuel cell-based single-stage grid integration with lyapunov control scheme. IEEE Syst. J. https://doi.org/10.1109/JSYST.2019.2948899
20. Priyadarshi, N., Bhaskar, M.S., Padmanaban, N., Blaabjerg, F., Azam, F.: New CUK–SEPIC converter based photovoltaic power system with hybrid GSA–PSO algorithm employing MPPT for water pumping applications. IET Power Electr. **13**, 2824–2830 (2020). https://doi.org/10.1049/iet-pel.2019.1154

21. Priyadarshi, N., Padmanaban, N., Holm-Nielsen, J.B., Bhaskar, M.S., Azam, F.: Internet of things augmented a novel PSO-employed modified zeta converter-based photovoltaic maximum power tracking system: hardware realisation. IET Power Electr. **13**, 2775–2781 (2020). https://doi.org/10.1049/iet-pel.2019.1121
22. Kamalapathi, K., Priyadarshi, N., Padmanaban, S., Holm-Nielsen, J.B., Azam, F., Umayal, C., Ramachandaramurthy, V.K.: A hybrid moth-flame fuzzy logic controller based integrated Cuk converter fed brushless DC motor for power factor correction. Electronics **7**, 288 (2018)
23. Priyadarshi, N., Padmanaban, S., Ionel, D., Mihet-Popa, L., Azam, F.: Hybrid PV-wind, micro-grid development using quasi-Z-source inverter modeling and control—experimental investigation. Energies **11**(9), 2277 (2018). https://doi.org/10.3390/en11092277
24. Priyadarshi, N., Ramachandaramurthy, V.K., Padmanaban, S., Azam, F.: An ant colony optimized MPPT for standalone hybrid PV-wind power system with single Cuk converter. Energies **12**(1), 167 (2019). https://doi.org/10.3390/en12010167
25. Priyadarshi, N., Sharma, A.K., Bhoi, A.K., Ahmad, S.N., Azam, F., Priyam, S.: A practical performance verification of AFLC based MPPT for standalone PV power system under varying weather condition. Int. J. Eng. Technol. **7**(2), 338–343 (2018)
26. Priyadarshi, N., Azam, F., Bhoi, A.K., Sharma, A.K.: A multilevel inverter-controlled photovoltaic generation. In: Bhoi, A., Sherpa, K., Kalam, A., Chae, G.S. (eds.) Advances in Greener Energy Technologies. Green Energy and Technology. Springer, Singapore (2020). https://doi.org/10.1007/978-981-15-4246-6_8
27. Priyadarshi, N., Azam, F., Bhoi, A.K., Sharma, A.K.: Dynamic operation of grid-connected photovoltaic power system. In: Bhoi, A., Sherpa, K., Kalam, A., Chae, G.S. (eds.) Advances in Greener Energy Technologies. Green Energy and Technology. Springer, Singapore (2020). https://doi.org/10.1007/978-981-15-4246-6_13
28. Priyadarshi, N., Azam, F., Bhoi, A.K., Sharma, A.K.: A proton exchange membrane-based fuel cell integrated power system. In: Bhoi, A., Sherpa, K., Kalam, A., Chae, G.S. (eds.) Advances in Greener Energy Technologies. Green Energy and Technology. Springer, Singapore (2020). https://doi.org/10.1007/978-981-15-4246-6_18
29. Priyadarshi, N., Azam, F., Bhoi, A.K., Sharma, A.K.: A closed-loop control of fixed pattern rectifier for renewable energy applications. In: Bhoi, A., Sherpa, K., Kalam, A., Chae, G.S. (eds.) Advances in Greener Energy Technologies. Green Energy and Technology. Springer, Singapore (2020). https://doi.org/10.1007/978-981-15-4246-6_25
30. Priyadarshi, N., Azam, F., Bhoi, A.K., Sharma, A.K.: A four-switch-type converter fed improved photovoltaic power system. In: Bhoi, A., Sherpa, K., Kalam, A., Chae, G.S. (eds.) Advances in Greener Energy Technologies. Green Energy and Technology. Springer, Singapore (2020). https://doi.org/10.1007/978-981-15-4246-6_29
31. Vardia, M., Priyadarshi, N., Ali, I., Azam, F., Bhoi, A.K.: Maximum power point tracking for wind energy conversion system. In: Bhoi, A., Sherpa, K., Kalam, A., Chae, G.S. (eds.) Advances in Greener Energy Technologies. Green Energy and Technology. Springer, Singapore (2020). https://doi.org/10.1007/978-981-15-4246-6_36
32. Azam, F., Priyadarshi, N., Nagar, H., Kumar, S., Bhoi, A.K.: An overview of solar-powered electric vehicle charging in vehicular adhoc network. In: Patel, N., Bhoi, A.K., Padmanaban, S., Holm-Nielsen, J.B. (eds.) Electric Vehicles. Green Energy and Technology. Springer, Singapore (2021). https://doi.org/10.1007/978-981-15-9251-5_5
33. Vardia, M., Priyadarshi, N., Ali, I., Azam, F., Bhoi, A.K.: Design of wind energy conversion system under different fault conditions. In: Bhoi, A., Sherpa, K., Kalam, A., Chae, G.S. (eds.) Advances in Greener Energy Technologies. Green Energy and Technology. Springer, Singapore (2020). https://doi.org/10.1007/978-981-15-4246-6_41
34. Choudhary, T., Priyadarshi, N., Kumar, P., Azam, F., Bhoi, A.K.: A fuzzy logic control based vibration control system for renewable application. In: Bhoi, A., Sherpa, K., Kalam, A., Chae, G.S. (eds.) Advances in Greener Energy Technologies. Green Energy and Technology. Springer, Singapore (2020). https://doi.org/10.1007/978-981-15-4246-6_38

35. Priyadarshi, N., Azam, F., Solanki, S.S., Sharma, A.K., Bhoi, A.K., Almakhles, D.: A bio-inspired chicken swarm optimization-based fuel cell system for electric vehicle applications. In: Bhoi, A., Mallick, P., Liu, C.M., Balas, V. (eds.) Bio-inspired Neurocomputing. Studies in Computational Intelligence, vol. 903. Springer, Singapore (2021). https://doi.org/10.1007/978-981-15-5495-7_1
36. Gautam, A.R., Singh, S., Fulwani, D.: DC bus voltage regulation in the presence of constant power load using sliding mode controlled dc-dc bi-directional converter interfaced storage unit. In: 2015 IEEE First International Conference on DC Microgrids (ICDCM), https://doi.org/10.1109/ICDCM.2015.7152050
37. Gao, W.B., Cheng, M.: On the quality control of variable structure systems. Contr. Decision **4** (4), 1–6 (1989) (in Chinese)
38. Hung, J., Gao, W., Hung, J.: Variable structure contrcl: a survey. Ind. Electr. IEEE Trans. **40** (1), 2–22 (1993)

Chapter 24
Overview of Bidirectional DC–DC Converters Topologies for Electric Vehicle and Renewable Energy System

S. Saravanan, P. Pandiyan, T. Chinnadurai, Ramji Tiwari, and N. Prabaharan

24.1 Introduction

Development in electric mobility has been increasing exponentially due to advanced technologies, noiseless operation and pollution-free property. The charging and discharging of the battery management especially gets attention for extending the life span of the batteries. Significant research has been carried out for power converters in the field of power electronics. The power flow in conventional converters is in a unidirectional way, whereas in bidirectional converters, power flow occurs in both the directions, i.e., forward and reverse directions. The applications of these converters are found in renewable energy harvesting applications like solar photovoltaic arrays, wind turbines, and fuel cells and smart grids, electric vehicles, uninterrupted power supplies and aerospace applications. The bidirectional configuration-based converters act as interfacing element between energy storage devices and power sources which shrink the size of the converter and enhance the performance of the overall system because the requirement of two individual converters is not required to perform the forward and reverse directions

S. Saravanan
Department of EEE, Sri Krishna College of Technology, Coimbatore, Tamilnadu, India

P. Pandiyan
Department of EEE, KPR Institute of Engineering and Technology, Coimbatore, Tamilnadu, India

T. Chinnadurai
Department of ICE, Sri Krishna College of Technology, Coimbatore, Tamilnadu, India

R. Tiwari
Department of EEE, Sri Krishna College of Engineering and Technology, Coimbatore, Tamilnadu, India

N. Prabaharan (✉)
School of Electrical and Electronics Engineering, SASTRA Deemed University, Thanjavur, Tamilnadu, India

N. Priyadarshi et al. (eds.), *DC–DC Converters for Future Renewable Energy Systems*, Energy Systems in Electrical Engineering,
https://doi.org/10.1007/978-981-16-4388-0_24

of power flow. The buck or boost converter is used based on the energy storage system location, and the corresponding control strategy is employed to adjust the current or voltage according to the system requirement [1].

A bidirectional DC-to-DC converter is employed in many applications where the direction of current in both the ways depending upon the mode of operation owing to its inherent features. With the purpose of starting and accelerating the vehicle to drive in an uphill location, an additional power is required in order to boost up the high-voltage DC bus. The required higher power can be achieved through secondary batteries installed in electric vehicles in which bidirectional DC-to-DC converter that provides the maximum amount of current from the batteries during motor starting stage. In contrast to unidirectional converter configurations, it can reverse the direction of the current and power flow to be precise, and the supplementary battery present in the BOEVs takes up regenerative energy returned from the motor in deceleration condition [2].

Smart grid and PHEVs charging station uses the bidirectional DC–DC converter. In this situation, bidirectional DC–DC converter is employed to charge the electric vehicle batteries from the connected grid side and fed back to the batteries of PHEVs to the grid side subject to energy demands. For this reason, the bidirectional DC–DC converters are needed with high reliability, higher efficiency and low-cost features to utilize in the charging station.

This converter is classified into two major groups, such as non-isolated configuration and isolated configuration. The transfer of power taking place without using magnetic isolation is known as non-isolated topology. Therefore, this topology does not have a transformer and is in need of galvanic isolation features like a high step-up voltage gain which is simple in construction and of lesser weight. The non-isolated topology is the best choice when weight and size are the key factors in specific applications. As opposed to the non-isolated converter, the isolated converter topology converts DC-to-AC voltage, and it is given to the higher frequency transformer. The output of the transformer is given to the rectifier to get the DC output voltage. Generally, the voltage transfer ratio of the isolated converter configurations is higher compared to the non-isolated converter topologies [3].

The control technique employed in a bidirectional converter is based on the topology and the issues occurred in the controller part in real-time scenario. The applications which do not need isolated configuration topology, non-isolated converter topology are preferred due to low cost and simple in construction and also possible to implement without using a transformer. The isolation between sources and load is needed in case of high-power applications. As a result, an isolated converter configuration is a more suitable choice. The unique features of this isolated converter topology are electrical isolation, high reliability, simple to implement the soft-switching control technique, safe operation and equipment protection. Despite choosing the specific configuration for the bidirectional DC-to-DC converters [4], higher efficiency as well as hybrid control techniques is needed to implement these converters.

The board categories of bidirectional DC-to-DC converter topology along with control strategies are discussed in this chapter. This review is useful for the researchers pursuing the field of bidirectional converter topology in diverse applications.

24.2 Non-isolated Bidirectional DC-to-DC Converters

Non-isolated bidirectional DC-to-DC converter is generally constructed using an anti-parallel diode with the switch and including a controllable switch with diode present in the unidirectional DC-to-DC converter. This non-isolated converter is categorized into boost, buck, cuk buck–boost and so on depending upon the operation and output requirement. In addition, interleaved multilevel and switched capacitor configurations are dedicated only for voltage boosting technique. Therefore, the non-isolated bidirectional configuration-based converters are divided further into eight clusters as follows.

24.2.1 Buck and Boost Derived Converter

The basic non-isolated bidirectional buck and boost DC-to-DC converter was reported depending upon on the unique boost and buck configuration [5]. Figure 24.1 shows the bidirectional configuration which is deduced from the unidirectional buck and boost configuration. On the other hand, bidirectional boost and buck converter can be formed by substituting the bidirectional switches with unidirectional switches available in the unidirectional converter. The converter which performs the conversion operation from low voltage (LV) level to high voltage (HV) level is known as boost converter, whereas buck converter performs the operation in converse way.

24.2.2 Buck–Boost Derived Converter

The technique adopted previously is applicable for designing bidirectional buck–boost converter by replacing bidirectional switches instead of unidirectional switches present in the unidirectional buck–boost converter configuration which is illustrated in Fig. 24.2. In conventional buck–boost converter topology, the voltage level can be decreased or increased as per the requirement [6]. In case of bidirectional buck–boost DC-to-DC converter gets benefited from the unique features of providing negative voltage output and power flow in both the directions.

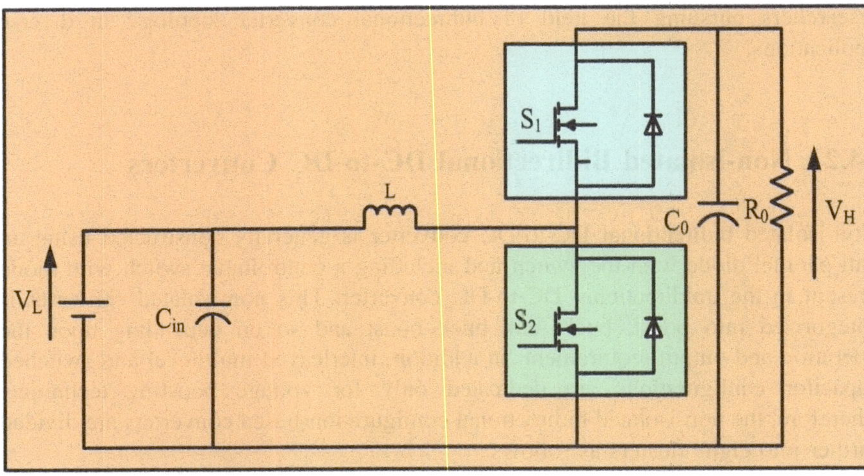

Fig. 24.1 Buck and boost derived bidirectional converter

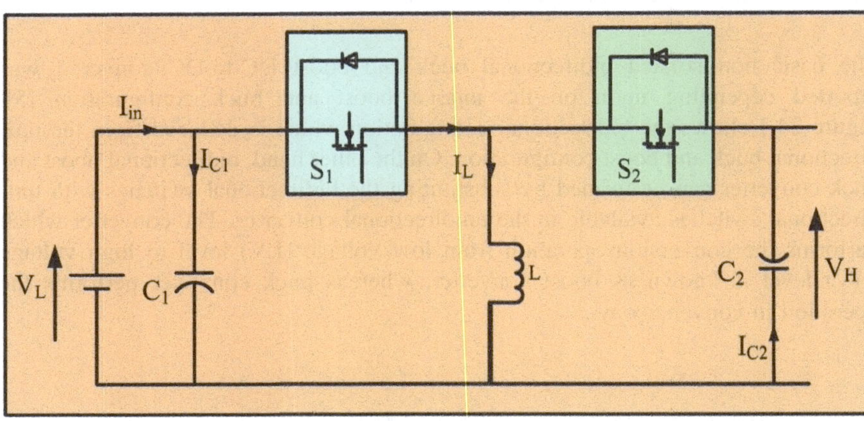

Fig. 24.2 Buck and boost derived bidirectional converter

24.2.3 Ćuk Derived Converter

Figure 24.3 illustrates the Cuk converter which has characteristics of continuous input and output current flow in both the directions by means of employing pair of bidirectional power switches in place of the diode and power switch combination of the regular circuit configuration. Some modifications have been implemented in the bidirectional non-isolated Cuk converter. In general, unidirectional Cuk converter with coupled inductor proposed to reduce the ripple content in the input and output

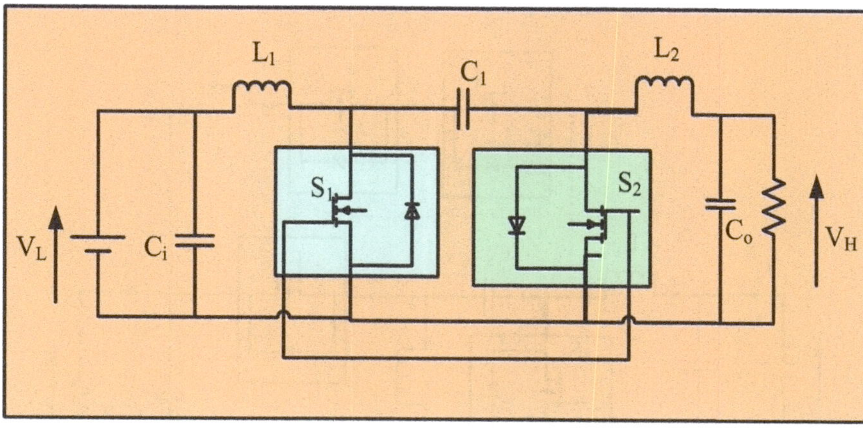

Fig. 24.3 Cuk derived bidirectional converter

current [7]. This method may be utilized to the bidirectional non-isolated Cuk converter configuration which has the potential features for advance analysis in the area of bidirectional Cuk converter with coupled inductor configuration.

24.2.4 Zeta and SEPIC Derived Converter

Zeta as well as Single-Ended Primary-Inductor (SEPIC) is the next level of DC-to-DC converter configuration that is constructed by means of reconfiguring the elements of Cuk converter topology to get positive output voltage. The SEPIC/ZETA bidirectional converter shown in Fig. 24.4 operates like a SEPIC converter when power flow takes place form low level to high level, whereas change in power flow direction (i.e., high level to low level) occurs the same converter works as ZETA converter. The highlighted auxiliary branch in the converter design proposed in [8] provides a power delivery path to the output from the input directly and mitigates the ripples in the current waveform.

24.2.5 Cascaded Converter

To enhance the voltage boosting capability as well as current stress of the converter, more than one converter is connected in cascade fashion depending on the requirement. The fundamental cascaded non-isolated bidirectional DC-to-DC converter is depicted in Fig. 24.5 which is dedicatedly meant for Electric Vehicle application. By closely observing this converter, it is formed by cascading the couple of two buck–boost bidirectional DC-to-DC converters [9]. However, to

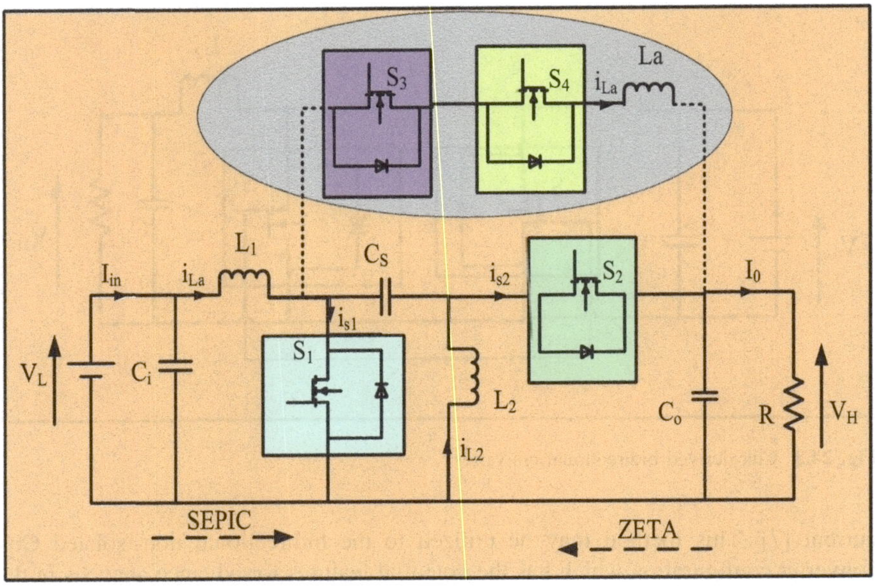

Fig. 24.4 Zeta and SEPIC derived bidirectional converter

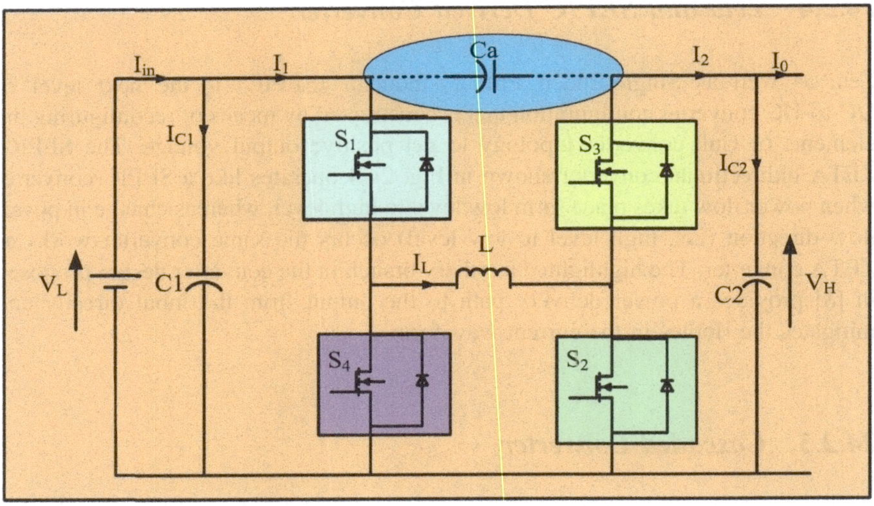

Fig. 24.5 Cascaded bidirectional converter

realize the fundamental buck–boost converter with more amount of components compared, this cascaded topology gets benefited with feature of high voltage transfer ratio along with the similar duty cycle of the switch. In addition, current stress on the switches, ripples and current stress on the capacitor, inductor and uncontrolled switches have been diminished which leads to operate this converters in higher power ratings.

24.2.6 Switched Capacitor Converter

The switched capacitor model can also be applied for improving the voltage boost capability of the converter. A bidirectional converter implemented with switched capacitor model to increase the voltage transfer ratio is depicted in Fig. 24.6. The research work presented in [10] evolved the bidirectional DC–DC converter topology implemented using unidirectional switched capacitor model. The absence of inductor in switched capacitor model gives the platform to reduce the weight in prototype design and also provides continuous current input by shunting same type of two strings realized by switched capacitor models and working together in anti-parallel manner. This can further be enlarged by incorporating the switched capacitor model.

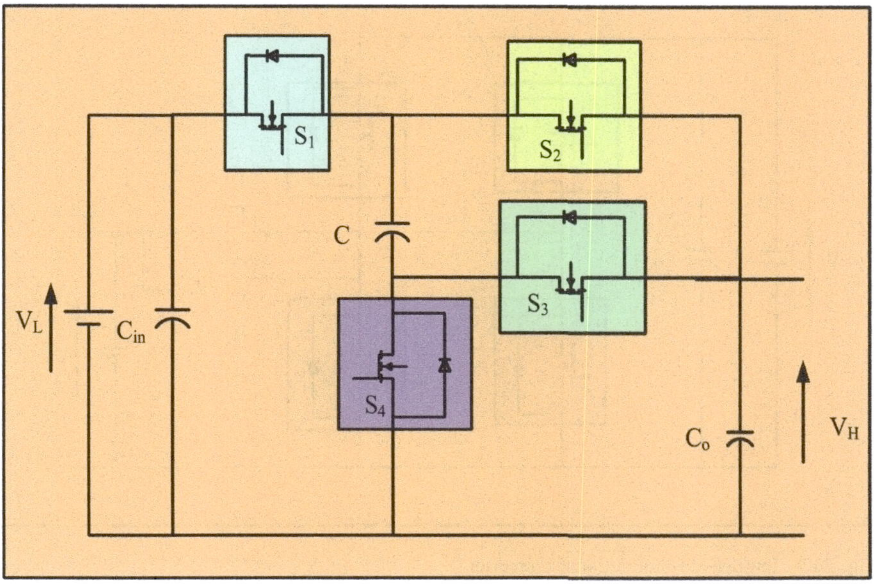

Fig. 24.6 Switched capacitor bidirectional converter

24.2.7 Interleaved Converter

The interleaved bidirectional DC-to-DC converter is depicted in Fig. 24.7 which has capability of reducing current ripple and switching frequency. Due to these advantages, it behaves as smaller electromagnetic interference filter. The research presented [11] a interleaved configuration-based bidirectional converter for automotive applications that consists of several stages which dominantly focused on filter size reduction, thermal management and better dynamic response. The interleaved converter is reported in [12] with coupled inductors in either reverse or direct manner to enhance the dynamic response of the converter and decreases the current ripple.

24.2.8 Multilevel Converter

The multilevel bidirectional DC-to-DC converter is illustrated in Fig. 24.8. A switching unit is employed at the same time as recurring pattern in every level in order to give high voltage transfer ratio. The proposed system in [13] implemented for dual-voltage automotive systems. The size and weight of this converter are significantly reduced due to the absence of the inductor.

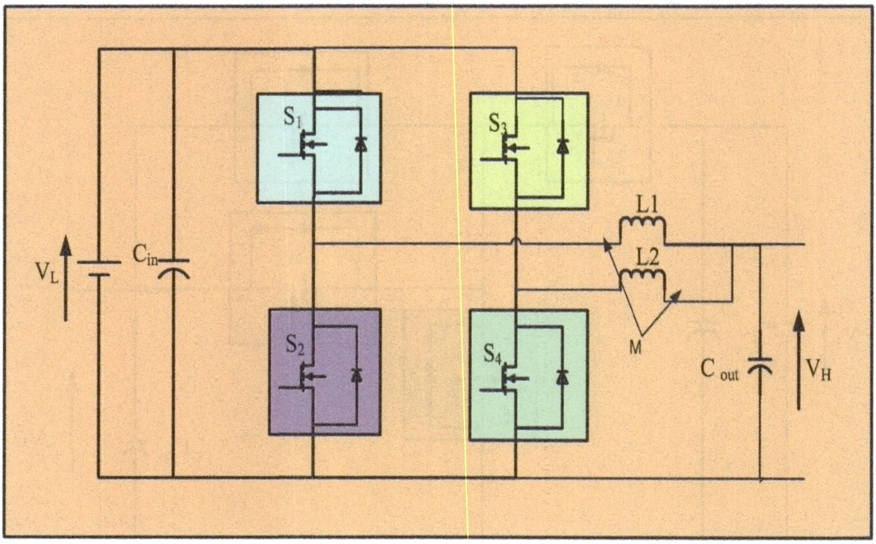

Fig. 24.7 Interleaved bidirectional converter

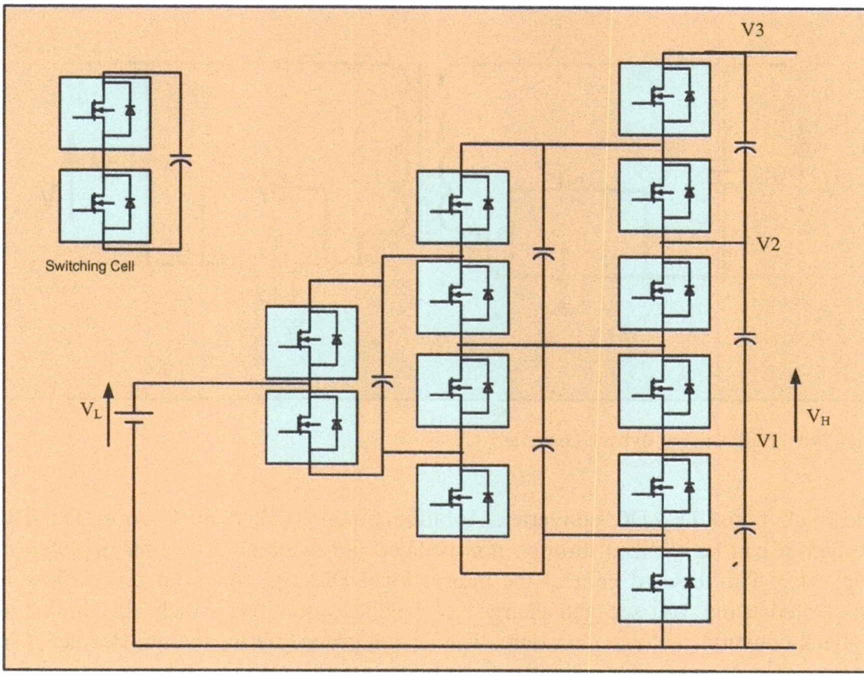

Fig. 24.8 Multilevel bidirectional converter

24.3 Isolated Bidirectional DC-to-DC Converters

Galvanic isolation is the most significant technique to establish a higher voltage gain boost capability by means of including an additional gain to the converter through the winding turns ratio in order to satisfy the output and input regulation [14]. This topology provides the prospects for realizing the multi-input and multi-output system and is also useful for the susceptible loads which are sensitive to faults as well as noisy signal. Furthermore, the safety is also considered. The isolated bidirectional DC-to-DC converter hit upon its usefulness in the area of BOEVs, PHEVs, aircraft and renewable energy resources.

24.3.1 Bidirectional Flyback Converter

The varieties of methods are available to improve the voltage boost-up capability in buck–boost converter to accomplish a higher voltage transfer ratio exclusive of isolation. On the other hand, magnetic isolation-based DC–DC converter specifically the flyback converter is constructed using transformer in place of inductor of

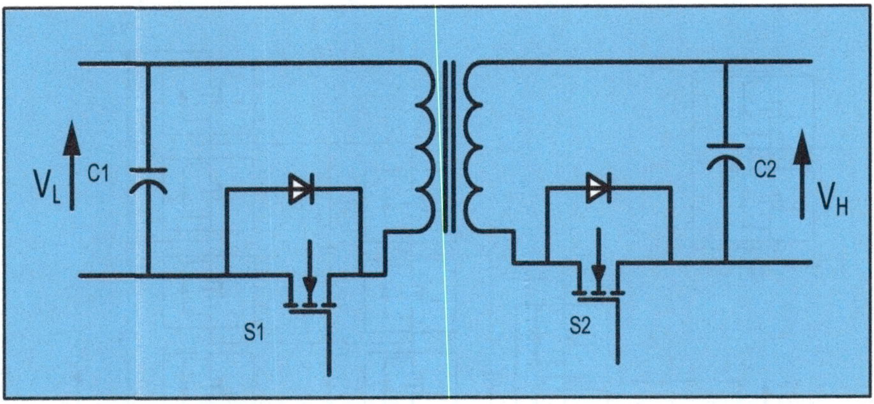

Fig. 24.9 Bidirectional flyback converter

the buck–boost DC–DC converter. A bidirectional isolated buck–boost DC–DC converter can be realized through non-isolated bidirectional converter depicted in Fig. 24.9. The forward gain of the bidirectional DC–DC converter power flow is calculated using volt-sec and charge sec balance equations which are similar to flyback converter voltage gain ratio. The design procedure for the transformer is to be considered, and snubber circuit is needed for suppressing the flyback transformer leakage current. Authors [15] proposed additional changes in this topology to increase the voltage gain.

24.3.2 Ćuk & SEPIC/Zeta Converter

By introducing the magnetic isolation in the non-isolated-based bidirectional DC–DC Cuk converter, the isolated bidirectional DC–DC Cuk converter is realized as shown in Fig. 24.10. In order to provide isolation among output and input parts with higher voltage transfer ratio which includes the transformer turns ratio and also supplies continuous input and output current [16], the coupling inductor linking the input and output directs toward the removal of ripple content in the input and output waveform which is significantly needed in renewable-based energy systems. Bidirectional SEPIC/ZETA converters were also realized in the similar way of approach.

24.3.3 Push–Pull Converter

Bidirectional push–pull converter is obtained from the unidirectional push–pull converter by incorporating the features of power flow takes place in both ways as

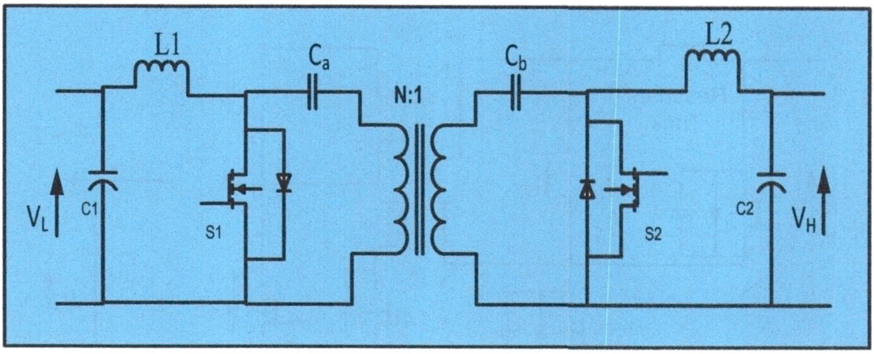

Fig. 24.10 Ćuk & SEPIC/Zeta converter

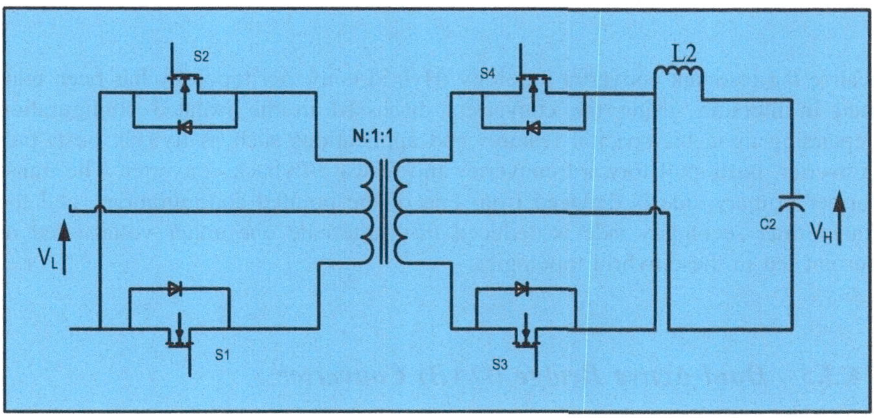

Fig. 24.11 Push–pull converter

shown in Fig. 24.11. In this converter, multiwinding transformer was used as in unidirectional push–pull converter. The authors [17] reported 3-ϕ bidirectional push–pull DC–DC converter applicable to high-power applications.

24.3.4 *Forward Converter*

The forward bidirectional DC–DC converter reported in [18] by considering the unidirectional forward converter is depicted in Fig. 24.12. The zero-voltage switching in this converter topology is accomplished using a clamped circuit. In addition, forward bidirectional DC–DC converter was proposed in [19] in which the leakage inductance of the transformer is utilized for creating a resonance in order to

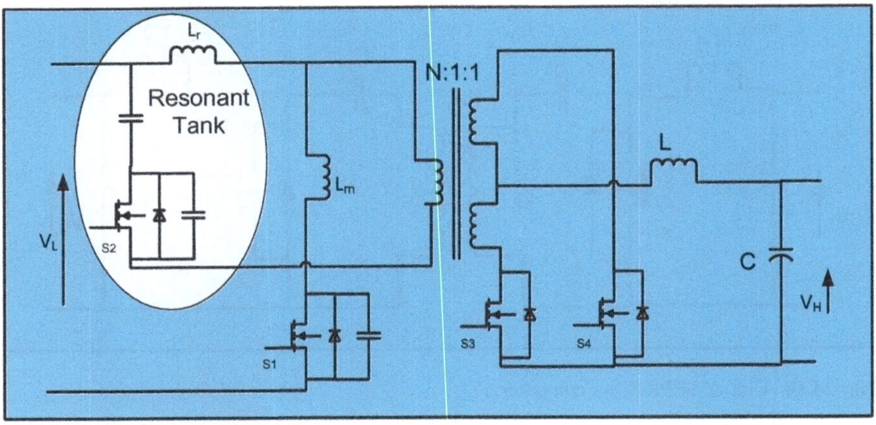

Fig. 24.12 Forward converter

realize the resonant converter topology. Hybrid converter topology has been realized in literature using the converters discussed in the isolated configuration depending upon the specific features and applications such as flyback push–pull converter, push–pull forward converter and forward-flyback converter. The transformer primary side is deduced from one of the pointed configurations, and the transformer secondary side is deduced from alternate one either voltage fed or current fed in these hybrid topologies.

24.3.5 Dual-Active Bridge (DAB) Converter

DAB converter is the well-known converter configuration that uses end-to-end bidirectional configurations which are isolated by means of a transformer with high-frequency feature. This DAB converter may be either full bridge or half bridge and voltage fed or current fed. The fundamental design for DAB converter consists of two full-bridge configurations and placed in both sides of the transformer. The number of switches employed in the converter is directly proportional to the power flow of this converter. The researcher reported [20] a work that has eight power switches with galvanic isolation which is most suitable configuration for automotive system realized with high voltage transfer ratio and also applicable for applications involving high power requirement. The control scheme employed converter gives highly efficient optimization. The basic circuit diagram of the derived DAB converter in Fig. 24.13 which consists of two stages. In the first stages, the DC–AC conversion takes place depending upon the desired specifications either a current-fed or voltage-fed full-bridge converter. A high-frequency step-up transformer amplifies the AC voltage level with galvanic isolation taken place in the second stage. A resonant circuit utilized in the converter along with the transformer

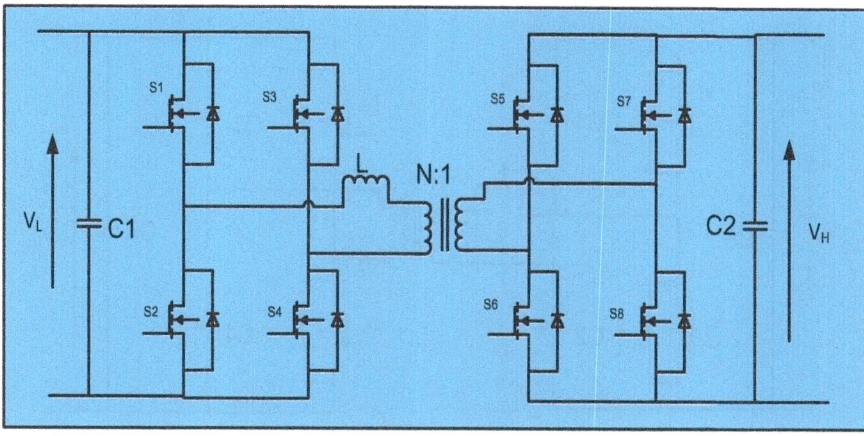

Fig. 24.13 Dual-active bridge converter

assembly helps in achieving either zero current switching or zero-voltage switching and consequently the efficiency gets improved [21]. Depending upon the specific requirement, either a current or a voltage-fed full-bridge converter does the AC-to-DC alteration in the third phase.

24.3.6 Dual Half-Bridge Converter

Half-bridge DAB consists of only four switches, and it is dedicated for low-power applications. Dual half-bridge DC-to-DC bidirectional converters along with voltage-fed configuration presented in [22] are depicted in Fig. 24.14. There is no inductor in the converter topology leads to no zeros in the right half of the s-plane. This makes the converter with minimum phase behavior and very easy to design the controller. Researchers reported the dual half-bridge bidirectional DC–DC converter which is developed with current-fed half-bridge configuration in primary side of the transformer as well as a voltage-fed half-bridge configuration in secondary side of the transformer. Another work in the dual half-bridge converter reported in [23] is exactly reverse configuration (i.e.) a voltage-fed configuration in primary side of the transformer and a current-fed configuration in the secondary side of the transformer. Current-fed configuration-based converter generates continuous current waveform which is suitable for electric vehicle applications. Authors implemented [24] interleaved bidirectional dual half-bridge configuration to enhance the voltage boost-up capability in their study to decrease the transformer ratio and current stress.

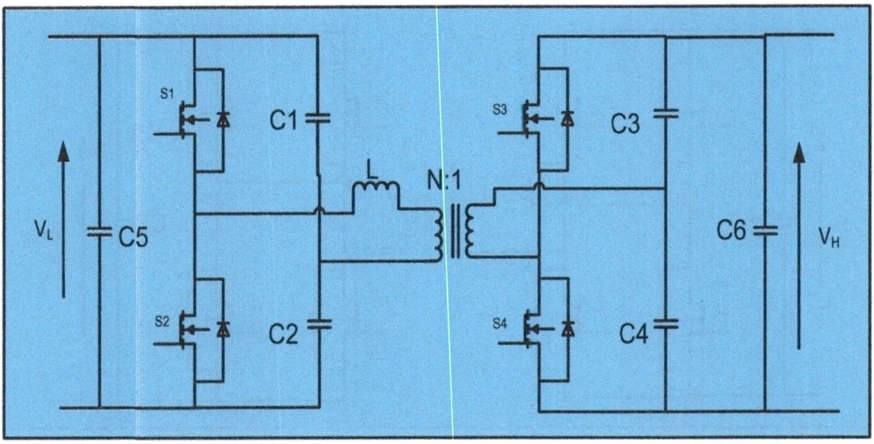

Fig. 24.14 Dual half-bridge converter

24.3.7 Half-Bridge-Full-Bridge Converter

In uninterrupted power supply (UPS) design, an isolated bidirectional converter used the voltage-fed half-bridge configuration in primary side of the transformer and voltage-fed full-bridge configuration in secondary side of the transformer shown in Fig. 24.15. This design makes converter with a smaller number of switches in comparison with DAB, and it leads to less complication in designing the controller for the same. This topology is very much preferable for combining pair of switch buck–boost converters in the half-bridge part to realize a whole UPS design. The authors [25] reported the full-bridge-half-bridge bidirectional DC–DC converter with impedance circuit which makes the system give better performance.

24.3.8 Multiport DAB Converter

The applications such as renewable energy integration system and electric vehicles require multi-input converters as reported in [26]. DAB-based isolated bidirectional multi-input bidirectional DC–DC converter implemented by means of multiwinding transformer configuration with decoupled power flow architecture is depicted in Fig. 24.16. The best usage of duty cycle as well as power flow control to maximize the system behavior is one of the significant research areas in multiport converters topology which is reported in [27].

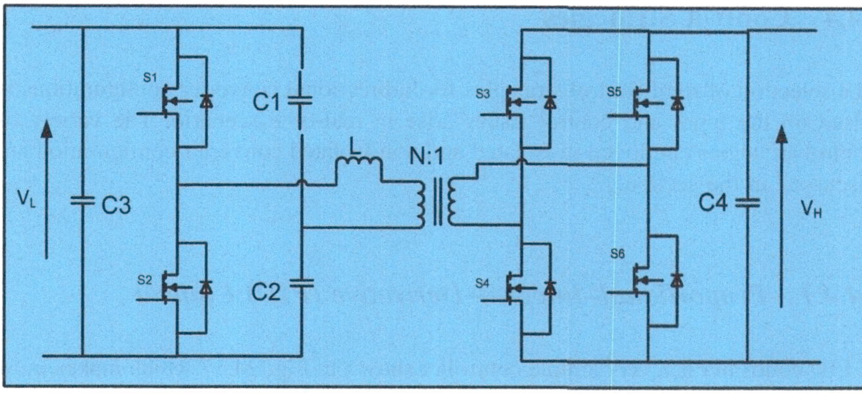

Fig. 24.15 Half-bridge-full-bridge converter

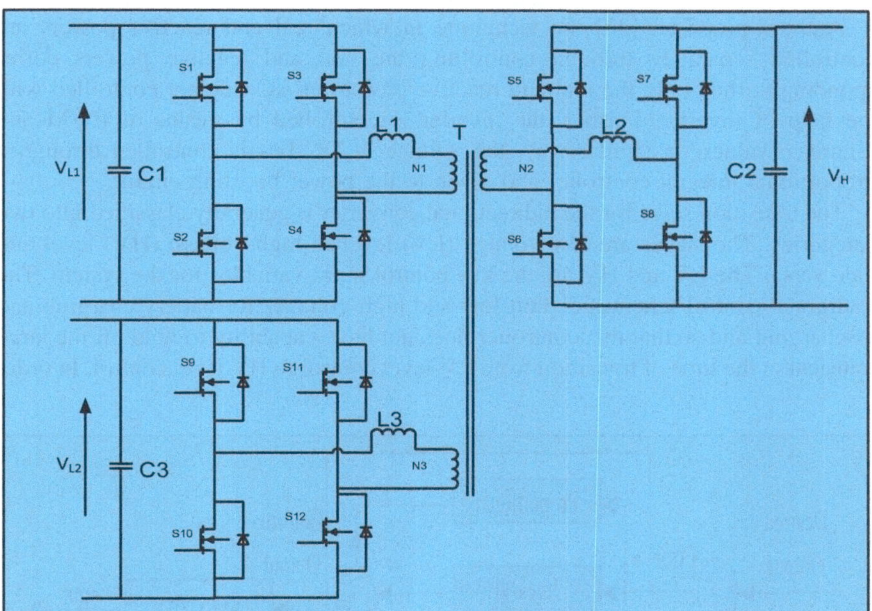

Fig. 24.16 Multiport DAB converter

24.4 Control Strategies

The selection of right control strategies for bidirectional converter configurations is based on the types and control issues arise in real-time scenario. The variety of control strategies employed in isolated and non-isolated converter configuration are discussed in the section.

24.4.1 Proportional–Integral–Derivative (PID) Control

A PID controller is a very simple controller shown in Fig. 24.17 which makes most of the researchers to select it as first choice for designing a control strategy. This PID controller is utilized in most of the converter configurations as well as variety of control problems and also used in combination along with another control strategy too. The major problem encountered in non-isolated bidirectional converter is power flow control in both directions.

Authors proposed [28] the technique in which real and reactive powers are controlled separately through controlling the real and reactive powers correspondingly; then only the real and reactive powers in AC side are controlled with the help of inverter. Further, the inverter is controlled by means of PWM and reference values. In similar way, the voltage in DC bus is controlled through a proportional integral controller and given to the power boosting circuit.

The transition stage for the bidirectional converter is generally classified into two categories. The stages are low voltage (LV) level to high voltage (HV) level and vice versa. The LV and HV are the key control input variables for the system. The traditional control schemes exploit low and high voltages for battery charging and discharging and so that this controller does not have capability to hold off the large transient at the time of transition from LV level control to HV level control. In order

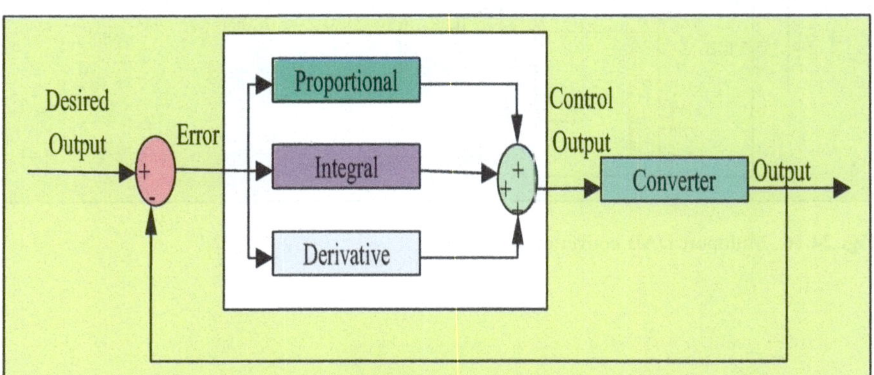

Fig. 24.17 PID controller

to reduce these problems, a PI control regulator is employed depending upon the pulse width modulation (PWM) technique which gives continuous power to the loads and diminishes the transition time as well as size of the capacitor present in the DC bus.

Multiple converter topologies provide high power efficiency with optimal integrated solution and low cost. To design a suitable control strategy for a non-isolated MIMO (Multi-input & Multi-output), multilevel bidirectional DC–DC converter is a challenging task which gives random number of voltage controllable nodes and power flow in both the directions. The job of the control strategy is to adjust the capacitor voltages of every module depending upon the specified reference values assigned for every nodal voltage. In view of that every row in the converter configuration operated along with an outside loop of PI regulator for regulating the voltage across the capacitor of the respective row path and likewise every unit of the bidirectional converter configuration operated with a higher bandwidth inside the loop of PI regulator. The inductor current is regulated by means of current controller employed in the corresponding module and set the duty cycle value for the required operation. The suitable way of inductor current regulation is used to protect the switching devices from the over current.

The procedure for stability analysis of step-down and step-up topology is similar. The local stabilization analysis performs the task by using eigenvalues of the linear system. On the other hand, the dynamics of the converter and control strategies proposed are nonlinear and to do stability analysis, the closed-loop control system ought to be linearized about the equilibrium point. The bode plot technique is used for stability analysis of transition of power flow takes place among the pair of step-down and step-up stages of the bidirectional converter. Therefore, required transfer function should be linear for analysis which can be implemented through state space averaging technique (SSA). The suitable control strategy must be designed with the help of stability analysis data obtained from bode plot.

24.4.2 Sliding Mode Control

The bidirectional converter consists of nonlinear elements which enable converter dynamic equation in nonlinear. One of the best ways to develop a control scheme dedicatedly for a nonlinear equation is that linearize the nonlinear equation about the optimum point by means of existing methods of linearization. Nevertheless, the estimation algorithm along with these techniques represents the improper system model. In linear model, the disturbance and perturbation are ignored. Hence, to achieve the desirable results as well as the system including the perturbation and disturbance existence, the controller should be designed with nonlinear strategies. Sliding Mode Controller (SMC) employed converter is illustrated in Fig. 24.18.

SMC falls under nonlinear control scheme which is more familiar for its unique features like fast, finite-time response, robust, insensitive to external perturbation

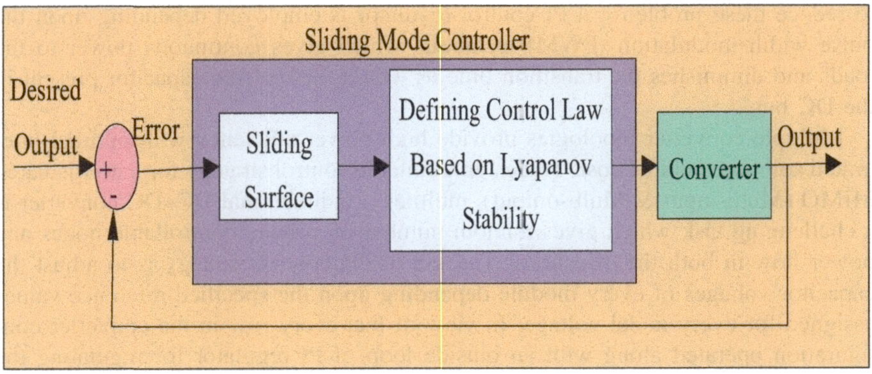

Fig. 24.18 Sliding mode controller

and parameter variation. This control strategy is applicable for both linear and nonlinear systems. The variable structure sliding mode controller to manage the bidirectional converter is employed for controlling the DC motor through rotor angular position.

Small-signal analysis depending upon the SSA model cannot forecast the regulator characteristics whenever a large signal present in the converter along with external perturbations. In order to overcome this issue, SMC is preferred due to its less sensitive to external perturbation and finite time convergence. However, this control strategy requires accurate parameter data as well as state information which makes the controller more complexity. Researchers reported [29] three variety of sliding surfaces to analyze the three particular switching states in a bidirectional DC-to-DC Cuk converter topology. This performance analysis reveals that whether the discontinuity surface is in linear grouping of the output voltage and current as well as negative magnetic coupling among the inductors; henceforth, the converter system will be less sensitive to the voltage output changes in steady-state condition.

Researchers in their work [30] proposed the applications of SMC for the bidirectional converter used in storage system for microgrid applications. The SMC is preferred due to energy resources as well as load demand fluctuations over a period of time. The SSA model based on microgrid systems is nonlinear; therefore, the equivalent load would also be nonlinear in nature. Sometimes, single control strategy is not sufficient enough to solve the problem. In that case, researchers go for hybrid versions to utilize the advantages of changing control strategy techniques. For example, the traditional method of cascade control technique uses pair of PI controllers in which outer loop controls the higher voltage side of the capacitor voltage with inner loop regulates the current flowing through inductor. In the view of such circumstances, some problems like severe variations occur in load as well as line, PID control strategy may not be the suitable for achieve the required performance. Therefore, PI controller in combination with a nonlinear fixed-frequency SMC to attain the required dynamic response and improve the system performance. In fuzzy-based sliding mode controller is developed for reproducing the energy for an ultra-capacitor in order to decrease the chattering problems occur in normal SMC. The combination of these two control strategies directs to robust controller even if the external perturbation is present and also diminishes the variations of the real-time response about the expected results.

24.4.3 Dynamic Evolution Control

In case of electric vehicle application, frequent acceleration and decelerations are required. In this view, a fastest dynamic response is the ultimate prerequisite for this type of applications. The fuel-cell-powered electric vehicle may not have tendency to offer a fastest dynamic response that is sorted out through ultra-capacitor usage as an additional power source [31]. The bidirectional converter is used to connect the ultra-capacitor to the fuel-cell-based electric vehicle. In this system, voltage drop is considerably minimized once the instant variation occurs in the load current. This type of nonlinear system prefers dynamic evolution control strategy as shown in Fig. 24.19. The fundamental idea behind this control strategy is to mitigate the dynamic steady-state error as well as forcing to chase the progress path in spite of disturbance present.

This type of control strategy and the dynamic characteristics pertaining to the control system work mainly depending upon the target equation with respect to time. The control strategy law related to this technique does not need accurate level of knowledge about the model parameters; therefore, it can balance the entire changes in the output and input voltages as well as variation in current flowing though the inductor which is an additional benefit of this control strategy which leads to produce enhanced system performance. From the obtained results, the developed controller is capable of responding to the fast-varying load conditions and returns to charge in the nominal level of voltage while power incurred due to fuel cell is larger compared to the load requirement or else vehicle is in braking mode of operation.

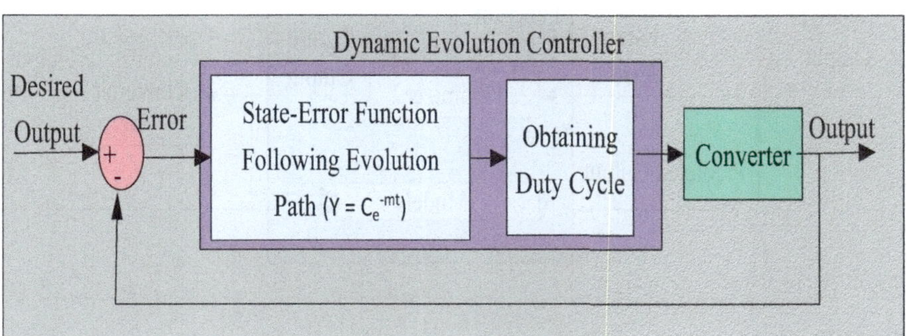

Fig. 24.19 Dynamic evolution controller

24.4.4 Model Predictive Control (MPC)

MPC strategy is derived from the predictive control scheme that utilizes the predefined function to build a system in which system parameters follow the set point values as illustrated in Fig. 24.20. This control strategy has its significant features like reference tracking characteristics, fast dynamic response and easy to implement using microprocessor. Researchers [32] proved that this control strategy is applicable for bidirectional converter used in battery-operated electric vehicles.

This control scheme requires exact discrete-time system model; then, only the forecasting and optimization stages are to be developed. In prediction block of the system, to implement the control strategy method, the measurement data is fetched from the preceding discrete-time block and the forecasted data is specified as a function of concurrent control variables in every converter switching conditions. In last, all predicted values are transferred into the optimization block. An optimization problem is solved in an online depending upon the predefined cost functions in the optimization stage and predicted values in each time step which guides to the optimal control action.

The linear-type MPC scheme needs a linear input–output equation to indicate the process flow. The main drawbacks in model predictive controller strategy are to produce good performance only when algorithm present inside the dynamic model should completely in linear or act liner according to the operational area. In order to eliminate this issue, the multi-MPC strategies were reported which employs multimodel control system to linearize the nonlinear model process of every model in local environment at various operating points. However, multiple MPC cannot be useful to carry out the nonlinear dynamic characteristics. The comprehensive dynamic control matrix utilizes the nonlinear system model to attain the accurate linear system model locally at each sampling time period. This method ensures that the linear system model to be changed in every optimization time interval to acquire

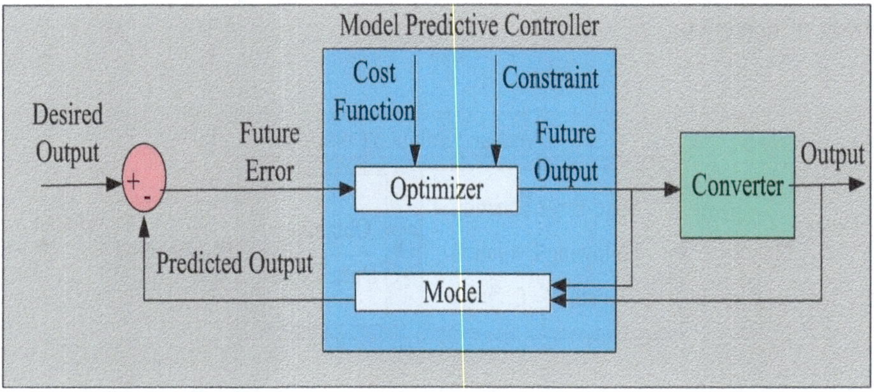

Fig. 24.20 Model predictive controller

the nonlinearity effects to be considered in every sampling step. Furthermore, the deviation occurs between nonlinear system model, and linear system model will be drastically reduced with the help of alternate algorithm. This method easily reduces the limitations caused by multiple MPC.

24.4.5 Fuzzy Logic Control (FLC)

FLC is preferred when the system is nonlinear, insensitive with parameter variation, inaccurate model, uncertainty and load disturbance. The time-variant and nonlinear behavior of converter switches creates a platform to model the single-stage converter dynamics in very tough manner. Furthermore, the charging and discharging circuits in a single-stage converter have a broad span of variation. FLC-based bidirectional converter is illustrated in Fig. 24.21.

Authors reported [33] two control strategies such as FLC and SMC for controlling the charging and discharging of battery of a DAB converter. In comparison with SMC, FLC can be implemented without enough knowledge about the system parameters, and less measurement is needed to design a controller. An artificial neural network (ANN) is an another variant of intelligent control strategy which is applicable for any systems owing to its learning behavior; therefore, it is well suited for nonlinear system. The major benefits are no need of thorough knowledge pertaining to the system, learning by training the large and complex earlier data enables to resolve the difficult problems in the easiest way. Authors implemented [34] an artificial neural network (ANN)-based controller to control the boost converter output voltage and enhance the system performance at the end of transient operating condition. The results obtained from the simulation in this control scheme

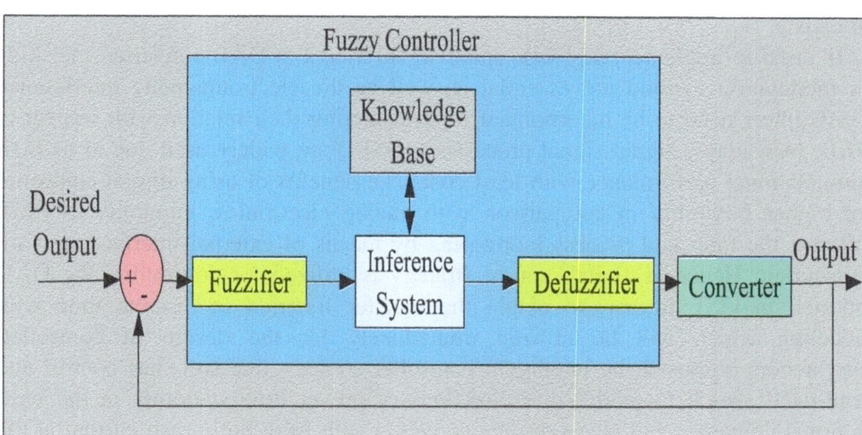

Fig. 24.21 Fuzzy logic controller

are similar to the working of a PI controller with a fastest dynamic response that reduces the overshoot.

Design of FLC requires skilled knowledge to derive the control strategy law; otherwise, it has influence on the system performance. In order to minimize the error, precise knowledge is highly appreciable. ANFIS (Adaptive-Network based Fuzzy Inference system) is employed to mitigate the expected error and adjust the controller parameters by means of gradient descent method and least square error estimation at the time of learning stage.

24.4.6 Digital Control (DC)

DC strategy-based controller produces on output which will be processed by the computer directly from the continuous-time error input signal. Once the data is processed, the discrete time signal is given to the digital controller. Then the output of the controller is sent to the bidirectional converter topology. The general block diagram of the digital control strategy is illustrated in Fig. 24.22. The digital control strategy is well-known controller scheme because of the existence of microcontrollers which process the inputs within microseconds. The authors reported [35] the bidirectional flyback converter using microcontroller. In this control scheme, the converter is utilized for mitigating the electromagnetic interference and switching loss occur in the capacitor without taking the sensor input from high-voltage side in charging/discharging phases of the load capacitance based on valley switching technique. The input voltage and VDS(LV) of the MOSFET are compared in this technique by means of high-speed comparator, and then it is given the high-speed microcontroller to identify the output coming from the comparator which charges and gives a fixed Ton (on-time) pulse which drives the specific actuator. The proposed control strategy provides fast charge/discharge speed and also higher efficiency.

In order to implement a precise small-signal model in DAB converter, the skills on modulation method are essential as well as the electromagnetic interference (EMI) filters need to be incorporated for considering their relation with respect to DAB. Nowadays, digital signal processors (DSPs) are widely used due to its high computational performance with least cost. The benefits of using digital controller are higher flexibility in comparison with analog electronics, immunity to EMI, improve the fault and process monitoring by means of external interface/network connection. Hence, a digital control strategy is utilized for controlling the DAB reported in [35], further, to obtain the transfer function in discrete time with modeling which will be utilized immediately for the design of controller. Researchers proposed the intelligent controller strategy like soft-start control and dead-band switch to modify the directions of power flow smoothly in the bidirectional converter and also keep the converter safe from the inrush current at the starting condition. This variant of controller gives a platform to control the power flow in both directions and enhance the conversion of power in efficient way in

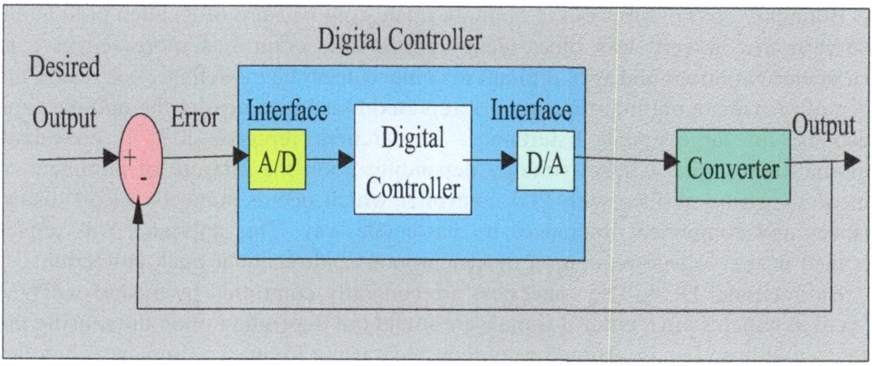

Fig. 24.22 Digital controller

low-voltage side distribution systems. This configuration can work with Zero-voltage switching in the primary side and soft commutation technique in the output rectifier side. Furthermore, this converter does not need any snubber and clamp circuits for mitigating the voltage stresses in the power semiconductor switches. The DSP processor was used for implementing all control algorithms related to commercial applications.

In addition, distribution system and power management need the integration of different load types, renewable energy resources as well as energy storage devices. The Field Programmable Gate Array (FPGA)-based DC strategy proposed in [36], in which a greater number of sources and loads are integrated by means of software reconfigurable-based power modules in connection with the DC bus. This technique gives more flexibility, enhanced reliability, simple to use with better probable energy usage by software reconfiguration done in each module. The concept of indistinguishable modules drastically diminishes the cost incurred for development as well as time and simplifies the system operation.

24.4.7 Boundary Control (DC)

A switching surface is defined depending upon the converter large-signal trajectories to intimate the control actions to the switches. The ideal switching surface has the features such as global stability, fast dynamics and better large-signal operation. In the boundary control strategy, time-varying topology-based switching converters are more preferable. Hysteresis control and SMC are the most employed boundary control method compared to other methods for the applications in power converters. These techniques give better stability and large-signal performance even though it fails to optimize effectively. A boundary control strategy in buck converters by means of second-order switching surfaces which can attain nearby optimum large-signal output and improve the velocity of the trajectory path along with the sliding switching surface.

Boundary control turns out to be more interesting because of its attempt to attain the response in very less time; ideal optimal-time control is more sensitive to parameter variations and very difficult to achieve accurate modeling. As a result, the idea of proximate optimal-time control reported to achieve nearby the optimal-time response in large-signal disturbance. Researchers reported [37] a proximate optimal-time digital control strategy depending upon the mixture of nonlinear or linear switching surface with PID controller which deems subjective load disturbances and component tolerances in reasonable way. This approach was implemented in the voltage regulation of synchronous bidirectional buck converter.

Bidirectional DC-to-DC converters are generally controlled by means of PWM signal in which switch control signals are found out depending upon the sensing the state variable and by inserting the compensators using frequency domain techniques and small-signal average models. Different types of large-signal-based techniques have been reported [38] to enhance the robustness and transient responses of converters; however, SMC strategy is the best method to characterize both large- and small-signal system conditions as well as provide robust output responses against disturbances and uncertainties. In addition, ripple specifications and transient performance can be modified easily and, in some applications voltage, and current overshoots may be removed.

24.5 Conclusion

This study gives the overview about bidirectional DC-to-DC converters starting from the topological arrangement as well as discussed in detail about the control strategy. The bidirectional DC–DC converters are basically non-isolated configuration in which unidirectional switches are replaced with the bidirectional switches in the fundamental converter. The different types of techniques are employed to obtain the objectives like enhancing the voltage boost capability of the converter by incorporating the multicell configurations and also utilizing advanced power semiconductor materials like GaN. In addition, the control strategies and switching schemes are also investigated for non-isolated and isolated configurations of bidirectional DC-to-DC converters.

References

1. Forouzesh, M., Siwakoti, Y.P., Gorji, S.A., Blaabjerg, F., Lehman, B.: Step-up DC–DC converters: a comprehensive review of voltage-boosting techniques, topologies, and applications. IEEE Trans. Power Electron., **32**(12), 9143–9178 (2017)
2. Tytelmaier, K., Husev, O., Veligorskyi, O., Yershov, R.: A review of non-isolated bidirectional dc-dc converters for energy storage systems. In: Proceeding II International Young Scientists Forum on Applied Physics and Engineering (YSF) - IEEE, pp. 22–28 (2016)

3. Chakraborty, S., Vu, H.N., Hasan, M.M., Tran, D.D., Baghdadi, M.E., Hegazy, O.: DC-DC converter topologies for electric vehicles, plug-in hybrid electric vehicles and fast charging stations: State of the art and future trends. Energies **12**(8), 1569 (2019)
4. Gorji, S.A., Sahebi, H.G., Ektesabi, M., Rad, A.B.: Topologies and control schemes of bidirectional DC–DC power converters: an overview. IEEE Access **7**, 117997–118019 (2019).
5. Matsuo, H., Kurokawa, F.: New solar cell power supply system using a boost type bidirectional dc-dc converter. IEEE Trans. Ind. Electron., **1**, 51–55 (1984)
6. Caricchi, F., Crescimbini, F., Noia, G., Pirolo, D.: Experimental study of a bidirectional DC-DC converter for the DC link voltage control and the regenerative braking in PM motor drives devoted to electrical vehicles. In: Proceeding of IEEE Applied Power Electronics Conference and Exposition-ASPEC'94, pp. 381–386 (1994)
7. Majo, J., Martinez, L., Poveda, A., de Vicuna, L.G., Guinjoan, F., Sanchez, A.F., Marpinard, J.C.: Large-signal feedback control of a bidirectional coupled-inductor Cuk converter. IEEE Trans. Ind. Electron. **39**(5), 429-436 (1992)
8. Song, M.S., Son, Y.D., Lee, K.H.: Non-isolated bidirectional soft-switching SEPIC/ZETA converter with reduced ripple currents. J. Power Electron. **14**(4), 649–660 (2014)
9. Lee, H.S., Yun, J.J.: High-efficiency bidirectional buck–boost converter for photovoltaic and energy storage systems in a smart grid. IEEE Trans. Power Electron. **34**(5), 4316–4328 (2018)
10. Chung, H.S., Chow, W.C., Hui, S.Y.R., Lee, S.T.: Development of a switched-capacitor DC-DC converter with bidirectional power flow. IEEE Trans. Circuits and Sys. I: Funda. Theory and App., **47**(9), 1383–1389 (2000)
11. Garcia, O., Zumel, P., De Castro, A., and Cobos, A.: Automotive DC-DC bidirectional converter made with many interleaved buck stages. IEEE Trans. Power Electron., **21**(3), 578–586,(2006)
12. Huang, X., Lee, F.C., Li, Q., Du, W.: High-frequency high-efficiency GaN-based interleaved CRM bidirectional buck/boost converter with inverse coupled inductor. IEEE Trans. Power Electron., 31(6), 4343–4352 (2015)
13. Peng, F.Z., Zhang, F., Qian, Z.: A magnetic-less DC-DC converter for dual voltage automotive systems. In: Conference record of the 2002 IEEE Industry applications conference. 37th IAS Annual Meeting (Cat. No. 02CH37344), vol. 2, pp. 1303–1310 (2002)
14. Gorji, S.A., Ektesabi, M., Zheng, J.: Isolated switched-boost push–pull DC–DC converter for step-up applications. Electron. Lett. **53**(3), 177–179 (2017)
15. Delshad, M., Farzanehfard, H.: A new isolated bidirectional buck-boost PWM converter. In: Proceeding 1st Power Electronic & Drive Systems & Technologies Conference (PEDSTC) - IEEE, pp. 41–45 (2010)
16. Ruseler, A., Barbi, I.: Isolated Zeta-SEPIC bidirectional dc-dc converter with active-clamping. In: Proceeding Brazilian Power Electronics Conference - IEEE, pp. 123–128 (2013)
17. Kwon, M., Park, J., Choi, S.: A bidirectional three-phase push–pull converter with dual asymmetrical PWM method. IEEE Trans. Power Electron., **31**(3), 1887–1895 (2015)
18. Lin, B.R., Chen, J.J., Lee, Y.E., Chiang, H.K.: Analysis and implementation of a bidirectional ZVS Dc–Dc converter with active clamp. In: Proceeding 3rd IEEE Conference on Industrial Electronics and Applications, pp. 382–387 (2008)
19. Khodabakhshian, M., Adib, E., Farzanehfard, H.: Forward-type resonant bidirectional DC–DC converter. IET Power Electron., **9**(8), 1753–1760 (2016)
20. Zhao, B., Song, Q., Liu, W., Sun, Y.: Overview of dual-active-bridge isolated bidirectional DC–DC converter for high-frequency-link power-conversion system. IEEE Trans. Power Electron. **29**(8), 4091–4106 (2013)
21. Jung, J.H., Kim, H.S., Ryu, M.H., Baek, J.W.: Design methodology of bidirectional CLLC resonant converter for high-frequency isolation of DC distribution systems. IEEE Trans. Power Electron. **28**(4), 1741–1755 (2012)

22. He, P., Khaligh, A.: Comprehensive analyses and comparison of 1 kW isolated DC–DC converters for bidirectional EV charging systems. IEEE Trans. Transp. Electr. **3**(1), 147–156 (2016)
23. Li, H., Peng, F.Z., Lawler, J.S.: A natural ZVS medium-power bidirectional DC–DC converter with minimum number of devices. IEEE Trans. Ind. App., **39**(2), 525–535 (2003)
24. Park, S., Song, Y.: An interleaved half-bridge bidirectional DC–DC converter for energy storage system applications. In: Proceeding 8th International Conference on Power Electronics-ECCE Asia - IEEE, pp. 2029–2034 (2011)
25. Chub, A., Vinnikov, D., Kosenko, R., Liivik, E., Galkin, I.: Bidirectional DC–DC converter for modular residential battery energy storage systems. IEEE Trans. Ind. Electron. **67**(3), 1944–1955 (2019)
26. Gorji, S.A., Ektesabi, M., Zheng, J.: Double-input boost/Y-source DC–DC converter for renewable energy sources. In: Proceeding IEEE 2nd Annual Southern Power Electronics Conference (SPEC), pp. 1–6 (2016)
27. Zhao, C., Round, S.D., Kolar, J.W.: An isolated three-port bidirectional DC–DC converter with decoupled power flow management. IEEE Trans. Power Electron. **23**(5), 2443–2453 (2008)
28. Cornea, O., Andreescu, G.D., Muntean, N., Hulea, D.: Bidirectional power flow control in a DC microgrid through a switched-capacitor cell hybrid DC–DC converter. IEEE Trans. Ind. Electron., **64**(4), 3012–3022 (2016)
29. Cao, J., Chen, Q., Zhang, L., Quan, S.: Sliding mode control of bidirectional DC/DC converter. In: Proceeding 33rd Youth Academic Annual Conference of Chinese Association of Automation (YAC) - IEEE, pp. 717–721 (2018)
30. Agarwal, A., Deekshitha, K., Singh, S., Fulwani, D.: Sliding mode control of a bidirectional DC/DC converter with constant power load. In: Proceeding IEEE First International Conference on DC Microgrids (ICDCM), pp. 287–292 (2015)
31 Samosir, A.S., Yatim, A.H.M.: Implementation of dynamic evolution control of bidirectional DC–DC converter for interfacing ultracapacitor energy storage to fuel-cell system. IEEE Trans. Ind. Electron. **57**(10), 3468–3473 (2010)
32. Pirooz, A., Noroozian, R.: Model predictive control of classic bidirectional DC–DC converter for battery applications. In: Proceeding 7th Power Electronics and Drive Systems Technologies Conference (PEDSTC) – IEEE, pp. 517–522 (2016)
33. Talbi, S., Mabwe, A.M., El Hajjaji, A.: Control of a bidirectional dual active bridge converter for charge and discharge of a Li-Ion Battery. In: Proceeding IECON 2015-41st Annual Conference of the IEEE Industrial Electronics Society, p. 000849–000856 (2015)
34. Dhivya, B.S., Krishnan, V., Ramaprabha, R.: Neural network controller for boost converter. In: Proceeding International Conference on Circuits, Power and Computing Technologies (ICCPCT) – IEEE, pp. 246–251 (2013)
35. Krismer, F., Kolar, J.W.: Accurate small-signal model for the digital control of an automotive bidirectional dual active bridge. IEEE Trans. Power Electron, **24**(12), 2756–2768 (2009)
36. Rodriguez, M., Stahl, G., Corradini, L., Maksimovic, D.: Smart DC power management system based on software-configurable power modules. IEEE Trans. Power Electron. **28**(4), 1571–1586 (2012)
37. Yousefzadeh, V., Babazadeh, A., Ramachandran, B., Alarcón, E., Pao, L., Maksimovic, D.: Proximate time-optimal digital control for synchronous buck DC–DC converters. IEEE Trans. Power Electron., **23**(4), 2018–2026 (2008)
38. Galvez, J.M., Ordonez, M., Luchino, F., Quaicoe, J.E.: Improvements in boundary control of boost converters using the natural switching surface. IEEE Trans. Power Electron., **26**(11), 3367–3376 (2011)

Chapter 25
Comparative Analysis of MPPT Techniques Using DC–DC Converter Topologies for PV Systems

S. Ravindra, A. Naveen Reddy, K. N. V. Sai Tejaswi, and K. Baby Shamili

25.1 Introduction

From the last decade, researchers are moving toward clean energy sources like solar and wind in order to reduce the environment pollution. However, the extraction of power from PV system is still a difficult task due to the poor conversion ratio of PV cells. Solar to electrical energy transformation is accomplished by photovoltaic system. Basic unit of PV system consists photovoltaic cell. Group of PV cells in parallel and series leads to solar modules, and again these modules are considered to be grouped in parallel or series to form solar arrays. Combination of solar cells in parallel enhances output-current and in series enhances output voltage.

The main parts of PV arrangement are PV array, converter and controller to control function of converter. So, the effective PV arrangement relies on array, inverter and MPPT. Also, reliability of PV panel is measured by the selection of material which depends upon the mechanized technology. Further effectiveness of converter is enhanced by considering appropriate control method. On account of nonlinear current–voltage characteristic (I-V), function of PV system keeps operating point on curve (I-V) where maximum power is shaped. The factors affecting the MPP are irradiance and temperature of cell. Irradiation can change quickly because of changing environmental conditions, for example, mists/stormy meetings. It is very necessary to tune the MPPT precisely under different environmental conditions to attain MPP.

DC–DC converter combined with MPPT allows PV network to attain maximum uninterruptable power, despite environmental conditions of temperature, solar radiation. There are different algorithms of MPPT control to enforce or to gain the performance.

S. Ravindra (✉) · A. N. Reddy · K. N. V.S. Tejaswi · K. B. Shamili
VVIT, Nambur, Guntur, Andhra Pradesh, India

N. Priyadarshi et al. (eds.), *DC–DC Converters for Future Renewable Energy Systems*, Energy Systems in Electrical Engineering,
https://doi.org/10.1007/978-981-16-4388-0_25

In this paper, describing many classical MPPT techniques, together with P&O [1] I&C [2] and FOCV [3], are investigated and explored. Modeling is considered through coding and simulating in MATLAB/Simulink. The exhibition of a PV framework relies upon in operating conditions of the solar cell comparably as array design. Solar irradiation level, temperature and load determine output voltage, current and power of PV array. The impacts of three factors must be accounted in design of PV array. Temperature variations and solar irradiation illumination levels should not unfavorably influence PV array output power to load. Moreover, these types of methods are not suggested for different operating conditions. For instance, mismatched PV modules because of aging where there are different local maxima in P–V module. These traditional algorithm experiences from tack in nearby maxima.

To improvise effectiveness of MPPT under different operative conditions, numerous endeavors have been finished to create MPPT strategies equipped for following Global MPP, for example, distributed MPPT strategy [4] and two-stage techniques [5]. Soft computing techniques such as artificial fish swarm [6], particle swarm (PSO) [7], gray wolf (GWO) [8] and cuckoo search [9] are explored. There are different DC–DC converter topologies that were established concerning to higher efficiencies, switching and control strategies and fault-tolerant designs and widely on renewable energy-based applications [10–16].

This paper proposes a meta-heuristic optimization algorithm dependent on social conduct of crow and named CSA to follow MPP for PV framework. The possibility of CSA is inspired from storing process of extra food in hiding places and restores it at required time. When distinguished with PSO, gray wolf, artificial fish warm and cuckoo search, the CSA likewise does not have algorithm explicit controlling boundaries.

The execution of CSA within tracking speed and efficacy is going to be estimated by PSO-MPPT validation of simulation results. The effective and efficient performance by CSA-MPPT methodology proves its viability of addressing MPPT.

25.2 Modeling of DC–DC Converters with DC Source

25.2.1 Modeling of Boost Converter with DC Source

Boost converter mainly consists of an inductor for storing energy temporarily, an electronic switch which has high switching frequency, a capacitor for reducing voltage ripple in output side and a diode. All these elements are connected as shown in Fig. 25.1.

The switch in converter should have to handle unidirectional current and voltage capability, and so mostly IGBT/MOSFET is used. Thus, for duty cycle 'δ', average output voltage can be computed using Eq. (25.1)

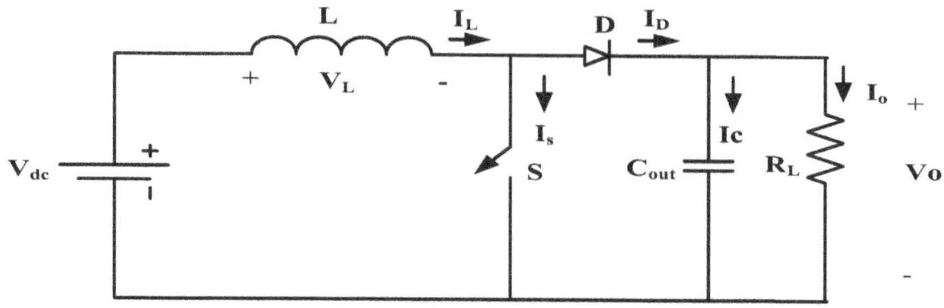

Fig. 25.1 Circuit diagram of boost converter

$$\frac{V_o}{V_{in}} = \frac{1}{1-\delta} \tag{25.1}$$

From the above equation, it seems that output voltage can be boosted to any value but due to practical considerations like diode drop, internal resistance of inductor and parasitic effects leads to limited boosting factor. In ideal network, output power is equal to input power of converter which yields.

$$P_o = P_{in}$$

i.e., $V_o I_o = V_{in} I_{in}$.

25.2.1.1 Inductor Design

The inductance value for boost converter for continuous current can be decided by using Eq. (25.2), unlike buck converter the input current in boost converter is continuous because inductor is connected in series with source.

$$L = \frac{R_L \delta (1-\delta)^2}{2 f_s} \tag{25.2}$$

25.2.1.2 Capacitor Design

In order to surge output voltage, the duty ratio of converter needs to be high means the device has to in on condition for most of time, so output will be disconnected from input. So, the chances of discontinuous operation will be high and to maintain less voltage ripple in output a large size might be required. The capacitor has to provide the current to load when switch is turned on. The filter capacitance that obtained is given by (25.3),

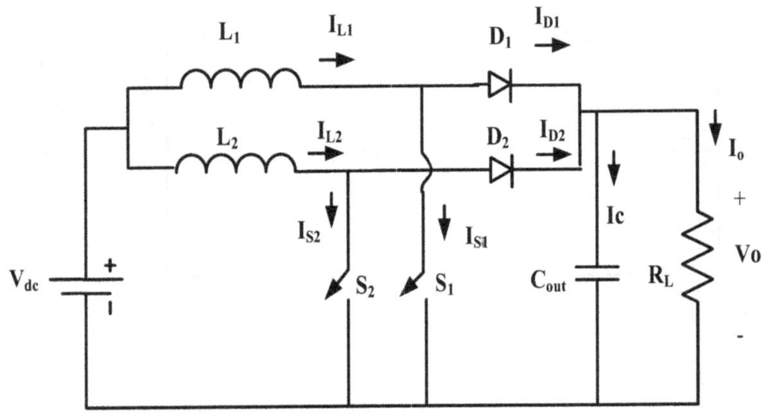

Fig. 25.2 Circuit diagram of two-phase IBC

$$C_{\text{out}} = \frac{\delta V_O}{V_r R_L f_s} \tag{25.3}$$

25.2.2 Modeling of IBC with DC Source

Interleaving means parallel operation, the drawbacks of CBC such as low-voltage gain, discontinuous operation, large size inductor and capacitor can be overcome by using IBC. Due to current sharing between inductor and switches, IBC has less complexity due to lower ratings of devices and passive components. The phase shift between the alternate circuits of IBC is given by $\frac{360}{n}$, where 'n' is number of parallel phases [10].

Figure 25.2 shows two-level IBC. L_1 and L_2 are inductors, S_1 and S_2 are two switches, D_1 and D_2 are two diodes, C_{out} is output filter capacitance.

The switches of IBC are working in continuous conduction mode with 180° out of phase. The two switching states are as follows:

i. When any one of the switches $S1$ or $S2$ ON, the corresponding inductor will store the energy via source which makes the diode to turn off.
ii. When the switch OFF, the stored energy in the inductor has to be dissipated so forces the corresponding diode to conduct and discharge through load.

25.2.3 Simulation Results of Conventional and IBC

The DC–DC boost converters are designed at $\delta = 0.5$, $V_r/V_O = 1\%$, $R_L = 10\ \Omega$ and $f_s = 100$ kHz. The CBC and IBC component values are given away in Table 25.1.

Table 25.1 CBC and IBC components

CBC components	Values	IBC components	Values
Capacitor C_{out}	50 μF	Capacitor C_{out}	50 μF
Inductor L	6.25 μH	Inductor L_1	6.25 μH
Commutation frequency f	100 kHz	Inductor L_2	6.25 μH
		Commutation frequency f	100 kHz

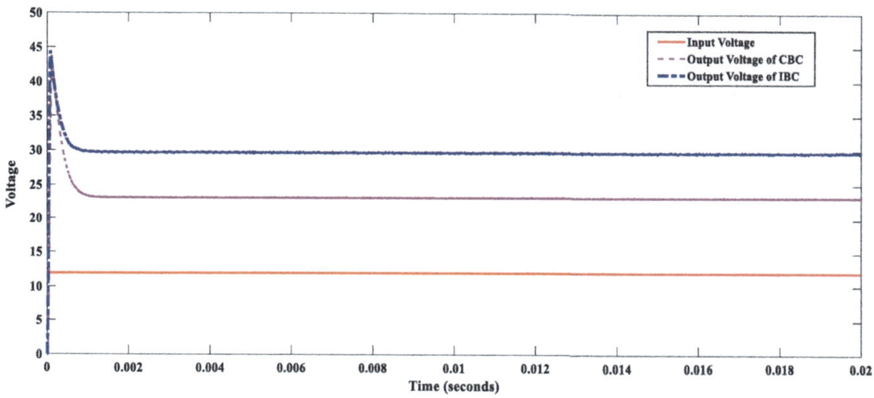

Fig. 25.3 Input and output voltages of CBC and IBC

Figure 25.3 shows the input and output voltages of CBC and IBC, and Fig. 25.4 shows the current in inductor, diode, capacitor and load of boost converter.

The currents in inductor, diode, capacitor and load currents of conventional boost convertor for this input are given Fig. 25.4.

The above results show that current flowing through inductor within 0 to +10A during switch ON and OFF position. The load current is almost constant after 0.001 s. The current in capacitor varies within –4A to +6A limits, and the output voltage of CBC is stable at 23.44 V after 0.001 s.

Similarly, the output voltage of IBC is stable at 30.44 V after 0.0005 s, which is shown in Fig. 25.5.

25.3 MPPT Techniques Used in Solar PV System

A solar system is interconnection of many PV modules, which are linked in series, parallel and grouping of series and parallel depends on power requirements. Imported one PV module such as Sun Power SPR-305E-WHT-D [11] and simulated in MATLAB. The specifications of such PV module are shown in Table 25.2.

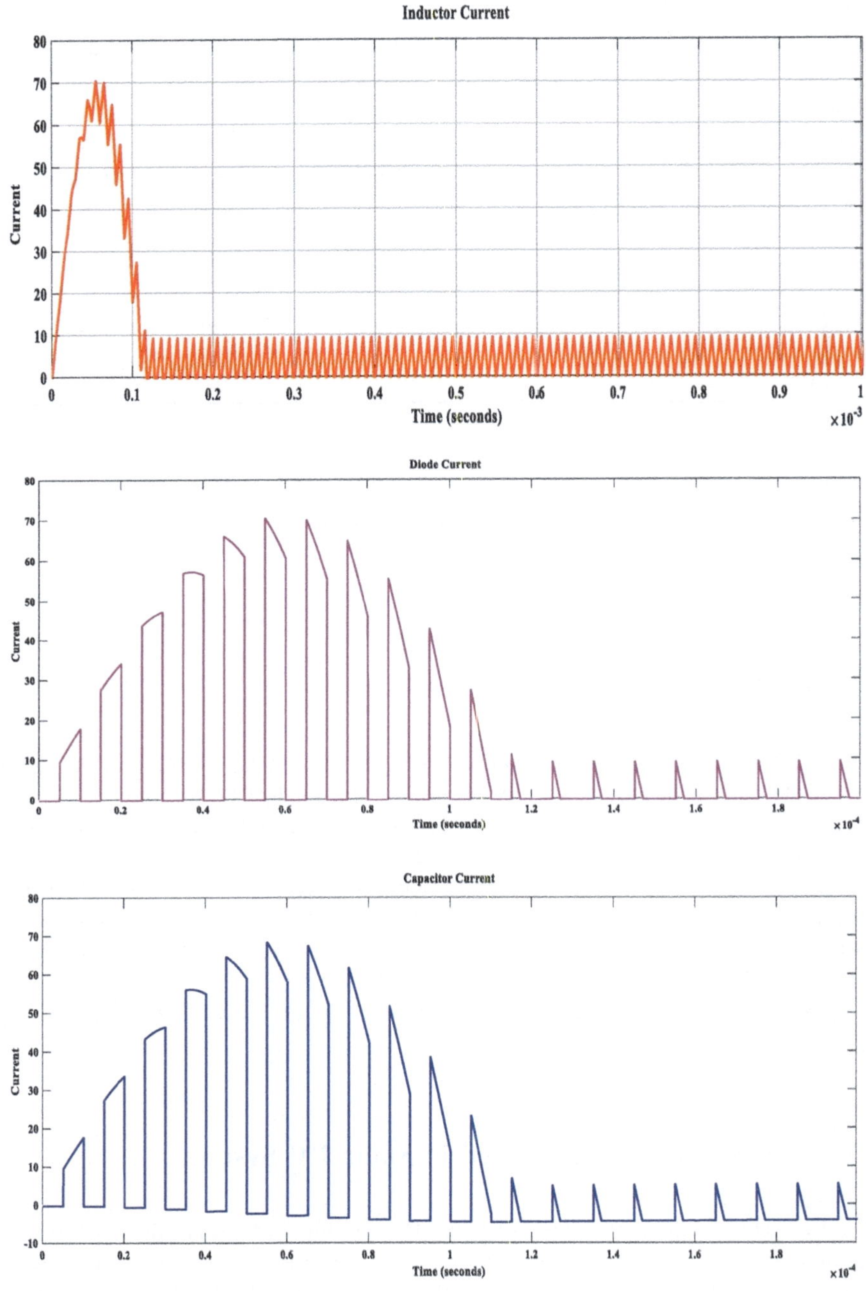

Fig. 25.4 Inductor, diode, capacitor and load currents of CBC

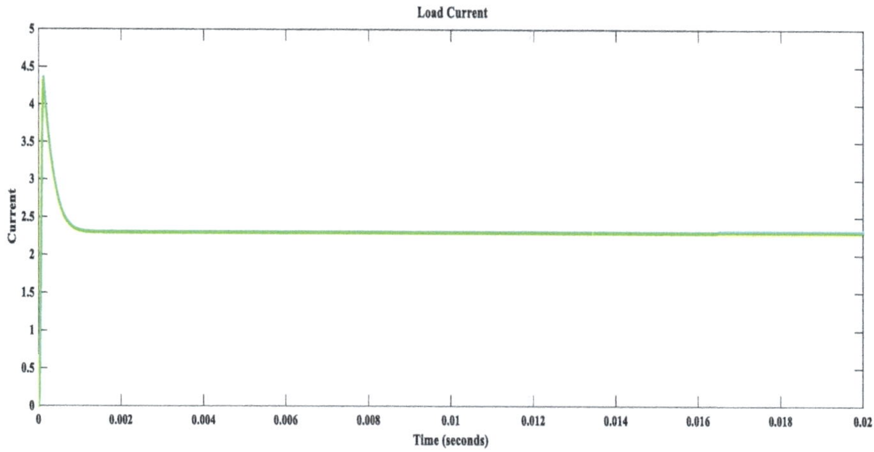

Fig. 25.4 (continued)

Under normal conditions (1000 W/m^2 irradiation and temperature of 25 °C), it can yield a maximal power of 305 W.

The variations of current–voltage and power–voltage under at different irradiance and temperature are shown in Fig. 25.6. It is observable that there are increasing current and power levels as increasing irradiation value, which is shown in Fig. 25.6a, and similarly increasing voltage and power levels as increasing temperature values as shown in Fig. 25.6b. Finally, it identified that only one MPP under uniform irradiance.

The block diagram consists of PV modules, power interface and load which is shown in Fig. 25.7. Here power interface is DC–DC converter; it is controlled by using MPPT controllers. It is simulated in MATLAB environment.

Several MPPT controllers are available to control DC–DC converter by extract MPP from PV module. Some of the MPPT controllers are listed below.

a. P&O MPPT controller method.
b. ICMPPT controller method.
c. FOCV method.
d. Crow search optimization algorithm.

25.3.1 P&O MPPT Controller

It is simple method to determine MPP on P–V curve. It requires majorly voltage and current of PV module. The operating on P–V curve is obtained from product of voltage and current of PV module and comparing with previous operating power. If

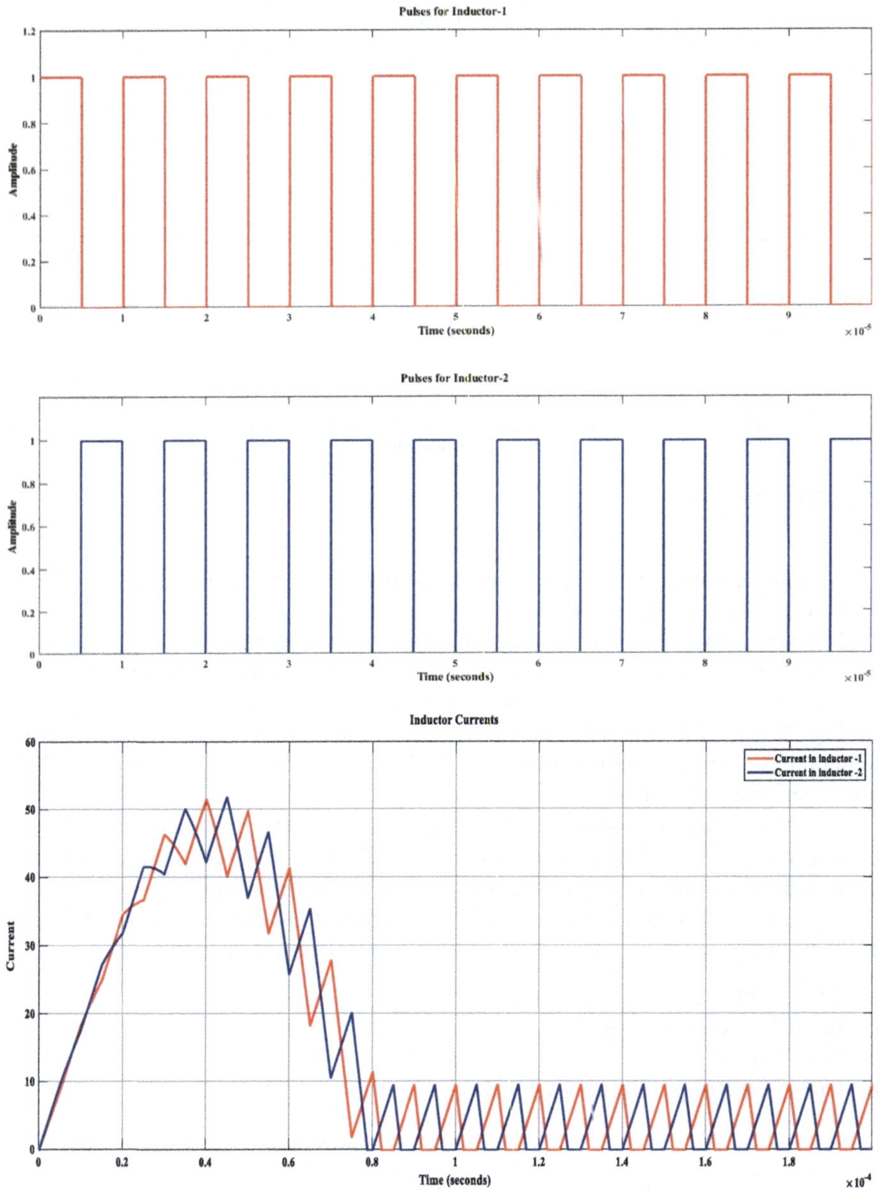

Fig. 25.5 Switching pulses and inductors current of IBC

Table 25.2 PV system parameters under standard condition

Parameters	Values
Maximum power (W)	305
Open-circuit voltage V_{oc} (V)	64.2
Short-circuit current I_{sc} (A)	5.96
Voltage at maximum power point V_{mp} (V)	54.7
Current at maximum power point I_{mp} (A)	5.58
Cells per module (Ncell)	96
Temperature coefficient of V_{oc} (%/deg.C)	−0.27269
Temperature coefficient of I_{sc} (%/deg.C)	0.0617

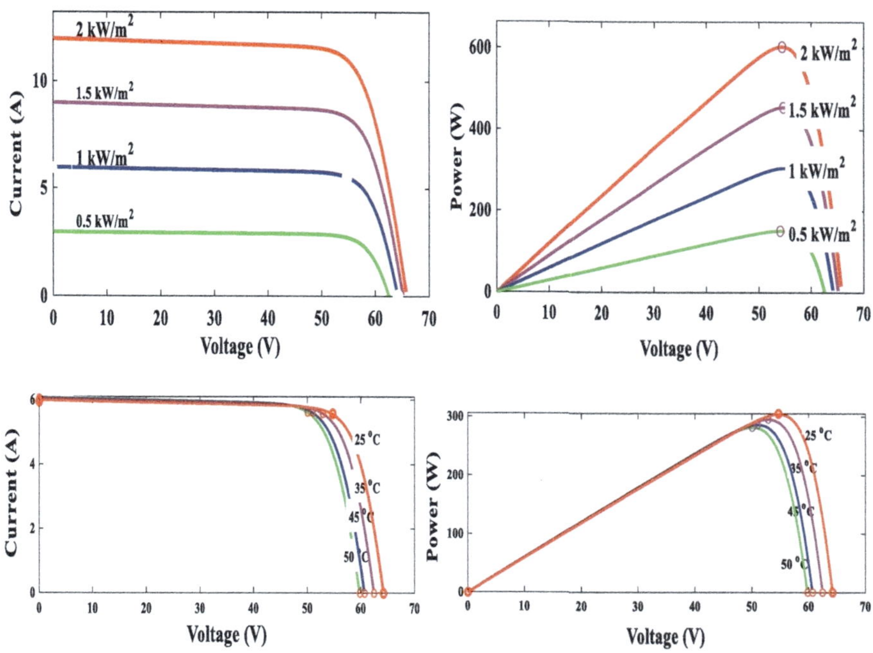

Fig. 25.6 I-V and P–V characteristics of PV module. **a** Variation levels of current and power at different irradiations. **b** Variation levels of voltage and power at different temperatures

the obtained operating point is more than previous power then move in this directions ($dP/dV > 0$), otherwise operating point is moved in backward direction. In this way, determine maximum operating point of PV module [12]. The following flowchart explained step by step procedure of P&O method, which is shown in Fig. 25.8.

A Simulink model is developed based on behavior of P&O MPPT controller from which observed input and output voltages of conventional as well as IBC at different irradiations and temperatures are shown in Table 25.3.

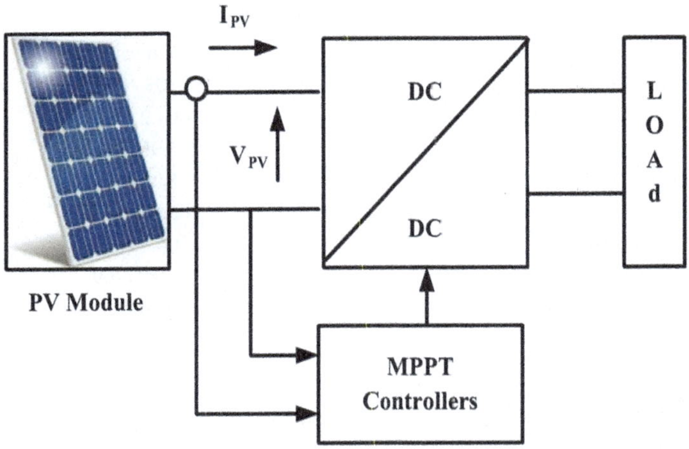

Fig. 25.7 Model of PV system with MPPT Controller

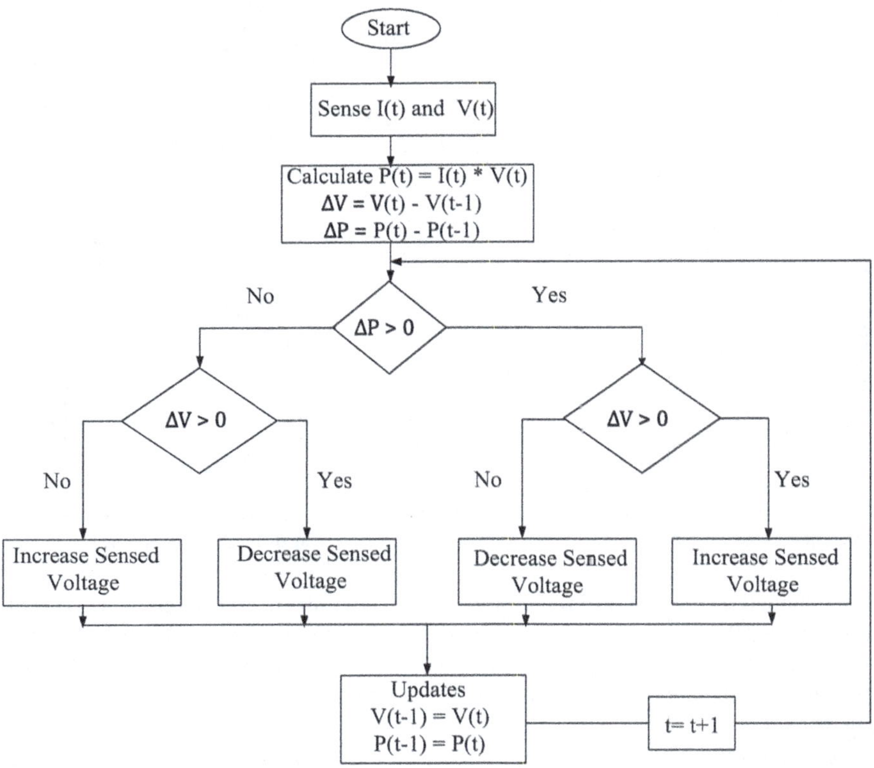

Fig. 25.8 Flowchart of P&O MPPT technique

Table 25.3 Input and output voltages at different irradiation and temperatures

S. No	Irradiance (W/m^2)	Temperature (Degrees)	CBC		IBC	
			Input voltage	Output voltage	Input voltage	Output voltage
1	2000	25	29.66	58.539	31.56	60.54
2	2000	35	29.74	58.59	31.84	60.99
3	2000	45	29.84	58.81	32.15	61.812
4	2000	50	30.11	59.41	32.17	61.99
5	1500	25	22.46	44.11	24.16	46.14
6	1500	35	22.59	44.39	24.591	46.41
7	1500	45	22.73	44.66	24.732	46.68
8	1500	50	22.8	44.81	24.79	46.811
9	1000	25	15.18	29.56	17.19	31.58
10	1000	35	15.267	29.75	17.27	31.76
11	1000	45	15.33	29.94	17.36	31.95
12	1000	50	15.401	30	17.41	32.03
13	500	25	7.827	14.85	9.828	16.87
14	500	35	7.872	14.94	9.873	16.96
15	500	45	7.9189	15.031	9.919	17.05
16	500	50	7.9419	15.072	9.942	17.09

25.3.1.1 Simulation Results

P & O simulation model is designed at particular irradiation [1000 W/m^2] and temperature [25 °C]. From designed model, the input and output voltages of CBC and IBC waves are shown in Fig. 25.9.

P&O is widely used method for tracking MPP on P–V curve. The main problem with P&O approach is that PV module terminal voltage is perturbed every cycle. Therefore, when MPP is attained, output power toggles around maximum, follow-on in power loss in PV system. The time intricacy of this technique is very less for computing maximum power but on reaching very close to MPP. But it does not stop at MPP and keeps on perturbing on both directions. For this reason, it has multiple local maxima at same point.

25.3.2 Incremental Conductance (IC) Method

The IC method [13] is superior algorithm of P&O. In this method, it tracks MPP on P–V curve even fast changes of atmospheric conditions. This method can be used to act directly on duty cycle of power interface. The MPP can be tracked by comparing the ICdI/dV to instantaneous one as described as follows (25.4).

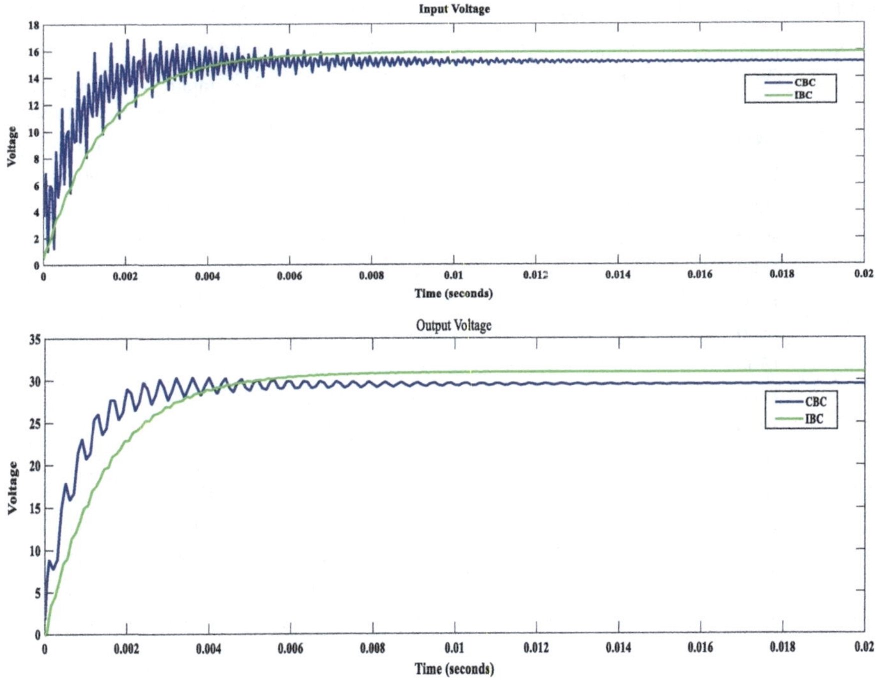

Fig. 25.9 Input and output voltages of P&O-based MPPT

$$\begin{cases} \mathrm{d}I/\mathrm{d}V = 0 \text{ at MPP} \\ \mathrm{d}I/\mathrm{d}V > -I/V \text{ at left hand of MPP} \\ \mathrm{d}I/\mathrm{d}V < -I/V \text{ at right hand of MPP} \end{cases} \quad (25.4)$$

The MPPT controller is controlled by duty cycle. The required duty cycle is calculated based on PV module output voltage. The new duty ratio value is changed in the system as per the controller sampling time.

The tracking accuracy of MPP depends on incremented or decremented step size. If incremented or decremented step sizes are large, reach MPP fast which means tracing accuracy is good, otherwise tracking accuracy will be less and searching MPP slowly occurs around MPP. Though, step size is smaller, increases accuracy tracking but at increase of searching time response.

The main drawback of this controller is more complex than P&O controller.

25.3.2.1 Flowchart of (IC) Method MPPT Algorithm

The following flowchart explained step-by-step procedure of IN method, which is shown in Fig. 25.10.

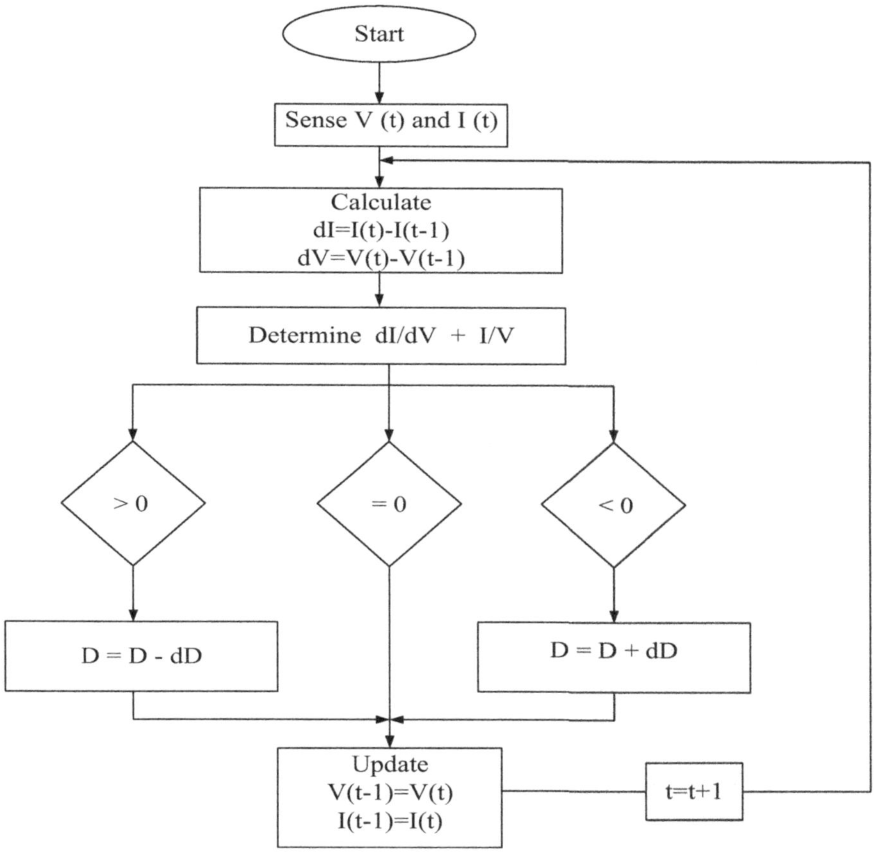

Fig. 25.10 Flowchart of ICMPPT algorithm

A simulation model is developed based on behavior of ICMPPT controller from which observed input and output voltages of conventional as well as IBC at different irradiations and temperatures are shown in Table 25.4.

25.3.2.2 Simulation Results

IC simulation model is designed at particular irradiation [1000 W/m^2] and temperature [25 °C]. From designed model, input and output voltages of CBC and IBC waves are shown in Fig. 25.11.

Table 25.4 Input and output voltages at different irradiation and temperatures

S. No	Irradiance (W/m^2)	Temperature (degrees)	CBC		IBC	
			Input voltage	Output voltage	Input voltage	Output voltage
1	2000	25	29.65	58.549	31.66	60.55
2	2000	35	29.82	58.72	31.84	60.91
3	2000	45	30.01	58.99	32.01	61.27
4	2000	50	30.11	59.44	32.1	61.45
5	1500	25	22.45	44.14	24.46	46.149
6	1500	35	22.589	44.41	24.59	46.401
7	1500	45	22.729	44.67	24.73	46.68
8	1500	50	22.789	44.812	24.79	46.82
9	1000	25	15.178	29.57	17.18	31.58
10	1000	35	15.26	29.76	17.27	31.77
11	1000	45	15.35	29.939	17.36	31.94
12	1000	50	15.411	30.01	17.41	32.03
13	500	25	7.827	14.86	9.828	16.87
14	500	35	7.873	14.95	9.874	16.96
15	500	45	7.92	15.04	9.919	17.05
16	500	50	7.9419	15.08	9.942	17.09

25.3.3 Fractional Open-Circuit Voltage (FOCV) Controller

The FOCV MPPT controller utilizes fact that PV module voltage corresponding to MPP exhibits a linear dependence with respect to module open-circuit voltage for different irradiation and temperature levels. The linear relation between V_{OC} and V_{MPPT} under various operating conditions is represented in the following way for FOCV [14, 15].

$$V_{MPPT}(t) = K_1 * V_{OC}(t) \tag{25.5}$$

where K_1 is propositional constant, and it depends on linear characteristics of PV module. The flowchart of the FOCV represents step-by-step procedure, which is shown in Fig. 25.12. The FOCV MPPT controller controlled by duty cycle. The value of duty cycle is increased or decreased based on V_{MPPT} computed from V_{OC} and actual voltage V_{Act}. The proportional constant K_1 is various between 0.7101 and 0.7802 [16].

25.3.3.1 Flowchart of FOCV

To know, behavior of FOCV MPPT flowchart, a simulation model is developed. The input and output voltages at different irradiations and temperatures are shown in Table 25.5.

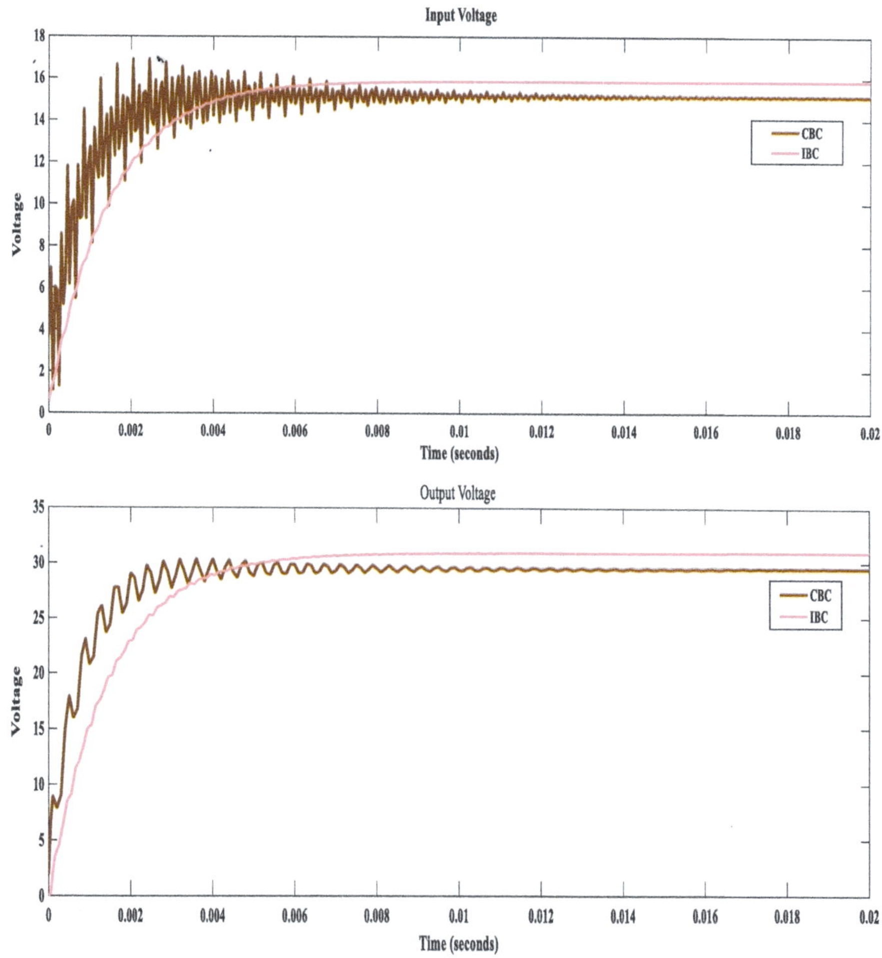

Fig. 25.11 Input and output voltages of IN-based MPPT

25.3.3.2 Simulation Results

Fast open-circuit voltage simulation model is designed at particular irradiation [1000 W/m^2] and temperature [25 °C]. From designed model, input and output voltages of CBC and IBC waves from observed are shown in Fig. 25.13.

25.3.4 Crow Search Algorithm

CSA is one of the most recent population-based evolutionary algorithms based on its activities in searching their food. A crow searching the places while others birds

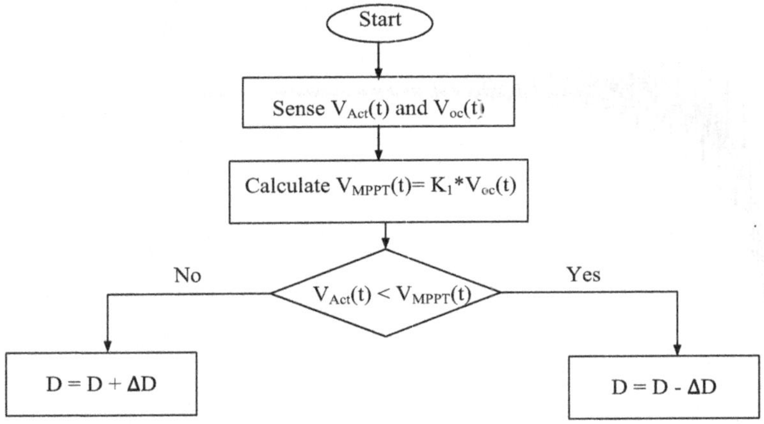

Fig. 25.12 Flowchart of FOCV MPPT algorithm

Table 25.5 Input and output voltages at different irradiation and temperatures

S. No	Irradiance (W/m^2)	Temperature (Degrees)	CBC		IBC	
			Input voltage	Output voltage	Input voltage	Output voltage
1	2000	25	29.66	58.567	31.68	60.57
2	2000	35	29.84	58.92	31.86	60.94
3	2000	45	30.01	59.11	32.03	61.29
4	2000	50	30.1	59.456	32.12	61.46
5	1500	25	22.42	44.19	24.48	46.18
6	1500	35	22.56	44.45	24.6	46.47
7	1500	45	22.69	44.73	24.77	46.69
8	1500	50	22.76	44.86	24.79	46.85
9	1000	25	15.18	29.58	17.19	31.59
10	1000	35	15.27	29.77	17.29	31.78
11	1000	45	15.36	29.94	17.38	31.96
12	1000	50	15.41	30.03	17.43	32.07
13	500	25	7.828	14.87	9.83	16.89
14	500	35	7.874	14.96	9.88	16.99
15	500	45	7.919	15.05	9.92	17.09
16	500	50	7.942	15.09	9.96	17.19

hide their foods and crow tries to pinch them when owner leaves its place. The performance of the CSA is assessed in terms of searching speed and efficiency against the PSO-based MPPT controller under different temperatures and irradiations.

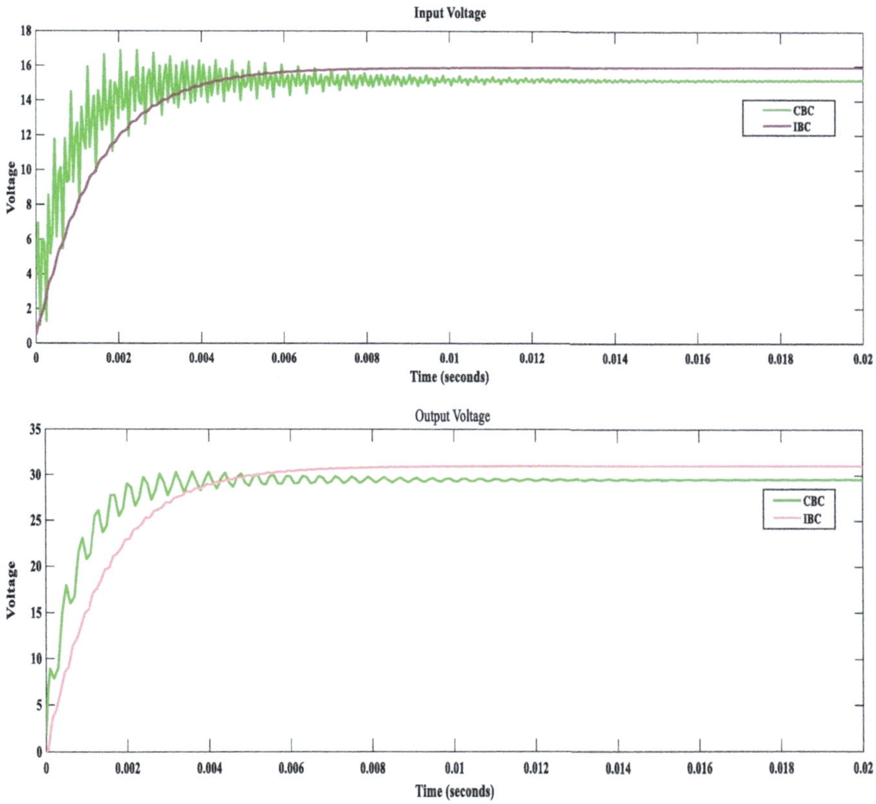

Fig. 25.13 Input and output voltages of FOCV-based MPPT

25.3.4.1 Crow-Based MPPT Controller

Crow-based MPPT controller determines the MPP on P–V curve. In this approach, consider 'N_C' crow flocks and i number of crows has position at iteration (iter) is x_i^{iter}. Crows hide their food at different places and are memorized. Crow moves to find the best food place in search which is defined as m_i^{iter}. The step-by-step procedure of CSA-based MPPT algorithm is explained through flowchart as shown in Fig. 25.14.

25.3.4.2 Simulation Results

The simulation results obtained for the adapted CSA-based MPPT controller at different irradiations and temperatures are listed in Table 25.6. From this table, it is identified that by increasing the irradiation and ambient temperature, both input and

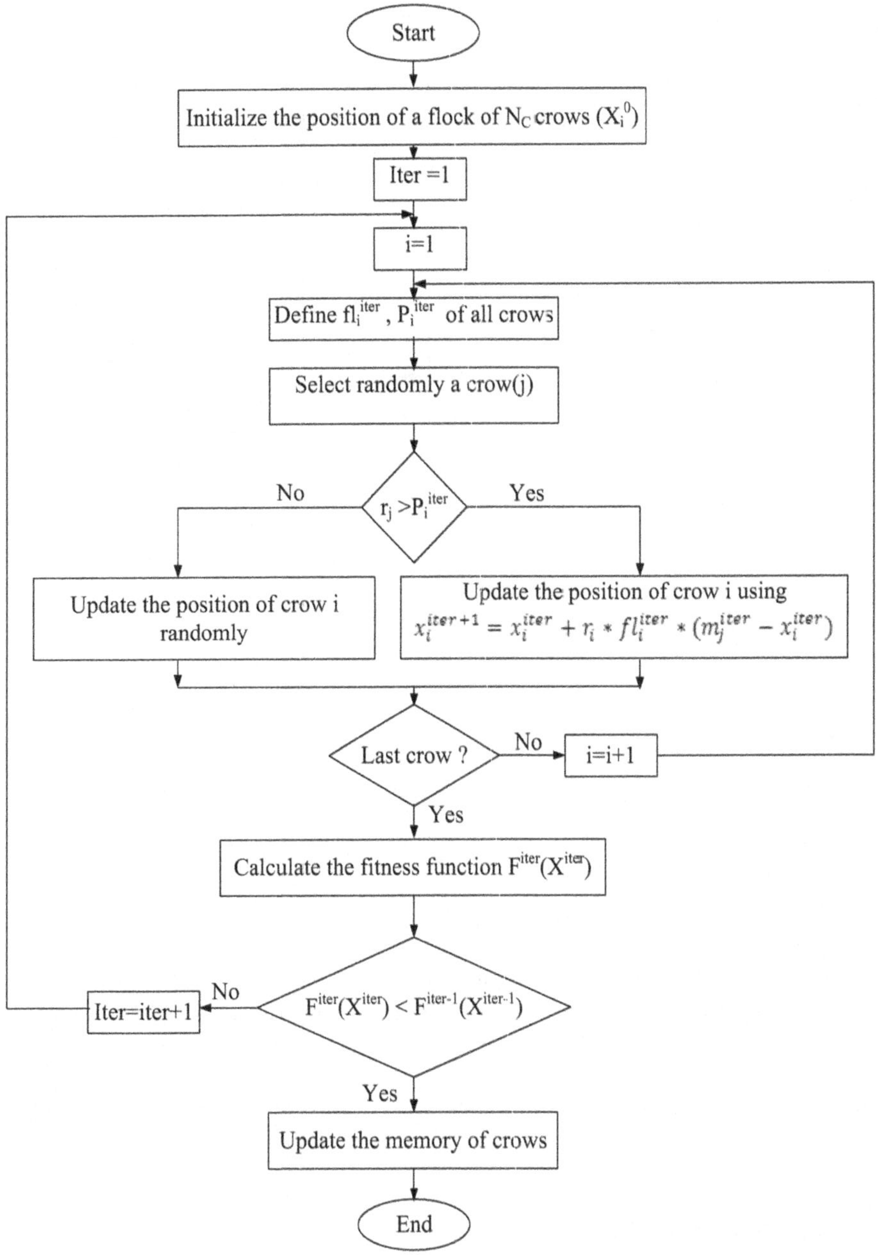

Fig. 25.14 Searching approach in CSA-based MPPT algorithm flowchart

Table 25.6 Input and output voltages at different irradiation and temperatures

S. No	Irradiance (W/m^2)	Temperature (Degrees)	CBC		IBC	
			Input voltage	Output voltage	Input voltage	Output voltage
1	2000	25	29.68	58.65	31.69	60.59
2	2000	35	29.88	58.99	31.88	60.99
3	2000	45	30.12	59.21	32.05	61.32
4	2000	50	30.14	59.471	32.15	61.47
5	1500	25	22.44	44.2	24.49	46.21
6	1500	35	22.58	44.47	24.61	46.5
7	1500	45	22.71	44.75	24.79	46.71
8	1500	50	22.77	44.86	24.81	46.88
9	1000	25	15.22	29.6	17.21	31.61
10	1000	35	15.28	29.79	17.31	31.8
11	1000	45	15.389	29.96	17.39	31.99
12	1000	50	15.43	30.06	17.45	32.09
13	500	25	7.838	14.89	9.88	16.92
14	500	35	7.876	14.99	9.91	17.01
15	500	45	7.921	15.09	9.95	17.11
16	500	50	7.945	15.13	9.99	17.22

output voltages increase. In addition, IBC has given higher voltages when compared to CBC in all conditions. The CSA-based MPPT algorithm shows its superiority in obtaining the voltages when compared to the classical methods.

CSA-based MPPT simulation model is designed at particular irradiation [1000 W/m^2] and temperature [25 °C]. From designed model, input and output voltages of CBC and IBC waves from observed are shown in Fig. 25.15.

25.3.5 Comparative Analysis of MPPT Techniques

The comparison of the adapted CSA-MPPT in terms of steady-state value, settling time, overshoot, rise time and peak value with classical methods is presented in Table 25.7. From this table, it is observed that when compared to CBC with CSA-MPPT and other methods, IBC with CSA-MPPT has given highest peak value with small overshoot and the highest steady-state value in small settling time (Fig. 25.16).

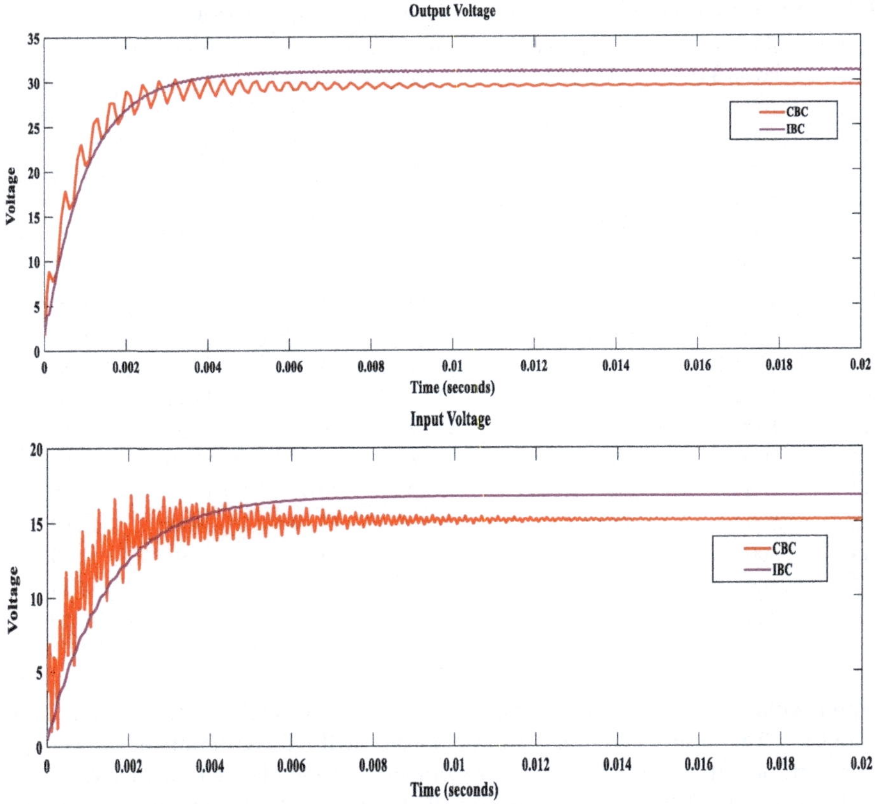

Fig. 25.15 Input voltage and output voltage of crow-based MPPT

Table 25.7 Comparison between conventional and interleaved boost convert

S. No	Parameters	CBC				IBC			
		P&O	IC	FOCV	CSA	P&O	IC	FOCV	CSA
1	Steady-state value (Watts)	86.92	87.62	87.73	87.85	88.57	88.59	88.65	89.01
2	Settling time, (10^{-2} s)	3.937	3.860	3.721	3.669	7.495	6.708	6.70	5.272
3	Overshoot (%)	4.505	4.150	3.981	3.505	5.851	3.646	3.251	2.577
4	Rise time (10^{-2} s)	2.4680	2.512	2.646	2.680	1.623	1.921	1.995	2.2
5	Peak value (Watts)	87.78	88.31	88.42	88.51	89.16	89.19	89.22	90.71

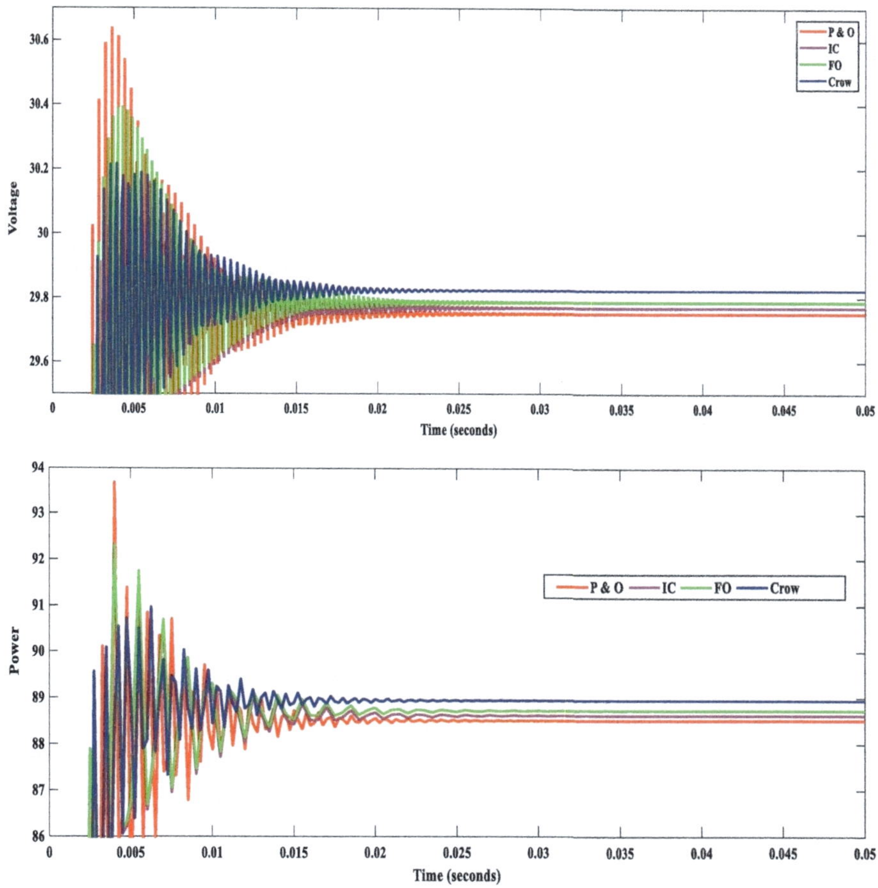

Fig. 25.16 Output voltage and output power of MPPT techniques

25.4 Conclusion

Conventional and interleaved boost converters are modeled at the beginning of this paper with DC and PV source to obtain maximum voltage. The input and output voltages are determined at different irradiance and temperature conditions from the two converters. The maximum power from the PV system is extracted by using adopted crow search algorithm in addition to the classical methods such as P&O, IC and FOCA. Later, the comparative results obtained from the two converters with CSA and other classical methods are presented. From this analysis, it is observed that the IBC has given better results than CBC with CSA and other classical approaches in obtaining input–output voltages under the weather changing and variable temperature conditions. The maximum power obtained from IBC with

CSA is 2.5% more than the CBC. Further, CSA is outperformed in obtaining lesser settling time, overshoot, rise time and required a smaller number of iterations compared with other evolutionary-based MPPT controllers and classical methods.

References

1. Devi, V.K., Premkumar, K., Beevi, A.B., Ramaiyer, S.: A modified perturb & observe MPPT technique to tackle steady state and rapidly varying atmospheric conditions. Sol. Energy **157**, 419–426 (2017)
2. Farayola, A.M., Hasan, A.N., Ali, A.: Implementation of modified incremental conductance and fuzzy logic MPPT techniques using MCUK converter under various environmental conditions. Appl. Solar Energy **53**, 173–184 (2017)
3. Bharath, K., Suresh, E.: Design and implementation of improved fractional open circuit voltage based maximum power point tracking algorithm for photovoltaic applications. Int. J. Renew. Energy Res. 7, 1108–1113 (2017)
4. Harish, D., Rao, G.V., Chowdary, D.: Three phase modular multilevel PV inverter with distributed MPPT for grid-connected applications. **5**, 118–124 (2017).
5. da Rocha, M.V., Sampaio, L.P., da Silva, S.A.O.: Comparative analysis of MPPT algorithms based on Bat algorithm for PV systems under partial shading condition. Sustain. Energy Technol. Assess.S. **40**, 100761 (2020).
6. Mao, M., Duan, Q., Duan, P., Hu, B.: Comprehensive improvement of artificial fish swarm algorithm for global MPPT in PV system under partial shading conditions. Trans. Inst. Meas. Control. (2017).
7. Yang, Z., Duan, Q., Zhong, J., Mao, M., Xun, Z.: Analysis of improved PSO and perturb & observe global MPPT algorithm for PV array under partial shading condition. In: Proceedings Control and Decision Conference (CCDC). 29th Chinese, pp. 549–553 (2017).
8. Rayapudi, S.R.: Enhanced grey wolf optimizer based MPPT algorithm of PV system under partial shaded condition. Int. J. Renew. Energy Dev. **1**, (2017).
9. Ahmed, J., Salam, Z.: A Maximum power point tracking (MPPT) for PV system using cuckoo search with partial shading capability. Applied Energy. **119**, 118–130 (2017)
10. Chandran, A., Reshmi, V., Mathew, B.K.: Comparative analysis of P&O and FLC based SEPIC boost converter for solar PV application. Third International Conference on Smart Systems and Inventive Technology (ICSSIT). Tirunelveli, India. pp. 616–621 (2020).
11. Bhattacharjee, S., Saharia, B.J.: A comparative study on converter topologies for maximum power point tracking application in photovoltaic generation. J. Renew. Sustain. Energy. (2014).
12. Alajmi, B.N., Marei, M.I., Abdelsalam, I.: A multiport DC–DC converter based on two-quadrant inverter topology for PV systems. IEEE Trans. Power Electron. **36**(1), 522–532 (2021).
13. Ribeiro, E., Cardoso, A.J.M., Boccaletti, C.: Fault-tolerant strategy for a photovoltaic DC–DC converter. IEEE Trans. Power Electron. **28**, 3008–3018 (2013).
14. Panigrahi, R., Mishra, S.K., Joshi, A.: Synthesizing a comprehensive set of converter topologies for a specified voltage gain. Energy Conversion Congress and Exposition (ECCE). IEEE, pp. 955–961 (2020).
15. Raghavendra, K.V.G., Zeb, K., Muthusamy, A., Krishna, T.N.V., Kumar, S.V.S., Kim, D.H., Kim, M.S., Cho, H.G., Kim, H.J.: A comprehensive review of DC–DC converter topologies and modulation strategies with recent advances in solar photovoltaic systems. Electronics. **9**, 1: 31 (2020).
16. Sayed, K., Gronfula, M.G., Ziedan, H.A.: Novel soft-switching integrated boost DC-DC converter for PV power system. Energies. **13**, 3:749 (2020).

The manufacturer's authorised representative in the EU is Springer Nature Customer Service Centre GmbH, Europaplatz 3, 69115 Heidelberg, Germany. If you have any concerns regarding our products, please contact ProductSafety@springernature.com

Printed and bound by CPI Group (UK) Ltd, Croydon, CR0 4YY
30/09/2026
02225294-0018